# CONSERVATION OF LEATHER
and related materials

# CONSERVATION OF LEATHER
## and related materials

Marion Kite • Roy Thomson

*Chairman,*
*The Leather Conservation Centre*

*Former Chief Executive,*
*The Leather Conservation Centre*

Routledge
Taylor & Francis Group

LONDON AND NEW YORK

First published 2006 by Butterworth-Heinemann

2 Park Square, Milton Park, Abingdon, Oxfordshire OX14 4RN
52 Vanderbilt Avenue, New York, NY 10017

*Routledge is an imprint of the Taylor & Francis Group, an informa business*

First issued in paperback 2020

Copyright © 2006 Taylor & Francis

All rights reserved. No part of this book may be reprinted or reproduced or utilised in any form or by any electronic, mechanical, or other means, now known or hereafter invented, including photocopying and recording, or in any information storage or retrieval system, without permission in writing from the publishers.

Notice:
Product or corporate names may be trademarks or registered trademarks, and are used only for identification and explanation without intent to infringe.

**British Library Cataloguing in Publication Data**
A catalogue record for this book is available from the British Library

**Library of Congress Cataloging-in-Publication Data**
A catalog record for this book is available from the Library of Congress

ISBN: 978-0-7506-4881-3 (hbk)
ISBN: 978-0-367-60635-0 (pbk)


Publisher's Note

The publisher has gone to great lengths to ensure the quality of this book but points out that some imperfections from the original may be apparent.

# Contents

| | | |
|---|---|---|
| Foreword | xi | |
| Dedications | xiii | |
| Acknowledgements | xv | |
| Contributors | xvii | |

**1 The nature and properties of leather** 1
*Roy Thomson*
References 3

**2 Collagen: the leathermaking protein** 4
*B.M. Haines*
2.1 The collagen molecule 4
2.2 Bonding within the molecule 6
2.3 Bonding between molecules 6
   2.3.1 Salt links 6
   2.3.2 Covalent intermolecular bonding 7
2.4 Fibril structure 8
2.5 Shrinkage temperature 9
References 10

**3 The fibre structure of leather** 11
*B.M. Haines*
3.1 The structure of mammalian skins 12
3.2 Variation of structure between animal types 12
   3.2.1 Mature cattle skins 12
   3.2.2 Calfskins 14
   3.2.3 Goatskins 14
   3.2.4 Sheepskins 14
   3.2.5 Deerskins 15
   3.2.6 Pigskins 15
3.3 Grain surface patterns 17
3.4 Suede surfaces 17
3.5 Variation in structure with location in the skin 17
3.6 Directional run of the fibres 19
3.7 The influence of fibre structure on leather properties, structure and tear strength 19
3.8 Structure and leather handle 20
3.9 Fibre weave and movement 20
References 21

**4 The chemistry of tanning materials** 22
*A.D. Covington*
4.1 Introduction 22
4.2 Vegetable tanning 23
4.3 Mineral tanning 27
   4.3.1 Chromium(III) salts 27
   4.3.2 Aluminium(III) salts 29
   4.3.3 Titanium(IV) salts 29
   4.3.4 Zirconium(IV) salts 30
4.4 Oil tanning 30
4.5 Aldehyde tanning 31
   4.5.1 Formaldehyde tanning 31
   4.5.2 Glutaraldehyde tanning 31
   4.5.3 Oxazolidine tanning 31
4.6 Syntans 32
   4.6.1 Auxiliary syntans 32
   4.6.2 Combination or retanning syntans 33
   4.6.3 Replacement syntans 33
4.7 Overview 34
References 34

**5 The mechanisms of deterioration in leather** 36
*Mary-Lou E. Florian*
5.1 Introduction 36
5.2 Agents of deterioration 37
   5.2.1 Introduction 37
   5.2.2 Acid hydrolysis 38
   5.2.3 Oxidation 38
   5.2.4 Metals and salts 40

|  |  | 5.2.5 | Heat | 41 |
|---|---|---|---|---|
|  |  | 5.2.6 | Water | 41 |
|  | 5.3 | Collagen | | 43 |
|  |  | 5.3.1 | Bonds in collagen: sites of deterioration mechanisms | 43 |
|  |  | 5.3.2 | Peptides | 43 |
|  |  | 5.3.3 | Amino acids in collagen | 44 |
|  | 5.4 | Vegetable tannins | | 47 |
|  |  | 5.4.1 | Introduction | 47 |
|  |  | 5.4.2 | Antioxidant ability of tannins | 47 |
|  |  | 5.4.3 | Analysis of tannins in aged leather – deterioration mechanisms | 47 |
|  | 5.5 | Other chemicals present due to fabrication and use | | 50 |
|  |  | 5.5.1 | Introduction | 50 |
|  |  | 5.5.2 | Fats, oils and waxes | 51 |
|  |  | 5.5.3 | Sulphur compounds and their acids | 51 |
|  |  | 5.5.4 | Acids in leather due to fabrication or use | 52 |
|  |  | 5.5.5 | Perspiration | 52 |
|  | 5.6 | Denaturation and shrinkage temperatures as a method of assessment for all tannages | | 52 |
|  | 5.7 | Summary | | 53 |
|  | Acknowledgements | | | 54 |
|  | References | | | 54 |
| 6 | **Testing leathers and related materials** | | | 58 |
|  | *Roy Thomson* | | | |
|  | 6.1 | Introduction | | 58 |
|  | 6.2 | Determination of raw material | | 58 |
|  | 6.3 | Determination of tannage type | | 58 |
|  |  | 6.3.1 | Ashing test | 58 |
|  |  | 6.3.2 | Spot test | 59 |
|  |  | 6.3.3 | Conclusion | 59 |
|  | 6.4 | Determination of degree of deterioration | | 59 |
|  |  | 6.4.1 | Organoleptic examination | 59 |
|  |  | 6.4.2 | Chemical tests | 60 |
|  | 6.5 | Conclusions | | 64 |
|  | References | | | 64 |
| 7 | **The manufacture of leather** | | | 66 |
|  | *Roy Thomson* | | | |
|  | 7.1 | Tanning in prehistoric and classical times | | 66 |
|  | 7.2 | Tanning in the medieval and post-medieval periods | | 68 |
|  | 7.3 | Tanning in the nineteenth century | | 73 |
|  | 7.4 | Tanning in modern times | | 77 |
|  |  | 7.4.1 | Pretanning | 77 |
|  |  | 7.4.2 | Tanning | 80 |
|  |  | 7.4.3 | Post-tanning | 81 |
|  | References | | | 81 |
| 8 | **The social position of leatherworkers** | | | 82 |
|  | *Robert D. Higham* | | | |
|  | References | | | 87 |
| 9 | **Gilt leather** | | | 88 |
|  | *Roy Thomson* | | | |
|  | 9.1 | Production and art historical aspects | | 90 |
|  | 9.2 | Conservation and restoration | | 91 |
| 10 | ***Cuir bouilli*** | | | 94 |
|  | *Laura Davies* | | | |
|  | 10.1 | The *cuir bouilli* technique | | 94 |
|  | 10.2 | Leather moulding techniques | | 94 |
|  | 10.3 | The origins of the *cuir bouilli* technique | | 94 |
|  | 10.4 | Changes undergone by the leather in the *cuir bouilli* process | | 97 |
|  | 10.5 | Conservation of *cuir bouilli* | | 97 |
|  |  | 10.5.1 | Stability | 97 |
|  |  | 10.5.2 | Damage caused by old treatments | 98 |
|  |  | 10.5.3 | Original treatments of *cuir bouilli* leather | 98 |
|  | 10.6 | Case study of the conservation of *cuir bouilli* leather | | 98 |
|  |  | 10.6.1 | Analysis and use of non-invasive xeroradiographic imaging | 99 |
|  |  | 10.6.2 | Removal of inappropriate surface coatings | 99 |
|  | Endnotes | | | 101 |
|  | References | | | 101 |
| 11 | **The tools and techniques of leatherworking: correct tools + skills = quality** | | | 103 |
|  | *Caroline Darke* | | | |
|  | 11.1 | Leatherworking tools | | 103 |
|  |  | 11.1.1 | The awl | 103 |
|  |  | 11.1.2 | The knife | 103 |
|  |  | 11.1.3 | The strop | 104 |
|  |  | 11.1.4 | The bone folder or crease | 104 |
|  |  | 11.1.5 | The steel rule | 104 |
|  |  | 11.1.6 | The dividers (compass) | 104 |
|  |  | 11.1.7 | The revolving hole punch | 105 |
|  |  | 11.1.8 | The hammer | 105 |
|  |  | 11.1.9 | The race | 105 |
|  |  | 11.1.10 | The clam | 105 |
|  |  | 11.1.11 | The edge shave | 105 |

|       |         | 11.1.12 | The crease iron | 106 |
|-------|---------|---------|-----------------|-----|
|       |         | 11.1.13 | The stitch marker | 106 |
|       |         | 11.1.14 | The pricking iron | 107 |
|       |         | 11.1.15 | The needle | 107 |
|       |         | 11.1.16 | Thread | 108 |
|       | 11.2    | Adhesives | | 108 |
|       | 11.3    | Reinforcements | | 108 |
|       | 11.4    | Techniques | | 109 |
|       |         | 11.4.1  | Skiving | 109 |
|       |         | 11.4.2  | Preparation | 109 |
|       |         | 11.4.3  | Sewing – stitch formation | 109 |
|       |         | 11.4.4  | Decorative stitching | 110 |
|       |         | 11.4.5  | Machine stitching | 110 |
|       |         | 11.4.6  | Decorative machine stitching | 110 |
|       |         | 11.4.7  | Seams and construction | 111 |
|       | Bibliography | | | 112 |

## 12  General principles of care, storage and display — 113
*Aline Angus, Marion Kite and Theodore Sturge*

- 12.1 Introduction — 113
- 12.2 Objects in use — 113
- 12.3 Display or storage — 114
- 12.4 Levels of treatment — 114
- 12.5 Handling by the public — 114
- 12.6 The 'finish' — 115
- 12.7 Preventive conservation — 115
  - 12.7.1 Environment — 115
  - 12.7.2 Pests — 116
  - 12.7.3 Storage and display — 116
- 12.8 Shoes — 117
- 12.9 Gloves — 117
- 12.10 Leather garments — 117
- 12.11 Luggage — 117
- 12.12 Saddles — 117
- 12.13 Harness — 118
- 12.14 Screens, wall hangings and sedan chairs — 119
- 12.15 Carriages and cars — 120
- 12.16 Conclusion — 120
- Reference — 120

## 13  Materials and techniques: past and present — 121
*Marion Kite, Roy Thomson and Aline Angus*

- 13.1 Past conservation treatments — 121
  - 13.1.1 Introduction — 121
  - 13.1.2 1982 Jamieson survey — 121
  - 13.1.3 1995 survey — 122
  - 13.1.4 2000 list — 123
  - 13.1.5 2003 Canadian Conservation Institute (CCI) survey — 124
- 13.2 Notes on treatments in use in 2004 – additional information — 124
  - 13.2.1 Introduction — 124
  - 13.2.2 Dry cleaning — 124
  - 13.2.3 Wet cleaning and solvent cleaning — 125
  - 13.2.4 Proprietary leather cleaners — 125
  - 13.2.5 Humidification — 125
- 13.3 Repair materials — 126
- 13.4 Adhesives — 126
- 13.5 Surface infilling materials and replacement techniques — 127
- 13.6 Moulding and casting materials and techniques — 128
- 13.7 Consolidation techniques — 128
- 13.8 Dressings and finishes — 128
- References — 129

## 14  Taxidermy — 130
*J.A. Dickinson*

- 14.1 A brief history — 130
- 14.2 Taxidermy terms — 131
- 14.3 Birds — 131
  - 14.3.1 Methods — 131
  - 14.3.2 Problems — 132
- 14.4 Mammals — 132
  - 14.4.1 Methods — 132
  - 14.4.2 Problems — 134
- 14.5 Fish — 135
  - 14.5.1 Methods — 135
  - 14.5.2 Problems — 136
- 14.6 Care — 137
  - 14.6.1 Light — 137
  - 14.6.2 Temperature — 137
  - 14.6.3 Relative humidity — 137
  - 14.6.4 Storage — 137
- 14.7 Preservatives — 140
- References — 140

## 15  Furs and furriery: history, techniques and conservation — 141
*Marion Kite*

- 15.1 History of fur use — 141
  - 15.1.1 Introduction — 141
  - 15.1.2 Background and history — 142
  - 15.1.3 Husbandry and harvesting — 145
  - 15.1.4 Some fashionable furs and dates — 146
- 15.2 Structure, morphology, dressing and making — 148
  - 15.2.1 Definitions and terminology — 148

|     |       | 15.2.2 | Brief history of fur-skin processing and dyeing | 148 |
| --- | ---   | ---    | ---                         | --- |
|     |       | 15.2.3 | Hair and fur fibres         | 149 |
|     |       | 15.2.4 | Keratin                     | 149 |
|     |       | 15.2.5 | Morphology of hair          | 150 |
|     |       | 15.2.6 | Fur-skin dressing           | 151 |
|     |       | 15.2.7 | Dyeing                      | 153 |
|     |       | 15.2.8 | Finishing                   | 154 |
|     |       | 15.2.9 | Pointing                    | 154 |
|     |       | 15.2.10| Making up into garments or accessories | 154 |
|     |       | 15.2.11| Plates and crosses          | 157 |
|     | 15.3  | Conservation and care       |     | 158 |
|     |       | 15.3.1 | Introduction                | 158 |
|     |       | 15.3.2 | Species identification      | 158 |
|     |       | 15.3.3 | Damage                      | 159 |
|     |       | 15.3.4 | Conservation methods        | 159 |
|     |       | 15.3.5 | Two case histories illustrating methods | 161 |
|     |       | 15.3.6 | Freezing tests of adhesives | 165 |
|     |       | 15.3.7 | Care of furs                | 166 |
|     | Endnotes |     |                             | 167 |
|     | References |  |                             | 168 |

## 16 The tanning, dressing and conservation of exotic, aquatic and feathered skins    170
*Rudi Graemer and Marion Kite*

|     |      |        |                          |     |
| --- | ---  | ---    | ---                      | --- |
|     | 16.1 | Exotic skins |                    | 170 |
|     |      | 16.1.1 | Introduction             | 170 |
|     |      | 16.1.2 | Origins and history of exotic leathers | 170 |
|     |      | 16.1.3 | Uses of exotic leathers  | 170 |
|     |      | 16.1.4 | Preparing the raw skins  | 171 |
|     |      | 16.1.5 | Tanning and dressing     | 171 |
|     |      | 16.1.6 | Conservation             | 172 |
|     |      | 16.1.7 | Conclusion               | 172 |
|     | 16.2 | Aquatic skins |                   | 173 |
|     |      | 16.2.1 | Fish skin preparation    | 174 |
|     |      | 16.2.2 | Structure and identification | 174 |
|     |      | 16.2.3 | Fish skin in ethnographic objects | 175 |
|     |      | 16.2.4 | Conservation             | 178 |
|     | 16.3 | Feathered skins and fashionable dress | 178 |
|     |      | 16.3.1 | Processing               | 178 |
|     |      | 16.3.2 | Conservation problems with bird skins | 181 |
|     | Endnotes |  |                              | 182 |
|     | References | |                            | 182 |

## 17 Ethnographic leather and skin products    184
*Sherry Doyal and Marion Kite*

|     |      |        |                          |     |
| --- | ---  | ---    | ---                      | --- |
|     | 17.1 | Introduction |                    | 184 |
|     | 17.2 | Ethics |                          | 184 |
|     | 17.3 | Uses   |                          | 184 |
|     | 17.4 | Tanning methods |                 | 185 |
|     | 17.5 | Construction techniques |         | 185 |
|     | 17.6 | Decoration |                      | 185 |
|     | 17.7 | Conservation |                    | 186 |
|     |      | 17.7.1 | Pre-treatment examination | 186 |
|     |      | 17.7.2 | Poisons – health and safety issues | 186 |
|     |      | 17.7.3 | Condition                | 187 |
|     |      | 17.7.4 | Cleaning                 | 187 |
|     |      | 17.7.5 | Solvent cleaning         | 188 |
|     |      | 17.7.6 | Reshaping                | 188 |
|     |      | 17.7.7 | Mounts/internal supports | 188 |
|     |      | 17.7.8 | Mending                  | 189 |
|     |      | 17.7.9 | Repair supports          | 189 |
|     |      | 17.7.10| Sewing                   | 189 |
|     |      | 17.7.11| Adhesives                | 189 |
|     |      | 17.7.12| Cosmetic repairs and infills | 190 |
|     |      | 17.7.13| Storage                  | 190 |
|     |      | 17.7.14| Display                  | 190 |
|     | References | |                         | 190 |
|     | Bibliography | |                       | 191 |

## 18 Collagen products: glues, gelatine, gut membrane and sausage casings    192
*Marion Kite*

|     |      |                              |     |
| --- | ---  | ---                          | --- |
|     | 18.1 | Animal glues and fish glues  | 192 |
|     | 18.2 | Skin glues and hide glues    | 192 |
|     | 18.3 | Parchment glue and parchment size | 193 |
|     | 18.4 | Rabbit skin glue             | 193 |
|     | 18.5 | Bone glue                    | 193 |
|     | 18.6 | Gelatine                     | 193 |
|     | 18.7 | Fish glue                    | 194 |
|     | 18.8 | Gut membrane                 | 194 |
|     | 18.9 | Sausage casings              | 195 |
|     | References |                        | 197 |

## 19 The manufacture of parchment    198
*B.M. Haines*

|     |      |                           |     |
| --- | ---  | ---                       | --- |
|     | 19.1 | Temporary preservation    | 198 |
|     | 19.2 | Soaking                   | 198 |
|     | 19.3 | Liming                    | 198 |
|     | 19.4 | Unhairing and fleshing    | 198 |
|     | 19.5 | Drying                    | 198 |
|     | Bibliography |                   | 199 |

## 20 The conservation of parchment    200
*Christopher S. Woods*

|     |      |                           |     |
| --- | ---  | ---                       | --- |
|     | 20.1 | Introduction              | 200 |
|     | 20.2 | Parchment production and use | 200 |
|     | 20.3 | Chemical, physical and deterioration characteristics | 203 |
|     | 20.4 | Display and storage       | 209 |

|   |   |   |   |
|---|---|---|---|
| 20.5 | Conservation treatments | 209 | |
| | 20.5.1 Mould and fumigation | 210 | |
| | 20.5.2 Cleaning methods | 210 | |
| | 20.5.3 Humidification and softening | 211 | |
| | 20.5.4 Consolidation of weak parchment | 215 | |
| | 20.5.5 Consolidation of inks and pigments | 216 | |
| | 20.5.6 Repairs and supports | 217 | |
| 20.6 | Conclusion | 220 | |
| Acknowledgements | | 220 | |
| Endnotes | | 220 | |
| References | | 221 | |

## 21 Conservation of leather bookbindings: a mosaic of contemporary techniques — 225

- 21.1 Introduction — 225
  *Randy Silverman*
- 21.2 Binding solutions to old problems — 225
  *Anthony Cains*
  - 21.2.1 Introduction — 225
  - 21.2.2 Klucel G — 226
  - 21.2.3 Application of Klucel G — 227
  - 21.2.4 Facing degraded leather — 227
  - 21.2.5 Technique — 227
  - 21.2.6 Treatment of the boards — 228
  - 21.2.7 Adhesives — 228
  - 21.2.8 Offsetting — 228
  - 21.2.9 Board attachment — 228
  - 21.2.10 Helical oversewing — 228
  - 21.2.11 The joint tacket — 229
  - 21.2.12 Drills — 229
  - 21.2.13 Making the needle drill bit — 229
- 21.3 Leather Conservation – bookbinding leather consolidants — 230
  *Glen Ruzicka, Paula Zyats, Sarah Reidell and Olivia Primanis*
  - 21.3.1 Introduction — 230
  - 21.3.2 ENVIRONMENT Leather Project — 230
  - 21.3.3 Consolidants — 230
- 21.4 Solvent-set book repair tissue — 232
  *Alan Puglia and Priscilla Anderson*
  - 21.4.1 Preparation of the repair tissue — 233
  - 21.4.2 Leather consolidation — 233
  - 21.4.3 Repair technique — 233
  - 21.4.4 Reversing solvent-set tissue repairs — 233
  - 21.4.5 Conclusion — 233
- 21.5 Split joints on leather bindings — 234
  *Don Etherington*
- 21.6 A variation on the Japanese paper hinge – adding a cloth inner hinge — 235
  *Bill Minter*
- 21.7 Split-hinge board reattachment — 235
  *David Brock*
- 21.8 Board slotting – a machine-supported book conservation method — 236
  *Friederike Zimmern*
  - 21.8.1 Introduction — 236
  - 21.8.2 The method — 237
  - 21.8.3 Treatment of the text block — 237
  - 21.8.4 Treatment of boards — 237
  - 21.8.5 Reattachment of text block and boards — 237
  - 21.8.6 The board slotting machine — 238
  - 21.8.7 Scientific analyses — 238
  - 21.8.8 Dyeing with reactive dyes — 239
  - 21.8.9 Conclusions — 241
  - 21.8.10 Acknowledgements — 241
- 21.9 A variation on the board slotting machine — 241
  *Bill Minter*

References — 242

## 22 The conservation of archaeological leather — 244
*E. Cameron, J. Spriggs and B. Wills*

- 22.1 Introduction — 244
  - 22.1.1 The archaeological context — 244
  - 22.1.2 Leather technology and material culture — 244
- 22.2 Wet leather — 245
  - 22.2.1 Condition — 245
  - 22.2.2 Preserving wet leather before treatment — 246
  - 22.2.3 Past treatments — 247
  - 22.2.4 Present-day conservation treatments — 248
- 22.3 Dry leather — 251
  - 22.3.1 Condition — 251
  - 22.3.2 On-site retrieval — 253
  - 22.3.3 Recording procedures — 254
  - 22.3.4 Present-day treatments — 256
- 22.4 Mineralized leather — 257
  - 22.4.1 Condition — 257
  - 22.4.2 On-site retrieval — 259
  - 22.4.3 Recording — 259
  - 22.4.4 Treatment — 259

| | 22.5 | Long-term storage of archaeological leather | 260 |
|---|---|---|---|
| | | 22.5.1 Storage requirements | 260 |
| | | 22.5.2 Condition assessments of treated leather | 260 |
| | | 22.5.3 Old collections/retreatments | 260 |
| | 22.6 | Purpose of treatment: a call for clarity | 260 |
| | 22.7 | Conclusion | 261 |
| | References | | 261 |
| **23** | **Case histories of treatments** | | **264** |
| | 23.1 | The Gold State Coach. 1762 | 265 |
| | | 23.1.1 Description | 265 |
| | | 23.1.2 The problems and the options | 265 |
| | | 23.1.3 Treatment | 265 |
| | 23.2 | Dog Whip – believed to be eighteenth century | 268 |
| | | 23.2.1 Description | 268 |
| | | 23.2.2 Treatment | 268 |
| | 23.3 | Fire Bucket | 271 |
| | | 23.3.1 Description | 271 |
| | | 23.3.2 Treatment | 271 |
| | 23.4 | Fireman's Helmet | 274 |
| | | 23.4.1 Description | 274 |
| | | 23.4.2 Treatment | 274 |
| | 23.5 | Leather Lion | 276 |
| | | 23.5.1 Description | 276 |
| | | 23.5.2 Treatment | 278 |
| | 23.6 | Sedan Chair | 279 |
| | | 23.6.1 Description | 279 |
| | | 23.6.2 Repairs | 279 |
| | | 23.6.3 Cleaning | 283 |
| | | 23.6.4 Gap filling and finishing | 284 |
| | 23.7 | Jewellery Box | 285 |
| | | 23.7.1 Description | 285 |
| | | 23.7.2 Treatment | 285 |
| | 23.8 | Dining Chairs | 287 |
| | | 23.8.1 Description | 287 |
| | | 23.8.2 The set of eight chairs for reupholstering | 287 |
| | | 23.8.3 The set of eight chairs repaired without removing the covers | 287 |
| | | 23.8.4 The four chairs where the covers were removed and conserved | 289 |
| | | 23.8.5 Overview | 290 |
| | 23.9 | Alum Tawed Gloves, having belonged to Oliver Cromwell | 293 |
| | | 23.9.1 Description | 293 |
| | | 23.9.2 Condition | 293 |
| | | 23.9.3 Treatment | 293 |
| | | 23.9.4 Future care | 294 |
| | 23.10 | Court Gloves | 296 |
| | | 23.10.1 Description | 296 |
| | | 23.10.2 Treatment | 296 |
| | 23.11 | Mounting of a Collection of Flying Helmets | 297 |
| | | 23.11.1 Description | 297 |
| | | 23.11.2 Mount instructions | 297 |
| | 23.12 | Leather Components from Panhard et Levassor Automobile. 1899 | 302 |
| | | 23.12.1 Description | 302 |
| | | 23.12.2 Condition | 303 |
| | | 23.12.3 Treatment | 304 |
| | | 23.12.4 Future care | 306 |
| | 23.13 | Altar Frontal. 1756 | 307 |
| | | 23.13.1 Description | 307 |
| | | 23.13.2 Treatment | 307 |
| | 23.14 | Gilt Leather Screen | 313 |
| | | 23.14.1 Description | 313 |
| | | 23.14.2 Treatment | 313 |
| | 23.15 | Gilt Leather Wall Hangings, Levens Hall | 315 |
| | | 23.15.1 Description | 315 |
| | | 23.15.2 Treatment | 316 |
| | 23.16 | Phillip Webb Settle. 1860–65 | 325 |
| | | 23.16.1 Description | 325 |
| | | 23.16.2 Treatment | 325 |
| | 23.17 | Gilt Leather Wall Hangings at Groote Schuur, Cape Town | 329 |
| | | 23.17.1 Description | 329 |
| | | 23.17.2 Condition | 329 |
| | | 23.17.3 Conservation treatment | 331 |
| | | 23.17.4 Future care | 333 |
| **Index** | | | **335** |

# Foreword

The first time I wished for a book like this was in 1957 when, as a member of the Victoria and Albert Artwork Room, I was asked to conserve sixteenth and seventeenth century gloves with beautiful embroidered cuffs. I knew little about leather. It was essential to learn about the methods of turning skins into leather and how they could be recognized. Available written information did not begin at the beginning.

It was then I met Dr Claude Spiers. Claude was a senior lecturer at the Leathersellers' Technical College in Bermondsey and he invited me to visit. There he showed me the vats in the floor where the skins were held in suspension in the various processing liquors and explained how tanning works. He then arranged a meeting with John Waterer; designer, antiquarian, author, historian and leather craftsman. John guided me through the conservation of the superfine tawed skins of the gloves and later wrote the chapter on leather for *Textile Conservation*, published by Butterworth in 1972. It was in the same year that his *Guide to the Conservation and Restoration of Objects made Wholly or in Part of Leather* was published for the International Institution for Conservation. These are still excellent introductions but *The Conservation of Leather and Related Materials* widens the scope to the benefit of collectors, conservators, curators and anyone with responsibility for the care of leather objects. It outlines the history and development of the different types of tanning and what makes each type of skin and each type of tanning suitable for particular purposes. Most importantly, it describes how to recognize skin patterns and treatments. Finally the case studies indicate the range of treatments available for the preservation of this often overlooked segment of our cultural heritage.

Karen Finch OBE

# Dedications

**John W. Waterer R.D.I., F.S.A., F.I.I.C., 1892–1977**

## 'FITNESS FOR PURPOSE'

This book is dedicated to John Waterer. Although John died in 1977, his lifelong involvement with leather was such that, without the interest, influence and enthusiasm he created it is doubtful whether this book could have been written. Much loved and respected, with an ever-ready smile, he epitomized Chaucer's words in the *Canterbury Tales* – 'To any kind of man he was indeed the very pattern of a noble Knight.'

John was born in South London in 1892 and after leaving school was invited in 1909 to join a well-known leathergoods company as an apprentice in their luggage department. Although John had very considerable career prospects as a talented musician, this proved, almost by chance, to be the stepping stone to his lifetime's work. After a break in the Navy during the Great War he rejoined his old company and became increasingly involved in the design and creation of the new 'lightweight' luggage, being increasingly demanded by the travelling public due to the evolution of the small inexpensive motor car and the slow but steady growth in air travel.

With the knowledge thus gained, in 1936 John joined S. Clarke & Co., a well-established but progressive travel goods manufacturer, as managing director. John was then able to fulfil his design flair but always with 'Fitness for Purpose' in his mind – a guiding principle throughout this life. After three exciting years came the Second World War. By then John was 47 years of age, happily married with a daughter and at the peak of his professional skill and ability.

The war years had a profound influence on John Waterer's life. With all its attendant problems, including bomb damage, S. Clarke & Co. continued making luggage but with part of its production given over to war work. With his ever-enquiring mind, John found time – possibly during the long hours of fire watching – to begin his research into the history of leather and its early uses. This led to a well-received lecture to the Royal Society of Arts in 1942 for which he subsequently received their Silver Medal. At the same time both the government and trade association set up committees to consider the best way forward in the immediate post-war years, little realizing that the years of difficulty and austerity would linger on until well into the 1950s. Here John preached his gospel: a vision of a better future where design and fitness for purpose would be paramount, overcoming the innate conservatism of manufacturers, by encouraging them to embrace the benefits that good design would bring to the manufacturing process.

All this led to the publication in 1946 of *Leather in Life, Art and Industry*. Although in later years John wrote many further well-researched books, this book set him up as an outstanding leather historian and authority and can truly be regarded as his magnum opus. If that was not enough, John was then instrumental in setting up the Museum of Leathercraft to enable others to see the use and evolution of leather over the ages, thereby fostering design and craftsmanship in the years to come.

John was by now conducting a worldwide correspondence on leather-related matters. In 1953 his total virtuosity resulted in his being elected to the faculty of Royal Designers for Industry. This appointment is considered the highest honour to be obtained in the United Kingdom in the field of industrial design and shows the high regard in which he was held by his contemporaries. In the same year he was also admitted to the Livery of the Worshipful Company of Saddlers, with whom he had a long, friendly and supportive association in the years that followed.

John remained as managing director of S. Clarke & Co. until the early 1960s, producing modern looking luggage designs which have stood the test of time. It was then by a turn of fate that Clarke's was acquired by the company he had joined way back in 1909! John was then 71 years 'young' but with undimmed enthusiasm and no concept of the meaning of retirement – it seems to have slipped his mind – which enabled him to give his increasing free time to further his research into leather history. This led to his realization that although there were many beautiful and historic leather artefacts there was little or no knowledge as to how they might be conserved for the benefit of future generations. After considerable research this led to his writing his *Guide to the Conservation and Restoration of Objects made Wholly or in Part of Leather*, first published in 1972, and his election as Fellow of the International Institute for Conservation.

His vision also led to the creation of the Leather Conservation Centre in 1978. The Centre is now housed in purpose-built premises in Northampton, through the generosity of the Worshipful Company of Leathersellers. John did not live to see this, but together with the Waterer/Spiers Collection, it is a fitting memorial to a very special and dedicated man whose like will not come again. The Waterer/Spiers Collection was the inspired decision of the Council of the Museum of Leathercraft, taken after John's death, to commission each year an article in leather to show the best in contemporary design, skill and workmanship. It was decided to conjoin his friend Claude Spiers – a leather chemist – who had been instrumental with John in setting up the museum during the Second World War. This annually growing collection now provides an outward and visible sign that leather design, excellence and workmanship, which John spent his life preaching and encouraging, still prosper.

Peter Salisbury

## Betty M. Haines MBE, B.Sc., F.R.M.S., F.S.L.T.C., 1925–2003

Betty Haines, whose name is known throughout the conservation world as a writer and teacher on all aspects of collagen, skin and leather science, died following a short illness while this book was being brought together.

Betty graduated from Chelsea College of the University of London in 1945 with a B.Sc. in Botany, Chemistry and Zoology. She joined the British Leather Manufacturers' Research Association in 1946 becoming one of a line of eminent lady scientists employed by them from its foundation in 1920 to the present day. Working in the Biology Department she applied her knowledge of protein science, bacteriology and entomology in the fields of hide and skin quality and the pretanning processes. In particular she developed the field of leather microscopy first using conventional light microscopes and later with the new electron microscopes.

One application of this microscopical expertise was with the identification of archaeological material and Betty's advice was sought by major museums throughout the UK. This led to collaboration with Dr Baines-Cope of the British Museum Research Laboratory which culminated in the publication of *The Conservation of Bookbinding Leather* in 1984.

It was in 1978 while this work was being undertaken that Betty was invited to join the Trustees of the newly formed Leather Conservation Centre. She was elected Chairman of the Technical Advisory Panel in 1984, Chairman of Trustees in 1987 and President from 1999.

During this period she contributed to summer schools and wrote a series of monographs for the Centre. She also lectured to students and gave papers at professional conferences and seminars both in the UK and abroad.

The chapters prepared by Betty for this volume will, sadly, be her last written contributions in a series of publications stretching over half a century. Her deep knowledge of leather and its conservation will, however, remain in the memories of those who were privileged to know or work with her.

Roy Thomson

# Acknowledgements

The editors wish to thank the many contributors to this volume for their hard work and patience during the editorial process. Particular appreciation is expressed to the Victoria and Albert Museum and the Leather Conservation Centre for permission to spend time on the preparation and editing of this work and to our respective colleagues there for their support.

We would like to thank Jodi Cusack and Stephani Havard at Butterworth-Heinemann and also Neil Warnock-Smith who was our first point of contact.

Thanks also must go to Carole Spring for her help in the preparation of the texts and to Stephen Kirsch for supplying an almost impossible to obtain image of a sewing machine used to sew furs and gloves.

We would both like to thank our respective spouses, John and Pat, for their unfailing help, encouragement and tolerance throughout this project.

Marion Kite
Roy Thomson

# Contributors

**Priscilla Anderson**

Priscilla Anderson was awarded a Batchelor of Arts *cum laude* majoring in the History of Art from Yale University in 1990. She also holds a Master of Library Science from the University of Maryland and a Master of Science in Art Conservation from the Winterthur/University of Delaware program. Following internships at the Wilson Library, University of North Carolina; the Walters Art Museum, Baltimore and the University of Maryland Libraries, she worked as a conservator/rare bookbinder at the Library of Congress. She is now a Special Collections Conservator at the Weissman Preservation Centre of the University of Harvard Library. She is a Professional Associate Member of the American Institute for Conservation.

**Aline Angus**

Aline Angus was educated in Scotland and has an honours degree in Ancient History and Archaeology from the University of Durham. She gained a Higher National Diploma in Conservation and Restoration at Lincolnshire College of Art and Design in 1992. She has worked on ethnographic collections at the Horniman Museum in London and the Royal Albert Museum in Exeter. She was at the Royal Museum in Edinburgh for three years preparing 18c and 19c objects for the new Museum of Scotland. She has spent seven years at the Leather Conservation Centre, Northampton.

**David Brock**

After studying at the University of Texas at Austin and being awarded a degree majoring in Photographic Studies at the Colombia College of Chicago, David Brock received his first instruction in hand bookbinding from Joan Flasch and Gary Frost at the Art Institute of Chicago in 1977. In the following year he began a six year apprenticeship with William Anthony in hand bookbinding and conservation. This was followed by five and a half years as a Rare Book Conservator at the Library of Congress. In 1990 David became a conservator in private practice and ran his own business for eight years, closing it in 1998 to assume his current position as Rare Book Conservator for Stanford University.

**Anthony Cains**

Anthony Cains was indentured to a London trade bookbinder in 1953. As part of his training he attended the London School of Printing where he received several prizes. During his National Service he studied under William Matthews at Guildford who recommended him to Douglas Cockerell and Sons where the foundation of his career in book and manuscript conservation was laid. He served both the British and American funded rescue teams after the Florence floods of 1966, being appointed Technical Director of the programme set up in the Biblioteca Nazionale Centrale Firenze. He was subsequently invited to design and establish a workshop in the Library of Trinity College Dublin which he ran until his retirement in 2002. He is a founding director and committee member of the Institute for the Conservation of Historic and Artistic Works in Ireland.

**Esther Cameron**

After reading Archaeology at Birmingham University, Esther Cameron trained in Archaeological Conservation at Durham University, gained a Masters degree and later went on to complete a doctorate at Oxford University. She has worked for the Wiltshire and Kent County Museums Services and for the Institute of Archaeology at the University of Oxford. She is now a freelance archaeological finds specialist working on a range of materials including leather. She is a Fellow of the Royal Society of Antiquaries of London and has served on the executive committees of the United Kingdom Institute for Conservation and the Archaeological Leather Group. She is a Trustee of the Leather Conservation Centre.

## Anthony Covington

Tony Covington is Professor of Leather Science at the British School of Leather Technology at University College Northampton. He is also Visiting Professor at Sichuan Union University, Chengdu, China and Nayudamma-Wahid Professor at Anna University, Chennai, India. He studied for Graduateship of the Royal Institute of Chemistry at Teesside Polytechnic and was awarded a doctorate at Stirling University in Physical Organic Chemistry. Before joining University College Northampton he carried out research at BLC the Leather Technology Centre for eighteen years. He is Past President of the Society of Leather Technologists and Chemists and of the International Union of Leather Technologists and Chemists' Societies. He is a Fellow of the Royal Society of Chemistry and the Society of Leather Technologists and Chemists.

## Caroline Darke

Caroline Darke graduated from St Martins School of Art with a National Diploma in Design (Fashion). Running her own business SKIMP she produced bags, belts, small leather goods and fashion accessories for major shops and stores in UK, USA, Europe and Japan. She has taught part time at Manchester College of Art, Guildford School of Art, St. Martins School of Art, Croydon College of Art and Brighton School of Art. From 1965–94 she was Associate Lecturer at London College of Fashion, from 1994–2000 Associate Lecturer and Accessories Co-ordinator at Cordwainers College and from 1995 MA Accessories course leader at Royal College of Art. In 2000 Caroline was appointed Course Director Professional Development Unit-Cordwainers at London's University of Arts.

## Laura Davies

Laura Davies graduated with a Fine Art Degree from Staffordshire University specialising in Sculpture. She then studied for a Masters degree at the Royal College of Art/Victoria and Albert Museum joint course in Conservation. During the three year duration of the course she was placed in the Applied Arts Conservation Department of the Museum of London for the practical content of the course where she gained experience with *cuir bouilli* objects. In 1999 she was awarded the Museums and Galleries Commission Student Conservator of the Year Award. After graduating she spent a year as an Objects Conservator at London's National Museum of Science and Industry. She is now a Sculpture Conservator at the Tate Gallery.

## James Dickinson

In 1968 James Dickinson was awarded a Carnegie UK/Museums Association bursary to study taxidermy. This enabled him to train at various UK, German and Swiss museums. In 1973 he was appointed Senior Conservator Natural History at the North West Museum Service, working on material from museums all over north of England. In 2001 he became the Conservation Officer Natural Sciences for the Lancashire County Museum Service. He is a Founder Member and former Chair of the Guild of Taxidermists. In 1990 he was appointed a Member of the Order of the British Empire for services to taxidermy. In 1991 he became a Fellow of the Museums Association.

## Sherry Doyal

In 1981 Sherry Doyal was awarded a City and Guilds Certificate with distinction in Conservation and Restoration Studies from the Lincoln College of Art. In 1984 she gained a post graduate Certificate in Upholstery Conservation from the Textile Conservation Centre and was subsequently engaged as a conservator of furnishing textiles and upholstery by the TCC, the Crown Suppliers, the Metropolitan Museum of Art and the Victoria and Albert Museum. From 1991–94 she was the National Trust House and Collections Manager at Ham House. From 1995 Sherry pursued her interest in ethnography and natural history conservation, first at the Horniman Museum and then Exeter City Museums.

From 1999 she combined a part time position as Natural Trust Conservator and latterly Regional Historic Properties Advisor with freelance enthnobotanical conservation. In February 2005 Sherry was appointed Deputy Head, Conservation and Collections Care at the Horniman Museum and Gardens, London. She is a Trustee of the Leather Conservation Centre.

## Don Etherington

Don Etherington began his career in conservation and bookbinding in 1951 as an apprentice after which he worked as a conservator for the British Broadcasting Corporation and Roger Powell and Peter Waters. Between 1967 and 1969 he was a training consultant at the Biblioteca Nazionale in Florence where he trained workers in book conservation practices after the 1966 flood. Between 1960 and 1970 he was a lecturer at Southampton College of Art in England where he developed a four year programme in bookbinding and design. From there he went to the Library of Congress in Washington DC where he served as a Training Officer and Assistant Restoration Officer. In 1980 Mr Etherington became Assistant Director and Chief Conservation Officer at the Harry Ransom Humanities Research Center at the University of Texas in Austin. In 1987 he joined Information Conservation, Inc. located in Greensboro, North Carolina where he created a new conservation division for the preservation of library and archival collections. He is now President of the Etherington Conservation Center, Greensboro, North Carolina. He is an Accredited Member of the Institute of Paper Conservation and Fellow of both the American Institute of Conservation and the International Institute of Conservation.

## Mary-Lou E. Florian

Mary-Lou Florian is Conservation Scientist Emerita and Research Associate at the Royal British Columbia Museum. She has a Bachelors and Masters degree in biology specialising in fungi, insects and plant anatomy. Her first introduction to conservation was as a Biologist at the Conservation and Restoration Research Laboratory at the National Gallery of Canada in the early 1960s. She later worked as a Senior Conservation Scientist in Environment and Deterioration Services at the Canadian Conservation Institute in Ottawa. In 1978 she went to the Royal British Columbia Museum in Victoria, British Columbia as a Conservation Scientist and retired as Head of Conservation Services there in 1991. In her present capacity as Research Associate at the Museum she is studying fungal stains and archaeological wood identification. She is a Lifetime Honorary Member of the American Institute of Conservation and besides other professional excellence awards has been awarded the 125th Commemorative Medal from the Governor General of Canada.

## Rudi Graemer

Rudi Graemer received his early education in Switzerland and in 1953 was awarded a First Class Diploma from the National Leathersellers College in London. His wide experience in technical management in the leather trade includes work in the UK, Switzerland, Australia and in the former Belgian Congo. He returned to the UK to work with the specialist reptile and exotic leather manufacturers, T. Kinswood and Co. in 1960 from where he retired as Managing Director in 1990.

## Betty Haines

See dedication page xiii.

## Robert D. Higham

Robert Higham qualified in leather technology at the National Leathersellers College, London, in 1959 and served in tannery technical management in Bolton, Galashiels and Edenbridge until 1969. In that year he became Technical Editor of *Leather*, the international journal for that industry, becoming Editor a few years later. In addition he carried out *ad hoc* consultancy work for several UN agencies. He moved to Aberdeen in 1980 to study for the Church of Scotland Ministry where he was awarded the degree of Batchelor of Divinity. During this period he continued as Consultant Editor of *Leather* and with consultancy for UNIDO. He retired from parish ministry in 2002 having served in Berwickshire and latterly the Isle of Tiree.

## Marion Kite

Marion Kite studied Textiles and Fashion at Goldsmiths College School of Art where she was awarded a Batchelor of Arts specialising in goldwork embroidery. She is a Senior Conservator in the Textile Conservation Section of the Victoria and Albert Museum having worked there since 1974 and where she developed a particular interest in the conservation of animal products and other unusual material incorporated into textiles and dress accessories. She served on the Directory Board of the International Council of Museum Committee for Conservation between 1993 and 1999 and as Treasurer between 1993 and 1996. She is a Fellow of the International Institute of Conservation and currently serves on the IIC Council. She is Chairman of the Council of Trustees of the Leather Conservation Centre and also sits on the Council of the Museum of Leathercraft. She is a Trustee of the Spence and Harborough collections of Gloves administered by the Worshipful Company of Glovers of London. She is an Accredited Conservator Restorer and a Fellow of the Royal Society of Arts.

## William Minter

Bill Minter was awarded a BSc in Industrial Technology in 1970 from the Stout State University in Menomonie. After completing seven years apprenticeship with the book conservator and fine bookbinder William Anthony, he set up his own workshop specialising in the binding and conservation of rare books and manuscripts which now operates from Woodbury, Pennsylvania. Included among his innovations for book conservation is the development of an ultrasonic welder for polyester fill encapsulation. He is a Professional Associate Member of the American Institution for Conservation and has served as President of both their Book and Paper and Conservator in Private Practice Groups.

## Olivia Primanis

After studying at the State University of New York at Albany majoring in English Literature and being awarded a Batchelor of Arts degree in 1973, Olivia Primanis began her training through an apprenticeship in hand book binding and book conservation at the Hunt Institute of the Carnegie Mellon University in Pittsburgh. Concurrently, she opened 'The Bookbinder' which offered artists' supplies and bookbinding services for individuals and institutions. In 1984 she moved to Los Angeles and continued her private practice of conservation bookbinding and teaching. Since 1990, Ms Primanis has held the position of Senior Book Conservator at the Harry Ransom Humanities Research Center at the University of Texas at Austin where she undertakes conservation treatments, teaches and participates in departmental administration. She serves on the Book and Paper Group Publication Committee of the American Institute for Conservation.

## Alan Puglia

Alan Puglia was awarded a Batchelor of Arts degree from the University of New Hampshire in 1986. Following studies in conservation at the George Washington University in Washington and the University of Texas at Austin, he was awarded the degree of Master of Library and Information Science and a Certificate of Advanced Study in Library and Archives Conservation. Having worked for a number of institutions in the field of book and archives conservation for ten years he was appointed Conservator for the Houghton Library Collections at Harvard University in 1999.

## Sarah Reidell

Sarah Reidell graduated from Bryn Mawr College and then studied for a Masters degree in Library and Information Science and a Certificate in Advanced Studies at the University of Texas in Austin. Having worked for a period as visiting conservator in France and Spain she undertook internships at the Center for American History in Austin, Harvard University Library and as Mellon Advanced Intern at the Conservation Center for Art and Historic Artifacts in Philadelphia. In 2003 she was appointed Conservator for Special Collections at the Harvard University Library.

## Glen Ruzicka

Glen Ruzicka was awarded the degree of BA at the Emory University in Atlanta in 1971. He then trained in rare book conservation at the Library of Congress where he worked for over ten years. From 1986 to 1988 he served as Head of the Preservation Department of the Milton S. Eisenhower Library, John Hopkins University, Baltimore. In 1988 he was appointed Chief Conservator of the Conservation Center for Art and Historic Artifacts in Philadelphia where he is now Director of Conservation. He is a Professional Associate Member of the American Institute for Conservation where he served as Chair of the Book and Paper Group. He is a member of the Board of Directors of the Pennsylvania Preservation Consortium and a member of the Historic Buildings and Collections Committee of Girard College, Philadelphia.

## Randy Silverman

Randy Silverman has worked in the field of book conservation since 1978 and was awarded a Masters degree in Library Science from the Brigham Young University in 1986. Having worked as conservator and preservation librarian at the Brigham Young University he was appointed as the Preservation Librarian at the University of Utah's Marriott Library in 1993. Mr Silverman initiated the passage of Utah's permanent paper law in 1995. He is a Professional Associate Member of the American Institute for Conservation and has served as co-chair of their Library Collections Conservation Discussion Group, as a member of the Institute's National Task Force on Emergency Response, and as President of the Utah Library Association. He is also an Adjunct Professor with Emporia State University (Kansas), the University of Arizona and the University of North Texas.

## James Spriggs

Jim Spriggs studied conservation at the Institute of Archaeology, University College London where he was awarded a Diploma in Archaeological Conservation. He is Head of Conservation at the York Archaeological Trust. His department specialises in the study and conservation of all types of archaeological material from excavations in York and elsewhere from both land based and marine environments. He is an Accredited Conservator Restorer, a Fellow of the Society of Antiquaries of London and of the International Institute of Conservation. He is a Founder Member of the York Consortium for Conservation and Craftsmanship.

## Theodore Sturge

Theo Sturge trained in conservation at the Institute of Archaeology, University College London in the 1970s. On leaving college he worked at Leicester Museum as Assistant Keeper, Antiquities Conservation, for 16 years. This was followed by six years as Senior Keeper, Conservation and Restoration at the Herbert Art Gallery and Museum in Coventry and six years as Senior Conservator at the Leather Conservation Centre, Northampton. In 2000 he set up his own studio specialising in leather conservation. He is an Accredited Member of the United Kingdom Institute for Conservation and a Fellow of the International Institute for Conservation.

## Roy Thomson

Roy Thomson was awarded the degree of BSc with Honours in the Chemistry of Leather Manufacture from the University of Leeds in 1960. He worked in research and technical services associated with the leather trades until 1968 when he was appointed Works Director responsible for technical and production management at the largest lambskin clothing leather tannery in the UK. In 1994 he was appointed Chief Executive at the Leather Conservation Centre from where he retired in 2004. He is an Accredited Conservator, Fellow of the Royal Society of Chemistry, a Fellow and Past President of the Society of Leather Technologists and Chemists and Fellow of the International Institute for Conservation. He is Past Chairman of the Council of the Museum of Leathercraft and Treasurer of the Archaeological Leather Group.

## Barbara Wills

Barbara Wills trained in conservation at Lincoln College of Art. She joined the Department of Conservation at the British Museum in 1979 and completed a Museums Association Certificate in Ethnographical Conservation in 1984. As Senior Conservator in the Organic Artefacts Section, she specialises in the treatment of leather, basketware and Ancient Egyptian material. She is an Accredited Member of the United Kingdom Institute for Conservation and has served on the committee of the Archaeological Leather Group for a number of years. She is a Trustee of the Leather Conservation Centre.

## Christopher S. Woods

Chris Woods gained a Post Graduate Diploma in Library and Archive Conservation from Colchester Institute following an Art History degree from

Sheffield Art College. He worked for fifteen years, first as Conservator and then Head of the Preservation Division for the Dorset Archives Services. He was appointed as Head of Collection Care and Conservation for Oxford University Library Service at the Bodleian Library in 2002 responsible for the care of the many and varied library and archive collections in the 40 Oxford University Library Service sites. He is an Accredited Conservator Restorer, a Fellow of the International Institute for Conservation and serves as Chairman of the United Kingdom Institute for Conservation.

## Friederike Zimmern

Following an apprenticeship in bookbinding in Hamburg, Friederike Zimmern studied at the Academy for Art and Design at Stuttgart and was awarded her Diploma in the Restoration and Conservation of Books, Paper and Archives in 1998. After working for restoration companies in Germany she obtained an advanced level internship at the Straus Center for Conservation at Harvard University Art Museums. In February 2002 she was appointed as the Head of the paper conservation workshop of the graphic art collection of the Hessisches Landesmuseum in Darmstadt.

## Paula Zyats

Paula Zyats studied at Temple University Tyler School of Art in Rome and Philadelphia College of Art and was awarded a Batchelor of Fine Art degree specialising in illustration in 1987. Having become involved with book conservation, she completed a Master of Science degree with a Certificate in Art Conservation from the Winterthur/University of Delaware Art Conservation Program. This involved internships served at Columbia University Libraries, the Library of Congress and the Folger Shakespeare Library with a Mellon Advanced Internship at the Conservation Center for Art and Historic Artifacts in Philadelphia. In 1998 she was appointed conservator at the CCAHA and in 2004 became Assistant Chief Conservator at the Yale University Library Preservation Department.

# 1

# The nature and properties of leather

*Roy Thomson*

Man and his early ancestors have exploited the unique properties of skin and leather for millennia and almost all human cultures have developed specialist techniques to utilize this readily available raw material for a wide variety of purposes. Indeed, tanning has been described as man's first manufacturing process. But what are the properties which make these skin-based products so special?

To begin with, leather is a sheet material with the area of each piece ranging from tens of square centimetres to six, seven or more square metres depending on the animal from which it was obtained. Until the development of woven textiles it was the only material available in sheets of this size.

Then there is the complex physical structure of skin and materials made from it. A close examination of the make-up of a piece of skin shows that it consists primarily of long thick fibres and fibre bundles interweaving in three dimensions within a jelly-like 'ground substance'. Other features such as hairs and hair roots, muscles, blood vessels and fat cells are present but it is this intricate, three-dimensional, woven structure that predominates and gives skin-based materials many of their unique physical qualities.

These properties include flexibility, a relatively high tensile strength with particular resistance to shock loads, resistance to tearing, puncturing and abrasion, low bulk density, good heat insulation and water vapour transmission. They also include mouldability, resistance to wind and liquid water, and an ability to be stretched and compressed without distorting the surface.

Many of these characteristics are common to both leather and other skin products but linguistic studies suggest that the various materials such as raw hide, oil-tanned pelt, alum-tawed skin and vegetable-tanned leather were differentiated from each other from early times. It was not until the late eighteenth century though that the actual nature of the tanning process was examined and the question posed as to how leather was different from these other materials.

A number of criteria have been put forward in an attempt to define what is a true leather (Bienkiewicz, 1983; Covington, 2001; Lollar, 1958; Reich, 1999). These will be considered.

A fundamental property of leather is that while a raw skin is subject to rapid bacterial degradation due in the main to the action of proteolytic enzymes, leather is resistant to such microbiological attack even if it is kept wet. There are, though, a number of techniques such as salt curing, drying, solvent dehydration and acid pickling which will impart temporary preservation against bacterial attack. This resistance to decay, however, is lost if the fibres are allowed to become wet. Similarly the effects of the treatments involved in the preparation of parchment or alum-tawed skins, both renowned for their longevity, are reversed by repeated immersion in water.

Skin-based materials are prepared by many indigenous peoples around the world by thoroughly impregnating the raw hide with fatty materials and then allowing it to dry out under carefully controlled conditions. The fats coat the individual skin fibres and fill the spaces between them. Even if the treated hides are then immersed in water, the presence of these water-repellent fats ensures that the fibres remain too dry for bacterial attack to take place. They therefore appear to satisfy the criteria of resistance to microbiological degradation. These products, which are found widely in ethnographic collections, have been termed pseudo leathers.

These pseudo leathers should not be confused with oil-tanned skins which are not treated with stable, water-resistant fats but with reactive, oxidizible oils often obtained from marine animals. These undergo various chemical changes during processing to liberate compounds with true tanning actions. Examples of these oil-tanned products include chamois wash leathers and the buff leather employed widely in the sixteenth and seventeenth centuries to make protective jerkins for the military.

Another characteristic attributed to leather is that whereas if a raw skin is allowed to dry out it is expected that it will become hard, horny, brittle and translucent, a true leather is said to dry to give a soft, flexible, opaque product.

It is true that if a raw skin is allowed to dry in an uncontrolled manner it is likely to give a product with the properties described. If, though, the rate of drying is regulated as with the production of the pseudo leathers described above, a soft opaque material results. Similarly if a dehaired skin is dehydrated by immersion in successive baths of a polar solvent such as acetone or one of the lower alcohols and the residual solvent evaporated, the resultant product will be soft, white, opaque and flexible. This flexibility will be enhanced by working the skin mechanically while it is still only just damp with the solvent. It will look and feel very similar to an alum-tawed or formaldehyde-tanned leather. These characteristics of solvent-dried skins are utilized in the solvent dehydration methods employed to conserve waterlogged archaeological leathers.

Parchment and vellum which are prepared by drying unhaired pelts under tension also exhibit many of the physical properties of a true leather.

The different properties of the various untanned products made in the past depended on the amount and type of oil used to treat the unhaired skin and the rate of drying. These properties enabled these materials to be used for such diverse purposes as mallet heads, textile machinery parts and the protective corners of basketwork skips. A modern successor to the latter is the use of rawhide to protect the corners and bottoms of baskets used by hot air balloonists. It is the unique combination of impact and abrasion resistance together with an elastic resilience which makes this age old material ideal for its modern purpose.

While leathers produced for gloving and clothing are soft and supple, those made for shoe soleing are firm and resilient. In the period when the technique of chrome tanning was being developed during the last quarter of the nineteenth century it was found that while a stable product could be made, this new type of leather was liable to dry out to give a hard, cracky, inflexible material, in many ways similar to untanned skin. It was only with the introduction of the fatliquoring process, which coated the tanned fibres with oils, that a material could be manufactured with the properties required for it to be recognized as a true leather.

If a piece of wet skin, tanned or untanned, is heated slowly it will reach a temperature at which it shrinks dramatically to about one third of its original area. This phenomenon has been likened to melting but is fundamentally different. The hydrothermal shrinkage of skin is irreversible and rather than being caused by a single physicochemical change is the cumulative result of a number of intermolecular processes.

The temperature at which this change takes place is termed the shrinkage temperature and the amount by which any process increases the shrinkage temperature of a skin has often been considered as a measure of its leathering ability.

The shrinkage temperature of a given sample of skin will depend on a large number of factors. These include the species and age of the animal from which the skin is obtained, what pretanning and tanning treatments the skin has undergone, the moisture content of the sample and the exact procedures employed in the determination. If, however, care is taken to carry out the measurement in a standardized manner, duplicate results within 1 or 2°C can be obtained. Using methods described in international standards, the following shrinkage temperatures are exhibited by typical commercial products:

| | |
|---|---|
| Raw mammalian skin | 58–64°C |
| Limed unhaired cattle hide | 53–57°C |
| Parchment | 55–64°C |
| Oil-tanned leather | 53–56°C |
| Alum-tawed skins | 55–60°C |
| Formaldehyde-tanned leather | 65–70°C |
| Alum-tanned skins | 70–80°C |
| Vegetable-tanned leather (hydrolysable) | 75–80°C |
| Vegetable-tanned leather (condensed) | 80–85°C |
| Chrome-tanned leather | 100–120°C |

Most of these results confirm that tannage enhances the shrinkage temperature. There is, however, an anomaly with oil-tanned skins such as chamois-tanned wash leathers or the brain-tanned 'elk skins' produced by Native American and other cultures. In these cases the stabilizing process does not increase the shrinkage temperature. These products exhibit all the characteristics of true leathers and what is more they retain these after frequent washing and drying in use. Oil-tanned leathers also exhibit another significant difference in their hydrothermal properties. When other skins and leathers shrink in hot water they turn into a rubbery material which dries to a hard, brittle, product. When oil-tanned materials shrink they retain their leathery handle and when dried retain their softness and flexibility to a significant extent. In addition, if a wet oil-tanned leather is heated above its shrinkage temperature and then immersed in cold water it can be stretched back to nearly its original size.

These exceptions to the various criteria proposed to define what is and what is not a true tannage have led to attempts to explain the conversion of skin into leather according to the mechanisms involved.

For many years it has been accepted that the cohesion of skin fibres is a result of the structure of the collagen protein molecules from which these fibres are formed. These have been shown to be held together by a combination of a few, relatively strong, covalent bonds and many weak hydrogen bonds. It has been thought that hydrothermal shrinkage occurs when the disruptive energy introduced by heating the sample exceeds the cohesive strength of the bonding within and between collagen molecules. Tannage has been thought to introduce extra chemical crosslinking bonds between adjacent collagen molecules which are resistant to microbiochemical attack. The nature and strength of these crosslinkages vary considerably depending on the type of tanning material employed. Vegetable tannage for instance is thought to introduce many extra hydrogen bonds between free amino side groups of the collagen protein and hydroxyl groups from the polyphenolic tannin molecules. Chrome tannage on the other hand is a result of side chain carboxylic groups on the protein molecule co-ordinating with the multinuclear chromium complexes present in chrome tanning liquors. The differences in the increase in shrinkage temperature brought about by the different tanning systems has been thought to be related to the combined strength of these crosslinking bonds.

Recent work has shown that the energy associated with the hydrothermal shrinkage is similar for all the different tannages irrespective of the temperature at which the shrinkage occurs. This has led to the concept of the formation of a supramolecular matrix around the collagen molecule during tanning and that it is the size and complexity of this matrix which determines shrinkage temperature. This mechanism does not preclude the presence or importance of crosslinking reactions occurring during tanning but it does explain why oil tannage can be considered to give a true leathering effect without increasing the shrinkage temperature.

Although indicating the complexity of the problem, the question of what exactly leather is has not been fully answered by the above discussion. However, a definition which appears to take into account the points raised is as follows.

Leather is a material produced from the skin of a vertebrate, be it mammal, reptile, bird, fish or amphibian, by a process or series of processes which renders it non-putrescible under warm moist conditions. A true leather retains this property after repeated wetting and drying. Leather usually dries out to give a relatively pliable, opaque product but it can be hard or soft, flexible or rigid, stiff or supple, thick or thin, limp or springy, depending on the nature of the skin used and the process employed.

It has been the aim of the tanner throughout the ages to manufacture a product with just the combination of properties demanded by the end user.

It should always be borne in mind that in a similar way to 'metal' or 'wood', leather is not a single material but a group of related products having many characteristics in common but each varying in its properties and reaction to conservation treatments.

## References

Bienkiewicz, W. (1983) *Physical Chemistry of Leather Making*. Malabar: Krieger, pp. 308–323.

Covington, A.D. (2001) Theory and Mechanism of Tanning. *J. Soc. Leather Technologists and Chemists*, **85**, 24.

Lollar, R.M. (1958) Criteria Which Define Tannage. In *Chemistry and Technology of Leather Vol. II* (O'Flaherty, F. et al., eds), pp. 1–27. New York: Reinhold.

Reich, G. (1999) The Structural Changes of Collagen During the Leather Making Processes. *J. Soc. Leather Technologists and Chemists*, **83**, 63.

# 2

# Collagen: the leathermaking protein

B.M. Haines

Collagen is the major protein from which skin is formed and its unique structure is fundamental to the leathermaking process. A knowledge of the nature of this protein is therefore required if the properties of leather are to be understood. The chemical and physical nature of collagen have been reviewed, among others, by Bailey (1992), Bailey and Paul (1998), Kennady and Wess (2003), Reich (1995), Ward (1978), and Woodhead-Galloway (1980).

## 2.1  The collagen molecule

The collagen molecule, like all proteins, is formed by the linking together of naturally occurring smaller units, amino acids.

All amino acids contain a carboxyl and an amino group, with a side chain denoted as 'R' (*Figure 2.1*). Amino acids differ only in the nature of their side chains.

With the simplest amino acid, glycine, the side chain is a single hydrogen atom (*Figure 2.2*). With other amino acids the side chains may be short or long, non-polar and therefore chemically inert, or polar and chemically reactive.

Non-polar side chains contain only carbon and hydrogen atoms. Polar side chains on the other hand contain oxygen, present as hydroxyl or carboxyl end

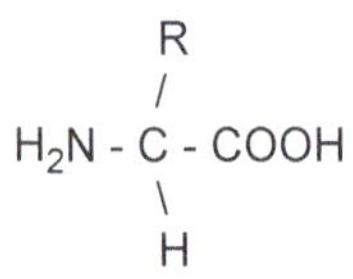

$$H_2N - \underset{\underset{H}{|}}{\overset{\overset{R}{|}}{C}} - COOH$$

**Figure 2.1**  The structure of amino acids.

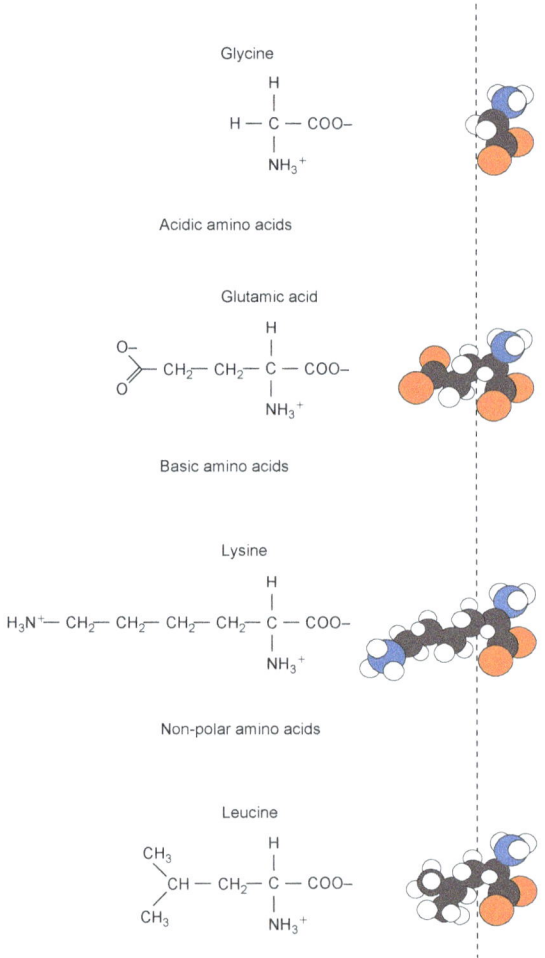

**Figure 2.2**  Structures of amino acids showing the different sizes of common side chains.

groups and therefore acidic in nature, nitrogen, present as amino or amide end groups and therefore basic in nature, or sulphur, present as mercaptan groups.

The amino acids are linked by covalent peptide bonding between the carboxyl group of one amino acid and the amino group of an adjacent amino acid. This formation of peptide bonds involves the loss of water in a condensation reaction (*Figure 2.3*). In this way numerous amino acids are linked to form a long chain, or protein backbone.

All proteins have identical backbones: the distinctive character of each protein lies in the particular sequence of amino acids along the chain.

Collagen is composed of about 20 different amino acids, linked to form a chain 300 nm long containing approximately 1000 units. Collagen is characterized by a high proportion of glycine (30%) and by the presence of the imino acids proline (10%) and hydroxyproline (10%) (*Figure 2.4*). Hydroxyproline is formed from proline after the backbone chain has been synthesized. Hydroxyproline is found only rarely in proteins other than collagen and so it is used to identify the presence of collagen in a sample or to determine the collagen content of a sample.

Many segments of the backbone chain of collagen consist of simple tripeptide repeats of glycine, X, Y, where X is frequently proline or hydroxyproline. The spatial shape, the ring structure, of proline and hydroxyproline twists the chain into a helical coil, a left-handed helix with three amino acids per twist with glycine occupying every third position along the chain (*Figure 2.5*).

The collagen molecule is made up of three such helical chains, which is the reason for the molecule being termed the triple helix. These three chains twist together to form a right-handed coil (*Figure 2.6*).

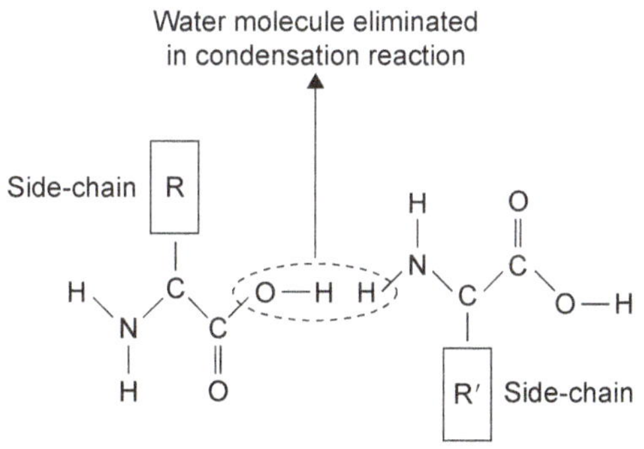

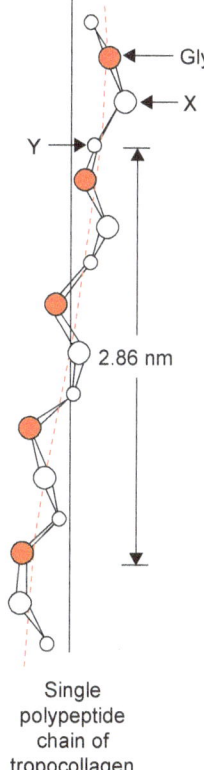

**Figure 2.3** The reaction between two amino acids to form a peptide.

**Figure 2.5** Single helical polypeptide chain of tropocollagen.

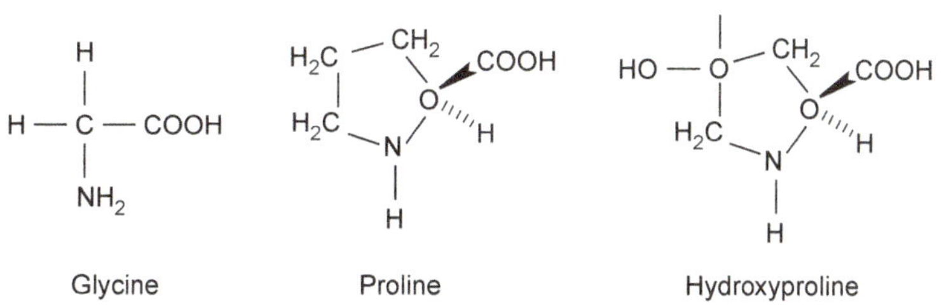

**Figure 2.4** The structures of the most common amino acids present in collagen.

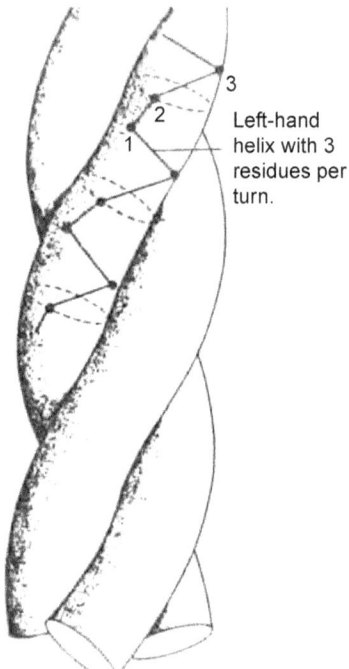

**Figure 2.6** Three tropocollagen helices coiled together to form the collagen molecule.

## 2.2 Bonding within the molecule

The coiling holds the three chains together but the triple helix is further stabilized by chemical crosslinks between the three backbone chains.

Hydrogen bonds form between the NH and the CO groups of adjacent chains (*Figure 2.7*). Hydrogen bonds are electrostatic and their stability depends on the distance separating the reactive groups, the greater the distance the weaker the bond. The three chains need to be closely packed within the molecule to allow hydrogen bonding to take place. Glycine has the smallest side chain and is present at every third position along the chain. The alignment of adjacent chains needs to be such that the small side chain of glycine projects into the centre of coiled molecule, the larger side chains projecting out.

In order to bring glycine into the required internal position in the helix, molecular models have shown that to achieve the maximum number of hydrogen bonds and to minimize the hindrance of larger side chains, each chain of the triple helix needs to be staggered by one amino acid with respect to its neighbour.

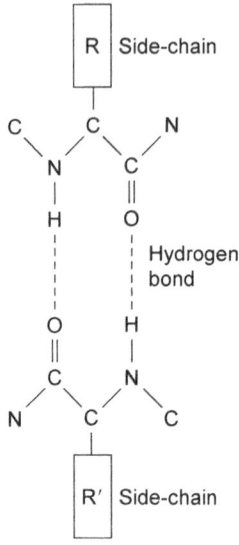

**Figure 2.7** Hydrogen bonding between adjacent peptides.

$$\text{Backbone} \qquad\qquad \text{Backbone}$$
$$| \qquad\qquad + \qquad - \qquad\qquad |$$
$$C-(CH_2)n-NH_2 \qquad O_2C-(CH_2)n-C$$
$$| \qquad\qquad\qquad\qquad |$$

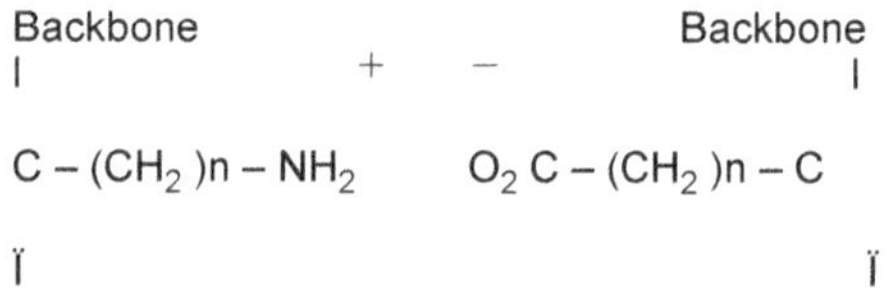

**Figure 2.8** Salt links between adjacent peptides.

## 2.3 Bonding between molecules

Many molecules pack together to form the fibril, which is the smallest collagen unit seen under the transmission electron microscope. The stability of the fibril depends on crosslinks formed between adjacent molecules, i.e. intermolecule bonding.

### 2.3.1 Salt links

Polar side chains project out from each molecule. When acidic and basic end groups of these side chains become aligned, electrostatic salt links can form (*Figure 2.8*).

The sequence of amino acids along each chain of the triple helix has been determined and this has shown there to be distinct alternate grouping of amino acids with either polar or non-polar side chains giving rise to domains along the chain that are either only polar or non-polar in reaction (*Figure 2.9*). When the collagen fibril is immersed in an electronoptically dense metallic stain for several

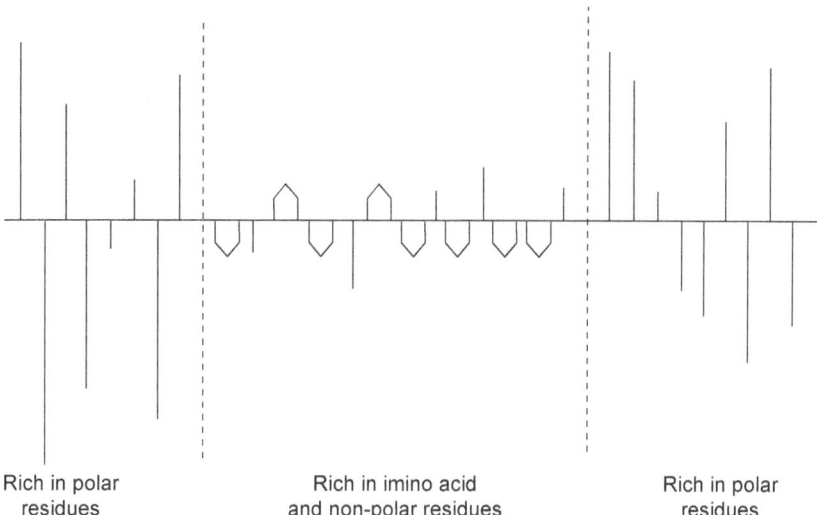

**Figure 2.9** Idealized scheme showing domains rich in polar or non-polar amino acids.

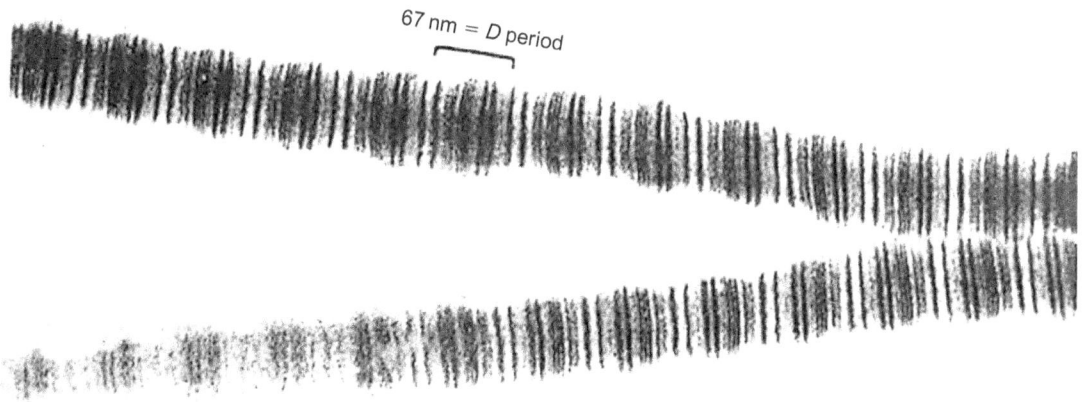

**Figure 2.10** Positively stained collagen fibril revealing reactive polar domains.

minutes (positive staining) the charged end groups of the polar side chains react with the metal. This enables their location to be revealed under the transmission electron microscope. Fibrils stained in this way exhibit a series of fine striations (*Figure 2.10*) which indicate that in fibril formation molecules are so aligned as to bring together those regions where polar side chains predominate. This allows the maximum salt links between adjacent molecules.

### 2.3.2 Covalent intermolecular bonding

If collagen fibrils are immersed in the metallic stain for only a few seconds (negative staining) this allows the stain to merely outline the surface features of the collagen such as cavities. A fibril stained in this way exhibits a regular dark banding at intervals of 67 nm (*Figure 2.11*). The collagen molecules are 300 nm long, that is 4.4 times that of the banding.

Since the sum of the band spacing does not coincide with the length of the molecule, a gap must therefore exist between the end of one molecule and the beginning of the next. Such gaps are accessible to the stain, hence the dark banding.

This banding has also led to the theory that adjacent molecules are displaced relative to one another by one quarter of their length, as shown

8  *Conservation of leather and related materials*

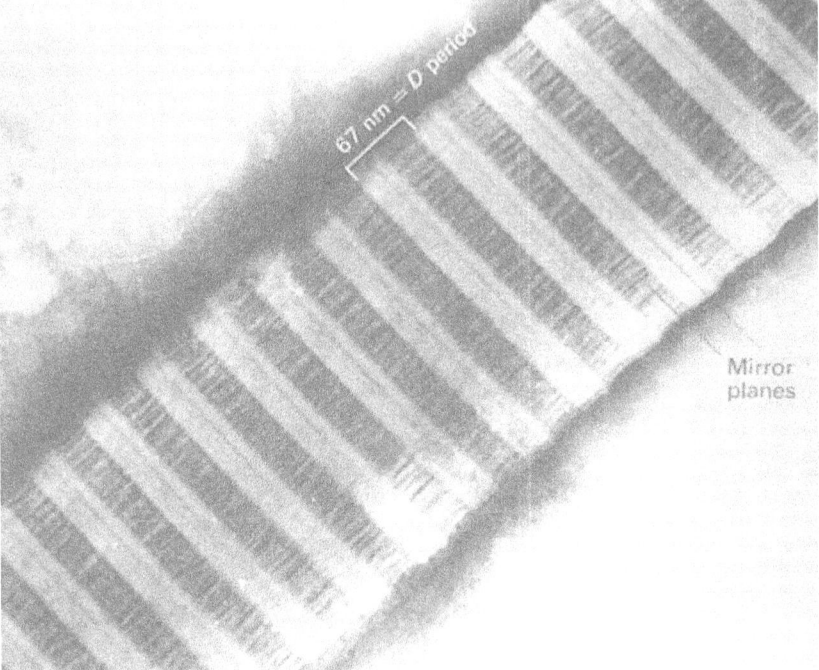

**Figure 2.11**  Negatively stained collagen fibril showing characteristic dark banding.

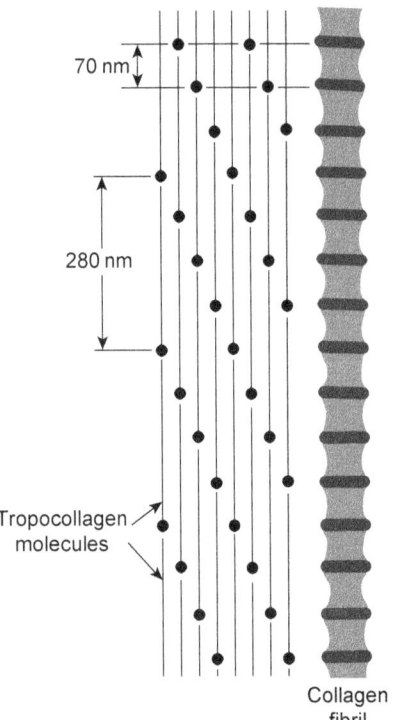

**Figure 2.12**  Diagram showing 'quarter stagger' alignment of adjacent collagen molecules.

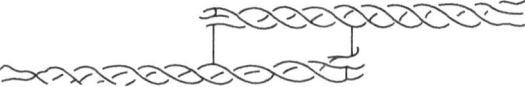

**Figure 2.13**  Diagram showing covalent bonding in the telopeptide region.

diagrammatically in (*Figure 2.12*). This alignment has been described as the 'quarter stagger' of the molecules, with overlaps of 27 nm.

It is at these overlaps that another form of cross-linking occurs. At each end of the triple helix there is a short non-helical region called the telopeptide. Covalent bonds form between the telopeptide region of one molecule and the helical region of an adjacent molecule (*Figure 2.13*). Such an array leaves no weak point along the fibril which could give way under stress.

### 2.4  Fibril structure

There have been various theories as to the spatial arrangement of the aligned molecules within the fibril. Some indication of the internal structure of

**Figure 2.14** Swollen fibril showing subfibrillar helical structure.

the fibril has been seen when collagen fibrils, highly swollen with alkali, were examined under the transmission electron microscope. The swollen fibril exhibited longitudinal striations with a helical twist (*Figure 2.14*). Therefore the fibril would appear to be made up of a series of helical coils, alternating in direction, beginning with the backbone chain of the molecule, then the triple helix and finally the grouping of molecules within the fibril. Such repeated coiling imparts strength to the fibril.

The collagen fibrils are remarkably consistent in diameter, irrespective of animal type or location within the skin. It is only at the extreme grain surface where fibrils smaller in diameter have been found.

Under the transmission electron microscope there appears to be a distinct grouping of fibrils into larger units or elementary fibres. The fibrils are held in these groups by helical coiling of the fibril.

The elementary fibres are in turn grouped into fibre bundles which then interweave through the skin in the manner already described.

## 2.5 Shrinkage temperature

One property of collagen is that it exhibits a sudden shrinkage in length when heated in water. Mammalian skin collagen that has received no chemical processing shrinks at about 65°C. There is little variation in this temperature with different mammalian species or different regions of the skin.

The reason for this shrinkage is that the backbone chains of the molecule exist in an extended form, held in this form by hydrogen bonding. When collagen is heated a point is reached at which the energy input exceeds that of the hydrogen bonding. There is then a sudden release from the extended form and the fibre shrinks to a rubber-like consistency. Only the remaining covalent and salt links hold the collagen molecules together and prevent the shrunken collagen from immediately going into solution.

Prolonged exposure to alkali, such as in the liming process, causes changes to certain amino acids; these changes in turn reduce the degree of hydrogen bonding within the collagen molecule. As a consequence, the shrinkage temperature of skins that have been limed falls to 60°C or even 55°C.

Ageing conditions that bring about hydrolytic or oxidative degradation of the collagen cause breaks in the backbone chain of the molecule and changes to the chemical composition of the side chains. These both lead to a reduction in shrinkage temperature.

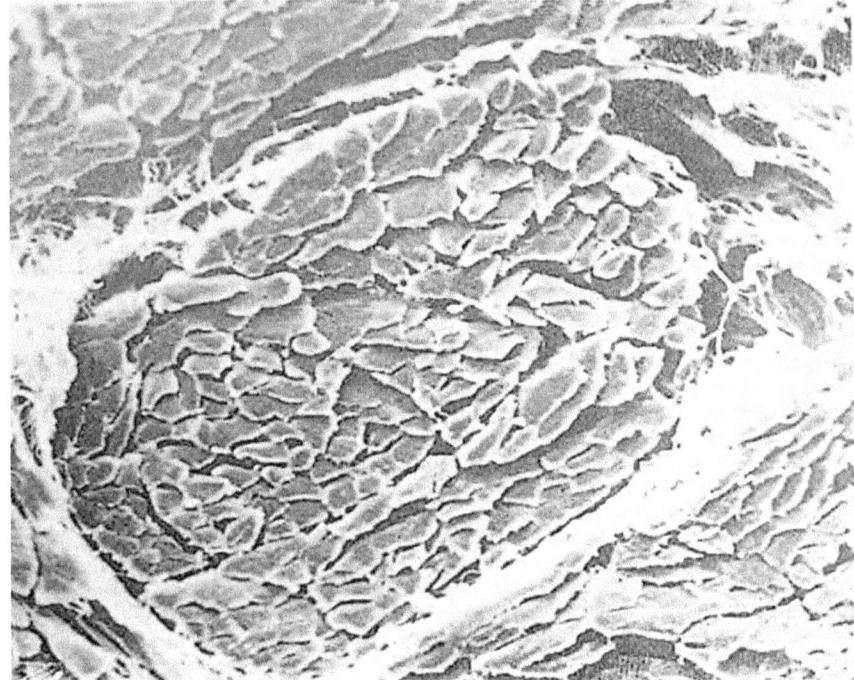

**Figure 2.15** Grouping of elementary fibres into fibre bundles.

Chemical cross-links introduced into the collagen by tanning agents raise the shrinking temperature depending on the type of tanning material and the nature of the process employed.

# References

Bailey, A.J. (1992) Collagen – Nature's Framework in the Medical, Food and Leather Industries. *J. Soc. Leather Technologists and Chemists*, **76**, 111.

Bailey, A.J. and Paul, R.G. (1998) Collagen: A Not So Simple Protein. *J. Soc. Leather Technologists and Chemists*, **82**, 104.

Kennady, C.J. and Wess, T.J. (2003) The Structure of Collagen within Parchment – A Review. *Restorator*, **24**, 61.

Reich, G. (1995) Collagen: A Review of the Present Position. *Leder*, **46**, 195.

Ward, A.G. (1978) Collagen 1891–1977: Retrospect and Prospect. *J. Soc. Leather Technologists and Chemists*, **62**, 1.

Woodhead-Galloway, J. (1980) *Collagen: The Anatomy of a Protein*. London: Edward Arnold.

# 3

# The fibre structure of leather

*B.M. Haines*

In making leather a raw, putrescible animal skin is converted into a dry, non-putrescible material with the handle and degree of flexibility required for its specific end use.

The skin of any vertebrate animal can be made into leather, and the one common characteristic of these skins is that they are primarily composed of the protein collagen. The molecules of collagen are extremely long in relation to their cross-section, and during their formation they become naturally orientated into fibrils and bundles of fibrils (*Figure 3.1*), which interweave in a three-dimensional manner through the skin.

This natural fibrous weave is preserved in the final leather and it is this fibrous structure that gives its unique physical properties of handle and ability to accommodate to the stresses and movement imposed during its use.

However, this fibrous skin structure varies considerably between skins of different species and types within species, thus giving the leather industry a wide variety in raw materials from which a careful

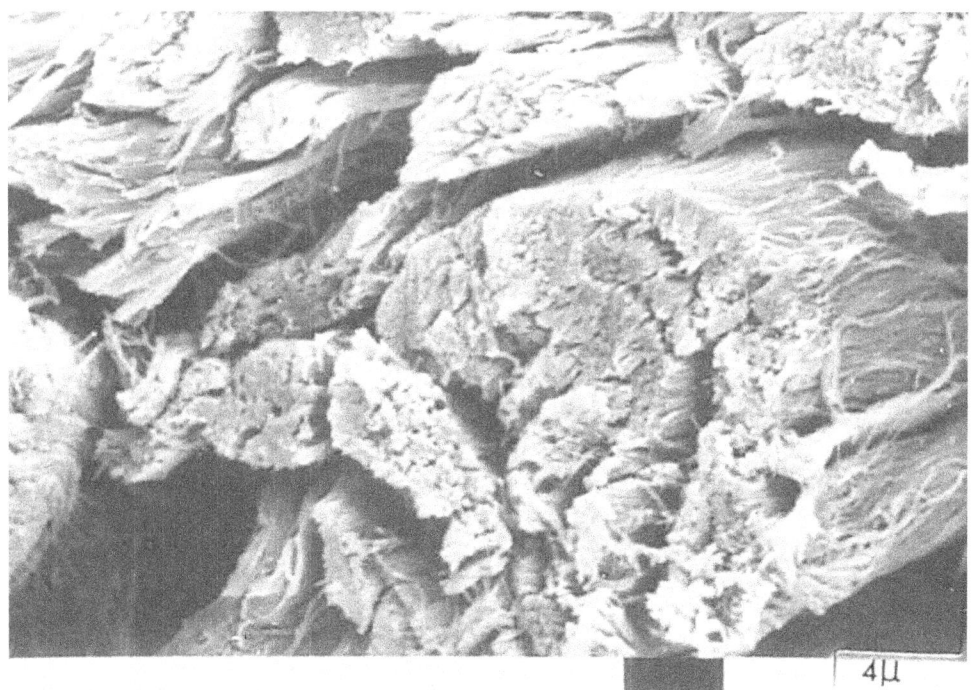

**Figure 3.1** Fibrils and fibril bundles of collagen.

selection has to be made in order to achieve the combination of mechanical and aesthetic properties required for specific end uses.

The diverse structures of skins and the effects on the properties of leathers made from them have been described by British Leather Manufacturers' Research Association (1957), Haines (1981, 1984, 1999) and Tancous (1986).

## 3.1 The structure of mammalian skins

Although the skin of any vertebrate animal, be it fish, reptile, bird or mammal, can be made into leather. The most commonly used skins have been, and are at the present time, those from cattle, calf, goats, sheep and to a lesser extent pig and deer. These being mammalian in origin they are covered by hair.

The mammalian skin has distinct layers (*Figure 3.2*): the layer which extends from the outer surface to the base of the hair roots is termed the grain layer and contains the hairs, sebaceous and sweat glands and numerous blood vessels. The collagen fibres become increasingly fine as they pass through this grain layer to the outer surface, which in life is covered by the epidermis.

In the underlying corium the fibre bundles are considerably larger and interweave at a higher angle relative to the skin surface. Towards the inner or flesh surface the fibres become finer and run in a horizontal plane to form a limiting or flesh layer, separating the skin from the underlying muscles.

In the early stages of leather processing the epidermis, together with the hair, is removed chemically. This exposes at the surface the compact interweaving of extremely fine fibrils that create the smooth, aesthetically pleasing grain surface of the leather.

## 3.2 Variation of structure between animal types

Each species has a distinctive skin structure: the skins vary in total thickness, dimensions of the corium fibre bundles and in the proportion of the total thickness occupied by the grain layer.

### 3.2.1 Mature cattle skins

The skins of mature cattle are generally between 4 and 6 mm thick with a proportion reaching 8 mm in thickness, and measuring 3.3 to 4.2 m$^2$ in area.

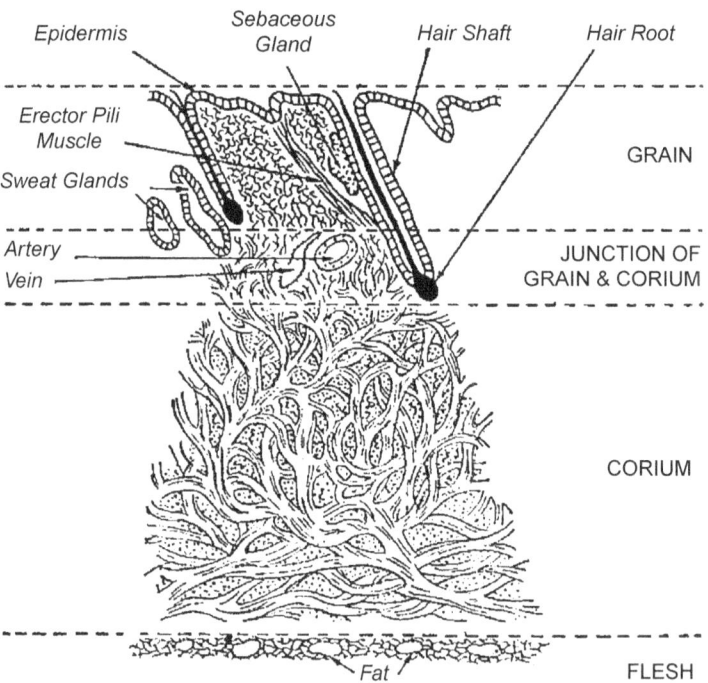

**Figure 3.2** Diagrammatic representation of structure of a typical mammalian skin.

The grain layer occupies about one sixth the total thickness (*Figure 3.3*). The hairs are straight, relatively coarse and spaced equidistant through the grain layer.

The corium fibre bundles are relatively large (0.1 mm in diameter) and interweave at a fairly high angle relative to the surface.

Such a skin is highly suited for sole leather, harness, saddlery, or mechanical belting leather but it is far too thick for shoe uppers or upholstery unless it is split into two layers. The upper or outer layer consisting of the grain layer and part of the underlying corium, the grain split, is used for shoe uppers or upholstery or case leather. This outer layer can be split to 1 mm thick for upholstery leather or from 1.3 to 2 mm for shoe uppers. Due to the coarseness of the corium fibres and the compactness of their weave, the leather tends to have a heavy handle and to lack drape. Consequently cattle skin is rarely used for clothing leather.

After splitting, the remainder of the corium forming the flesh split is used in various ways. Some splits are taken in the untanned state for the production of sausage casings, but most are used for making coarse suede leather, the split surface being abraded to form the suede nap. Due to the large size of the corium fibres the nap raised is coarse and less suited to fashion footwear or clothing. Generally flesh splits are made into coarse suede shoes and boots or industrial gloves.

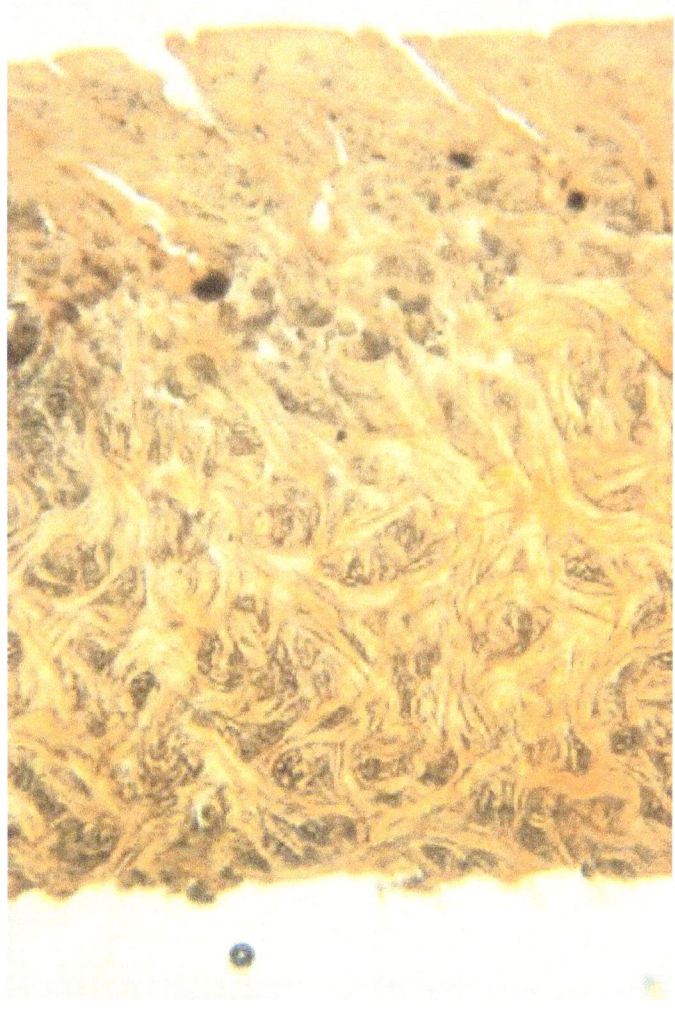

**Figure 3.3** Cross-section of leather made from mature cattle hide.

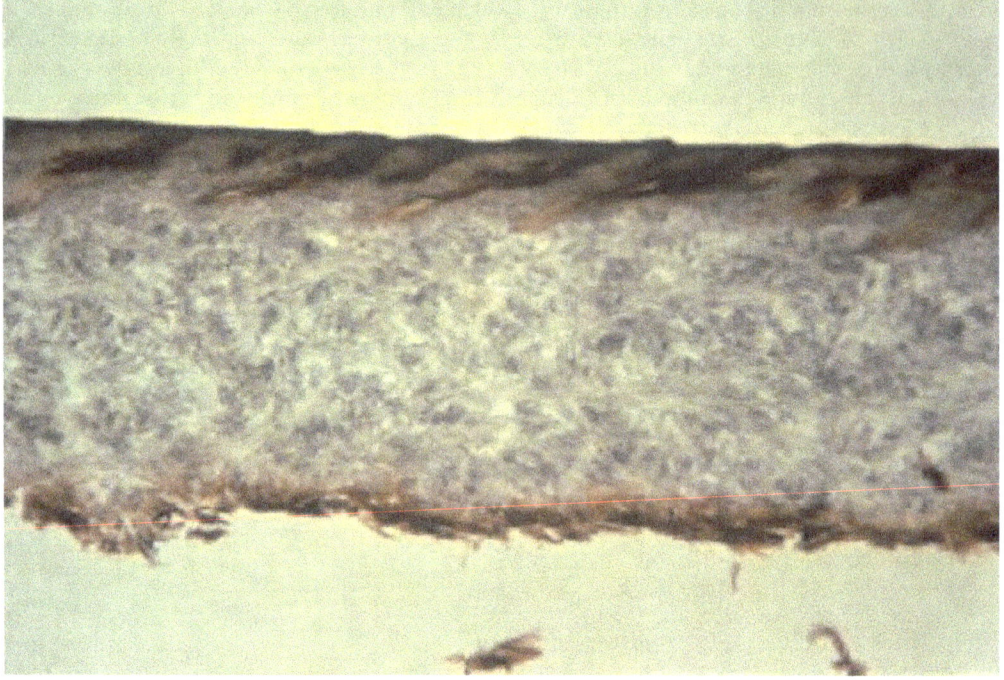

**Figure 3.4** Cross-section of leather made from young calfskin.

### 3.2.2 Calfskins

A calfskin is a miniature version of the adult cattle skin. The proportion of grain layer to total thickness is similar (approximately one sixth) but the skin thickness and fibre bundle size are dependent on the age of the animal, both increasing with age.

The skin of a young calf aged 1 month (*Figure 3.4*) is about 1 mm thick and 0.5 to 0.7 m² in area. The corium fibre bundles are fine and interweave compactly at a medium angle.

The skin of an animal aged 6 months is about 1.3 mm thick and 1 m² in area, the corium fibre bundles are thicker and interweave at a higher angle than in the younger animal.

At 12 months, when the calf is almost fully grown, the skin will be about 3 mm thick and with an area of 2.7 m². At this stage the fibre structure closely resembles that of the mature animal.

A young calfskin yields a leather which although only 1 mm thick, is strong due to the compact interweaving of the fine corium fibre bundles and the grain surface is extremely fine. This makes the leather ideally suited to fashion footwear, handbag, bookbinding and fine case leathers. Calf leather is unsuitable for clothing as the weave is too compact for the softness and drape required from clothing leather.

### 3.2.3 Goatskins

These skins range between 1 and 3 mm in thickness and measure 0.5 to 0.7 m² in area. The grain layer occupies about one third the total thickness (*Figure 3.5*). The hairs are a mixture of coarse and fine straight hairs widely spaced through the grain layer. This spacing allows for the smooth interweaving of the corium fibres into the grain layer and there is no discontinuity between the two layers. The corium fibre bundles are relatively fine and interweave compactly at a medium angle. This structure makes goat leather highly suitable for shoe uppers and particularly for bookbinding. The compactness of the fibre structure makes the leather less suitable for clothing leather as it lacks softness and drape. It is only with very young kid skins that the fibre structure is sufficiently open for the leather to be used in gloving.

### 3.2.4 Sheepskins

There are several types of sheep, each with a different skin structure.

The hair sheep, indigenous to tropical countries, is a relatively small animal, yielding thin skins (0.8 mm) of 0.4 to 0.5 m² in area. The fine corium

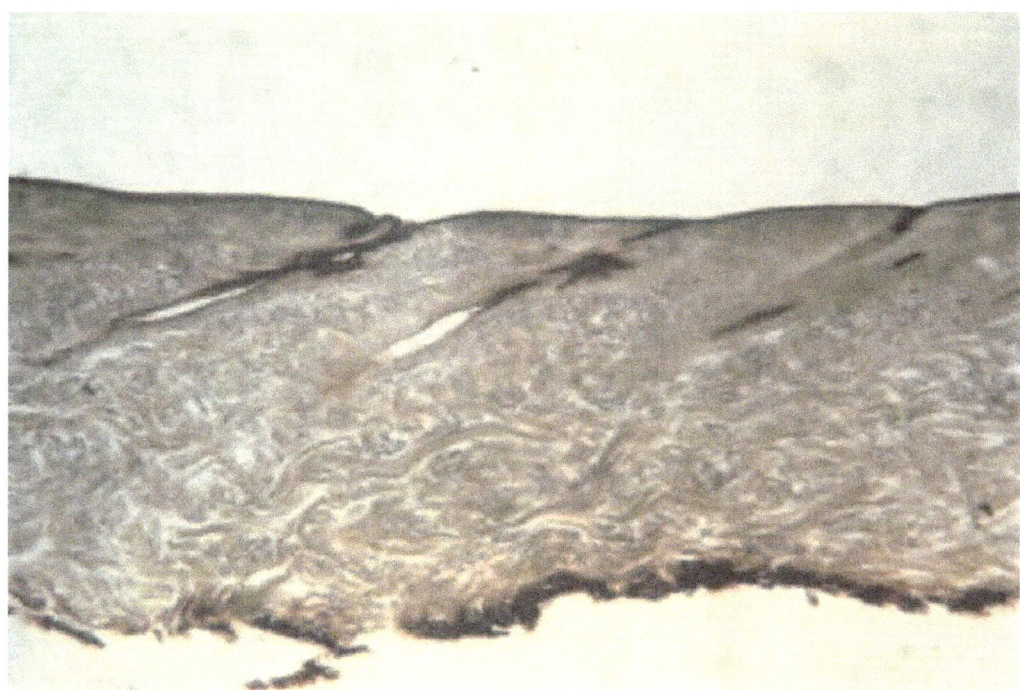

**Figure 3.5** Cross-section of leather made from goatskin.

fibres interweave fairly compactly with no looseness between the grain and corium layers. Such skins are ideally suited for gloving.

In woolbearing sheep, native to Britain, the skins are thicker (2 to 3 mm) and 0.5 to 0.6 m² in area. To accommodate the density of the wool fibres the grain layer occupies at least half the total skin thickness (*Figure 3.6*).

The density and curl of the wool fibres within the grain layer limits the space through which the corium fibres can interweave into the grain layer and there is a tendency to looseness at the junction between the two layers. In addition, natural fat tends to be stored in a layer of fat cells at the junction between grain and corium. These fat cells interrupt the fibre weave still further and after the fat has been removed during leather processing, the collapsed fat cells add to the looseness in this region.

The corium fibres are fine and less compactly interwoven than in the goat or calf skin. This allows the leather to be softer and drapeable; qualities required of clothing leather.

The coarse-woolled domestic hill sheep has a lower wool density with less tendency to looseness at the junction of grain and corium. These skins are primarily used for grain or nappa clothing leather.

The fine-woolled sheep is best suited to the production of woolskin clothing where the wool is retained and the flesh surface of the skin is sueded.

For the production of chamois leather the grain layer is split off and only the underlying corium layer is used.

### 3.2.5 Deerskins

These skins range between 2 and 3 mm in thickness and measure 0.9 to 1.3 m². The shallow grain layer occupies one sixth the total thickness. The corium fibre bundles are somewhat coarse and rather loosely interwoven, yielding a leather that tends to be stretchy. This property was well suited to the production by oil tannage of buff clothing leather.

### 3.2.6 Pigskins

Pigskins differ in structure from the other skins described in that there is no distinct grain layer, the hair penetrating the full thickness of the skin (*Figure 3.7*). Throughout the skin the fibres interweave in a particularly compact and distinctive basket weave type of pattern, yielding a tight-structured leather best suited for case leather and bookbinding.

16  *Conservation of leather and related materials*

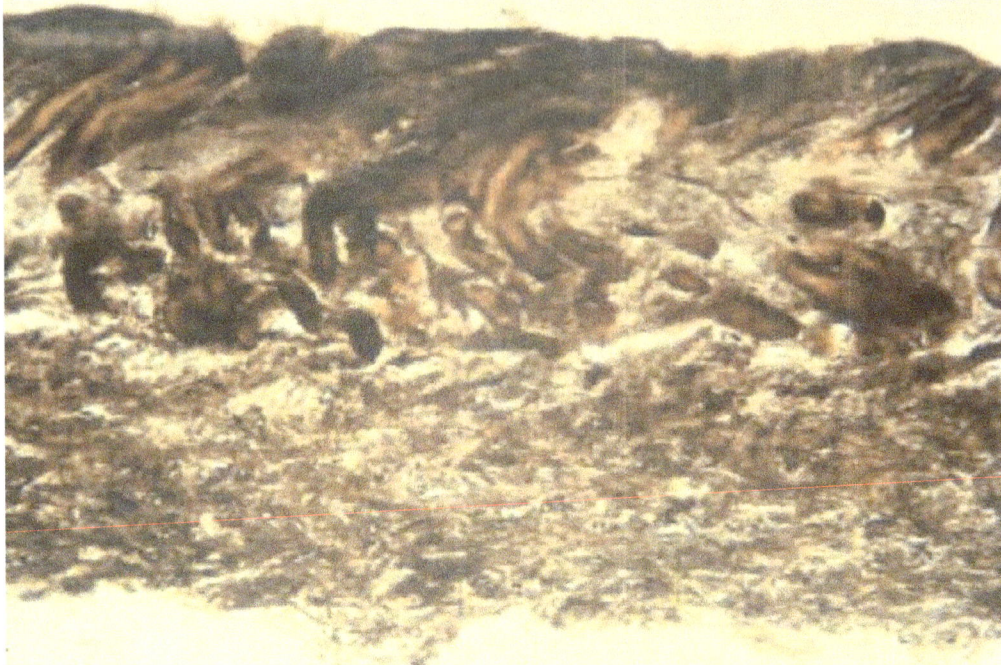

**Figure 3.6**  Cross-section of leather made from woolled sheepskin.

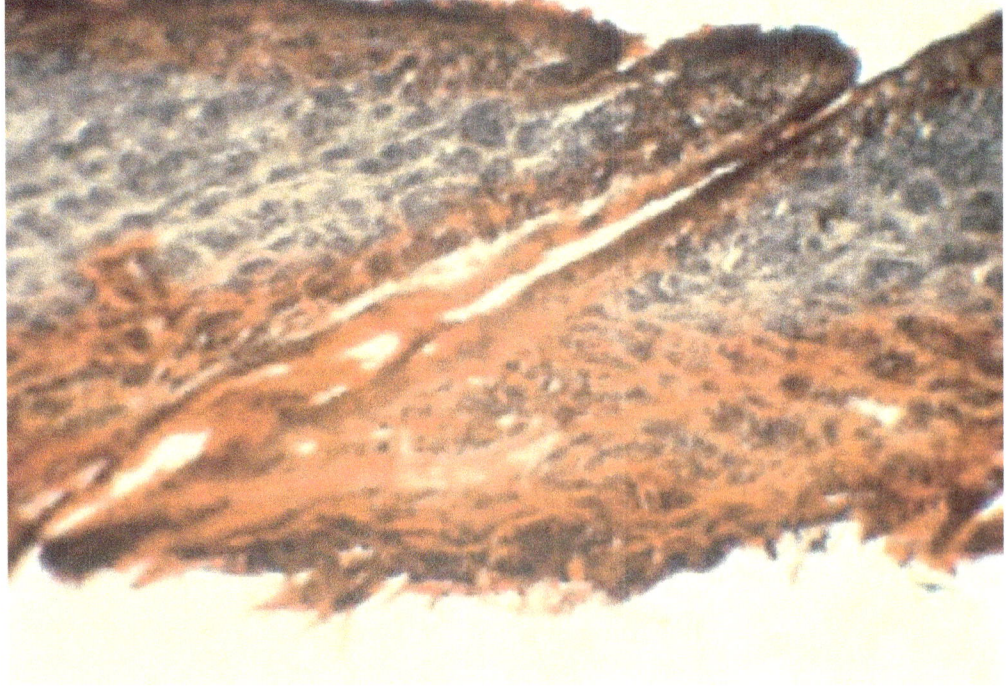

**Figure 3.7**  Cross-section of leather made from pigskin.

The rather coarse grain surface pattern makes pig leather less suitable for fashion footwear.

## 3.3 Grain surface patterns

When in processing the hairs are removed chemically from the skin, the empty hair follicles at the grain surface can be seen to be arranged in distinctive patterns, characteristic of each animal type. For example, in calfskin the follicles are of equal size and arranged in regular rows (*Figure 3.8*). As the animal matures, while the sizes of the follicles and the distances between them increase, the overall pattern is maintained (*Figure 3.9*).

In goatskin there are regular alternating rows of large and fine follicles (*Figure 3.10*) whereas in woolled sheepskins the fine follicles are arranged in groups (*Figure 3.11*).

Such follicle patterns can be used to identify the animal origin of leather artefacts.

## 3.4 Suede surfaces

The fineness of nap that can be raised is determined by the size of the constituent fibre bundles of the skin. If the grain surface is sueded then the fine fibrils of the grain surface yield a particularly fine nubuck nap. Where the nap is raised on the split corium surface of cattle skin the large corium fibre bundles give rise to a coarse open nap. The naturally finer corium fibre bundles of a goat or sheep permit a finer nap to be raised at the flesh surface.

## 3.5 Variation in structure with location in the skin

As the skin develops and has to adapt to meet the demands of the animal in life, so the natural fibre weave changes with location on the original animal.

Along the line of the backbone the skin is thickest and the weave most compact and dense. These backbone features are particularly marked in animals with a mane such as goats.

In the central butt region which originally covered the back of the animal, the fibre weave is particularly compact with the fibres interweaving at a high angle. In the belly region, the weave is looser with fibre bundles running at a far lower angle of weave. These structural differences result in the leather in the belly region being weaker, softer and more stretchy. Due to the looser weave in the belly the grain surface has

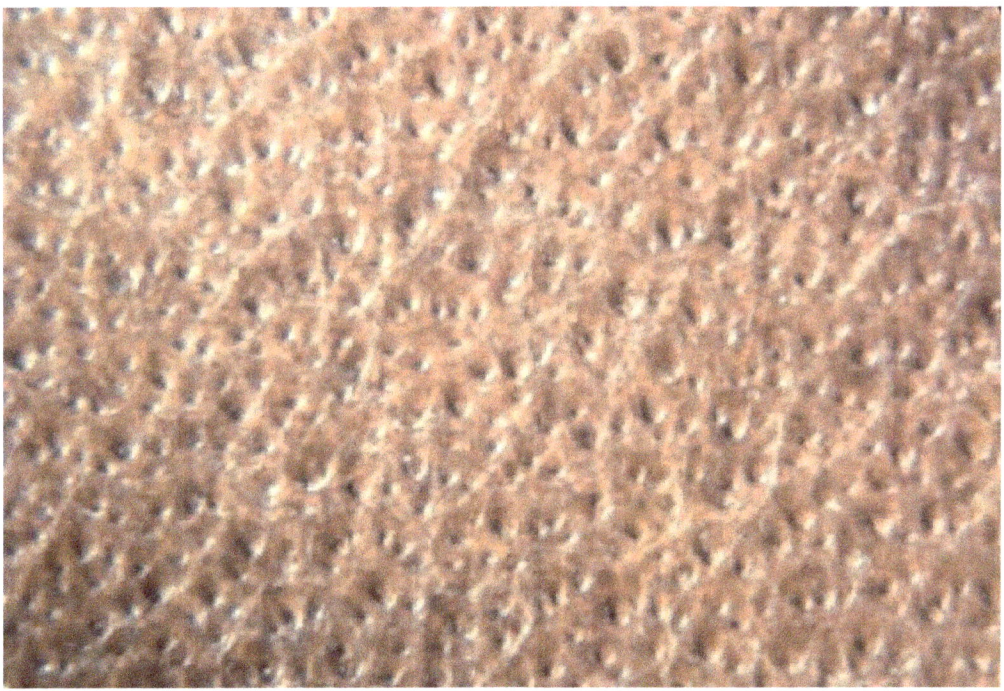

**Figure 3.8** Hair follicle pattern of calfskin leather.

**Figure 3.9** Hair follicle pattern of mature cattle hide leather.

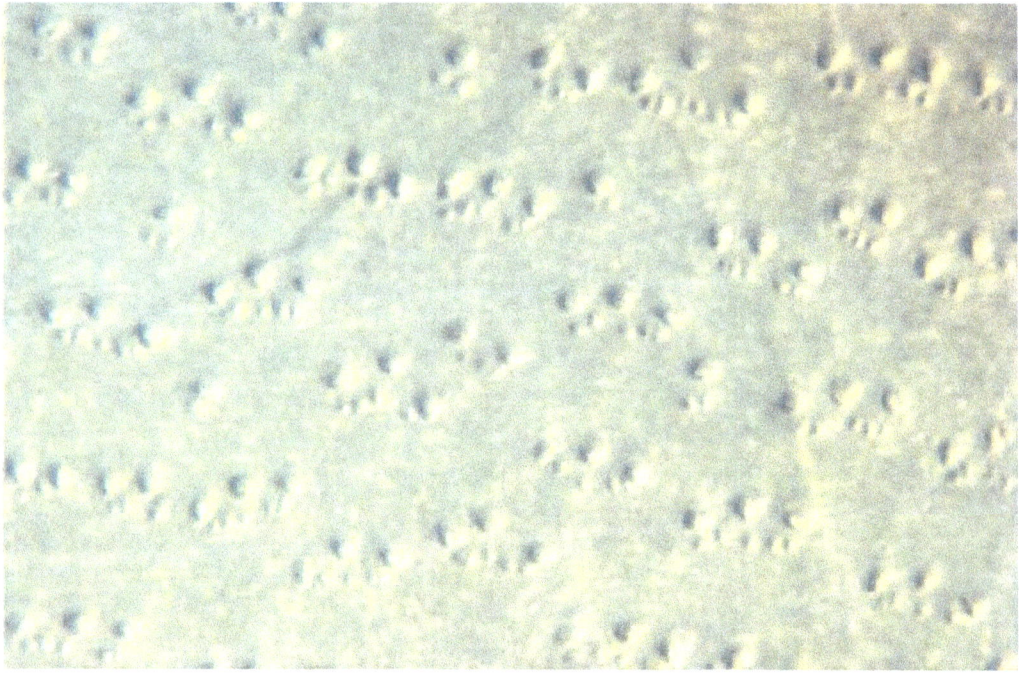

**Figure 3.10** Hair follicle pattern of goatskin leather.

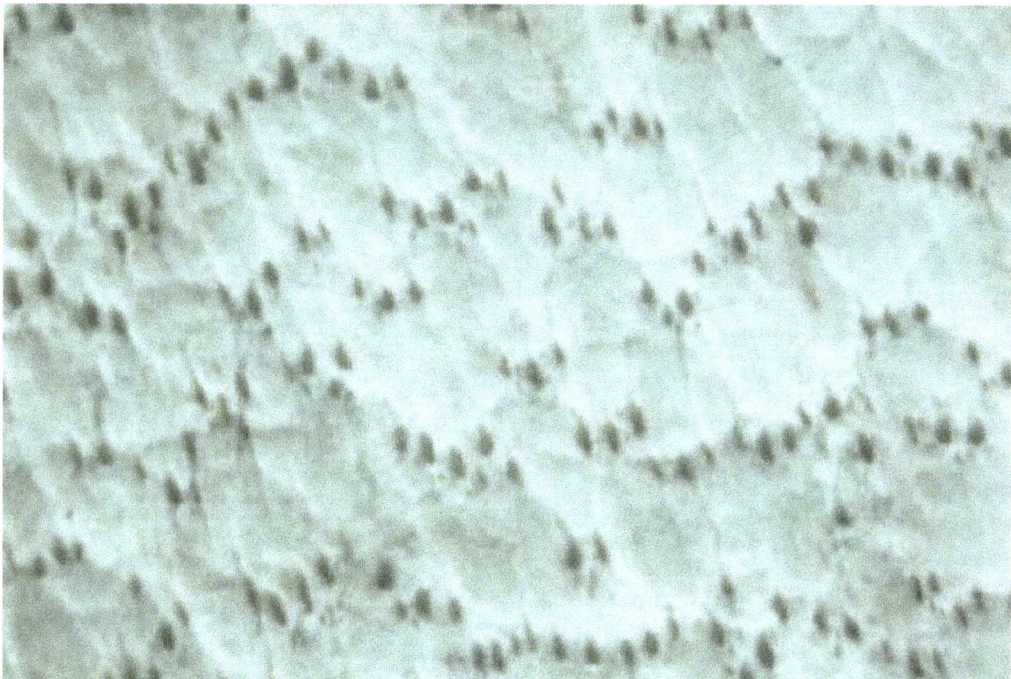

**Figure 3.11** Hair follicle pattern of sheepskin leather.

a greater tendency to form coarse wrinkles when the leather is flexed. The loosest weave is found in the axillae regions.

These location differences in fibre structure are common to most animal types.

## 3.6 Directional run of the fibres

The three-dimensional weave of the fibre bundles is not random throughout the skin. There is, parallel to the grain surface, a predominating direction in which the majority of the fibres run (*Figure 3.12*).

This arises when the collagen fibrils are being laid down in the developing foetus. The fibrils tend to become aligned along the main lines of tension as when the backbone lengthens and the legs extend. The hairs' growth follows the same lines.

This directional run profoundly affects the physical properties of the skin and of the final leather. For example, the strength of the leather is greater and the extensibility less in the direction parallel with the main run of the fibres. The reason for this is that when measuring the strength of leather or skin, it is standard practice to take the mean of two tests, one made in the direction parallel to the backbone, the other at right angles.

If the backbone line is not known then two tests can be made: one parallel with the run of the hairs, the other across the run of the hairs.

The directional run governs the way in which leather is cut in use. Shoe uppers are cut, toe to heel, in line with the backbone, the direction of greater stretch. Similarly the bindings of books are cut with the greater stretch along the spine of the book.

## 3.7 The influence of fibre structure on leather properties, structure and tear strength

The strength of a leather is dependent not only on its total thickness but on the proportion that is corium tissue and the frequency with which the corium fibres interweave and cross over each other, a prerequisite for strength.

To illustrate this point three leathers of 1mm substance are compared:

(a) A 1 mm leather prepared from a young calfskin where the full thickness of the original skin is retained will have a grain layer of about 0.16 mm deep and the fineness of the corium fibres permits frequent interweaving through the corium. Such a leather will tear at about 8 kg.

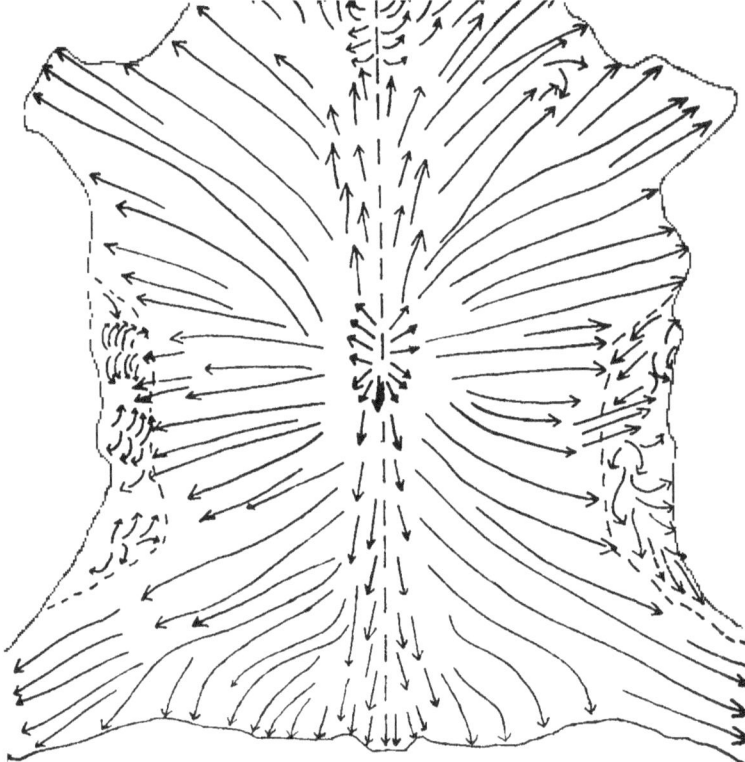

**Figure 3.12** Direction of fibre run through a skin.

(b) A 1 mm leather produced from a woolled sheepskin will have a grain layer of 0.5 mm depth and the corium fibres although fine will be more loosely interwoven than in the calfskin. This leather can be expected to tear at about 4 kg.

(c) When an upper leather is prepared by splitting a layer from a mature cattle skin the natural weave is disturbed and the tear strength is highly dependent on the depth of corium present. A 1 mm leather split from a mature cattle skin may have a grain layer as deep as 0.6 mm. The large size of the corium fibres restricts the frequency with which they can interweave in the remaining 0.4 mm of corium tissue. Such a leather will have a tear strength of 3 kg.

It is common practice in, for example, bookbinding to reduce the substance of leather by paring away corium tissue from the flesh side. In so doing the reduction in leather strength will be far greater than the reduction in substance. For example, a goat binding leather that before paring measures 1.4 mm and tears at 7.5 kg: after paring to 0.6 mm the leather would tear at about 1.2 kg. The paring removes almost all of the corium tissue but whereas the thickness had been reduced by about a half, the strength was reduced by five sixths.

## 3.8 Structure and leather handle

Although the tanner is bound by the natural structure of the skin, his skill lies in bringing about, by chemical and mechanical processing, limited changes to the structure in order to produce a range of leathers with differing properties.

For example, if in the final dry leather the processing allows fine spaces to remain between the fibrils and between fibre bundles, then the leather will have a full, soft and flexible handle. Conversely if the processing causes the fibrils and fibres to adhere on drying, then the leather will be firmer and far less flexible.

## 3.9 Fibre weave and movement

When fine spaces remain between fibrils and fibres they are free to move over each other within the fibre bundle and within the fibre weave as a whole. This permits the leather to accommodate to stretching, compression or creasing.

When a leather is stretched the angle of weave falls and the structure becomes more compact. Similarly if the leather is compressed the angle of weave falls and the weave becomes more compact as spaces within the structure are reduced.

If the leather processing has not allowed sufficient spaces to remain within the fibre weave then on stretching the leather, unable to accommodate to the stress, will break and on compression the leather will be cut.

One of the attractive features of leather is the fine creasing of the outer grain surface that occurs when leather is curved grain inwards. The folds in the grain surface are formed by the underlying weave being able to open out and extend into the fold or to compress in the adjoining crease line.

The fineness of the surface folds also depends on uniformity of structure throughout the leather. Any change to a firmer or looser layer will create coarser and unattractive surface folds. This is all important when laminating leather. The backing material needs to be equally flexible as the leather and the method of bonding should not restrict movement of the leather fibres, particularly the extension of the flesh fibres. Failing in this can cause an unattractive coarse wrinkling of the grain surface.

## References

British Leather Manufacturers' Research Association (1957) *Hides, Skins and Leather Under the Microscope*. Egham: BLMRA.

Haines, B.M. (1981) *The Fibre Structure of Leather*. Northampton: The Leather Conservation Centre.

Haines, B.M. (1984) The Skin Before Tannage – Procter's View and Now. *J. Soc. Leather Technologists and Chemists*, **68**, 57.

Haines, B.M. (1999) *Parchment*. Northampton: The Leather Conservation Centre.

Tancous, J.J. (1986) *Skin, Hide and Leather Defects*. Cincinnati: Leather Industries of America.

# 4

# The chemistry of tanning materials

*A.D. Covington*

## 4.1 Introduction

The chemical nature of collagen allows it to react with a variety of agents, often resulting in its conversion to leather (Covington, 1997), when the putrescible material becomes resistant to microorganisms – the definition of tanning. Untanned skin can be very stable; witness the longevity of parchment. However, when wetted and redried untanned skin, even parchment, becomes hard and horny, useless for most purposes. Part of the benefit of tanning and lubricating leather is the ability to resist fibre sticking upon drying. Of the changes in appearance and properties that are a consequence of tanning, one of the more important is the increase in hydrothermal stability. This can be measured by the conventionally measured shrinkage temperature, $T_s$. Shrinking is a kinetic process and, as such, can be treated thermodynamically. Such thermodynamic studies indicate that breakdown of the tanning interaction is not the cause of shrinking; indeed $^{27}$Al NMR studies demonstrate that even the weak, hydrolysable aluminium tannage is not reversed during shrinking (Covington *et al.*, 1989).

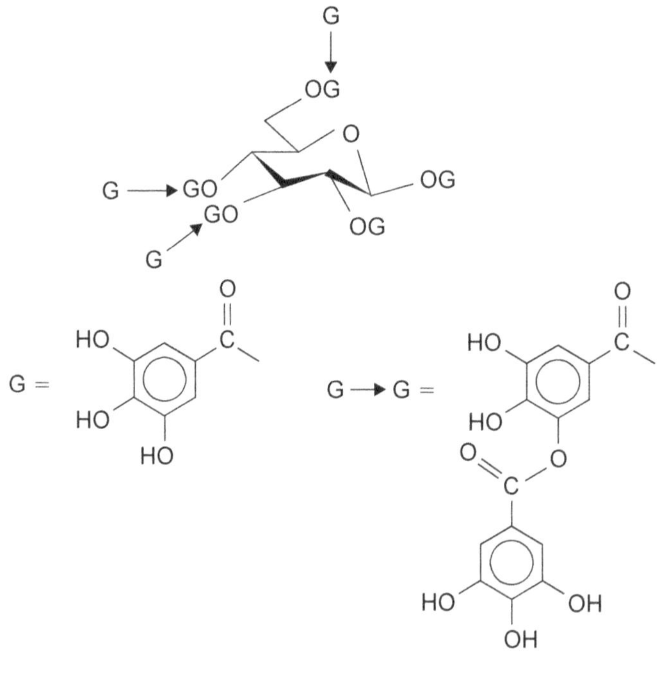

**Figure 4.1** The structure of Chinese gallotannin (tannic acid).

The reaction which is manifested as heat shrinking is a breakdown of the hydrogen bonding in collagen or leather; that is, regardless of the tanning process, the shrinking reaction is the same.

This begs the question: where does hydrothermal stability come from? That is, if the tanning process only modifies the shrinkage temperature, without changing the shrinking mechanism, what causes the $T_s$ to rise? The answer may have something to do with the size of the co-operating unit in the shrinking process: the larger the unit, the slower are the kinetics and the higher is the shrinkage temperature (Covington, 1998). It has been found that the size of the co-operating unit in raw collagen, with shrinkage temperature 60°C, is 20 residues, but aluminium- and chromium-tanned collagens with shrinkage temperatures of 73 and 107°C have co-operating units containing 71 and 206 amino acid residues respectively.

The nature of these co-operating units is not clear; they may function through the natural covalent crosslinks and additional hydrogen bonding structure elements in collagen, supplemented or modified by the tanning effects of hydrogen bonding or covalent reaction at polar groups on the amino acid side chains or through multiple interactions at the peptide link itself. It has universally been assumed that the stabilizing effects of tanning reactions are founded in the creation of new crosslinks, giving rise to new structure within the collagen triple helix. However, recently it has been proposed that stabilization is based on modifying the structure around the triple helix (Covington, 2001a). In raw collagen, this supramolecular structure is bound water, nucleated at the hydroxyproline moieties in the amino acid sequence (Berman et al., 1995). Tanning reactions change the water matrix, causing it to confer different degrees of stability, which depend on the nature of the interactions between the matrix and the triple helix (Covington and Song, 2003). It is the ability of tanning agents to fit into or displace the water structure and bind the matrix covalently to collagen that affects the magnitude of the change in the shrinkage temperature of the collagen. Thus, tanning reactions may be highly complex on the molecular level.

Current tanning technology is dominated by chromium; it was introduced about 130 years ago and by the turn of the nineteenth century had begun to replace the traditional tannages, which were based on plant polyphenols (so-called vegetable tannins). Today, about 90% of the world's leather production is chrome tanned, the remainder is tanned with vegetable tannins, mostly for leathergoods or shoe soles.

The range of options available to the modern tanner will now be considered.

## 4.2 Vegetable tanning

Many plant materials contain polyphenols which can be used in tanning. To be effective, the molecular mass must be in the range 500–3000; lower molecular mass fractions in the tannin are referred to as non-tans and higher molecular mass species are gums, which are sugar complexes with polyphenols. The properties that vegetable tannins confer to the leather are as varied as the many sources from which they are obtained. Tannins are classified as follows: hydrolysable or pyrogallol tannins, subclassified as gallotannins (examples are Chinese gallotannin or tannic acid, sumac, tara) or ellagitannins (examples are myrabolam, chestnut, oak) and condensed or catechol tannins (examples are mimosa, quebracho, gambier).

Hydrolysable tannins are sugar derivatives, based on glucose, but may contain larger polysaccharides. Gallotannins are characterized by glucose esterified by gallic acid (*Figure 4.1*), where esterification may occur directly with the glucose ring or as depside esterification of bound gallic acid.

Ellagitannins have sugar cores, esterified not only with gallic acid, but also with ellagic acid (*Figure 4.2*) and chebulic acid (*Figure 4.3*).

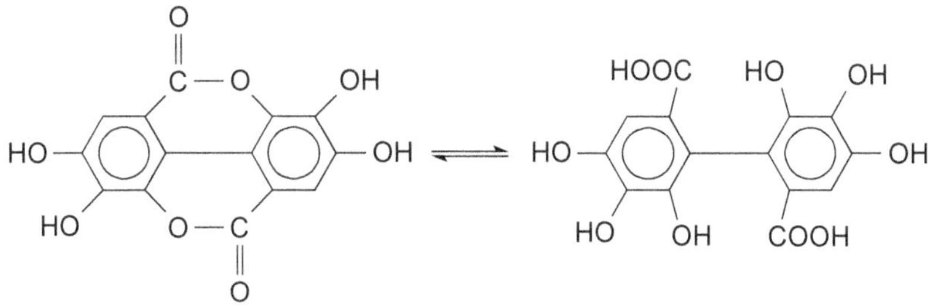

**Figure 4.2** The structure of ellagic acid.

Examples of the structures of hydrolysable tannins are chebulinic acid (*Figure 4.4*), chebulagic acid (*Figure 4.5*) from myrabolam or vescalagin and castalagin (*Figure 4.6*) from chestnut.

The presence of the trihydroxyphenyl moiety gives this group its name, the pyrogallol tannins.

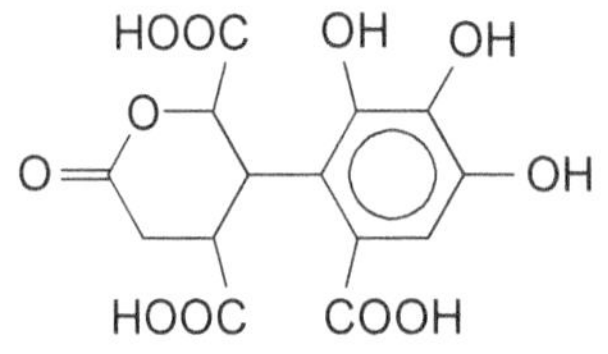

**Figure 4.3** The structure of chebulic acid.

Hydrolysable tannins typically raise the shrinkage temperature of collagen to 75–80°C. The presence of the trihydroxy group allows complexation with metal ions, to make so-called semi-metal tannage. The remarkable property of the reaction is that the shrinkage temperature is raised as high as 120°C. Many metals exhibit this effect, but the preferred salts are aluminium(III), see below.

As the name suggests, these tannins tend to break down by hydrolysis, depositing the esterifying acids in the fibre structure of the leather; this is called 'bloom'. Because of the presence of so many hydroxy groups in the hydrolysable tannins, they are highly reactive (astringent) as tanning agents. Advantages of this class of polyphenols are that they can be pale coloured, e.g.

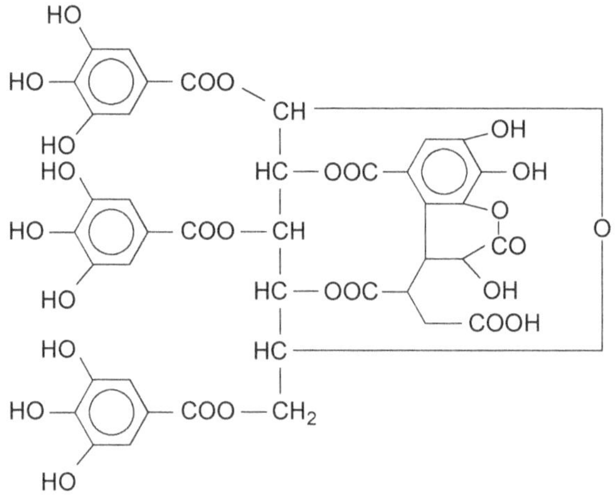

**Figure 4.4** The structure of chebulinic acid.

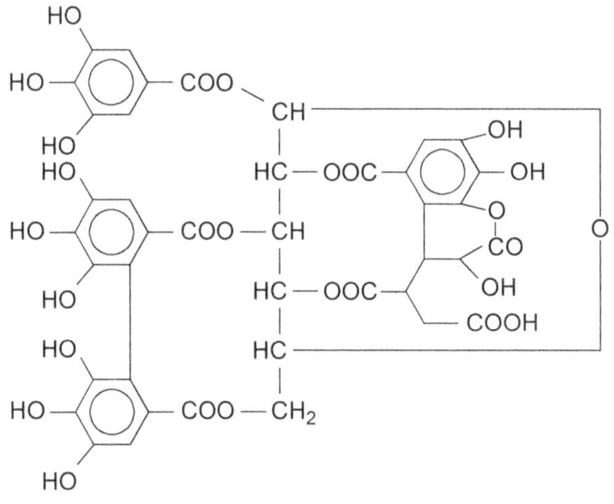

**Figure 4.5** The structure of chebulagic acid.

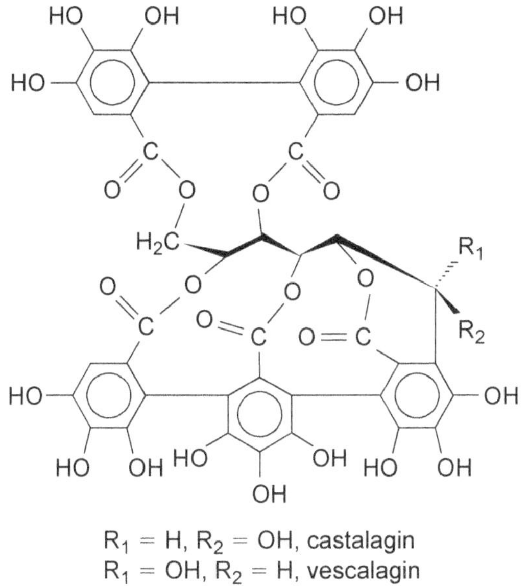

$R_1$ = H, $R_2$ = OH, castalagin
$R_1$ = OH, $R_2$ = H, vescalagin

**Figure 4.6** The structures of chestnut tannins.

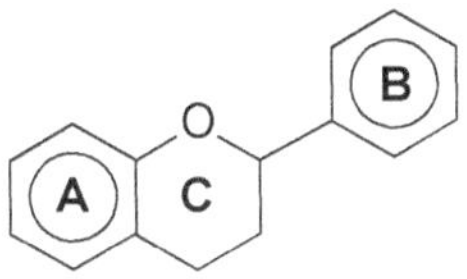

**Figure 4.7** The flavonoid ring system of condensed polyphenols.

tara, and they are typically lightfast, i.e. they do not readily darken on exposure to light.

Condensed tannins are based on the flavonoid ring system (*Figure 4.7*).

The A ring usually contains phenolic hydroxy groups and the presence of the C ring makes both rings reactive to forming carbon-carbon bonds, to create flavonoid polymers: the B ring does not exhibit the same reactivity, it often contains the dihydroxyphenyl moiety, hence the alternative name for this group of compounds is the catechol tannins. The flavonoid structure can be hydroxylated in different ways, giving rise to different types of polyphenols: the more common patterns are shown in *Figure 4.8*.

In nature, the commonest classes are the procyanidins and the profisetinidins. The dihydroxy group in the B ring is not reactive towards metal salts, so these tannins do not undergo the semi-metal process. The very much less common prodelphinidins and prorobinetinidins can form semi-metal complexes and consequently produce high shrinkage temperature.

The condensed tannins are illustrated in the structures of mimosa (*Figure 4.9*) and quebracho (*Figure 4.10*) tannins.

The condensed tannins do not undergo hydrolysis, instead they may deposit a precipitate, an aggregate of polyphenol molecules, called 'reds' or phlobaphenes. Unlike the hydrolysable tannins, the

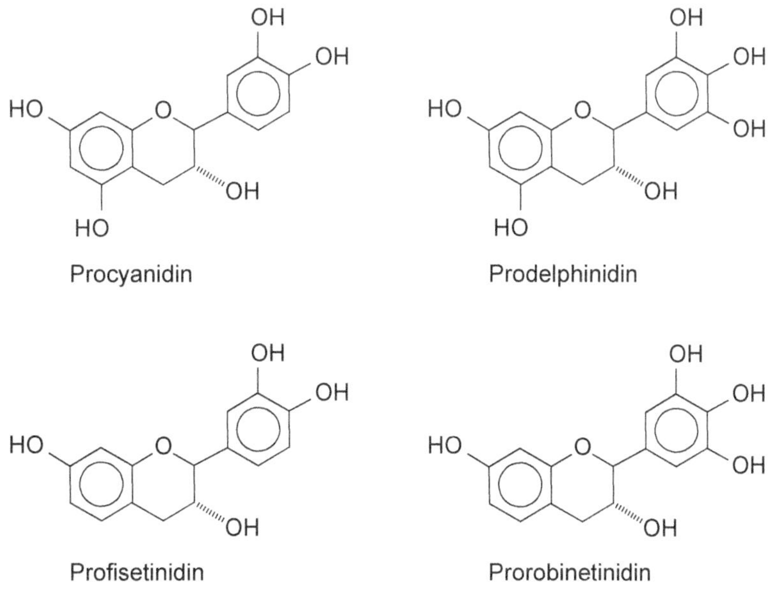

Procyanidin

Prodelphinidin

Profisetinidin

Prorobinetinidin

**Figure 4.8** Hydroxylation patterns for classes of condensed tannins.

condensed tannins redden markedly upon exposure to light; this is understandable in terms of their linked ring structure, conferring the ability to undergo oxidative crosslinking. Condensed tannins typically raise the shrinkage temperature of collagen to 80–85°C.

All vegetable tannins react with collagen primarily via hydrogen bonding at the collagen peptide links (*Figure 4.11*), but it is known that polyphenols also fix to amino and carboxylic acid groups on side chains (depending on pH).

Condensed tannins have an additional mechanism for reaction: it has been suggested that this additional interaction is a covalent reaction between the protein and aromatic carbon in the tannin molecules via quinoid structures and this more stable interaction with collagen would account for the higher shrinkage temperatures achievable by these vegetable tannins. The presence of hydroxyl groups in the flavonoid ring system makes the A ring susceptible to reaction with aldehydic agents (see below). This means they can be crosslinked in a way analogous to metals crosslinking hydrolysable polyphenols and similarly the combination tannage produces the same high shrinkage temperature. The catechol B ring plays no part in the reaction, but a pyrogallol structure in the B ring is reactive like the A ring and contributes to the effectiveness of the reaction.

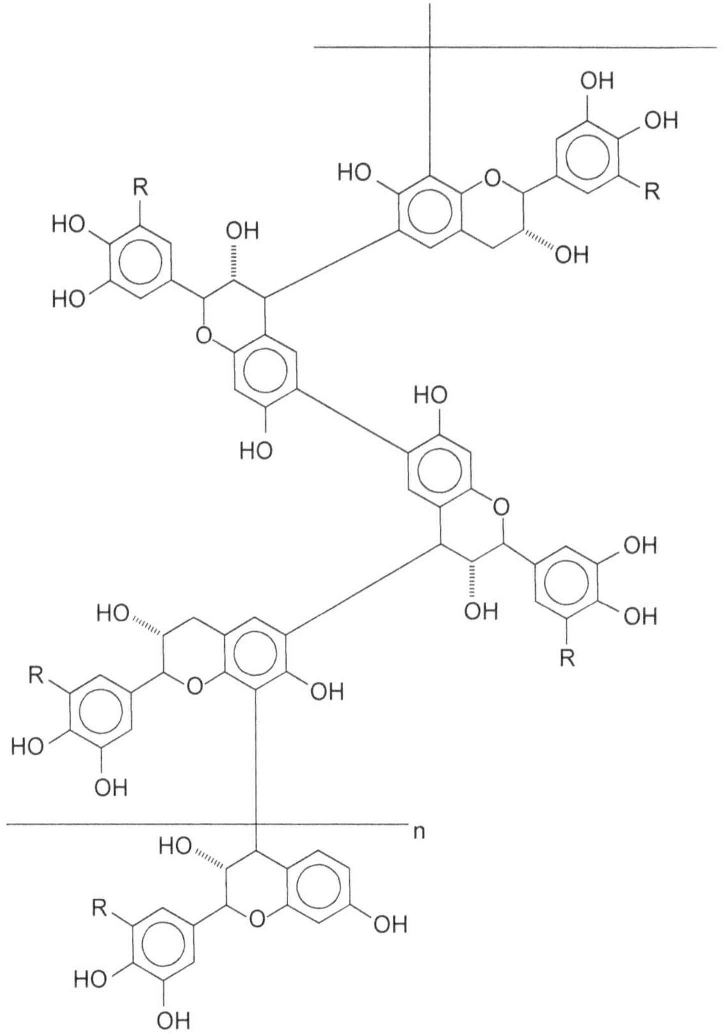

**Figure 4.9** The structure of mimosa tannin, a commercial tannin, with an uncommon prorobinetinidin structure.

## 4.3 Mineral tanning

A review across the Periodic Table of the tanning effects of simple inorganic compounds reveals that many elements are capable of being used to make leather. But, if the practical criteria of effectiveness, availability, toxicity and cost are applied, the number of useful options is much reduced. In all cases, the benchmark for comparison is tannage with chromium(III), when $T_s > 100°C$ is easily achieved, it is readily available, it is relatively cheap and there are minimum associated health hazards or environmental impact.

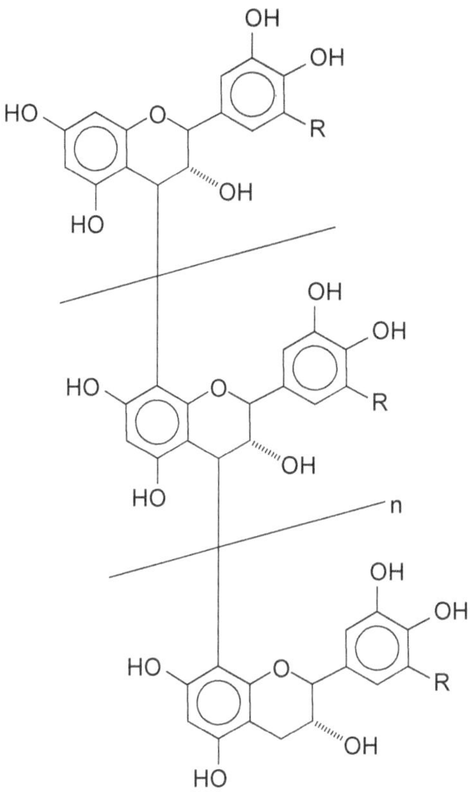

From the Periodic Table: Groups 1, 2, 6, 7, 8 do not tan, in Groups 3, 4, 5 elements of the first period do not tan. The remainder have weak tanning powers, but only elements of the second period are of practical interest: of those, aluminium has the best effect, silicates and polyphosphates have auxiliary functions in tanning. From the transition elements only titanium(III) or (IV), zirconium(IV), chromium(III) and iron(III) have practical possibilities. The lanthanides, as individual elements or mixtures, have moderate tanning properties, but no greater than any alternative in the transition periods. The actinides are ruled out.

So, in the whole Periodic Table there are only five elements plus the lanthanides that might find application in tanning, but only four of those play a significant role in the modern industry.

### 4.3.1 Chromium(III) salts

There is a fortuitous coincidence of reactivities in chrome tanning. The reaction between the tanning material and the collagen molecule occurs at ionized carboxy groups: aspartic and glutamic acid side chain carboxys have $pK_a$ values of 3.8 and 4.2 respectively, providing a reaction range at pH 2–6. Chromium(III) forms basic salts in the range pH 2–5, although in practice the useful range is pH 2.7–4.2.

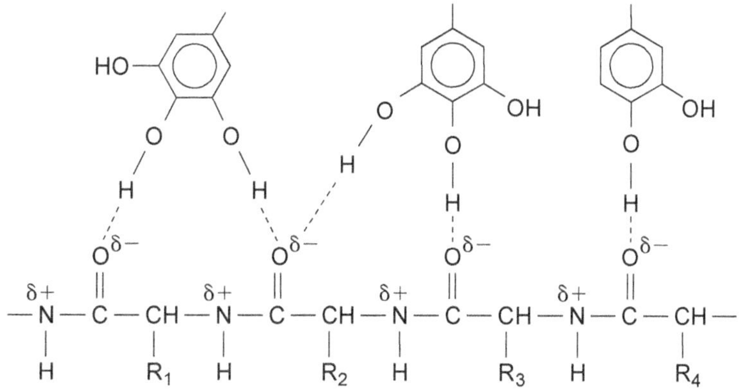

**Figure 4.10** The structure of quebracho, a typical profisetinidin tannin.

**Figure 4.11** Model of hydrogen bonding between polyphenols and peptide link in protein.

In that useful range, the number of chromium atoms in the molecular ion increases from 2 to 4 and the availability of ionized collagen carboxyls increases from 6 to 47% of the total number. Chrome tanning is usually initiated at pH 2.5–3.0 using 33% basic chromium(III) sulphate. During the course of the tanning process, the pH is raised to 3.5–4.0, causing both the number of reaction sites on collagen to increase and the chrome species to increase in size. Starting the process under conditions of low reactivity of both collagen and chrome favours fast penetration of chrome into the substrate, but slow reaction; increasing the pH increases the reactivity of the collagen, although not the reactivity of the chrome (Covington, 2001a), resulting in reduced penetration rate. Chrome fixation is accelerated by elevated temperature and pH: in general the higher the chrome content of the leather, the higher the shrinkage temperature, although the industrial requirement is to achieve the highest shrinkage temperature from the least amount of chrome. To obtain a continuing balance between reaction rate and penetration rate is part of the tanner's art: this is not simple, especially if the skin is thick.

The chemistry of chrome tanning is essentially based on the formation of co-ordination compounds and the changes that occur in chrome species are complex. The formation of olated species was first suggested by Bjerrum. Co-ordinated sulphate is present in the solid tanning salt, but it is rapidly hydrolysed in solution: it plays no part in the bound complex, but it is an important component in stabilizing the tanning matrix (*Figure 4.12*) (Covington 2001b; Covington *et al.*, 2001).

It has been generally assumed in leather science that the reaction which determines the powerful tanning effect and high hydrothermal stability conferred by chromium(III) is through crosslinking at the carboxylate side chains. Indeed, Gustavson (1953) claimed to calculate that about 10% of bound

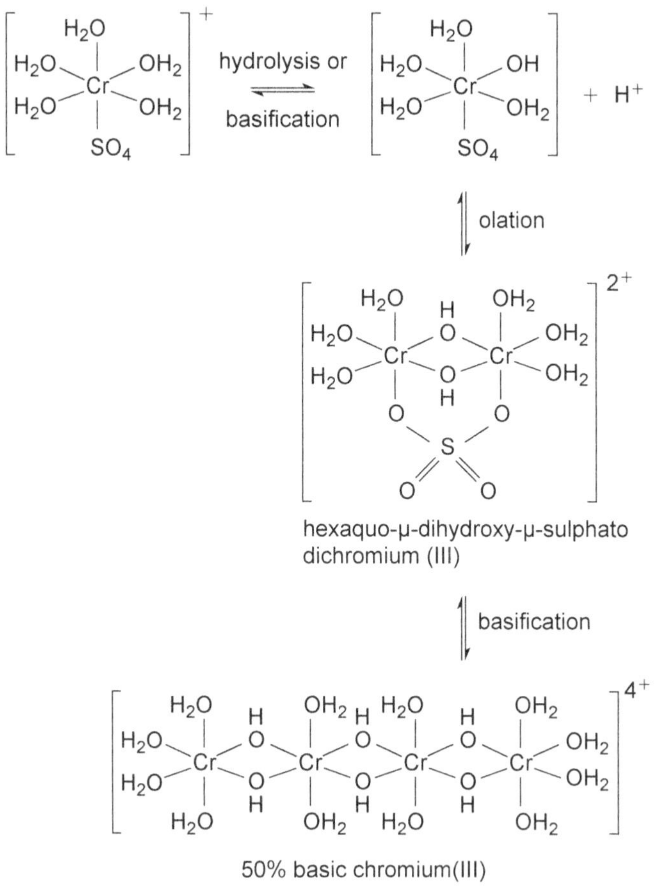

**Figure 4.12** The changes in chromium(III) species during basification.

chrome is reacted in this way. However, the argument has been shown to be flawed and incorrect. Recent research has indicated that crosslinking does not need to be invoked as the mechanism of chrome tanning (Covington and Song, 2003).

The co-ordination of ligands to chromium(III) to modify the properties of the salt is routinely exploited as 'masking'. Monodentate ligands, especially formate, may be applied at different ligand-to-metal ratios, taking up co-ordination sites and hence reducing the reactivity, increasing the ability to penetrate hide. In addition, such masking can increase the pH at which the salt precipitates; in these circumstances the final pH of the tannage may be elevated beyond that of unmasked tannage, thereby enhancing the reactivity of the collagen. In this way, the reaction rate can be accelerated, but without the same effect of increasing the size of the chrome species, which hinders penetration. Masking with bidentate ligands, capable of crosslinking chromium ions, causes an increase in molecular size and changes the hydrophilic–lipophilic balance of the complex. Dicarboxylates containing two or more methylene groups perform this function: phthalate is the salt of choice, when the mechanism of enhanced fixation depends on making the complex more hydrophobic, weakening the hydrating effect of the solvent, therefore driving the complex onto the collagen. Whichever masking salts are used, they are usually added to the tanning bath: this means the masking reaction proceeds at the same rate as the tanning reaction, and because the reactions are identical, both are the formation of carboxy complexes with chromium(III).

Chrome-tanned leather is highly versatile, largely due to the low level of tanning agent needed to achieve the desired stability. This means that the variety of retanning materials which might be applied to the part-processed leather can produce a wide range of final products. Indeed, from any one chrome-tanned cattle hide, it is possible (at a pinch) to produce a sole leather or combat boot upper leather or softee shoe upper leather or upholstery leather or garment leather, all as full grain or suede.

### 4.3.2 Aluminium(III) salts

The use of potash alum in leathermaking is almost as old as leathermaking itself; it occurs in nature and it is known that the Egyptians used it 4000 years ago, because written recipes survive. Throughout tanning history, alum was often used in conjunction with vegetable tannins, typically to enhance the colouring reactions, in which the natural dyes would form intensely coloured lakes with the basified metal species.

Used by itself, alum interacts only weakly with collagen, scarcely raising the shrinkage temperature and having little leathering effect. However, in a mixture of water, salt, flour (to mask the aluminium ion and fill the fibre structure) and egg yolk (the lecithin content is an effective lubricant), skin can be turned into a soft, white, leathery product, traditionally used in the past for gloving. But, even in this case, the shrinkage temperature is not raised (hence, it is possible to discriminate between leathering and tanning) and the aluminium salt can be washed out of the leather if it gets wet; for these reasons, this process is called 'tawing', to distinguish it from tanning.

The reaction sites for aluminium(III) are the collagen carboxyls, but unlike chromium(III) to which it bears a superficial resemblance in a tanning context, aluminium(III) does not form defined basic species nor does it form stable covalent complexes with carboxy groups; that interaction is predominantly electrovalent, accounting for the ease of hydrolysis. The reaction can be optimized for tanning by modifying aluminium sulphate with masking salts, such as formate or citrate, and basifying the tannage to pH 4, close to the precipitation point. In this way, reversibility of tannage is minimized and shrinkage temperatures as high as 90°C can be achieved. Basic aluminium(III) chlorides are also well known in leathermaking and several commercial tanning formulations are available. As solo tanning agents, they are slightly superior to salts based on the sulphate. However, the leathering effect of aluminium(III) is inadequate, producing firm leather, which may dry translucent due to the fibre structure resticking. Therefore, as tanning agents, aluminium(III) salts have limited value.

### 4.3.3 Titanium(IV) salts

In tanning terms, titanium(IV) salts are slightly superior to Al(III). Empirically, the chemistry is dominated by the titanyl ion $TiO^{2+}$ but the actual species are chains of titanium ions bridged by oxygen atoms. However, the co-ordinating power is weak with respect to carboxy complexation, so the interaction is more electrostatic than covalent.

The traditional use for titanium(IV) in tanning was in the form of potassium titanyl oxalate, to retan vegetable-tanned leather for hatbanding, a product for which demand fell away in the latter half of the twentieth century. Titanium solo tanning is only moderately effective, because large quantities are required to achieve the highest shrinkage temperatures, ~90°C, and this causes the leather to be overfilled, although remaining soft. An advantage of tanning with titanium(IV) is that it is a colourless tannage and therefore makes white leather. Hence, it has found application in tanning sheepskins with the wool on for rugs.

### 4.3.4 Zirconium(IV) salts

The development of zirconium tannage is relatively recent but it soon gained industrial acceptance. From its position in the Periodic Table, Zr(IV) might be expected to display similar tanning properties to Ti(IV). The tanning power exceeds that of Al(III) and Ti(IV), but in no way matches Cr(III). While the tanning effects of Zr(IV) and Ti(IV) are similar, the chemistries of their salts are different. Zirconium(IV) salts are characterized by eight co-ordination and high affinity for oxygen, resulting in a tetrameric core structure (Figure 4.13); the basic unit of structure is four Zr ions at the corners of a square, linked by diol bridges, above and below the plane of the square.

By hydrolysis or basification, the tetrameric units can polymerize, by forming more diol or sulphato bridges. In this way, zirconium species may be cationic, neutral or anionic and large ions can form. So, tanning may involve all the polar side chains of collagen, those bearing carboxy, amino or hydroxy groups. Hydrogen bonding via the hydroxy groups in the Zr(IV) species is an important feature of the tanning reaction; together with the filling effect by the big molecules, the overall tanning effect is somewhat similar to tanning with plant polyphenols, hence zirconium tanning has been referred to as the inorganic equivalent of vegetable tanning.

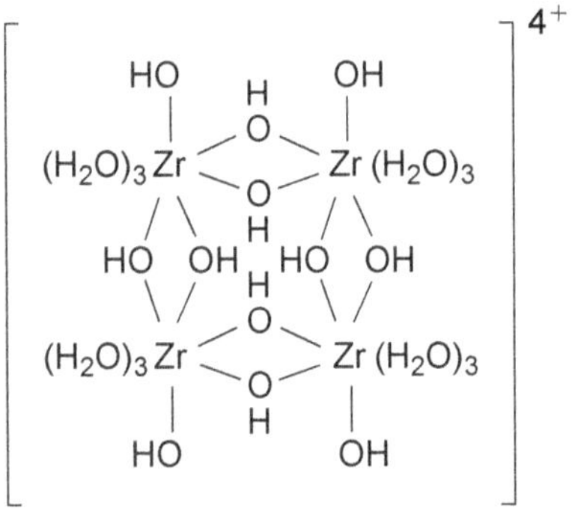

**Figure 4.13** The structural unit of basic zirconium(IV).

## 4.4 Oil tanning

The familiar wash leather (chamois or 'chammy') is tanned with unsaturated oil and the preferred agent is cod liver oil. Useful tanning oils contain fatty acids, either free or as glyceride derivatives, which are polyunsaturated. The degree of unsaturation is critical, because if there is too little unsaturation the oil will not oxidize readily and therefore function only as a lubricant, if there is too much unsaturation the oil will crosslink itself and harden with oxidation, like linseed oil.

In this tannage, dewoolled (fellmongered) sheepskins are processed in the normal way to the pickled state, when they are then swollen by the osmotic effect in water, so they can be split more easily. The flesh splits are then treated with the cod oil in rotating drums. Blowing warm air into the vessel serves two functions; the skins are dried a little, to aid oil penetration, and autoxidation of the oil is initiated, which is the basis of the process.

The actual nature of the tannage is not known, except for the following observations (Sharphouse, 1985):

1. The unsaturation of the oil decreases as the process progresses.
2. Peroxy derivatives are formed.
3. Hydroxy function appears.
4. Acrolein, $CH_2$=$CH.CHO$, is produced.

It is thought that the tannage may be due in part to an aldehydic reaction and to polymerization of the oil; the presence of the latter effect could account for the difference between the characteristics of oil- and aldehyde-tanned leathers. The situation is further complicated by the observation that oil tanning hardly raises the shrinkage temperature of collagen; so oil tanning is a leathering process rather than a typical tanning process, based on the accepted criteria of tanning. This is an extreme form of a stabilizing matrix, relying for its effect on physical rather than chemical interaction with the triple helix. It also has unusual consequences.

The most remarkable feature of oil-tanned leather is its hydrophilicity, surprising considering the chemical nature of the tanning agent. A well-tanned chamois leather is expected to take up at least 600% water on its dry mass and to be hand wrung to 180%. Also this must be repeatable after drying. In use, no grease must be exuded to cause smearing. The hydrophilicity probably arises from the effect of the matrix, holding the hydrophilic protein structure apart, allowing it to absorb a lot of water. Another strange property is the Ewald effect, in which heat shrunk leather will regain about 80% of its dimensions if plunged into cold water. This presumably can happen because the matrix allows the separated chains to reregister. The phenomenon has been exploited to mould oil-tanned leather into required shapes, sometimes called tucking.

A synthetic version of oil tanning is to use a sulphonyl chloride which reacts predominantly with the amino groups on collagen:

$$\text{Collagen—NH}_2 + \text{C}_{16}\text{H}_{33}\text{SO}_2\text{Cl} \rightarrow$$
$$\text{Collagen—NH—SO}_2\text{C}_{16}\text{H}_{33} + \text{HCl}$$

Clearly this cannot be a crosslinking reaction, so it is not surprising that the shrinkage temperature is not raised by this tannage, but there is a powerful leathering effect and the product exhibits similar properties to oil-tanned leather.

It was pointed out that acrolein is produced during the oil tanning reaction: indeed, acrolein itself can be used to make a product similar to oil-tanned leather. But, although it is not used itself for this purpose (for toxicity reasons) it is used indirectly, as a component of wood smoke. The traditional method of preserving hides and skins used by the plains dwellers, such as the Native Americans and the Mongols, is to use brains tanning. In this process, the animal brain is partly cooked in water, so it can be mashed into a paste, which can be worked into the pelt. The leathering effect turns the skin to a soft, open-structured leather, buckskin, largely due to the lubricating power of the phospholipids of the brain. The Sioux Indians have a saying, 'every animal has enough brains to tan its own hide'. The leathering effect is serviceable, as long as the pelt is not rewetted, because then it will harden on drying due to the fibres resticking. To make the leather resistant to wetting, the solution is to smoke it over a wood fire; the multiplicity of free radical and other reactions does not adversely affect the handling qualities and the tannage is made permanent.

## 4.5 Aldehyde tanning

### 4.5.1 Formaldehyde tanning

The archetypal aldehyde tannage is with formaldehyde, probably most familiar in preserving biological specimens or in embalming. Reaction occurs primarily at amino groups, shown empirically as follows:

$$\text{Collagen—NH}_2 + \text{HCHO} \rightarrow$$
$$\text{Collagen—NH—CH}_2\text{OH}$$

The N-hydroxymethyl group is highly reactive and crosslinking can occur at a second amino group:

$$\text{Collagen—NH—CH}_2\text{OH} + \text{H}_2\text{N—Collagen} \rightarrow$$
$$\text{Collagen—NH—CH}_2\text{—HN—Collagen}$$

In this way, the shrinkage temperature can typically be raised to 80–85°C. However, the crosslinking is relatively inefficient, probably because the formaldehyde species are not monomeric, as shown in the equations. Among the species formed in solution is paraformaldehyde, $\text{HOCH}_2(\text{CHOH})_n\text{CHO}$. The presence of polyhydroxy species and their reaction with skin produces a white, spongy, hydrophilic leather, although the absorptive property of oil-tanned leather is not matched.

### 4.5.2 Glutaraldehyde tanning

Of the many mono- and multifunctional aldehydes which might be used for tanning (and all can be made to work), only glutaraldehyde and its derivatives have found commercial acceptance, with the possible exception of the more expensive starch dialdehyde. There would appear to be a number of possible tanning reactions of glutaraldehyde: the crosslinking options seem wider than for simple aldehydes, but the result is the same, a maximum shrinkage temperature of 85°C, because in the same way that formaldehyde is not a simple species in solution, glutaraldehyde can form polymerized species (*Figure 4.14*).

The terminal hydroxy groups of the polymer are active and capable of reacting with amino groups. The polymer itself can interact with the collagen peptide links by hydrogen bonding via the alicyclic oxygens and so the leather is given its spongy, hydrophilic character.

Tanning with glutaraldehyde itself confers a yellow–orange colour to the leather, which is undesirable. Several attempts have been made to modify the chemistry, to prevent colour development, including making the monobisulphite addition compound or hemiacetals, but none has been totally successful.

### 4.5.3 Oxazolidine tanning

An alternative to aldehyde tanning, but which retains the essential reactions, is to use oxazolidines, reported about 20 years ago. These compounds are alicyclic derivatives of an amino alcohol and formaldehyde. Under hydrolytic conditions, the rings can open, to form an N-hydroxymethyl compound which can react with one or more amino sites, in an effective though acridly odiferous tannage (*Figure 4.15*).

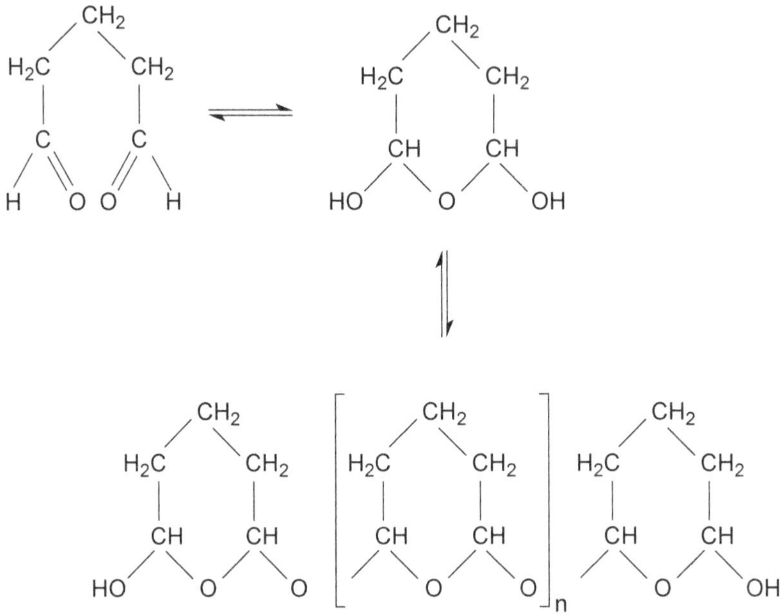

32  *Conservation of leather and related materials*

**Figure 4.14** The reactivity of glutaraldehyde.

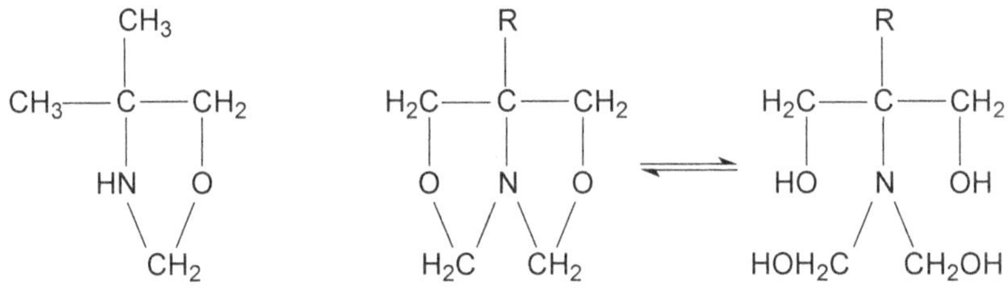

4,4-dimethyl-1,3-oxazolidine

5-(ethyl or hydroxyethyl)-1-azo-3, 7-dioxabicyclo (3,3,0) octane

1,3-dihydroxy-2-(ethyl or hydroxyethyl)-2-N, N-di (hydroxymethyl) propane

**Figure 4.15** The structures of oxazolidines.

## 4.6 Syntans

The term syntan means synthetic tanning agent. This class of tanning agents was introduced early in the last century, with the purpose of aiding vegetable tanning, although the range of reactivities currently available means that they may serve several different functions. They are classified into three types, according to their primary properties. This can be expressed diagrammatically, as relative trends in properties (*Figure 4.16*).

### 4.6.1 Auxiliary syntans

These compounds are frequently based on naphthalene and are synthesized by the 'Nerodol' method, i.e. the base material is sulphonated to a high degree and then polymerized, typically by formaldehyde (*Figure 4.17*); the products are usually relatively simple chemical compounds.

The presence of the sulphonate groups means that these compounds can interact strongly with the amino side chains of collagen at pH < 6:

$$\text{Collagen—NH}_3^+ \text{---} {}^-\text{O}_3\text{S—Syntan}$$

In this way, reaction sites for vegetable tannins can be blocked (see vegetable tanning) promoting penetration through the hide cross-section. At the same time, they serve to solubilize the aggregated phlobaphenes

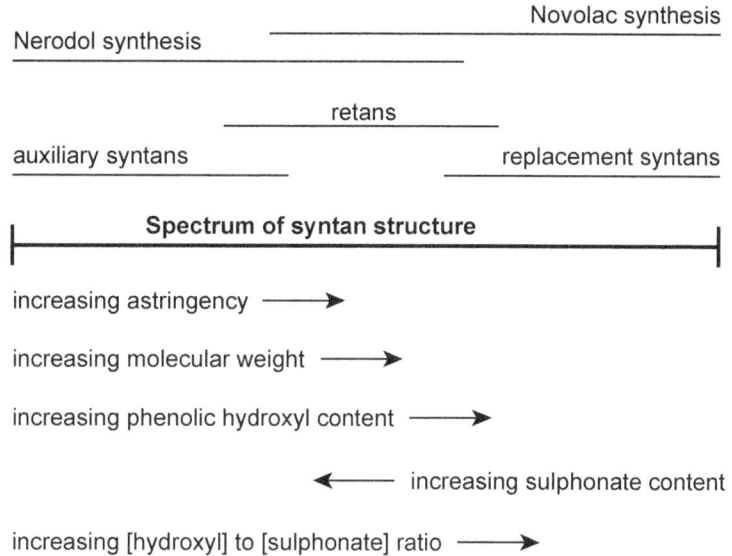

**Figure 4.16** Classification of syntans.

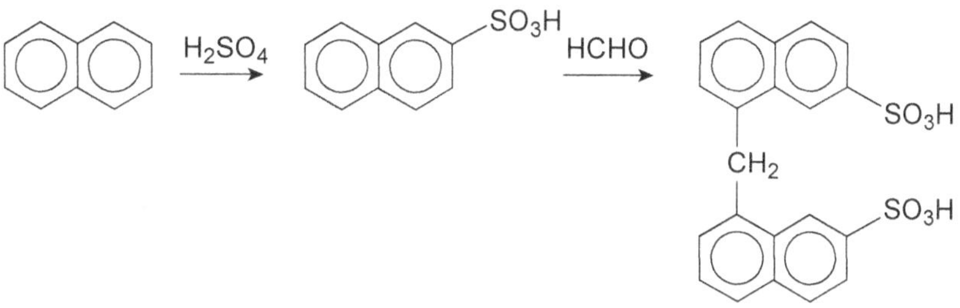

**Figure 4.17** The Nerodol synthesis of syntans.

of condensed tannins, thereby reducing reaction with the hide surfaces.

### 4.6.2 Combination or retanning syntans

These syntans are usually based on simple phenolic compounds, they are synthesized by the 'Novolac' method, i.e. the base material is polymerized, typically with formaldehyde, and then the product may be partially sulphonated (*Figure 4.18*).

The products are more complex than the auxiliary syntans, having higher molecular masses, and may be crosslinked in two dimensions. Their enhanced tanning functionality means that they can confer hydrothermal stability and their larger molecular size means that they can have a filling effect. Because they are relatively small polymers, with consequently weak tanning power, these syntans work best as retanning agents; they are applied after main chrome tannage, to modify the handling properties of the leather.

### 4.6.3 Replacement syntans

By increasing the tanning power of syntans, the agents may be classified as replacement syntans, by which it is meant that they could replace vegetable tannins. These syntans can be used for solo tanning, because their properties of tanning are comparable with plant polyphenols. Again, there is no clear distinction between the retanning syntans and the replacement syntans, the difference lies in the degree of the effects. Base materials for syntans can range from the simple to the relative complicated (*Figure 4.19*).

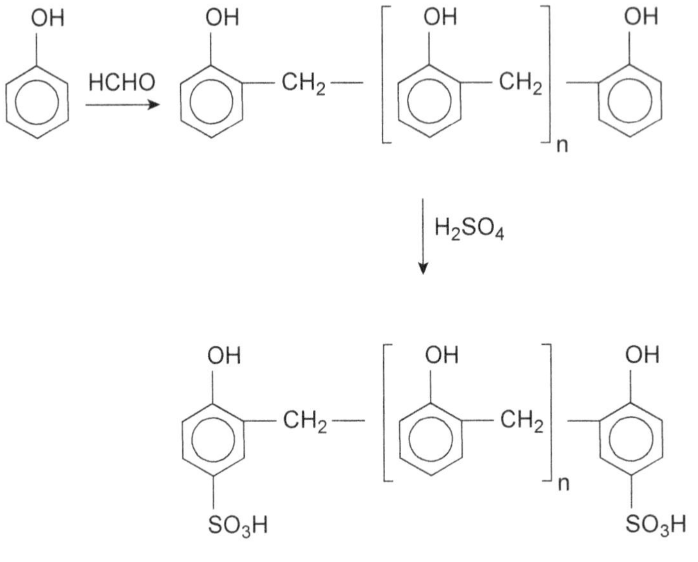

**Figure 4.18** The Novolac synthesis of syntans.

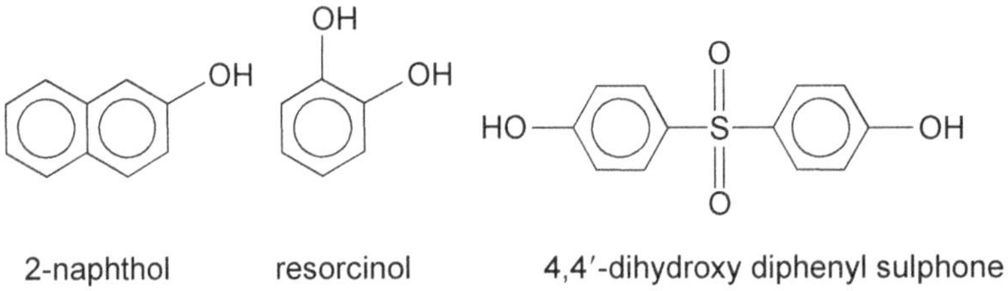

**Figure 4.19** Examples of base materials for replacement syntans.

In addition, the bridging groups may be more diverse, including dimethyl methylene, ether, and urea. They rely less on sulphonate groups for their reactivity, but synthesis by the Novolac method may incorporate some sulphonic acid functionality.

The replacement syntans vary in their effects on leather, but can produce properties similar to vegetable tannins, including raising the shrinkage temperature to 80–85°C. They are still used to prepare hide to receive vegetable tannins, though they can be used in their own right, to make leather that is more lightfast than vegetable-tanned leather; a common use is for making white leather.

## 4.7 Overview

Leathermaking has traditionally been viewed as a craft, often as a dark art. That has changed. Since the early part of the twentieth century, leather science has had an increasing impact on the understanding of the processes involved in the preparation of hides and skins for leather and the subsequent tanning reactions, followed by retanning, dyeing, lubricating, drying and finishing the leather.

It is clear that the deeper our understanding is of the scientific principles that underpin the technologies involved in leathermaking, the better will be our potential to conserve leathers and to improve leathers of the future. Improvements in the environmental impact of tannery operations and in the performance of the products will ensure the continued place of leather in the range of modern materials.

## References

Berman, H.M., Bella, J. and Brodsky, B. (1995) Hydration Structure of a Collagen Peptide. *Structure*, **3**(9), 893.

Covington, A.D. (1997) Modern Tanning Chemistry. *Chem. Soc. Rev.*, **26**(2), 111.

Covington, A.D., Hancock, R.A. and Ioannidis, I.A. (1989) Novel Studies on the Mechanism of the Aluminium Tanning Reaction. *Proc. IULTCS Congress*, Philadelphia, USA.

Covington, A.D. (1998) New Tannages for the New Millennium. *Amer. Leather Chem. Assoc.*, **93**(6), 168.

Covington, A.D. (2001a) Theory and Mechanisms of Tanning: Current Thinking and Future Implications. *J. Soc. Leather Technol. Chem.*, **85**(1), 24.

Covington, A.D. (2001b) Chrome Tanning Science. *J. Amer. Leather Chem. Assoc.*, **96**(12), 467.

Covington, A.D. and Song, L. (2003) Crosslinking? What Crosslinking? *Proc. IULTCS Congress, Cancun, Mexico*.

Covington, A.D. *et al.* (2001) Leather Tanning Studies Using Extended X-ray Absorption Fine Structure. *Polyhedron*, **20**, 461.

Gustavson, K.H. (1953) Uni- and Multipart Binding of Chromium Complexes by Collagen and the Problem of Crosslinking. *J. Amer. Leather Chem. Assoc.*, **48**(9), 559.

Sharphouse, J.H. (1985) Theory and Practice of Modern Chamois Leather Production. *J. Soc. Leather Technol. Chem.*, **69**(2), 29.

# 5

# The mechanisms of deterioration in leather

*Mary-Lou E. Florian*

## 5.1 Introduction

This chapter describes the mechanisms of leather deterioration. Not just the changes in the material and the causes of deterioration but how and where, in the chemical make-up of the leather, the deterioration occurs. To understand the mechanisms of deterioration two components must be known: first, the agents of deterioration and their environmental parameters for reactivity and, second, the bonds in the material where the deterioration occurs. This still is not enough, as it only illustrates the potential reactions. For the final analysis the deteriorated leathers must speak for themselves.

Thus this chapter will start with a review of the deteriorating agents. This will be followed by a discussion of the reactivity of some of the various chemical constituents within the leather structure, and finally a review of the analyses of old, artificially aged, and new leathers to determine what deteriorating mechanisms have actually occurred.

Skin has survived under special conditions for centuries, e.g. the ice man, the bog man, the Dead Sea scrolls, and Egyptian mummies. What is common in their environments that has allowed the skin to survive is that all have been held in conditions that have prevented deterioration, by either reduced moisture, reduced temperature or astringent environments and all without significant environmental fluctuations. In all, the rates of reactions were reduced either by too little or too much water or by reduced temperature. It is important to note that all were without light. As we proceed we will see the significance of these environments to the deterioration of leathers.

Leathers have a very complex mixture of chemical structures. They vary according to the chemicals used in hide pretreatment, the chemicals that occur naturally in the tannin extracts that are used in the tanning process and the chemicals used for the various finishing operations. They also vary in the chemicals they come into contact with in use and the chemicals they absorb from their varying environments. Basically what conservators have to deal with is an active chemical reaction. Thus each type of leather may have some specific mechanism of deterioration because of its own particular make-up. But because of the common denominator present in all leathers, the collagen protein – the component that gives it integrity – there must be many deterioration mechanisms in common with all leathers. To make a logical presentation, vegetable-tanned leather will mainly be discussed because it presents the major problems of leather deterioration and is the commonest leather in heritage collections. Also the greatest amount of research has been conducted on these. Many of the same mechanisms of deterioration that occur in vegetable-tanned leather also occur in leathers of other tannages.

Over the past 20 years the major research into the deterioration of leathers has been conducted by the British Leather Confederation formerly the British Leather Manufacturers' Research Association (BLMRA) in Northampton, UK; the Canadian Conservation Institute in Ottawa, Canada; the Leather Conservation Centre in Northampton, UK; and by the co-operative work of the STEP Leather Project Group in the European Community. The published results of this co-operative work will be discussed in this chapter. Much of the co-operative work was carried out on the historical leather samples from the British Long Term Storage Trials (BLTST) started in 1935 as a source of naturally aged vegetable-tanned leather (Haines, 1991a). These trials were undertaken using over 300 volumes which were bound in leathers prepared in various ways. Pairs of books bound with these leathers were deposited for natural ageing, one in the British Library in London, the other in the National Library of Wales in Aberystwyth. The goal was to determine if the sulphur dioxide present in the polluted air in London, but not in Wales, was the cause of leather deterioration. These samples are still being studied with new scientific instrumentation in hopes of understanding the causes and mechanisms of deterioration on a physical and molecular level. These are not the only naturally aged leathers used for research, but they are still an excellent reference collection because of the detailed documentation of the materials and their environmental history. For more information on the British Long Term Storage Trials see BLMRA (1932, 1945, 1954, 1962, 1977).

At the Leather Conservation Centre in Northampton, UK, much of the recent research on leather deterioration, started in 1986, was done to determine the effect of controlled environmental conditions on artificially aged leathers of different tannages (Calnan, 1991). The purposes were to find a suitable leather for conservation treatments, define a suitable environment for storage, evaluate conservation treatments, and study the mechanisms involved in ageing. Haines (1991b) and Chahine (1991) extended the work to examine the role of the acidic environmental pollutants, sulphur dioxide and nitrogen dioxide, in the ageing of new leathers, by using artificial ageing chambers at BLMRA and in France respectively, which also allowed adsorption and desorption studies to be undertaken (Chahine and Rottier, 1997a).

The STEP project, which was started in 1991, involved conservation and research laboratories in Denmark, France, England, Holland, and Belgium. The goals of the project were to identify and quantify the chemical and physical changes which occurred in historic and BLTST naturally aged vegetable-tanned leathers caused by environmental factors including air pollution; to establish the parameters and conditions of an artificial ageing method which will bring about the same changes in new leathers as in these naturally aged leathers and to establish a standard test method to determine the resistance of new leathers to artificial ageing and the effects of leather conservation methods on this resistance.

Having the three sources of leather (naturally aged historical samples, the BLTST leathers and artificially aged leathers) the STEP Project, with the expertise, knowledge and state of the art equipment of the European participant institutions, was able to do a comparative, comprehensive and detailed analysis of the causes and mechanisms of deterioration of, mainly, vegetable-tanned leathers.

It is difficult to know how to proceed without putting the cart before the horse. Should the mechanisms of deterioration be discussed before the structure of leather or vice versa? In this chapter a general overview of the agents and mechanisms of deterioration will be considered first, followed by an examination of the chemical structure of leather and the relevant details that assist in the interpretation of the actual analytical data from aged leathers.

## 5.2 Agents of deterioration

### 5.2.1 Introduction

For the past 150 years, leather chemists have been speculating on the cause of leather deterioration. The history of the interpretation of the deterioration of vegetable-tanned leathers is reviewed in the literature (Larsen, 1995; Haines, 1991a; Calnan, 1991). From 1900 to the 1990s, many components of the tannin/collagen complex and aspects of the environment were considered as factors of deterioration.

In reviewing the literature on all aspects of the deterioration of the tannin/collagen complex it is apparent that there are two main causes, hydrolysis and oxidation. These reactions are both influenced by the environment within the leather, e.g. water, heat, light, pH, gases, etc. Both may act in the leather on the tanning materials and other chemicals used in its fabrication, as well as on the amino acids and peptides of the collagen protein. The two main causes are, however, quite different in their mechanisms and results. Recent studies have suggested that the hydrolytic breakdown in vegetable-tanned leather is chiefly due to the actions of sulphur dioxide and nitrogen dioxides present in acidic air pollution and that oxidative breakdown is due to the effects of high energy radicals from light, oxygen, oils, and the self-perpetuating reactions within the altered internal chemical structure. It has been shown that acid hydrolysis from acidic air pollution has an inhibitory action on the oxidative decomposition of collagen (Larsen, 1995).

Tanning agents themselves break down under oxidative and acid hydrolytic conditions and the breakdown products formed can promote both the oxidative and hydrolytic breakdown of collagen. Thus there are interactions on many levels.

The reactions of various deteriorating agents on synthetic polymers are described by Reich and Stivala (1971). Many of the reactions are similar for protein polymers such as collagen.

### 5.2.2 Acid hydrolysis

Water in the liquid form contains a few molecules that have formed positive hydronium ($H_3O^+$) and negative hydroxyl ($OH^-$) ions. The hydronium ions both act as hydrogen bond breakers and cause the breaking of bonds within ionic structures. The breakage of bonds by hydronium ions is termed hydrolysis.

When an acid, e.g. hydrochloric acid (HCl), is dissolved in water it dissociates to form hydrogen ions ($H^+$) and chloride ions ($Cl^-$). The hydrogen ions react with water molecules to form hydronium ions ($H_3O^+$) which cause hydrolysis. When an acid is involved in hydrolysis in this way it is called acid hydrolysis.

Sulphuric acid in leather, originating from the adsorption of sulphur dioxide from industrial air pollution, is commonly considered as the primary agent of acidic hydrolysis in historical leathers. In sunlight, sulphur dioxide is transformed to sulphur trioxide. The sulphur trioxide is adsorbed by tannins in the leather. There it is hydrated into sulphuric acid, which dissolves in the moisture present in the leather producing active hydronium ions. These hydronium ions, in turn, break the links between the amino acids in the collagen polymer chain. It should be noted that sulphuric acid is only one part of the internal acidic environment in the leather. There are organic acids and acidic breakdown products from tannins and amino acids inherently present within the structure which add to the hydrogen ion concentration and acid hydrolysis potential.

The collagen protein is colloidal in that it does not ionize in water but swells. A concentration of hydronium ions in the water can break bonds selectively within the polymeric structure leading to loss of structural integrity. Ultimately this results in the breakdown of the protein into a gelatin colloidal solution. Heat, moisture and low pH increase the rate of these hydrolytic reactions.

In synthetic poly films the rate of hydrolysis for a given film thickness is dependent on the relative humidity. This must also apply to leather. The water in leather which is involved in chemical reactions is the multilayer or free water (see section 5.2.6) which moves in and out of the fibre structure as water vapour with changes in relative humidity.

Although the most common form of hydrolytic deterioration is acid hydrolysis, different hydrolytic reactions can occur under acidic, basic or neutral conditions.

### 5.2.3 Oxidation

#### 5.2.3.1 *Introduction*

Oxidation involves the loss of electrons from, and an increase in positive valence of, a chemical compound. Often oxidation is described as a reaction in which oxygen combines chemically with another molecule, but there are many oxidation reactions where the loss of an electron occurs in which oxygen itself is not involved. When oxygen combines with other elements it is the oxidizing agent, but when two compounds react to cause a loss of an electron by one, the substance which loses the electron is called an oxidizing agent. Oxidation and reduction occur simultaneously (redox reactions) with the transfer of electrons between substances. The substance oxidized, the one which gains in positive valence, loses electrons: the substance reduced, the one which decreases in positive valence, gains electrons.

Oxidation, the loss of electrons, can occur due to the effects of oxygen, light, heat or the presence of high energy free radicals.

#### 5.2.3.2 *Free radicals*

(For details see Bernatek *et al.*, 1961; Scott, 1965; Eisenhauer, 1968; Reich and Stivala, 1971.)

##### 5.2.3.2.1 *Introduction*

In most oxidative reactions free radicals are involved. Radicals are highly reactive groups of atoms containing unpaired electrons such as N., NO. or $O_2$. (the subscript point notates the odd electron). They may be positively charged, negatively charged or neutral. Their reactivity is influenced by temperature and by chemical concentration.

The main reactions that produce free radicals which are associated with leather deterioration are caused by: radiation, which includes some visible light as well as UV; air pollutants such as ozone, nitrogen oxides, and sulphur dioxide; peroxides from oxidizing agents; products of autoxidation of lipids and charged amino acids and tannins and their breakdown products. Heckly (1976) states that free radicals are produced by a reaction between oxygen and a variety of freeze-dried biological materials including tissues and microorganisms. This has relevance to leathers which have been freeze-dried. He also reports that although heat denaturation destroys the protein structure, the cofactors of bio-oxidation remain undamaged, and they could still react with molecular oxygen. This suggests that oxidation can also occur in dry materials.

The great array of sources of radicals suggests that they are possibly the most common cause of the deterioration of organic materials and Pryor (1976) considers that all oxidation of organic molecules involves free radicals and intermediates. Examples of each source will follow.

#### 5.2.3.2.2 *Light – photolysis, photoxidation*

As mentioned in the introduction, the skin of the bog man and mummies, etc. were in environments without light. What is the importance of this?

The optical radiation spectrum extends from the far ultraviolet through the visible region into the near infrared, from the wavelengths of 200 nanometers (nm) to about 1400 nm (McNeill, 1992). The bonds in macromolecules have energies between 300 and 500 kJ mole$^{-1}$. Visible and infrared light have energies too low to break these bonds, but ultraviolet light with wavelengths lower than 400 nm are sufficiently strong to break them. In order for ultraviolet light to cause photolysis, it must first be absorbed, this requires a chromophore. Chromophores are present, for example, in the red anthocyanins in condensed tannins and in amino acids with aromatic side chains, such as phenylalanine, tyrosine and tryptophane.

Ionizing radiation produces cationic radicals and electrons as primary species and these decay and react further to yield a host of charged and neutral reactive free radicals. These radicals can undergo various electron transfer reactions without oxygen, i.e. they can transfer the energy to another molecule or dissipate the energy by breaking the bonds or producing heat.

Usually the energy from visible light photons that is absorbed by chemical compounds is reflected as light energy or converted to heat energy. In light sensitive materials, such as dyes, the energy causes molecular changes, expressed as loss of colour. Most polymers which are light sensitive at room temperature will be below their glass transition temperature ($T_g$). This is often equated with the shrinkage temperature of collagen ($T_s$) (see section 5.6). In other words, the shrinkage temperature of visible light-sensitive, strongly deteriorated, hydrated collagen can be as low as 20°C, whereas for normal undamaged collagen it is around 65°C.

With polymers, photolysis is capable of cleaving the C–C bonds in the backbone of the polymer chain, and causing chain scission, crosslinking, and the production of small molecular weight fractions, including momomers (Reich and Stivala, 1971).

If oxygen is involved, the light reaction is called photoxidation. This occurs when a photon which is absorbed into a material reacts with an oxygen molecule forming an activated oxygen free radical which in turn reacts with water to form a peroxide which is a strong oxidizing agent.

Photoxidation does extensive damage to polymers, for example resulting in loss of mechanical strength, embrittlement, cracking, crazing, solubility changes, lowering of pH, and colour change. The causes of this damage are chain scission, changes in the amorphous/crystalline ratio, crosslinkage, density change and the production of new functional groups and acidic products. Analogous reactions occur to the collagen protein chain.

#### 5.2.3.2.3 *Autoxidation of lipids*

The oxidation of organic materials by self-perpetuated radicals is called autoxidation.

In the process of manufacturing leather, a lubricant is essential. In addition, during use, leathers are commonly surface treated with dressings that contain fats or oils. The most common fats and oils are triglycerides, which are molecules made up of a combination of one glycerol molecule to which are attached three fatty acids. The fatty acids are either saturated or unsaturated.

In saturated fatty acids each carbon atom is linked to adjacent carbon atoms by a single bond and the remaining two bonds are reacted with hydrogen atoms. The saturated fatty acids commonly do not combine with oxygen but will react under severe conditions at a much slower rate than unsaturated fatty acids.

In unsaturated fatty acids some of the carbon atoms are linked by double bonds and each carbon atom is attached to only one hydrogen atom. All unsaturated fatty acids are particularly susceptible to oxidation. The conversion of the unsaturated oils to a gummy or hard surface is the result of autoxidation by a free radical chain reaction (Mills and White, 1987). The free radicals and peroxides formed in these reactions can also oxidize adjacent proteins. This is a major problem in the food industry where the reactions occur at room temperatures and in materials with low moisture content (Karel and Yong, 1981). They certainly must be a source of free radicals in leather.

This reaction between lipids and molecular oxygen increases with degree of unsaturation and in lipid autoxidation there are many complex reactions involved and chain reactions may occur which make them self-perpetuating. Oxygen, ozone, nitrogen dioxide, sulphur dioxide and metals are potent catalysts of autoxidation of unsaturated fatty acids.

#### 5.2.3.2.4  *Atmospheric pollutants – ozone, sulphur dioxide, nitrogen dioxide and carbon dioxide*

Ozone is a strong oxidizing agent. It is the most reactive component of air pollutants. It can cause main-chain scission in polymers. The concentration of ozone in air, 0.01–0.02 ppm, is sufficient to initiate autoxidation of unsaturated fatty acids. Ozone can react with many types of organic molecule to yield radicals.

Unsaturated fatty acids or oils are particularly sensitive to oxidation by ozone. They produce intermediary ozonides which, in turn, produce aldehydes and hydrogen peroxide, which interact to form acids. In living systems, ozonides can bring about disulphide crosslinkage with sulphur containing amino acids. In collagen the amino acid methionine contains sulphur and theoretically could be subject to similar oxidative changes.

As has been described previously, sulphur dioxide, in the presence of sunlight, is converted to sulphur trioxide on the surface of leather. In the presence of oxygen, sulphur trioxide may form sulphur dioxide and ozone, a strong oxidizing agent.

Tannins in leather act as sinks for sulphur dioxide. Wouters (1994) showed that, under similar conditions, leathers tanned with condensed tannins adsorbed twice as much sulphur dioxide as those prepared with hydrolysable tannins. Unlike hydrolysable tannins, the condensed tannins because of their polymerized–condensed state, form an impermeable surface in which sulphur dioxide can concentrate. Absorption of sulphur dioxide depends on ambient concentration of the gas and the velocity of deposition onto the leather surface. Reich and Stivala (1971) discuss some aspects of the interaction between sulphur dioxide and polymer films. The mechanism of the reactions of sulphur dioxide with polymers in the presence of ultraviolet radiation is difficult to ascertain because of the many side reactions which they can undergo. Thus in the presence of oxygen, ozone may form along with sulphur trioxide. Sulphur trioxide reacts with water to produce sulphuric acid which dissociates to form the damaging hydronium ion.

Electron spin resonance spectroscopy experiments (Rasmussen *et al.*, 1999) show that sulphur dioxide combined with oxygen causes oxidation of nitrogen atoms resulting in nitroxide radicals and possibly –CO. radicals which cause hydrolysis of the free peptide bond.

Nitrogen dioxide is a potent catalyst of the peroxidation of polyunsaturated fatty acids. Nitrogen dioxide reactions are complicated. In sunlight and water it produces nitrogen trioxide and oxygen radicals. This activated oxygen reacts rapidly with organic molecules to produce further radicals which cause chain scission. The ethylene group present in unsaturated fatty acids is readily attacked by nitrogen dioxide, generating further free radical species.

Carbon dioxide is the most common of air pollutants and is present in the largest amount. With sunlight and moisture it accelerates the production of the hydroxyl radical HO. from nitrogen dioxide. There is little reference to the role of carbon dioxide in deterioration of leather.

### 5.2.4  Metals and salts

Ash analysis of tannin extracts (Okell, 1945) show the presence of copper, iron, aluminium, manganese, zinc, calcium, magnesium, sodium, and potassium. Electron dense X-ray (EDX) analysis of samples of 12 contemporary leathers shows the presence of iron in all 12 samples along with the same variety of the above metals (Florian, 1984). In leather processing metal salts are used for a wide range of reasons – from biocides to whitening. We often consider generally that metals act as catalysts for oxidizing reactions, but the role of metals is complex. Magnesium and iron increase the rate of photolysis whereas zinc and aluminium slow it down and at the same time iron and aluminium can act as antioxidants. As deteriorating agents it is those metals which act as catalysts of oxidation which are the main concern.

Several metals such as cobalt, copper, and iron, which undergo univalent redox reactions, can participate as catalysts in the promotion of autoxidation and peroxidation of polyunsaturated fatty acids. Iron is the most active promoter of lipid peroxidation and is common as a contaminant in the fabrication of leather and in vegetable tannin extracts and is readily chelated by organic compounds.

The role of iron in causing hydrolysis of collagen is discussed by Stambolov (1996). Under aerobic conditions the iron acts as a catalyst to cause oxidation of the air pollutant sulphur dioxide to form sulphuric acid which, through hydrolysis, causes solubility of the collagen. Florian (1987) has reviewed the literature on the effects of metal ions on collagen in reference to archaeological objects excavated from a marine environment.

Ciferri (1971) discusses the role of different metal salts and salt concentrations on the shrinkage temperature of collagen crosslinked with aldehydes. The following ions are listed in decreasing order of activity: $Cl^-$ $N_3^-$ $Br^-$ $SCN^-$ $K^+$ $Mg^{2+}$ $Na^+$ $Cs^+$ $Li^+$ $Ca^{2+}$, with $Ca^{2+}$ causing the lowest shrinkage temperature. Common salt used in the tanning industry often contains calcium chloride, $CaCl_2$. Thampuran et al. (1981) suggest that protein degradation observed with long storage of salted hides was probably caused by calcium and magnesium chloride salts.

The reason for the varied effects of the different salts on shrinkage temperature is due to the different degree of adsorption or ionic bonding of each salt to the protein molecules; the greater the adsorption, the greater the effect. Some of these salts may be present in leather in high concentrations as a result of the reduction in moisture content due to ageing.

### 5.2.5 Heat

Supplying external heat to a chemical compound is simply transferring kinetic energy to it. This externally applied energy causes an increase in molecular movement which increases the rate of collision between adjacent molecules and thus increases the rate of reaction. In a polymer it increases the molecular movement until it is so extreme that the bonds holding the chain together can no longer resist the movement and break. Thermal degradation can lead to separation of the three protein chains in a triple helix, such as is found in collagen. An example is the denaturing of proteins using heat to determine the shrinkage temperature. It is a sudden reaction and can be compared to a zipper breaking all at once.

External heat and the moisture content of materials and the air around them are interconnected. The long-term effects of moderate heat fluctuations cause loss of the ability of the leather to adsorb water vapour from the air. After many fluctuations the leather loses its regain ability and becomes hard and brittle. The internal chemical compounds become concentrated because of the reduced amounts of water present. When leather loses or gains moisture a rapid change in the internal temperature of the leather will occur due to the latent heat of vaporization or condensation (Florian, 1984).

In general a leather which is highly degraded, not hard but fibrous and porous, will lose or gain moisture more rapidly than one which is not degraded. The reasons for this are the changed physical and chemical state such as losses in degree of tannage and coherence, the increase in permeability of the fibre structure, and an increase in the presence of hygroscopic chemicals (Calnan and Thornton, 1997).

### 5.2.6 Water

#### 5.2.6.1 Types of water in leather

At normal temperatures and relative humidities there are two types of water present in permeable organic materials such as leather: multilayer water (also called free water) and molecularly bound water.

Multilayer free water is located between the network of interwoven collagen bundles. It is present as monolayers bound to strong and weak binding sites on the protein and as multilayers in the interfibrillar spaces. It is weakly bonded by hydrogen bonds and van der Waals bonds between and on the surfaces of the fibrils, and to themselves, pushing the polymers apart and increasing their mobility. The amount of multilayer free water varies between 0.07 g and 0.25 g water/g of dry collagen depending on relative humidity (Horie, 1992). Free water moves in and out of the fibre structure with changes in relative humidity whereas bound water is retained attached to the protein molecules. This free, multilayer water can enter into, and increase the rate of, chemical reactions such as hydrolysis and oxidation. Its loss will cause stiffness but it is reversible with increases in relative humidity as long as the physical structure, the regain ability, of the leather has not been destroyed. The amount of free water varies with the physical state of the leather, such as porosity and chemicals present.

In describing the water relationship in food in systems containing chelating agents and other antioxidants Karel and Yong (1981) suggest that free water, at the monolayer and bilayer, is required to permit solubilization and activity of these agents, which are used to lower the rate of oxidation.

Bound water is present in the form of an integral part of the collagen structure and affects its physical and chemical properties. Bound water is considered to be that water which is not available to dissolve electrolytes because it does not move by diffusion or according to the osmotic gradient. In the case of collagen, the amount of bound water is about two molecules per tripeptide, i.e. up to 0.07 g water/g of dry collagen, at an equilibrium relative humidity of around 10% (Nomura et al., 1977). Removal of bound water alters the intra- and intermolecular bond arrangement and causes irreversible stiffness. Freeze-drying can cause removal of the bound water and it is also reported to cause denaturation of the interfibrillar proteins.

### 5.2.6.2  Water activity

Certainly, one aspect of the deterioration of vegetable-tanned leather, which is often overlooked, is due to the water in the leather itself. Water in food is always characterized by its vapour pressure ratio relative to the vapour pressure of pure water and is expressed as water activity, $a_w$. This means that if a solute is dissolved in the water its vapour pressure is changed as well as properties such as rate of evaporation and availability to microorganisms. Water activity at $a_w = 0.5-0.6$ accelerates oxidation by introducing swelling of macromolecules and exposing additional catalytic sites. At higher water activity ($a_w = 0.75-0.85$), oxidation will be retarded by dilution.

Another aspect of water relationships deals with deliquescence of specific compounds. Some salts in leathers can deliquesce and form condensed water to enhance deterioration. This is called sweating and has nothing to do with temperature and relative humidity relationships. Water content will also dictate the pH of the complex system. With natural loss of water as a result of ageing, the pH may drop due to an increase in concentration of any acids present.

### 5.2.6.3  Hysteresis

Stiffness or brittleness in leather is commonly caused by fluctuations in moisture content as a result of temperature and relative humidity changes. As material goes through these fluctuations the moisture gain and loss can be graphed into two curves, adsorption and desorption, collectively called the hysteresis curve (*Figure 5.1*) or isotherms (Florian, 1998). The curves show that the amount of adsorption of water is lower than desorption. Over many cycles the two curves become one straight line because of a decrease in the ability of the material to adsorb water, which reflects the relative accessibility of hydrophilic groups within the collagen. This results in the leather becoming stiff. This may be due to the realigning of the polymers and their hydrogen bonding to each other using sites that were originally used to bond water.

### 5.2.6.4  Colloidal nature of collagen

Another aspect of water is its role in the colloidal nature of collagen (see Florian, 1987 for details).

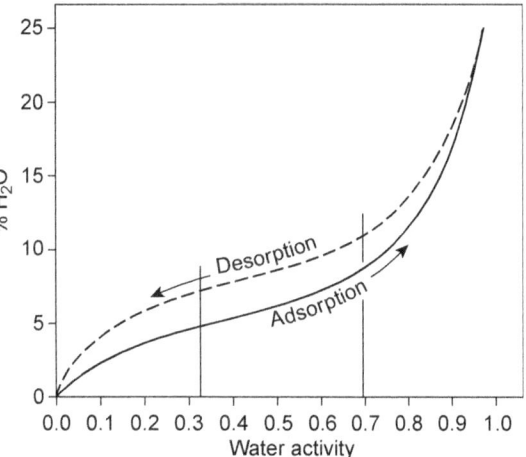

**Figure 5.1**  Schematic representation of moisture sorption hysteresis loop. Water activity can be converted to RH for ease of interpretation. The graph shows that the rate of water vapour adsorption is faster than the rate of desorption. Thus at 0.5 water activity or 50% RH there is more water in materials on the desorptive curve (from Florian, 1998).

This will be discussed under the description of collagen reactivity. There is a link between moisture content, shrinkage temperature and the chemical state (Odlyha *et al.*, 1999a). We often forget that loss of moisture in materials means concentration of solutes which can increase the rate of and initiate chemical reactions.

### 5.2.6.5  Relative humidity changes

Relative humidity is the amount of water vapour in a specific volume of air, at a specific temperature and pressure, relative to the full amount that that volume of air could hold at saturation. Relative humidity changes mean varying amounts of water vapour becoming available for diffusion into or out of organic materials. The moisture present in materials moves according to a diffusion gradient but is also controlled by the water bonding in the material. In materials treated with humectants, these hygroscopic materials chemically bind water molecules into the material taking them out of the diffusion process. This allows more water to enter until all the hygroscopic bonds are satisfied and diffusion has reached an equilibrium between the material and the environment. The movement of water in and out of material is an exothermal and endothermal reaction. Thus the fluctuations over a long-term scale enhance heat damage. Also, the presence of water often enhances the rate of, or is required for, chemical reactions. Water also plays a role in the formation of high energy radicals in photolysis, enhancing light damage.

In summary, the role of free and multilayer water is to act as a solvent for hydrolytic chemical reactions and its presence often increases the rate of reaction. Decreases in water content may increase the concentration and thus increase the rate of reaction of other chemical compounds present. In some cases, such as autoxidation of fats, water slows the reaction. With biological systems the vapour pressure or water activity is the limiting feature for growth. Even if there is a large amount of water present, it may have a low vapour pressure or water activity because of the solutes in it and is not available for fungal growth. This must also play a role in chemical activity in materials. Water bound to molecules such as humectants and salts may not be available for chemical reactions.

## 5.3 Collagen

(See also Chapter 2.)

### 5.3.1 Bonds in collagen: sites of deterioration mechanisms

Hydrolysis of the polypeptide chains in collagen involves the breaking of the peptide (C–N) linkages forming smaller polymers or releasing free amino acids. Oxidative breakdown of the collagen on the other hand is restricted to specific amino acid residues and to certain tripeptide segments situated in charged areas of the collagen peptide chains. Oxidation of the amino acid involves altering the side chains with the formation of breakdown products and finally ammonia. Collagen derives its strength from the higher structures of superhelices. Crystalline structures give a stability, and the tight packing prevents access to deteriorating chemicals. Loss of crystallinity therefore makes it vulnerable to chemical attack. Young (1999) researched the use of loss of crystallinity as a measure of deterioration.

It is important to realize that deterioration can occur at any one of the levels of organization of the collagen molecule independent of each other and may result in changes in the physical–chemical characteristics of the material. At every level of organization there are both hydrogen and covalent bonds holding the complex structure together. It is the breaking of these bonds which causes the denaturation and ultimately dissolving of the collagen. The bonds have different strengths, and it takes equal or greater energy (strength) to break them. This energy comes from the agents of deterioration discussed in the previous section, e.g. the electrical charge of chemical ions such as the hydronium ions of water in acids hydrolyses, and the high energy radicals from light, peroxides or oxidizing agent reactions and from heat. Where and how are these bonds broken?

### 5.3.2 Peptides

#### 5.3.2.1 Introduction, polypeptides to peptides

In proteins, all the bonds between the amino acids in the main chain are called peptide linkages. If these bonds are broken it results in fragments of the protein polypeptide called peptides, i.e. amides formed from two or more amino acids.

In the peptide, by convention, the amino acid with the free amino group is called the N-terminal amino acid or residue and the amino acid with the free carboxyl group is called the C-terminal residue. Degradation of a polypeptide will increase the number of peptides which in turn increases the number of N and C terminal amino acids. The increase in number of N-terminals in aged leather has been used as a measure of the production of peptides which itself is a measure of deterioration.

#### 5.3.2.2 Peptide formation in deteriorated leather

Larsen (1995) reviewed the literature on peptide formation in deteriorated leather. He reports that both hydrolysis and oxidation produce peptides and that the rate of hydrolysis increases as the reaction proceeds, probably due to autocatalysis by the carboxyl end groups. Bowes (1963) suggests that during deterioration only partial hydrolysis of the collagen occurs and, thus, free amino acids are not present, only peptides. Larsen (1995) also reports that splitting occurs at the telopeptide regions and that in 9 out of 10 peptides produced by the breakdown of the collagen molecule, arginine is in the second position, either before or after the splitting point. Arginine is a hydrophobic residue with a basic amino group in its side chain. It is more sensitive to heat oxidation after short-term exposure to hydrolysis, suggesting that the arginine becomes more exposed for oxidation. The opposite occurs after prolonged hydrolysis, suggesting that acid pollution inhibits oxidation.

In the STEP Project Vilmont (1992, 1993) and Juchauld and Chahine (1997) have analysed aged leathers for the presence of N-terminals. Larsen (1995) reviews the literature on N-terminal analysis and states that in all essentials the STEP results are in accordance with those reported in the literature reviewed for leathers of different tannage (Bowes, 1963; Bowes and Raistrick, 1964; and Deasy and Michele, 1965). Vilmont (1993) found an increase in N-terminal residues in artificially aged samples but not a significant increase in historic leathers, suggesting that hydrolysis is not the predominant agent in natural ageing.

Larsen (1995) compared the results of the Vilmont (1993) report to those of Bowes (1963) on N-terminals and states that even a few chain breaks may lead to severe damage of the collagen structure and the formation of the N-terminal peptides with molecular weights down to 10 000. This size is in agreement with those found in gelatine and hydrolysis of collagen and leather.

Larsen (1995) concludes that the preliminary results of the Vilmont (1992, 1993) research show that there is a fine agreement between leathers containing 2% or more of sulphate and the number of N-terminal residues formed by acidic cleavage of the collagen chain and that both the qualitative identification of the N-terminal amino acids formed and the quantitative determination of N-terminals are important for a complete clarification of the extent of leather deterioration.

### 5.3.3 Amino acids in collagen

(See Chapter 2, Hey, 1973 and Munro and Allison, 1964 for details.)

#### 5.3.3.1 *General characteristics of amino acids*

In understanding the potential deterioration mechanisms in leather, a knowledge of the various amino acids present is important, not only because of their loss, but also the presence of their breakdown products such as ammonia and organic acids. Also, as will be shown, the presence of a specific ratio of basic to acidic amino acids in deteriorated leather is an indicator of the degree of deterioration.

*Figure 5.2* shows the relative amounts of amino acids in calf collagen, untanned goatskin and calfskin leather. The high amount of glycine, hydroxyproline and proline is characteristic of collagen.

Because amino acids are amphoteric, the predominant form of the amino acid present in a solution

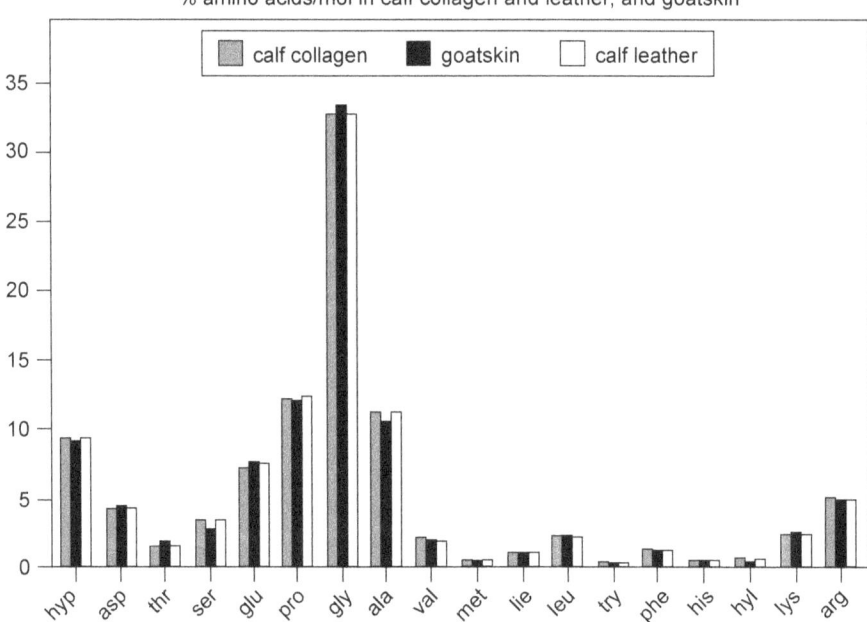

**Figure 5.2** % amino acids in new calf collagen and leather and goatskin. The three sources of collagen show similar amounts of the different amino acids. The high amount of glycine naturally occurs because glycine is in every third position on the polypeptide chain. Raw data from Larsen *et al.* (1994).

depends on the nature of the amino acid and the pH. In a strong acid solution all amino acids are present primarily as cations; in strong basic solutions they are present as anions. At some intermediate pH, called the isoelectric point (pI), the concentration of the dipolar ion is at maximum and the concentration of the anions and cations are equal. Thus the isoelectric point is the pH at which an amino acid carries no net ion charge.

In solutions they are electrolytes with the ability to form acidic and basic salts and thus act as buffers over at least two ranges of pH.

#### 5.3.3.2 Side chains of amino acids (Mills and White, 1986)

Each amino acid has its own properties because of its side chain. According to the nature of the side chain (R-group), the amino acids can be classified into neutral, acidic and basic groups The isoelectric points of the amino acids reflect the neutral, acidic or basic nature of the side chains. Because of the hydroxyl group in the side chain, the amino acids serine, threonine and tyrosine are significant hydrogen binders as is hydroxyproline.

The reactive groups on the side chains of amino acids are capable of a wide range of chemical alterations leading to a large variety of degradation, synthetic and transformation products.

#### 5.3.3.3 Amino acid changes

##### 5.3.3.3.1 Deterioration mechanisms

The common reactions of amino acids are deamination (loss of amine group), and decarboxylation (loss of carboxyl group). There is also the conversion of one amino acid to another called transamination. All three types of reactions result in changes in the amino acids and the formation of products which suggest specific types of deterioration, e.g. amines, ammonia, organic acids, $\gamma$-aminobutyric acid, $\alpha$-amino adipic acid, $\beta$-alanine.

In oxidative decarboxylation, the carboxylic acid group is lost and the amino acid converted to an amine, e.g. glutamic acid is converted to the amine glutamine. As the word deamination suggests, there is a loss of the amine group and the formation of the corresponding $\alpha$-keto acid, such as $\alpha$-ketoglutaric acid. The fate of the amino acid damaged by oxidative deamination and non-oxidative decarboxylation is the same, that is the conversion of the amino acid to the $\alpha$-keto acid, and then to ammonia. When analysing the amino acid profile in old and aged leathers Larsen et al. (1994) observed an increase in aspartic acid compared with the reference, new leathers. This increase could be due to transamination. Another example is the transformation of glutamic acid to alanine.

In ageing studies of living skin it has been observed that the degradation of histidine, arginine, proline and hydroxyproline by a series of reactions, transamination, dehydrogenation, oxidation etc., all result in the formation of glutamic acid. This could also occur in ageing of leather producing an increase in the relative amount of glutamic acid in old leathers.

Thus changes in the kinds and relative amounts of amino acids and the presence of the deterioration products mentioned above, suggest deterioration of the free amino acids or the amino acids at the ends of peptides in leather.

As leather deteriorates the ammonia derived from breakdown of amino acids is thought to react with any sulphuric acid present and form ammonium sulphate, thus the presence of sulphates has also been used as an indicator of hydrolytic deterioration. The greater the oxidative breakdown of the amino acids to ammonia, the greater is the buffering power against sulphuric acid. Thus oxidation can reduce hydrolytic activity.

Other characteristics of amino acids, important in leather deterioration mechanisms, are their ability to bind metals of many kinds and to act as adsorbers of ultraviolet and infrared radiation. Both these characteristics can enhance deterioration of other components of the leather such as fats and tannins.

The effect of ultraviolet damage depends on the structure of the amino acids. Methionine, histidine, tryptophane, tyrosine, and cysteine are the principal amino acids affected, both as free amino acids or in peptides. No breaking of peptide or disulphide bonds occurs. Methionine is oxidized at all pH values (Foote, 1976).

When amino acids are irradiated in air in the dry solid state, the following reactions occur: decarboxylation, with the formation of the corresponding amine; deamination and formation of the corresponding $\alpha$-keto acid; the formation by dehydrogenation of amino acids with $\beta,\gamma$-, or $\gamma,\delta$-unsaturated bonds ($\gamma$-butyric acid). The end product is ammonia.

The products of photolysis of glycine, the commonest amino acid in collagen, may be ammonia, acetic acid, glyoxylic acid and formaldehyde.

Photolytic products of the amino acids can be obtained at room temperature (Greenstein and Winitz, 1961). The rate of their production is significantly increased by the addition of divalent metals such as magnesium or iron, but slowed down by the addition of zinc or aluminium ions (Lewis, 1991).

In the decarboxylation and deamination processes, leucine, glutamic acid and phenylalanine yield radicals. *Figure 5.3* shows variable vulnerability of some amino acids to the breakdown to ammonia by ultraviolet light.

The above information is necessary to understand the significances of the changes in amino acid types and amounts in aged leather.

#### 5.3.3.3.2 Amino acid changes in deteriorated leather

Larsen *et al.* (1997a) has shown that after the natural or artificial ageing of vegetable-tanned leathers, there are changes in the relative amount of some amino acids. *Figure 5.4* shows the increases and decreases in amounts of the specific amino acids in aged leather compared to new leather. They found lower values for the basic amino acids: lysine, arginine and hydroxylysine, accompanied by an increase in the acidic amino acids; aspartic and glutamic acid as well as changes in neutral amino acids and the formation of several breakdown products such as γ-aminobutyric acid and β-amino acids.

Changes in the ratio of the basic and acidic amino acids present were expressed calculated as changes in the B/A ratio. The B/A ratio is the sum in mol% of basic (B) amino acids to the sum in mol% of acidic (A) amino acids. The B/A ratio (Arg + Hyl + Lys)/(Asp + Glu) was shown (Larsen, 1995; Larsen *et al.*, 1997) to be a good measure for the average degree

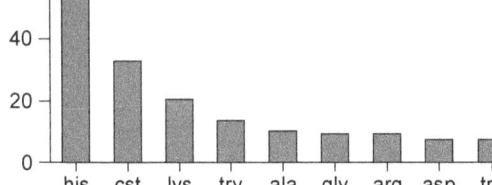

**Figure 5.3** % total α-NH$_3$ formed on deamination of some amino acids after ultraviolet exposure (2 hours under a mercury lamp) (after Greenstein and Winitz, 1961).

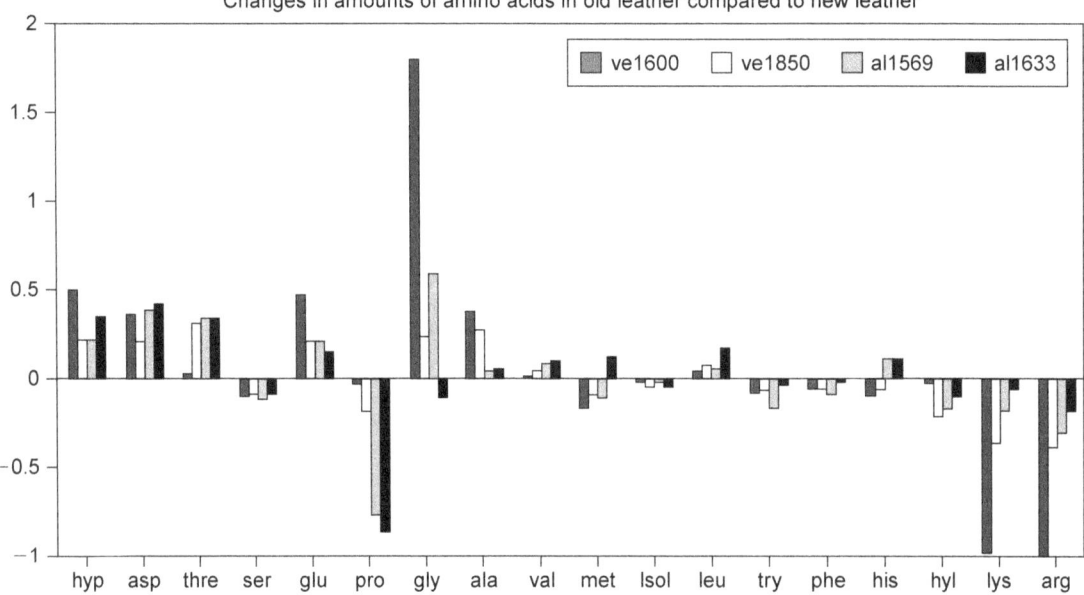

**Figure 5.4** The amounts of change, positive or negative, as compared to new leather, in the relative amounts of the amino acids in naturally aged, vegetable-tanned (ve) and alum-tawed (al) historic leathers. 0 = the baseline for new leather. Those amounts of the specific amino acids above 0 show an increase in amount, and those below 0 a decrease from the new leather due to natural ageing. Raw data from Larsen *et al.* (1994).

of oxidation. The isoelectric points of these amino acids are shown in *Figure 5.5*. In new leather the B/A value is around 0.69 whereas in old leather and artificially aged leather it is below 0.50. This ratio was used to determine the best methods of fabrication of leather to be used in conservation.

The following is an excerpt from Larsen and Vest (1999) which gives the mechanisms and reasons for the amino acids changes.

> *Lysine, arginine, hydroxyproline and proline are sensitive to oxidation. In the oxidation of these amino and imino acids, breakdown products are formed in the form of amino acids, Lysine to Glu, Arg to Glu, Pro to β-ala and Arg to β-ala. The formation of β-ala is only possible by the oxidation of the α-carbon atom in Pro and Arg which leads to cleaving of peptide chains. Many of the breakdown products are amino acids with acidic side chains. This means that the balance of charge between the collagen side chains is altered and the isoelectric point of the collagen is shifted further towards the acidic area, whereby the stability of the leather is weakened. This oxidative breakdown can be measured by amino acid analysis, since the form of deterioration opposite to the hydrolytic breakdown involves the modification of the amino acids in the collagen. This causes a change in the relative amounts of different amino acids in old leather or collagen relative to intact leather and collagen. The different tannages do not significantly change the relative amounts of amino acids on analysis.*

## 5.4 Vegetable tannins

### 5.4.1 Introduction

(General references for the tannins include: Bickley, 1991; Harborne, 1989; Haslam, 1989; Porter, 1989a, b; Porter and Hemingway, 1989; Hemingway, 1989; Haslam, 1979; Roux *et al.*, 1975; Ishak, 1974; Ribéreau-Gayon, 1972; White, 1958.)

### 5.4.2 Antioxidant ability of tannins (Jurd and Geissman, 1956; Lentan, 1966)

Tannins have strong antioxidant activity. The strength of this activity depends on the number and position of phenolic hydroxyl groups (Greenhow and Shafi, 1975). The hydrolysable tannins appear to have greater antioxidant ability than the condensed tannins. Antioxidants work in several ways: in the removal of free radicals, as metal ion deactivators (sequestering, chelation), as UV light deactivators, and as peroxide decomposers.

The antioxidant ability of tannins can be significantly reduced by oxidizing agents, such as UV light, hydrogen peroxide and ozone. In the analysis of metals in water, ozonation under acid conditions below pH 2 is used to release all metal ions sequestered by tannins. This conceivably would occur in deteriorated (pH 2) vegetable-tanned leather, releasing the metal ions to catalyse autoxidation processes of tannins, collagen and oils. In the history of the technology of the tanning process, citrate, oxalate and tartrate salts and sodium pyrophosphate have been used to sequester any iron contamination. The antioxidant ability of tannins may also be reduced by methylation. Tannins could conceivably become methylated during fumigation with a methylating agent such as methyl bromide (Florian, 1985).

### 5.4.3 Analysis of tannins in aged leather – deterioration mechanisms

Wouters (1994) undertook analysis of extracts from new, naturally aged and artificially aged leathers using HPLC (high power liquid chromatography). The amount of tannin and monomers of the phenolic acids present in the extracts was measured by optical density OD/100 mg. The HPLC profiles show the typical changes on ageing of the extracts from leathers tanned

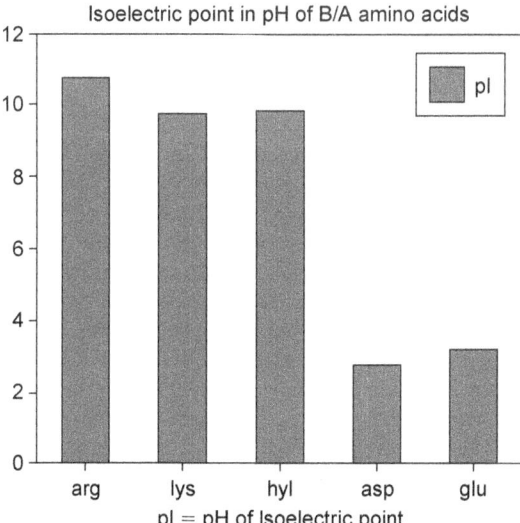

**Figure 5.5** The isoelectric point in pH of the amino acids that were used to calculate the Basic/Acidic (B/A) amino acid ratio B (Arg + Hyl + Lys)/A (Asp + Glu). Raw data from Larsen (1995) and Larsen *et al.* (1996).

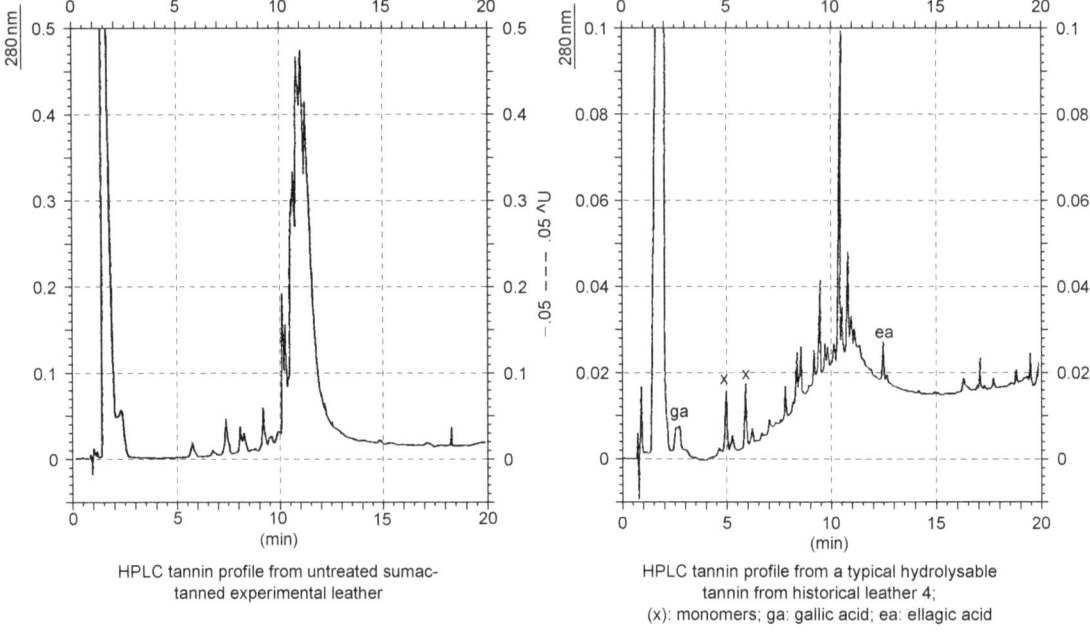

**Figure 5.6** HPLC hydrolysable tannin profile from untreated and a historical leather (from Wouters, 1994).

with hydrolysable and condensed tannins. *Figure 5.6* shows the changes in the chromatographic profiles of the extracts from hydrolysable-tanned leather before and after ageing. It shows the presence of new monomers in the extract from aged leathers and an increase in gallic and ellagic acid monomers. The change in extracts from leathers made with condensed tannins (not shown) was a chemical shift of the main peaks towards more hydrophobic compounds. Both showed increases in ellagic acid. Gallic acid can be copolymerized to ellagic acid by mineral acids such as sulphuric acid. This may explain the increased presence of ellagic acid in deteriorated leathers with a low pH.

*Figure 5.7* shows the optical density of the extractable tannin, its total monomer content and the amount of the monomers, gallic and ellagic acid present in the extracts from the different leathers.

*Figure 5.7A* indicates that with all the naturally and artificially aged leathers produced with hydrolysable tannins there is less tannin extracted after ageing than with the new leather. This suggests loss of tannins due to their breakdown into end products such as organic acids and sugar. The condensed tannins do not show a specific pattern of change. This could be explained by differences in the original tannin extracts because of methods of fabrication; it is difficult to work with such heterogeneous materials.

Because of the above, Wouters and Claeys (1997) state that a measurement of the total amount of tannin extracted cannot be used to determine deterioration of leather.

*Figure 5.7B* shows that ageing causes an increase in the amount of monomers extracted from all samples compared with new leather as a result of oxidation of the tannins. Extracts from aged samples of leathers prepared with condensed tannin contained a greater percentage of monomers than those tanned with hydrolysable tannin, suggesting greater deterioration due to oxidative and hydrolytic reactions.

In *Figure 5.7C*, gallic acid is naturally present in both tannin types. While the relative amounts of gallic acid are not significant, the presence of the gallic acid monomers is. The figure shows a comparable increase of these acids present in extracts prepared from aged leathers made from both types of tannin as compared to the amount present in new leather.

*Figure 5.7D* shows the amount of ellagic acid in the tannin extract. Gallic acid can be dimerised by mineral acids, such as sulphuric acid to ellagic acid. This is a possible reason for the large amount of ellagic acid in the naturally aged leathers produced with condensed tannins.

As expected, Figures 5.7A–D show, increases in monomer contents due to both oxidation and hydrolysis of the tannin.

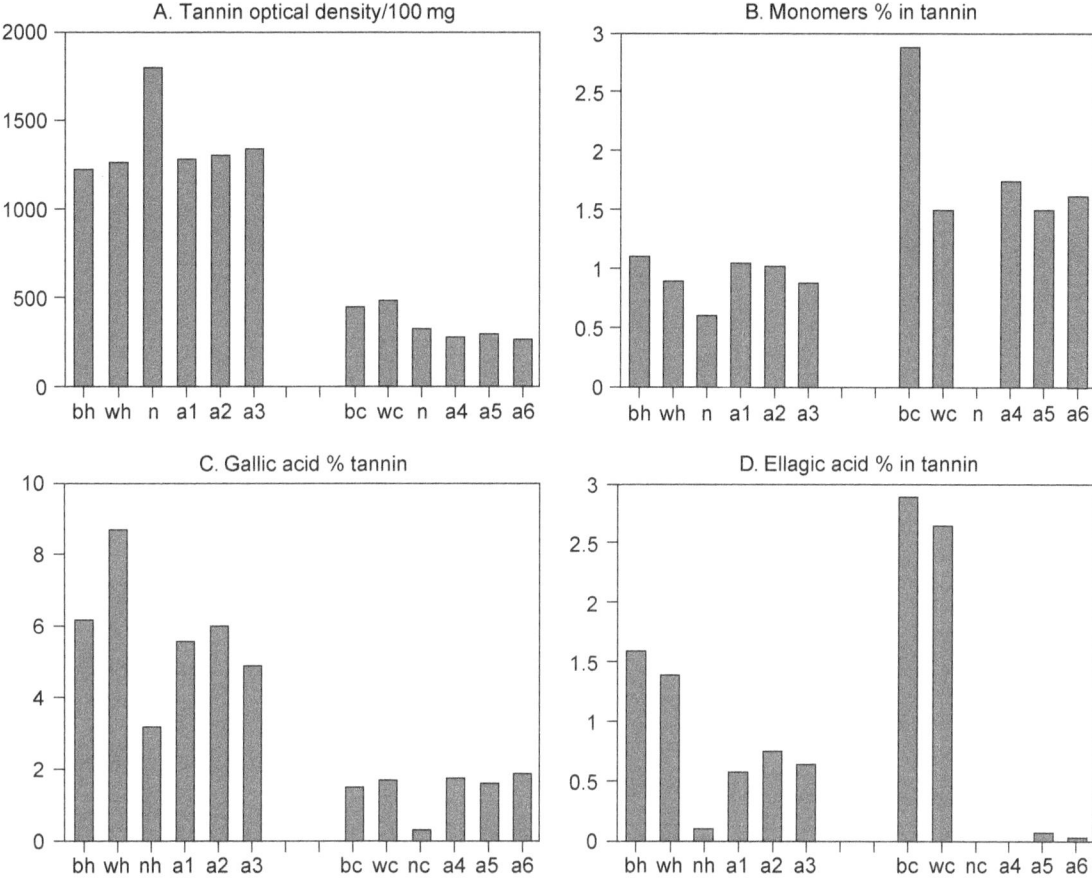

**Figure 5.7** The tannin analysis and their breakdown products in naturally aged leather long-term trial and new and artificially aged leather. Raw data from Wouters (1994). The amounts of extractable tannin and monomers were analysed by HPLC and their amounts determined by optical density. The group on the left in each chart is hydrolysable tannins and the group on the right is condensed tannins.

**h** = hydrolysable tannin; **c** = condensed tannin; **b** = naturally aged leather from bindings of books placed in the British Library in 1932; **w** = as b, except books placed in Wales Library, 1932; **n** = new leather; **a** = leather artificially aged (6th ageing trial); **1 and 4** = 2 weeks' ageing with $O_3$, plus 2 weeks with $SO_2$ (45 ppm), $NO_2$ (10 ppm), NO (10 ppm), 40°C, 35% RH; **2 and 5** = 4 weeks' ageing with $O_3$, plus 4 weeks with $SO_2$, $NO_2$, 40°C, 35% RH; **3 and 6** = 1–2 days at 120°C plus 4 weeks $SO_2$, $NO_2$, 40°C, 35% RH.

In summary (Wouters and Claeys, 1997), it can be concluded that the absolute or relative amounts of monomer products which are produced during natural and artificial ageing can be used as a measure of deterioration.

Gallic acid and ellagic acids are both monomers but were considered separately. The amount of gallic acid present increased during ageing in acid environments in both condensed- and hydrolysable-tanned leathers. The authors suggested their analysis is useful to determine deterioration. The presence of ellagic acid in condensed tannins suggests its formation from gallic acid due to the presence of strong acids such as sulphuric acid. The data is also significant in that it shows that the changes that occur in naturally aged leather also occur in the artificially aged leathers.

The variations in the amount of monomers present in extracts from BLTST-aged leathers due to plant species used as the source of the tannin extract and methods of fabrication are illustrated in *Figure 5.8*. The figure shows similarities in the results from several samples of leathers made using the same species, i.e. oak and quebracho, but significant differences in the

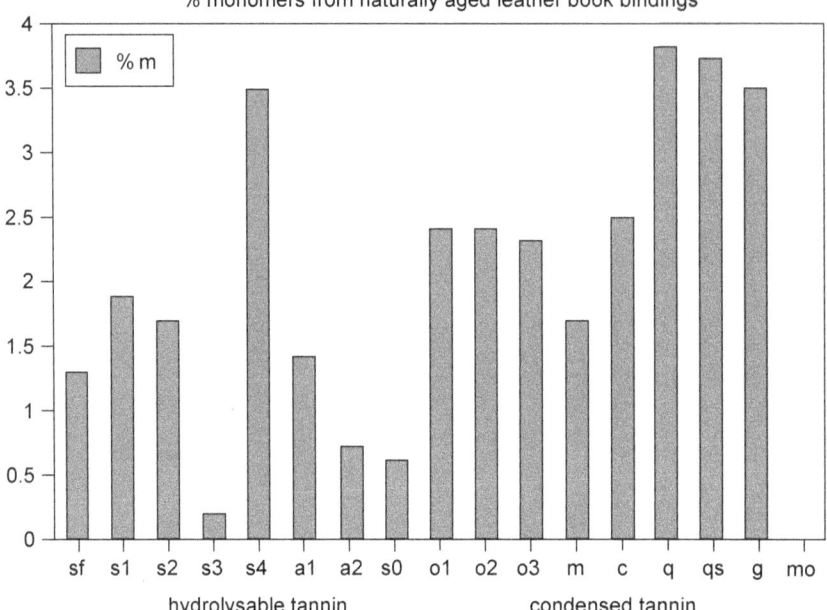

**Figure 5.8** Comparison of analysis of % monomers in extracted tannins from different species used in the British long-term tests in book bindings from books placed in the British Library in 1932. The ageing environments were the same for all tannins. Raw data from Larsen (2000). **sf** = sulphited sumac; **s** = sumac; **a** = acacia; **n** = new; **o** = oak; **m** = mimosa; **q** = quebracho; **g** = gambier; **%m** = % monomers in extractable tannin.

samples prepared from sumac. The results suggest differences are due to the species used but there are also differences from one leather to another as well due to different methods of fabrication. Overall both groups show significant changes with aged leather compared to the new leather, but the two groups of tannins show different deterioration patterns. The condensed-tanned leathers show a higher amount of monomers in the tannin extracted than that obtained from the hydrolysable-tanned leathers. The average and range of amount of monomers extracted from aged leathers tanned with hydrolysable tannins, 2.4% (1.9–3.9), and from those tanned with condensed tannins, 5.25% (3.5–6.8), show, despite the species differences, that a trend is present. This illustrates the greater deterioration of the condensed tannin.

## 5.5 Other chemicals present due to fabrication and use

### 5.5.1 Introduction

Leather is not just a product of collagen and tannins, it is a complex mixture made up of collagen, tannin extracts which themselves contain a complex mixture of different types of polyphenolic tannins, non-tannins (sugars, low molecular weight phenols, etc.), metal ions, organic and mineral acids, salts and sulphur compounds. Besides these two major mixtures there are other materials used in its fabrication. Tanned leather may be subjected to a variety of chemical and mechanical processes or finishing operations such as fat liquoring, staining, dyeing, graining or embossing, plating, boarding, enamelling or abrading, etc. (Thomson, 1991a, b).

Products employed during use to increase its serviceability, such as saddle soap, oils and phenolic biocides, and adsorbed gases (sulphur dioxide, nitrous oxide) may also be present. These products may prevent or enhance deterioration. The presence of some of these materials may give misleading information in analyses of old leathers, for example the phenols from biocides may be mistaken as coming from the tannins, and sulphur compounds from fat liquors and dyes as coming from air pollution.

Contaminants arising from use, such as perspiration, are also a source of deteriorating chemicals. Only the relevant common chemicals involved in deterioration will be discussed.

### 5.5.2 Fats, oils and waxes

The role of fats and oils as agents of deterioration is in producing high energy radicals which catalyse oxidation of proteins (see section 5.2.3.2.3).

Fats, oils and waxes are introduced after tannage, to lubricate the leather fibres and prevent them from sticking together as the leather dries.

Sulphated (incorrectly called sulphonated) oils are oils treated with sulphuric acid to make them miscible in water (Tuck, 1983). These oils have an ester-type bond between the oil and the combined sulphur trioxide which is not stable and is easily hydrolysed releasing sulphuric acid. Sulphited oils made by treating unsaturated fatty acids esters with sodium bisulphite form the true sulphonated oil with carbon–sulphur linkage. These are very stable, especially in the presence of electrolytes such as salts.

Analysis of the amounts of fat in vegetable-tanned leathers has been done by Wouters and Claeys (1997), and Wouters et al. (1997), but their implication in deterioration was not pursued.

### 5.5.3 Sulphur compounds and their acids

#### 5.5.3.1 Introduction

The topic of sulphur compounds (sulphites, sulphides, sulphates, sulphuric and sulphurous acids) in the tannin liquors and extracts is indeed complex. Again the information is simplified and presented as a first step in studying the deterioration of the tannin/leather complex.

The sulphur compounds come from various sources. Sulphuric acid and sulphide may come in with pretreated hides. Organic sulphonated materials and inorganic chemicals may be added to the tanning liquor to maintain the necessary acidic pH for tanning, solubility of components and control of viscosity. It should be noted that hydrolysable tannins are not usually sulphonated in the final tanning process, and that most of the inorganic water soluble sulphur compounds will be washed out. But the organic sulphonated materials (the sulphonated condensed tannins and ligno-sulphonates and syntans) may remain as bound sulphur. On degradation of these sulphonated organic materials in the leather, the sulphur must be released.

In deteriorated leather at pH 2, there is a possible sulphurous acid content along with sulphuric acid from sulphur dioxide pollution, which could cause hydrolysis of protein and tannins. Sulphur dioxide as an agent of deterioration is discussed in section 5.2.3.2.4.

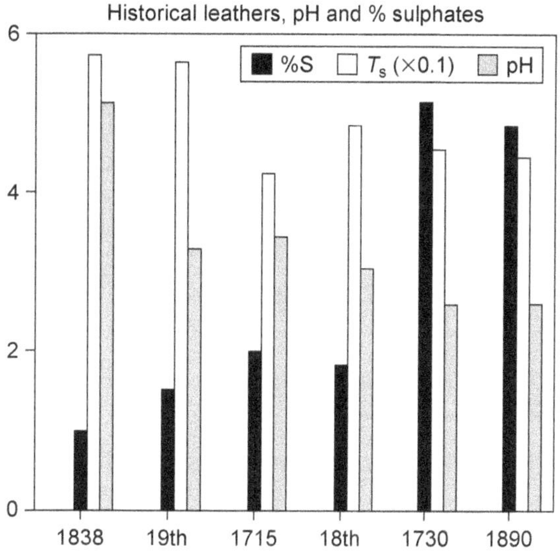

**Figure 5.9** Historic leathers: %S = percent sulphate in 2200–1000 g of leather; $T_s$ = shrinkage temperature expressed as ×0.1 for ease of comparison. Raw data from Larsen (2000).

#### 5.5.3.2 Changes in sulphates and pH in deteriorated leather

In leathers the presence of sulphates has been considered as an indication of the presence of sulphuric acid. Wouters and Claeys (1997) showed that with aqueous extracts from artificially aged vegetable-tanned leathers, as the pH becomes more acidic the sulphate content increases. On the other hand they observed that some badly degraded historical leathers (see *Figure 5.9*, 1715 and eighteenth century), with low shrinkage temperature, which had not been exposed to industrial sulphur dioxide, had a normal pH, but a high sulphate content. Because of this they stated that the pH measurement alone cannot be a good measure for deterioration, and suggested that the total concentration of sulphate is also needed.

Wouters (1994) described how the amounts of sulphate and nitrate present in a leather can be used to calculate the quantity of nitric acid or sulphuric acid which would be required to give these amounts. Using these quantities, a theoretical pH was calculated assuming that no other relevant material was present pH(c). If this pH(c) was different from the measured pH (pH(m)), the difference is considered to be due to sulphate present as ammonium sulphate. High levels of ammonium sulphate are considered to be indicative of severe deterioration in the past. The amount of sulphate present in hydrolysable-tanned and condensed-tanned leather inherently varies because the condensed tannins act as a sulphur dioxide sink. Sulphur dioxide is deposited at twice the rate in condensed- than hydrolysable-tanned leathers.

*Figure 5.9*, besides showing that badly degraded historical leathers can have a normal pH, shows the inverse relationship between pH and sulphate, that is as the pH decreases the sulphates increase. This confirms the work of Wouters (1997) who suggested that pH alone is not a parameter to assess deterioration.

### 5.5.4 Acids in leather due to fabrication or use

Okell (1945) lists acids that are likely to be met in tannery practice. These include hydrochloric, sulphuric, sulphonic, oxalic, sulphurous, formic, lactic, acetic, ellagic, gallic, and carbonic. Burton (1948) reports that the acidity of a tan liquor was considered to be the result of free hydroxyl groups of the pyrogallol (hydrolysable) tannins and possibly from catechol (condensed) tannins; phenolic hydroxyl groups of the pyrogallol and catechol tannins and lignins; natural acidity due to gallic, lactic, acetic, carbonic, oxalic, citric, tartaric, and phosphoric acids present in natural tanning materials; uronic acid and pectins; added formic, boric, sulphurous and sulphuric acid.

How many of these end up in the leather is an unknown. Their role in deterioration mechanisms is their involvement in acid hydrolysis and as a part of the acidic environment of the leather. pH changes in leathers is discussed in section 5.5.3.2.

### 5.5.5 Perspiration

Perspiration is considered as an important deteriorating agent of leathers in use. Bowes (1963) reports on extensive research on its effects on different types of tannage. It is the combination of moisture, heat and perspiration that cause the problem in, for example, shoes. The chief constituents of perspiration are sodium chloride (0.3–0.5%), lactic acid (0.1–0.3%), amino acids (0.05%) and urea (0.05%). The sodium lactate formed under moist conditions causes losses of chromium crosslinkage and reduces the shrinkage temperature (Haines, 1991c). In vegetable-tanned leather, tests using synthetic perspiration showed that these tannins reduced the effects of the perspiration as compared to chrome-tanned leather. The main change in the vegetable-tanned leather was the darkening caused by released phenols, due to the increase in pH.

## 5.6 Denaturation and shrinkage temperatures as a method of assessment for all tannages

(For details see Chahine and Rottier, 1997b; Larsen et al., 1993, 1997b; Larsen, 1995; Young, 1990, 1998, 1999.)

Denaturation is simply a term which implies the physical change of the hydrated collagen complex. It is commonly defined as the transition from the triple helix to a randomly coiled form that has been caused by any one of the many deterioration agents that could affect leather. These include alcohols, acids, heat, radiation, oxidizing and reducing agents, detergents, etc. Boghosian et al. (1999) using Raman spectroscopy did not observe random coiling, which suggests more research is necessary.

Our present understanding is that denaturation is the result of the loss of collagen's higher structural features due to disruption of the hydrogen bonding and other secondary forces that hold the molecule together. The temperature at which fully hydrated collagen or leather shrinks under defined conditions is called the shrinkage temperature ($T_s$).

The $T_s$ of undeteriorated mature collagen is around 65°C but it can be as low as 30°C in deteriorated collagen. The $T_s$ for new vegetable-tanned leather is from 70 to 90°C depending on the tannin type. Again with severely deteriorated leathers this can be as low as 30°C. In chrome-tanned leather the weak hydrogen bonds keeping the collagen polypeptides together are replaced by strong covalent chromium chemical bonds, thus the shrinkage temperature is around 100°C.

(For details of the methods used for determining $T_s$ see Young, 1990, 1998, 1999; Larsen, 1995, Larsen et al., 1997b; Larsen and Vest, 1999; Chahine and Rottier, 1999; Bowden, 1999; Odlyha et al., 1999b.)

$T_s$ is used as a measure of deterioration of collagen or leathers. It is apparent from the information already presented that hydrolytic and oxidative mechanisms occur in the deterioration of both tannins and collagen and that these are influenced by environmental parameters. As vegetable-tanned leathers deteriorate, the extractable vegetable tannins show a decrease and the amount of monomers present increases. The hydrolytic breakdown is measured by an increase in sulphate content, by a reduction in the B/A ratio of amino acids and by changes in hydrogen ion concentration (*Figure 5.10*). The total effect of oxidation and hydrolysis is reflected in the shrinkage temperatures.

Using data for monomers and sulphate contents, B/A and [H$^+$], Larsen (1995) has shown by common

| $O/T_s$ | $P/T_s$ | B/A | % T/monomers | % sulphate | $(H^+) \times 10^4$ | pH |
|---|---|---|---|---|---|---|
| *Historical leathers* | | | | | | |
| 32.9 | 31.2 | 0.50 | 5.85 | 3.60 | 11.04 | 2.96 |
| 43.8 | 47.3 | 0.54 | 6.18 | 3.46 | 14.75 | 2.83 |
| 52.4 | 54.2 | 0.58 | 5.92 | 2.18 | 12.61 | 2.90 |
| 61.4 | 61.6 | 0.60 | 3.88 | 1.84 | 10.05 | 3.00 |
| 70.0 | 67.7 | 0.61 | 3.50 | 5.60 | 26.30 | 2.58 |
| 81.9 | 81.8 | 0.60 | 1.63 | 0.45 | 0.89 | 4.05 |
| *Unaged leathers S = hydrolysable tannage, M = condensed tannage* | | | | | | |
| 76.0 S | 74.9 | 0.69 | 0.6 | 0.1 | 6.92 | 3.16 |
| 79.0 M | 78.2 | 0.69 | 0.0 | 0.2 | 5.50 | 3.26 |

**Figure 5.10** The observed $T_s$ (O/$T_s$/°C) and predicted $T_s$ (P/$T_s$/°C) and the variable used for the prediction of the $T_s$. The values are the mean values of 44 (only half the original data (Larsen, 1995, table 1, p. 124) is included in the table) historical leather samples. The means are calculated by grouping and averaging the values within overlapping intervals of 5°C. The pH values are calculated from the average hydrogen ion concentration (after Larsen, 1995).
**O/$T_s$/°C** = observed shrinkage temperature, the temperature at which hydrated corium fibres shrink, according to Larsen (2000).
**P/$T_s$/°C** = predicted shrinkage temperature. The prediction of $T_s$ is preformed as a common multiple regression based on the statistical classical least-squares method (see Larsen, 2000).
**B/A value** = the ratio of the amounts of basic (B) and acidic (A) amino acids.
**% sulphate** = sulphate as dry mass/%.
**%T/monomers** = tannin monomers%.
**pH** = the acid/alkaline nature of the aqueous extract of soluble components.
**[H$^+$]** = [H$^+$] − $10^4$ calculated from the pH value.
Measurements of pH, sulphate and tannin were analysed according to procedures of Wouters and Clayes (1997).

multiple regression based on classical least-squares method, that it is possible to predict the shrinkage temperature of a deteriorated vegetable-tanned leather. This shows clearly that shrinkage temperature changes are effected by the interaction of all four factors.

*Figure 5.10* compares the observed $T_s$ (O/$T_s$) and predicted $T_s$ (P/$T_s$).

As mentioned in the introduction, the emphasis in this chapter has been on vegetable tanned leather, but the deterioration mechanisms are similar in leathers of different tannages. Assessment of deterioration of parchment by $T_s$ has been demonstrated by Odlyha et al. (1999), Chahine and Rottier (1999), Bowden (1999) and Rasmussen et al. (1999). In addition, Bowes (1963), Haines (1991a) and Larsen (1995) show that oxidation and hydrolytic activity are the causes of deterioration in mineral-and alum-tanned leather and Ciferri (1971) discusses the role of salts on shrinkage temperature of aldehyde-tanned collagen. There are, of course, differences of degree of deterioration, chrome tannage is vulnerable to detannage because of perspiration and alum-tanned leather has the most resistance to deterioration provided it is kept dry.

## 5.7 Summary

It is now possible by using as little as 50 mg of leather to determine its composition and deterioration (Wouters and Claeys, 1997). The analytic procedure determines the denaturation of collagen polymer or shrinkage temperature, the amount, type, and ratio of degradation products of extractable tannins, the amount and composition of protein materials, the concentration of anions such as sulphate and nitrate, the pH of the aqueous extract, the relative amount of volatile materials and amounts and composition of lubricants and mineral ashes.

The results of these analyses have made it possible to determine the deteriorating mechanisms. Oxidation and hydrolysis are the main deteriorating agents. Sulphur dioxide is considered the main source of the acidic environment that causes hydrolysis. High energy free radicals from light, ozone, sulphur dioxide and specific chemical reactions must be important in initiating oxidation. Other aspects of the environment such as heat, moisture and the presence of acidic gases are also part of the group of agents of deterioration or play a role in increasing the rates of oxidation and hydrolysis. The presence of chemicals such as sulphur compounds and metals within the leather and which make up a part of the internal chemical environment may alter the hydrogen ion concentration and rates of reactions.

Analysis has shown that the effect of these agents on tannins cause them to become insoluble or to break down into their monomers and to finally degrade into products such as sugar and phenols. The latter may be broken down by further oxidation into organic acids. Equivalent effects occur with the collagen protein. It may be hydrolysed into peptides and these into amino acids. The amino acids can be broken down by oxidation into other amino acids through transamination, oxidative decarboxylation or deamination. Finally the amino acids may be oxidized into organic acids or ammonia. Knowing this and being aware of the variability in leather due to environmental and use history as well as differences in fabrication, it seems overwhelming and almost impossible to be able to fully assess the deterioration state of a particular piece of leather. However, through the research efforts of the Leather Conservation Centre and the STEP and ENVIRONMENT Leather Project groups, it has been shown that it is possible to make a logical assessment of leather deterioration and to determine some of the mechanisms of deterioration.

As was already mentioned, leather chemists have been speculating on the deterioration process of leathers for the past 150 years, it is probable that they will still be speculating in another 150 years.

There are many new approaches today with new instrumentation and non-destructive analyses (Boghosian et al., 1999; Odlyha et al., 1999a, b; Rasmussen et al., 1999; Young, 1999 and others) still in initial stages of work. They are promising as they give a more detailed analysis without disruption of the complex product of a series of complicated chemical reactions. This seems logical because in such a complex chemical reaction as deteriorating leather, any chemical analytical intervention can alter the products of deterioration present.

The new unintrusive methods of analysis may help determine the role of radicals in oxidation, which must be a major cause of deterioration. The water relationships between the chemical environment in the leather and its reactivity on the tannins and collagen are important. How much internal water is necessary for reactions to occur? The sulphur complexes present in the leather are also confusing. Certainly sulphur dioxide has been shown to cause hydrolytic activity in leather in a polluted environment, but what is the role of the other sulphur compounds inherently in the complex chemical mixture?

More recently there have been critical reviews (Larsen and Chahine, 2003; Thomson, 2001, 2002, 2003a, b) on the appropriate artificial ageing protocol that should be used for research. The STEP group goal was to determine the processes of deterioration and devise an artificial ageing protocol which would give similar chemical changes to proteins, vegetable tannins and leather to those observed in historical, naturally aged leathers. They proposed that this protocol could be used to standardize research. The ENVIRONMENT Group devised a similar protocol but using a lower temperature for a longer period of time than the STEP protocol for assessing tannages appropriate for archival use (Larsen et al., 1997). The CRAFT Group, a consortium of specialist bookbinders and leather chemists, used yet another protocol for assessing bookbinding leathers which omitted the oxidizing heating stage (Thomson 2001, 2002, 2003a, b). It may be that these differences in test procedures express the need for different protocols reflecting the tremendous variation in leather and the environments in which they are held over the years. All this suggests that more intensive research is needed.

But without the goal of the 'Golden Egg' as Larsen (1995) termed it, there would not be the incentive to continue to strive to find it.

## Acknowledgements

I want to thank Dr René Larsen, Scientific Co-ordinator of the STEP Project, and Rector of the School of Conservation, Royal Danish Academy of Fine Arts, Copenhagen, for his generous assistance in supplying the STEP literature and help in the interpretation of research details as well as his encouragement. I also want to thank Dr Gerry Poulton, Professor of Organic Chemistry at the University of Victoria, for his kind assistance in reviewing chemical aspects of this chapter.

Any remaining inaccuracies or inadequate expression of fact or concepts are of course wholly my responsibility. I know well that the risk of error is high when generalizing from so many and such varied studies by so many different researchers.

## References

Bernatek, E., Moskeland, J. and Valen, K. (1961) Ozonolysis of Phenols II. Catechol, Resorcinol and Quinol. *Acta Chemica Scandinavica*, **15**, 471–476.

Bickley, J.C. (1991) Vegetable Tannins. In *Leather, its Composition and Changes with Time* (C. Calnan and B. Haines, eds), pp. 16–23. Northampton: The Leather Conservation Centre.

Boghosian, S., Garp, T. and Nielsen, K. (1999) Study of the chemical breakdown of collagen and parchment by Raman spectroscopy. In *Preprints of the Advanced Study Course 1999*, pp. 73–88. Copenhagen: Royal Danish Academy of Fine Arts.

Bowden, D. (1999) Studies in the Thermophysical Behaviour of Parchment Using Surface Sensing Techniques. In *Preprints of the Advanced Study Course 1999*, pp. 161–168. Copenhagen: School of Royal Academy of Fine Arts.

Bowes, J.H. (1963) *Deterioration of Leather Fibres*. Final Technical Report to US Dept of Agriculture, Agricultural Research Services Project UR-E29-(60)-2, July 1958–June 1963, p. 63.

Bowes, J.H. and Raistrick, A.S. (1964) The Action of Heat and Moisture on Leather. V. Chemical Changes in Collagen and Tanned Collagen. *Journal of the American Leather Chemists Association*, **29**, 201–215.

British Leather Manufacturers' Research Association (1932–77) Laboratory Reports, 1977, 58; 1962, 41; 157; 1954, 33, 208; 1945, 24, 104; 1932, 11, 276.

Burton, D. (1948) Vegetable Tanning – I. The Determination of the Different Kinds of Acids and Salts in Tan Liquors. *Journal of the Society of Leather Trades' Chemists*, **32**, 362–376.

Calnan, C.N. (1991) Ageing of Vegetable Tanned Leather in Response to Variations in Climatic Conditions. In *Leather, its Composition and Changes with Time* (C. Calnan and B. Haines, eds), pp. 41–50. Northampton: The Leather Conservation Centre.

Calnan, C.N. and Thornton, C. (1997) Determination of Moisture Loss and Regain. In *Environment Leather Project* (R. Larsen, ed.), pp. 17–22. Copenhagen: Royal Danish Academy of Fine Arts.

Chahine, C. (1991) Acidic Deterioration of Vegetable Tanned Leather. In *Leather, its Composition and Changes with Time* (C. Calnan and B. Haines, eds), pp. 75–79. Northampton: The Leather Conservation Centre.

Chahine, C. and Rottier, C. (1997a) Artificial Ageing. In *Environment Leather Project* (R. Larsen, ed.), pp. 33–38. Copenhagen: Royal Danish Academy of Fine Arts.

Chahine, C. and Rottier, C. (1997b) DSC Measurement. In *Environment Leather Project* (R. Larsen, ed.), pp. 129–144. Copenhagen: Royal Danish Academy of Fine Arts.

Chahine, C. and Rottier, C. (1999) Studies of Change in the Denaturation of Leather and Parchment Collagen by Differential Scanning Calorimetry. In *Preprints of the Advanced Study Course 1999*, pp. 151–158. Copenhagen: Royal Danish Academy of Fine Arts.

Ciferri, A. (1971) Swelling and Phase Transition of Insoluble Collagen. In *Biophysics of the Skin*, **I** (H.R. Eldon, ed.), pp. 101–151. Chichester: Wiley-International.

Deasy, C. and Michele, Sr. S.C. (1965) A Study of Oxidative Degradation of Gelatine and Collagen by Aqueous Hydrogen Peroxide Solutions. *Journal of American Leather Chemists Association*, **60**, 665–674.

Eisenhauer, H.R. (1968) The Ozonization of Phenolic Wastes. *Water Pollution Control Fed. Journal*, 1887–1899.

Florian, M.-L.E. (1984) Conservation Implications of the Structure, Reactivity, Deterioration and Modification of Proteinaceous Artifact Material. In *Protein Chemistry for Conservators* (C.L. Rose and D.W. von Endt, eds), pp. 61–88. American Institute of Conservation.

Florian, M.-L.E. (1985) A Holistic Interpretation of the Deterioration of Vegetable Tanned Leather. *Leather Conservation News*, **2**(1), 1–5.

Florian, M.-L.E. (1987) Deterioration of Organic Materials Other Than Wood. In *Conservation of Marine Archaeological Objects* (Colin Pearson, ed.), pp. 21–54. London: Butterworths.

Florian, M.-L.E. (1998) *Heritage Eaters; Insects and Fungi in Heritage Collections*. London: James & James.

Foote, C.S. (1976) Photosensitized Oxidation and Singlet Oxygen: Consequences in Biological Systems. In *Free Radicals in Biology*, **II** (W.A. Pryor, ed.), pp. 85–133. New York: Academic Press.

Greenhow, E.J. and Shafi, A.A. (1975) The Determination of Polyfunctional Carboxylic Acids and Phenols, Including Vegetable Tannins by Catalytic Thermometric Titrimetry. *Proc. Anal. Div. Chem. Soc.*, **12**(11), 286–288.

Greenstein, J.P and Winitz, M. (1961) *Chemistry of the Amino Acids*, **1**. New York: John Wiley and Sons.

Haines, B. (1991a) Natural ageing of leather in Libraries. In *Leather, its Composition and Changes with Time* (C. Calnan and B. Haines, eds), pp. 66–74. Northampton: The Leather Conservation Centre.

Haines, B. (1991b). Deterioration Under Accelerated Acidic Ageing Conditions. In *Leather, its Composition and Changes with Time* (C. Calnan and B. Haines, eds), pp. 80–87. Northampton: The Leather Conservation Centre.

Haines, B. (1991c) The Structure of Collagen. In *Leather, its Composition and Changes with Time* (C. Calnan and B. Haines, eds), pp. 5–9. Northampton: The Leather Conservation Centre.

Harborne, J.B. (1989) *Plant Phenolics, Methods in Plant Biochemistry*, **1** (P.M. Dey and J.B. Harborne, eds), New York: Academic Press.

Haslam, E. (1979) Vegetable Tannins. In *Recent Advances in Phytochemistry*, **12**. *Biochemistry of Plant Phenolics* (T. Swain, J.B. Harborne and C.F. Van Sumere, eds), pp. 475–523. New York: Plenum Press.

Haslam, E. (1989) Gallic Acid Derivatives and Hydrolysable Tannins. In *Natural Products of Woody Plants*, **1** (J.W. Rowe, ed.), pp. 399–438. New York: Springer-Verlag.

Heckly, R.J. (1976) Free Radicals in Dry Biological Systems. In *Free Radicals in Biology*, **II** (W.A. Pryor, ed.), pp. 135–158. New York: Academic Press.

Hemingway, R.W. (1989) Biflavonoids and Proanthocyanidins. In *Natural Products of Woody Plants*, **II**: *Chemicals Extraneous to Lignocellulosic Cell Wall* (J.W. Rowe, ed.), pp. 571–651. New York: Springer-Verlag.

Hey, D.H. (1973) Amino Acids, Peptides and Related Compounds. *Organic Chemistry Series 1*, **6**. London: Butterworths.

Horie, C.V. (1992) Preservation of Natural Macromolecules. In *Polymers in Conservation* (N.S. Allen, M. Edge and C.V. Horie, eds), pp. 32–36. Cambridge: The Royal Society of Chemistry.

Ishak, M.S. (1974) Some Aspects on the Chemistry of Phlobaphenes from *Acacia Nilotica* Bark. *Egypt. J. Chem.*, **17**(5), 699–703.

Juchauld, F. and Chahine, C. (1997) The Analysis of N-terminal Residues in Hide Collagen. In *Environment Leather Project* (R. Larsen, ed.), pp. 75–85. Copenhagen: Royal Danish Academy of Fine Arts.

Jurd, L. and Geissman, T.A. (1956) Absorption of Spectra of Metal Complexes of Flavonoid Compounds. *Journal of Organic Chem.*, **21**, 1395–1401.

Karel, M. and Yong, S. (1981) Auto-oxidation B Initiated Reactions in Foods. In *Water Activity: Influence on Food Quality* (L.B. Rockland and G.R. Steward, eds), pp. 511–529. New York: Academic Press.

Larsen, R. (1994) Summary Discussion and Conclusion. In *STEP Leather Project* (R. Larsen, ed.), pp. 165–180. Copenhagen: Royal Danish Academy of Fine Arts.

Larsen, R. (1995) *Fundamental Aspects of the Deterioration of Vegetable Tanned Leathers*. Unpublished Ph.D. thesis. The Royal Danish Academy of Fine Arts, Copenhagen, p. 35.

Larsen, R. (2000) Experiments and Observations in the Study of Environmental Impact on Historical Vegetable Tanned Leathers. *Thermochimica Acta*, **365**, 85–99.

Larsen, R. and Vest, M. (1999) The Study of Oxidative Breakdown and Identification of Collagen Materials by Amino Acid Analysis. In *Preprints of the Advanced Study Course 1999*, pp. 11–20. Copenhagen: School of Royal Academy of Fine Arts.

Larsen, R., Vest, M. and Nielsen, K. (1993) Determination of Hydrothermal Stability (Shrinkage Temperature) of Historical Leather by the Micro Hot Table Technique. *Journal of the Society of Leather Technologists and Chemists*, **77**, 151–156.

Larsen, R., Vest, M., Nielsen, K. and Jensen, A.L. (1994) Amino Acid Analysis. In *STEP Leather Project* (R. Larsen, ed.), pp. 39–58. Copenhagen: Royal Danish Academy of Fine Arts.

Larsen, R., Vest, M., Poulsen, D.V., Kejser, U.B. and Jensen, A.L. (1997a) Amino Acid Analysis. In *Environment Leather Project* (R. Larsen, ed.), pp. 39–68. Copenhagen: Royal Danish Academy of Fine Arts.

Larsen, R., Vest, M., Poulsen, D.V. and Kejser, U.B. (1997b) Determination of Hydrothermal Stability by Micro Hot Table Method. In *Environment Leather Project* (R. Larsen, ed.), pp. 145–166. Copenhagen: Royal Danish Academy of Fine Arts.

Larsen, R. and Chahine, C. (2003), Comments on John Arthur Wilson Lecture. *Journal of the American Leather Chemists Association*, **98**, 360–363.

Letan, A. (1966) The Relation of Structure to Antioxidant Activity of Quercetin and Some of its Derivatives. **II**. Secondary Metal-complexing Activity. *Journal of Food Science*, **3**, 395–399.

Lewis, D.M. (1991) Some Aspects of the Photochemistry of Fibrous Proteins. In *Leather, its Composition and Changes with Time* (C. Calnan and B. Haines, eds), pp. 60–65. Northampton: The Leather Conservation Centre.

McNeill, I.C. (1992) Fundamental Aspects of Polymer Degradation. In *Polymers in Conservation* (N.S. Allen, M. Edge and C.V. Horie, eds), pp. 14–31. Cambridge: Royal Society of Chemistry.

Mills, J.S. and White, R. (1987) *The Organic Chemistry of Museum Objects*. London: Butterworths.

Munro, H.N. and Allison, J.B. (1964) *Mammalian Protein Metabolism*, **1**. New York: Academic Press.

Nomura, S., Hiltner, A., Lando, J.B. and Baer, E. (1977) Interaction of Water with Native Collagen. *Biopolymers*, **16**, 231 pp.

Odlyha, M., Foster, E.M., Cohen, N.S. and Caupana, R. (1999a) Studies of the Changes in Thermophysical Behavior of Leather and Parchment Samples Using Thermomechanical (TMA and DMTA), Non-invasive Dielectric and Thermogravimetric (TGA) Techniques. In *Preprints of the Advanced Study Course 1999*, pp. 169–194. Copenhagen: Royal Danish Academy of Fine Arts.

Odlyha, M., Cohen, N.S., Campana, R. and Aliev, A. (1999b) Study of Chemical Changes in Leather and Parchment by Solid Phase Nuclear Magnetic Resonance Spectroscopy (NMR). In *Preprints of the Advanced Study Course 1999*, pp. 103–117. Copenhagen: Royal Danish Academy of Fine Arts.

Okell, R.L. (1945) Acids and Salts as a Control Factor in Tannery Practice. *Journal of the Society of Leather Trades' Chemists*, **29**(56), 56–74.

Porter, L.J. and Hemingway, R.W. (1989) Significance of the Condensed Tannins. In *Natural Products of Woody Plants*, **I**, *Chemicals Extraneous to Lignocellulosic Cell Wall* (J.W. Rowe, ed.), pp. 988–1027. New York: Springer-Verlag.

Porter, L.J. (1989a) Condensed Tannins. In *Natural Products of Woody Plants*, **II** (J.W. Rowe, ed.), pp. 651–690. New York: Springer-Verlag.

Porter, L.J. (1989b) Tannins. In *Plant Phenolics, Methods in Plant Biochemistry*, **1** (P.M. Dey and J.B. Harborne, eds), pp. 389–420. New York: Academic Press.

Pryor, W.A. (1976) *Free Radicals in Biology, vols 1–5*. New York: Academic Press.

Rasmussen, S.B., Hansen, D.B. and Nielsen, K. (1999) Detection of Radicals in Collagen and Parchment Produced by Natural and Artificial Deterioration,

Electron Spin Resonance Spectroscopy (ESR) and its Implications for Artificial Ageing and Test of Conservation. In *Preprints of the Advanced Study Course 1999*, pp. 89–101. Copenhagen: Royal Academy of Fine Arts.

Reich, L. and Stivala, S.S. (1971) *Elements of Polymer Degradation*. New York: McGraw-Hill.

Ribéreau-Gayon, P. (1972) *Plant Phenolics*. New York: Hafner.

Roux, D.G., Ferreira, D., Hundt, H.K.L. and Mahan, E. (1975) Structure, Stereochemistry and Reactivity of Natural Condensed Tannins as Basis for Their Extended Industrial Application. *App. Polymer Symposium*, **28**, 335–553.

Scott, G. (1965) *Atmospheric Oxidation and Anti-oxidants*. New York: Elsevier.

Stambolov, T. (1996) *Manufacture, Deterioration, and Preservation of Leather. A Literature Survey of Theoretical Aspects and Ancient Techniques*. ICOM.

Thampuran, K.R.V., Vijayaramayya, T., Ghosh, D. et al. (1981) The Use of Lyotropic Salts in the Rapid Tanning of Heavy Leather. *Leather Science*, **28**(12), 442–443.

Thomson, R.S. (1991a) A History of Leather Processing from the Medieval to the Present Time. In *Leather, its Composition and Changes with Time* (C. Calnan and B. Haines, eds), pp. 12–15. Northampton: The Leather Conservation Centre.

Thomson, R.S. (1991b) Surface Coatings and Finishes. In *Leather, its Composition and Changes with Time* (C. Calnan and B. Haines, eds), pp. 34–38. Northampton: The Leather Conservation Centre.

Thomson, R.S. (2001) Bookbinding Leather; Yesterday, Today and, Perhaps, Tomorrow. Wolstenholme Memorial Lecture 2000. *Journal of the Society of Leather Technologists and Chemists*, **85**, 66–71.

Thomson, R. (2002). Conserving Historical Leathers: Saving our Past for the Future. *Journal of American Leather Chemists Association*, **97**, 307–320.

Thomson, R. (2003a) Towards a Longer Lasting Leather: A Summary of the CRAFT Leather Project. *Leather Conservation News*, **17**(1), 1–9.

Thomson, R. (2003b) Letter to the Editor. *Journal of the American Leather Chemists Association*, **98**, 363–365.

Traub, W. and Piez, K.A. (1971) The Chemistry and Structure of Collagen. In *Advancements in Protein Chemistry*, **25**, 243–352.

Tuck, D.H. (1983) *Oils and Lubricants Used on Leather*. Northampton: The Leather Conservation Centre.

Vest, M. and Larsen, R. (1999) Studies in Shrinkage Activities of Leather and Parchment by Micro Hot Table Method (MHT). In *Preprints of the Advanced Study Course 1999*, pp. 143–150. Copenhagen: Royal Danish Academy of Fine Arts.

Vilmont, L. (1992) Determination of the N-terminal Groups in Vegetable Tanned Leathers by HPLC-Derivation Procedure and Preliminary Qualitative Analysis. In *ICOM Committee for Conservation Leathercraft Working Group Interim Symposium, London*, pp. 10–15. London: ICOM.

Vilmont, L. (1993) N-terminal Group Analysis. In *The STEP Leather Project, Second Progress Report* (R. Larsen, ed.), pp. 105–114.

White, T. (1958) Chemistry of Vegetable Tannins. In *Chemistry and Technology of Leather*. (F. O'Flaherty, W.T. Roddy and R.M. Lollar, eds), pp. 98–160. New York: Reinhold.

Wouters, J. (1994) Tannin and Ion Analysis of Naturally and Actinically Aged Leathers. *STEP Leather Project* (R. Larsen, ed.), pp. 91–106. Copenhagen: Royal Danish Academy of Fine Arts.

Wouters, J. and Claeys, J. (1997) Analysis of Tannins, Sulphate, Fat, Moisture and Ash: Evolution and a Significance of Parameters to be Followed for Measuring the Degradation of Vegetable Tanned Leather. In *Environment Leather Project* (R. Larsen, ed.), pp. 87–94. Copenhagen: Royal Danish Academy of Fine Arts.

Wouters, J., van Bos, M., Poulsen, D.V., Claeys, J. and Oostvogels, A. (1997) Analysis of Tannins, Sulphate, Fat, Moisture and Ash of Leather, Treated or Produced for Conservation. In *Environment Leather Project* (R. Larsen, ed.), pp. 103–112. Copenhagen: Royal Danish Academy of Fine Arts.

Young, G.S. (1990) Microscopical Hydrothermal Stability Measurements of Skin and Semi-tanned Leather. *Preprints of the ICOM Committee for Conservation 9th Triennial Meeting, Dresden, 1990*, **II**, pp. 626–630. Los Angeles: ICOM.

Young, G.S. (1992). Loss of Infrared Linear Dichroism in Collagen Fibers as a Measure of Deterioration in Artifacts of Skin and Semi-tanned Leather. In *Materials Issues in Art and Archaeology*, **III** (P.B. Vandiver, J. Druzik and G.S. Wheeler, eds), pp. 859–867. Pittsburgh: Materials Research Society.

Young, G.S. (1998) Thermodynamic Characterization of Skin, Hide and Similar Materials Composed of Fibrous Collagen. *Studies in Conservation*, **43** (2), 65–79.

Young, G.S. (1999) *The Application of Thermal Microscopy, Differential Scanning Calorimetry and Fourier Transform Infrared Microspectroscopy to Characterize Deterioration and Physicochemical Change in Fibrous Type I Collagen*. Unpublished Ph.D. Thesis, University of London.

# 6

# Testing leathers and related materials

*Roy Thomson*

## 6.1 Introduction

Leather is not a single substance but rather a group of related materials with many characteristics in common. Other skin-based products such as rawhide, parchment and the so-called pseudo leathers also share many of these properties. In this manner, the term 'leather' is analogous to the word 'metal' which covers a wide range of related materials. Just as different metals vary widely in their properties, each type of leather will have its own unique range of characteristics. This will depend on the raw material employed and the processes used in its manufacture. The properties of a particular piece of leather will also be affected by the type and degree of deterioration which has taken place over its lifetime. These properties will determine which methods can and, more importantly, should not be used during conservation. The application, for instance, of aluminium alkoxide in non-polar solvents, a technique developed for the chemical stabilization of red rotted leathers, to ethnographic pseudo leathers will cause irretrievable damage. It is necessary, therefore, to determine its exact nature before embarking on any treatment programme.

The examination of leathers and related materials falls into three main groups:

(a) The determination of the type of raw material.
(b) The determination of the tanning process employed.
(c) An assessment of the type and degree of deterioration that has occurred.

## 6.2 Determination of raw material

The determination of the species of animal used by a microscopical examination of hair follicle patterns and the fibre structure has been described in Chapter 3. Work is being undertaken to analyse DNA extracted from skin-based objects. It might well be possible to develop such procedures for untanned materials and successful results have been reported with oil-tanned chamois leathers (Langridge, 2004). It is less likely, however, that successful methods will be found for use with vegetable- or mineral-tanned leathers as the crosslinking mechanisms involved in the tanning processes will probably interfere with the extraction procedure.

## 6.3 Determination of tannage type

Determination of the type of tannage is relatively straightforward. An initial visual and manual examination is often enough to give the required information and has the advantage of being non-destructive. With experience, the colour and structure of a skin will reveal whether it is untanned or whether it has been treated by a vegetable, alum or chrome process.

### 6.3.1 Ashing test

The type of tannage can sometimes be determined by burning a small sample. Chrome-tanned leather will give a green ash, alum-tanned materials a white ash and vegetable, oil and untanned samples will combust completely. Many untanned products, however, contain appreciable quantities of common salt or calcium compounds which give white residues that can be confused with those of aluminium salts. In any case, it is necessary to burn the carbon completely from the charred leather before it is possible to examine the ash fully. This requires surprisingly high temperatures and long times if valid conclusions are to be reached. It is for this reason that a range of spot tests or more complex instrumental analytical techniques have been developed (Reed, 1972).

### 6.3.2 Spot test

Originally spot tests were carried out on the objects themselves, ideally in an inconspicuous area, but this left a contaminated, coloured mark. This might once have been thought of as an advantage as the spot is a permanent record that the test has been carried out and results obtained. However, any permanent contamination of an object can never be considered good practice. Spot tests have therefore been modified for application to single fibres or small groups of fibres taken from the leather. These include tests to determine the presence of vegetable tannins and whether these are of the condensed or hydrolysable type, as well as the presence of aluminium compounds. Details of these are as follows.

#### 6.3.2.1 *Detection of vegetable tannins*

Place two samples, each consisting of fibres from the material under examination, one at each end of a microscope slide. Moisten each sample with a drop of distilled water. Add one drop of a 1% aqueous solution of ferric chloride to one sample. The presence of vegetable tannins is indicated by the development of a blue–black or green–black colouration.

#### 6.3.2.2 *Detection of aluminium*

Place two samples, each consisting of several fibres from the material under examination, one at each end of a microscope slide. Moisten each sample with a drop of 2M aqueous ammonium hydroxide solution. Add one drop of a 0.1% solution of sodium alizarin sulphonate solution in 90% ethyl alcohol to one sample. After five minutes, remove excess reagent using filter paper and add several drops of 2M aqueous acetic acid solution. The presence of aluminium compounds is indicated by a red colouration which does not turn yellow on acidification.

#### 6.3.2.3 *Detection of condensed tannins*

Place two samples, each consisting of several fibres from the material under examination, one at each end of a microscope slide. Moisten one sample with a 1% solution of vanillin in 90% ethanol. Remove excess reagent with a filter paper and add one drop of concentrated hydrochloric acid to both samples. The development of a deep red colour indicates the presence of condensed tannins.

#### 6.3.2.4 *Detection of hydrolysable tannins*

Place two samples, each consisting of several fibres from the material under examination, one at each end of a microscope slide. Add one drop of 2M aqueous sulphuric acid to one sample. Wait three minutes. Remove excess acid with filter paper. To the same sample, add one drop of a 0.7% solution of rhodomine in 99% ethanol. Wait five minutes. Remove excess reagent with a filter paper. Add one drop of a 2M aqueous potassium hydroxide solution to both samples. The development of a red colouration indicates the presence of gallic acid derived from the acid hydrolysis of hydrolysable tannins.

### 6.3.3 Conclusion

Of these, the most widely used has been the test for the presence of vegetable tannins (Driel Murray, 2002). Care must be taken in the interpretation of results. This is particularly relevant with leather recovered from archaeological sites as vegetable tannins and aluminium salts are commonly found in soils, leading to false positive results.

More complex techniques requiring specialist analytical instruments include atomic adsorption or X-ray fluorescence spectroscopy for the detection of aluminium and potassium associated with the use of alum, calcium possibly arising from the pretanning liming process or chromium. They also include the use of high performance liquid chromatography on extracts from the leather to determine the presence and type of vegetable tanning materials (Wouters, 1992).

## 6.4 Determination of degree of deterioration

As with the determination of the type of tannage, an organoleptic examination of a leather object will yield a considerable amount of information on the amount of deterioration it has suffered. Different senses can be involved independently or, often subconsciously, in combination.

### 6.4.1 Organoleptic examination

Simply looking at an object, with or without magnification, will reveal whether the leather surface is coherent or friable. This immediately gives a reasonable indication as to whether the material is in a good, fair or poor condition. The colour of a leather, particularly a vegetable-tanned leather, will indicate whether it has been affected by the form of acidic degradation known as red rot. As the name suggests, this manifests itself as an orange–red discolouration.

Handling the leather and gently rubbing the surface will confirm whether it is breaking up. It will also give an indication as to whether the leather is flexible or stiff. Larsen and his co-workers have developed this subjective type of handling assessment into a more systematic, objective method capable of giving numerical values which can then be replicated by different testers (Larsen et al., 1994, 1996).

The smell of vegetable-tanned leather can also give an indication as to its state of deterioration. Red-rotted leather has a characteristic acrid odour which pervades the stacks of many major libraries. It is difficult to describe but once it has been experienced, it is instantly recognizable.

The taste of acidic leather is also characteristic but this method of testing leather is not to be recommended.

### 6.4.2 Chemical tests

In addition to these non-invasive, organoleptic examination methods to determine the condition of leather objects, chemical tests have been developed. Hendricus van Soest and his colleagues from the then Central Research Laboratory in Amsterdam have described a series of tests they considered necessary to undertake before deciding on a conservation programme. These include the determination of pH and the sulphate, fat and moisture contents (Soest et al., 1984). They suggest samples should be taken from at least three different places on an object to obtain a representative set of results. This, they state, can require up to 30 g of material. Wouters (1992) pointed out that limitations exist on the amount of sample that can be taken from an historic object. He describes an analytical scheme for the determination of volatile matter (primarily moisture content), shrinkage temperature, pH and pH difference figure, the contents of anions such as sulphate chloride, nitrate and phosphate, tannin type and content, fat content and the amount of such cations as aluminium, chromium and iron present, all using a sample as small as 200 mg. Due to the heterogeneity of leather, particularly aged leather, he admits that large deviations between the results given by different samples from the same object might be expected. He states, however, that this lack of absolute accuracy is preferable to obtaining more reproducible figures which can be produced from larger samples. It should be noted that this analytical scheme requires access to specialist equipment such as anion and reverse phase, high performance liquid chromatographs and atomic absorption spectroscopes, together with experience in using such instruments on a regular basis.

It may be useful to concentrate on the most important tests that can be carried out on historic leathers by bench conservators and on how the results obtained should be interpreted.

#### 6.4.2.1 Shrinkage temperature

If a piece of leather or other collagen-based material is thoroughly wetted, placed in water and gradually heated, a temperature will be reached at which the sample will suddenly and irreversibly shrink to about one third of its original area. This temperature is called the shrinkage temperature and will depend on the raw material, the methods of tannage, the amount and type of deterioration the leather has undergone during its lifetime and details of how the shrinkage temperature was determined (Balfe and Humphreys, 1948; Nayndamma, 1958). The difference between the shrinkage temperature of an historical skin product when it was new and the figure obtained now is considered to be an indication of the various chemical changes which have occurred to the collagen/tanning material complex during its lifetime. These chemical changes are together responsible for the physical changes which take place such as losses in strength and flexibility. Measurement of the shrinkage temperature, together with a knowledge of the probable initial shrinkage temperature of the product will therefore give an understanding of the level of deterioration it has undergone.

Details of the standard method for the determination of shrinkage temperature have been standardized as Test IUP 16 of the International Union of Leather Technologists and Chemists Societies (Williams, 2000). This method requires a sample of 50 mm × 3 mm. It is unlikely that such a sample could be cut from any historic artefact. Conservators have therefore used smaller samples and procedures using single fibres under the microscope have been developed (Young, 1990; Larsen et al., 1993) based on earlier work by leather chemists (Nageotte and Guyon, 1930; Salcedo and Highberger, 1941; Nutting and Borasky, 1948, 1949).

The hydrothermal shrinkage of leather has been likened to melting, except that it is irreversible. As with melting, the shrinkage phenomenon is a transition from one phase to another and, as with other phase transitions, this shrinkage is associated with energy transfer. These energy changes can be measured using techniques such as differential scanning colorimetry, also known as differential thermal analysis. This technique was first used to examine the shrinkage of collagen in the 1960s (Witnauer and Wisnewski, 1964) and had been applied to the examination of the effects of natural and artificial ageing of leathers in particular by Chahine and her co-workers (1991, 1992).

Differential scanning colorimetry studies indicate that the temperature at which the phase transition occurs is close to the shrinkage temperature measured by conventional means. They also show that the transition associated with hydrothermal shrinkage is endothermic, that is there is an overall absorption of energy by the sample. Ageing, whether natural or artificial, causes a reduction in the shrinkage temperature. It also results in a reduction of the amount of energy required to shrink the leather.

This has practical implications. Not only will wet, partially degraded leathers shrink irreversibly at relatively low temperatures, that is at ambient summer temperatures in temperate climates, but the amount of heat required to cause damage is much less. It must be remembered, however, that the shrinkage is hydrothermal, not just thermal. The amount of water present is critical to the transition. An undegraded vegetable-tanned leather, for instance, will contain about 15% moisture when equilibrated at 65% relative humidity. Although it would shrink at about 80°C when wet, it will withstand temperatures well in excess of 100°C in this equilibrated condition (Bienkiewicz, 1983). It is therefore possible to use treatments involving heat on new or partially degraded leather objects provided that they are relatively dry. These procedures include the use of hot melt adhesives or fillers (see Chapter 22) and pest eradication systems such as the Thermo Lignum process (Thomson, 2002).

### 6.4.2.2 pH

In 1923, Brønsted and Lowry proposed their theory regarding the nature of acids and bases. This included the concept of strong and weak acids and bases. This proposes that when they are dissolved in water, strong compounds, such as hydrochloric and or sodium hydroxide dissociate completely into their constituent ions:

$$HCl \rightarrow H^+ + Cl^-$$
$$NaOH \rightarrow Na^+ + OH^-$$

Weak compounds, on the other hand, such as acetic acid or ammonia only dissociate partially on solution:

$$CH_3COOH \leftrightarrow H^+ + CH_3COO^-$$
$$NH_3 + H_2O \leftrightarrow NH_4^+ + OH^-$$

It has been shown that the degree of dissociation for any compound is constant at a given temperature.

The concentration of hydrogen or hydroxyl ions in a solution is therefore determined not only by the amount of acid or alkaline compound present, but also by the degree to which the particular compound has dissociated. By convention, this concentration of hydrogen or hydroxyl ions is expressed as the pH value. This has been defined as the negative logarithm to the base of 10 of the concentration of hydrogen ions present in a solution:

$$pH = -\log_{10}[H^+]$$

Because water dissociates only very weakly to give equal quantities of hydrogen and hydroxyl ions, it can be calculated that the pH value of pure water at ambient temperatures is 7.0. It can also be shown that adding acid reduces this figure to a minimum of 0 and adding alkali increases it to a maximum of 14 (Phillips, 1999).

It will be noted that the dissociation of a compound and the formation of ions is a solution effect. The measurement of a pH of a solid such as leather is therefore problematical. As a result it is usual to measure the pH of an aqueous extract. It will also be noted that the result obtained depends on the concentration of the ions in the extract which itself is determined by the amount of water used to extract a given quantity of solid. In order to obtain reproducible results, it is therefore necessary to fix the ratio of sample to water. This has been defined internationally as 5 g of leather to 100 ml of water (Standard Test IUC 11). It is rarely possible to obtain samples this large from historic objects but similar results can be obtained from smaller samples provided this ratio is maintained. Using standard glass electrodes, it is possible to reduce the sample size to 0.25 g extracted with 5 ml water and 'one drop' electrodes are now obtainable which can measure solutions prepared from samples as small as 0.025 g extracted in 0.5 ml water.

In the scheme for leather analysis described by Wouters (1992) 1 g of leather is extracted with 50 ml of water. As he points out, results given using this procedure would be expected to be up to 0.4 units higher than those obtained with the standard method.

The form of deterioration known as red rot is considered to be caused by the action of strong acids on vegetable-tanned leathers especially on those processed with condensed tannins. These strong acids, particularly sulphuric acid, could have been added to the leather during processing or to have been formed *in situ* from sulphur dioxide absorbed from polluted atmospheres. A number of analytical methods have been developed to determine the amount of strong acid present in a leather. However, it has been found empirically that if a standard extract from a vegetable-tanned leather has a pH in excess of 3.2 it was unlikely to contain damaging quantities of strong acids and to develop red rot.

Conversely, a leather having a pH of 2.8 or less is likely to be suffering actively from this kind of deterioration.

By extension, it has sometimes been thought that any leather having a pH of 3.2 or more is unlikely to be in a deteriorated condition. This is not the case. As was pointed out in the 1930s, leathers which are strongly degraded by the 'conjoint effects of oxidation and acidity' can break down to liberate ammonia (Atkin and Thompson, 1937). This reacts with any strong acid present with a resultant formation of ammonium sulphate and increases in pH. Wouters has noted that a number of historic leathers, which despite having a low shrinkage temperature indicating deterioration, have a relatively high pH. These, though, have a relatively high sulphate content. He suggests that at one stage in their lifetime, these leathers had a low pH due to the presence of sulphuric acid but at a later stage, the acid was neutralized. He has developed a procedure for determining a calculated pH assuming that all the sulphate present in the leather at one time existed as free sulphuric acid (Wouters and Claeys, 1966).

While there is no doubt that the pH of a leather will be a major factor as to whether it will deteriorate rapidly or not, it is not the only or even the predominant one. In 1975, Raistrick surveyed leather from the Long Term Deterioration Trials undertaken by the British Leather Manufacturers' Research Association (Raistrick, 1977). He showed that the initial pH values measured in the 1930s bore no relationship to the amount of deterioration observed in the 1970s. Indeed the leather with the lowest pH of 2.6 in 1931 had not deteriorated to any extent whereas that with the highest initial pH of 6.5 had decayed severely.

### 6.4.2.3 pH difference

If a solution of a strong acid is titrated against a solution of a strong alkali (that is small, equal aliquots of

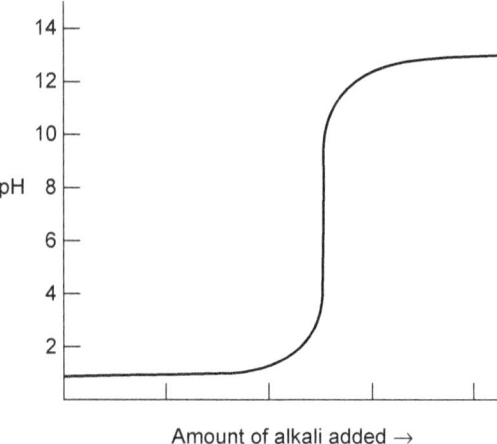

**Figure 6.1** Strong acid–strong alkali.

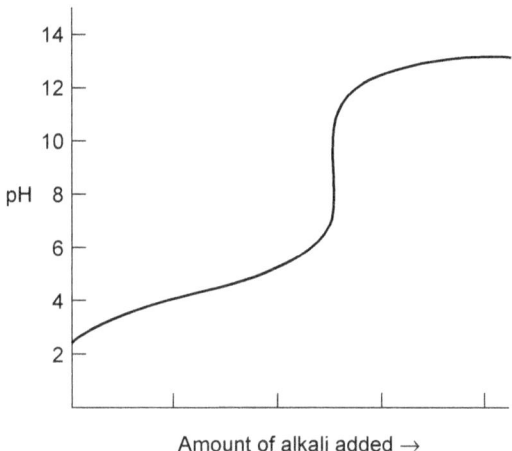

**Figure 6.2** Weak acid–strong alkali.

the alkali solution are added to a given quantity of the acid solution) and the changes in pH are noted, the following will be observed. Initially, there will be little change from the original low value. This will be followed by a sudden increase. The pH will then level off at a high value. This is shown in *Figure 6.1*.

If a similar solution of weak acid is titrated against the solution of a strong alkali, a different situation emerges. Initially the titration curve will increase gradually. Then, the rate of increase jumps suddenly. The titration curve then follows that observed with the strong acid. The point at which this change occurs is related to the degree of dissociation of the weak acid. This is shown in *Figure 6.2*.

As has been defined above, the pH is related logarithmically to the base 10 to the concentration of hydrogen ions present. Therefore if a given quantity of a solution of a strong acid having a pH value of 3–5 is diluted tenfold, by definition its pH value will increase by one unit. On the other hand, if a solution with a similar pH value contains only a weak acid, it will be seen that the reduction will be less. This phenomenon has been utilized to determine whether a leather contains appreciable quantities of strong acid. An aliquot of the aqueous extract used to determine the pH of the leather is diluted tenfold, the new pH is determined and the difference between the two figures calculated. If this difference is greater than 0.7 it is considered that dangerous quantities of strong acid are present. If it is less than 0.6 harmful acids are absent (Innes, 1948).

It must be noted that as with the initial pH figure, the absence of strong acids in a sample of historic leather at the time of examination does not indicate that they have never been present. Nor does the presence of strong acids mean that the leather will inevitably deteriorate. As Innes put it, 'vegetable tanned leathers which contain sulphuric acid after storage are not always rotted' and this is confirmed by Raistrick (1977) who showed that the amount of deterioration of a leather over 40 years was independent of its initial difference figure.

### 6.4.2.4 Sulphate content

The sulphate content of a leather has generally been determined gravimetrically by adding barium chloride to an aqueous extract and weighing the amount of barium sulphate precipitated (Rowley, 1999). Many leathers only contain small quantities of sulphate so to obtain meaningful results large samples are required. Alternative methods have therefore been suggested which employ much smaller samples. These include titrimetry with the aid of ion selective lead electrodes (Hallebeek, 1992) and anion chromatography (Wouters, 1992).

The presence of sulphate in vegetable-tanned leathers has been linked to the presence, now or in the past, of sulphuric acid. Indeed, as has been described above, sulphate contents have been used to determine the 'calculated pH' of historic leathers. It should be noted, however, that this calculation assumes that all the sulphate in the leather has derived from sulphuric acid. It also assumes that all strong acid in the leather has arisen from sulphuric acid.

A wide variety of sulphur compounds, which may or may not be converted to sulphuric acid, have been employed increasingly in the manufacture of leather from the mid-nineteenth century.

These include sodium sulphide for unhairing, sodium sulphite and related compounds as bleaches and stabilizing agents in the vegetable tanning process, sodium sulphate associated with chrome tanning compounds or added to many dyestuffs as a diluent, sulphated and sulphonated oils used for fatliquoring and a wide range of sulphonated aromatic compounds employed as synthetic tanning agents, the syntans. Care should therefore be taken before assuming that sulphate present in a leather is necessarily deleterious. Indeed the presence of large quantities of neutral sulphate in vegetable-tanned leather added in the form of the sodium salt has been reported to give 'considerable protective power' against the effects of acidic atmospheric pollution (Innes, 1948).

### 6.4.2.5 Fat content

Oils and fats are added to leather during its manufacture to prevent the fibres from sticking together as it dries thereby imparting the desired degree of softness and flexibility. These can be applied as a paste to the warm, wet leather and allowed to impregnate the fibre structure over an extended period of time in the currying process. Alternatively, they are applied in the form of an emulsion in the fatliquoring process which was introduced at the end of the nineteenth century.

As leathers age they can become firm, hard and cracky. At the same time, the amount of material that can be extracted using organic solvents is reduced. It has therefore been assumed that the fatty materials added during processing have been lost. This concept of 'feeding' the leather with new oil and fat mixtures to replace this lost material and restore the original properties has therefore grown up (Waterer, 1972). It was noted early on that adding too much new fatty material could result in an even greater amount of damage. Methods were therefore developed to determine the amount of fatty materials still present in the leather with a view to calculating how much new oil and fat needed to be applied to achieve a desired final oil content. Initially, methods involved extracting samples of the leather with organic solvents such as petroleum ether (Soest et al., 1984), n-methane (Wouters, 1992) or dichloromethane (Wouters, 1994) in a Soxhlet apparatus. As with the other determinations, the size of sample required is rather large. An alternative method has therefore been investigated using gas chromatography/mass spectrometry techniques. These, potentially, could give some idea of the type of oil present in the leather, the type and degree of deterioration which it has undergone as well as the quantity of fatty material still free to be extracted (Bos et al., 1996).

### 6.4.2.6  Moisture content

This is determined by equilibrating a sample of the leather in an environment of 20°C and 65% RH, weighing it, drying it for 24 hours at 102°C, cooling it in a desiccator and weighing it again. The material lost is presumed to be moisture.

New leathers would be expected to contain about 14% moisture under these conditions. Old, deteriorated leather usually contains less, down to 10%. In addition, it is generally considered that deteriorated leathers absorb and desorb moisture to a smaller extent and at a slower rate than new leathers. Recent studies (Hallebeek, 1994) have shown that this is not necessarily true particularly if the degraded leathers contain hygroscopic breakdown products.

## 6.5  Conclusions

As with other materials, it is generally accepted that historic leather objects presented for conservation should be examined thoroughly before treatment proposals are drawn up and any practical work undertaken. Questions to be answered include:

(a) What was the leather made from?
(b) How was it manufactured?
(c) How much and what type of deterioration has it suffered?

In addition, conservators, because of their specialist experience and knowledge, are often asked their opinion on historical technological and other aspects regarding the object.

Any study of an object can include a non-invasive, organoleptic examination and/or a more intrusive series of chemical or physical tests. In principle, providing the necessary information can be gained without changing or damaging the object this is the route to follow.

If, for instance, the object is a pair of ladies' white gloves from the mid-nineteenth century, the leather is likely to be alum tawed and should be treated as such. If it comes from the early twentieth century, it might have been tanned using formaldehyde or one of the early syntans. An experienced conservator with a knowledge of the changes in the tanning processes that took place over this period would realize that the methods employed for treatment would be exactly the same for all three types of leather. It would not therefore be necessary or ethically valid to take samples to carry out tests.

Equally, if a leather bookbinding exhibits the colour, surface texture and smell of a red-rotted vegetable-tanned leather, is it necessary to take samples to determine the pH and difference figure?

These are the type of questions a practising conservator needs to consider on a daily basis. Nevertheless, any conservator considering a leather object should be aware of the test methods available so that they can apply them when required.

## References

Atkin, W.R. and Thompson, F.C. (1937) *Proctor's Leather Chemists' Pocket Book*, 3rd edition, p. 273. London: E. & E.N. Spon.

Balfe, M.P. and Humphreys, F.E. (1948) The Shrinkage Temperature of Leather and the Effects of Heat and Moisture on Leather. In *Progress in Leather Science 1920–1945*, pp. 415–425. London: British Leather Manufacturers' Research Association.

Bienkiewicz, K. (1983) *Physical Chemistry of Leather Making* pp. 243–252. Malabar, Fl.: Krieger.

Bos, M. van (1996) Quantitative and Qualitative Determination of Extractable Fat from Vegetable Tanned Leather by GC-MS. In *ENVIRONMENT Leather Project* (R. Larsen, ed.), pp. 95–102. Copenhagen: Royal Danish Academy of Fine Arts.

Chahine, C. (1991) Acid Deterioration of Vegetable Tanned Leather. In *Leather, its Composition and Changes with Time* (C. Calnan and B. Haines, eds), pp. 75–87. Northampton: Leather Conservation Centre.

Chahine, C. and Rottier, C. (1992) Changes in Thermal Stability During Artificial Ageing with Pollutants: A DSC Study. In *Postprints of the ICOM Committee for Conservation Leathercraft Group Interim Symposium*. pp. 6–10, London: ICOM.

Driel Murray, C. van (2002) Practical Evaluation of a Field Test for the Identification of Ancient Vegetable Tanned Leathers. *Journal of Archaeological Science*, **29**, 17–21.

Hallebeek, P.B. (1992) Moisture Uptake/Release and Chemical Analysis. In *Postprints of the ICOM Committee Leathercraft Group Symposium, London 1992*, pp. 19–21. ICOM.

Hallebeek, P.B. (1994) Moisture Uptake/Release and Chemical Analysis. In *STEP Leather Project* (R. Larsen, ed.), pp. 107–116. Copenhagen: Royal Danish Academy of Fine Arts.

Innes, R.F. (1948) The Preservation of Vegetable Tanned Leathers Against Deterioration. In *Progress in Leather Science 1920–1945*, pp. 426–450. London: British Leather Manufacturers' Research Association.

Langridge, D. (2004) A Goat in Sheep's Clothing? *BLC Journal*, September, 3–5.

Larsen, R. et al. (1993) Determination of Hydrothermal Stability (Shrinkage Temperature) of Historical Leather by the Micro Hot Table Technique. *Journal of the Society of Leather Technologists and Chemists*, **77**, 151–156.

Larsen, R. et al. (1994) Materials. In *STEP Leather Project* (R. Larsen, ed.), pp. 11–30. Copenhagen: Royal Danish Academy of Fine Arts.

Larsen, R. et al. (1996) Fibre Assessment. In *ENVIRONMENT Leather Project* (R. Larsen, ed.), pp. 113–120. Copenhagen: Royal Danish Academy of Fine Arts.

Nageotte, J. and Guyonne, L. (1930) *American Journal of Pathology*, **6**, 631–638.

Nayndamma, Y. (1958) Shrinkage Phenomena. In *The Chemistry and Technology of Leather* (F. O'Flaherty et al., eds), 28–65. New York: Reinhold.

Nutting, G.C. and Borasky, R. (1948) *Journal of the American Leather Chemists' Association*, **43**, 96–101.

Nutting, G.C. and Borasky, R. (1949) *Journal of the American Leather Chemists' Association*, **44**, 831–838.

Phillips, P. (1999) Acids, Bases and Salts. In *Leather Technologists' Pocket Book* (M. Leafe, ed.), pp. 353–369. Withernsea: Society of Leather Technologists and Chemists.

Raistrick, A. (1977) *Long Term Evaluation of Bookbinding Leathers with a Critique of the Peroxide Test*. British Leather Manufacturers' Research Association Laboratory Report LR58. Egham: BLMRA.

Reed, R. (1972) *Ancient Skins, Parchments and Leathers*, pp. 265–287. London: Seminar Press.

Rowley, G. (1999) The Analysis of Leather. In *Leather Technologists' Pocket Book* (M. Leafe, ed.), p. 261. Withernsea: Society of Leather Technologists and Chemists.

Salcedo, I.S. and Highberger, J.H. (1941) *Journal of the American Leather Chemists' Association*, **36**, 271–283.

Soest, H.A.B. van et al. (1984) Conservation of Leather. *Studies in Conservation*, **29**(1), 21–31.

Thomson, R.S. (1995) The Effects of the Thermo Lignum Pest Eradication Treatment on Leather and Other Skin Products. *Leather Conservation News*, **11**, 11–12.

Thomson, R.S. (2002) Conserving Historical Leathers: Saving Our Past for the Future. *Journal of the American Leather Chemists' Association*, **97**, 307–320.

Waterer, J. (1972) *A Guide to the Conservation and Restoration of Objects Made Wholly or in Part of Leather*. London: G. Bell.

Williams, J.M.V. (2000) IULTCS (IUP) Test Methods. *Journal of the Society of Leather Technologists and Chemists*, **84**, 359–362.

Witnauer, L.P. and Wisnewski, A.J. (1964) Absolute Measurement of Shrinkage Temperature by Differential Thermal Analysis. *Journal of the American Leather Chemists' Association*, **59**, 598–612.

Wouters, J. (1992) Evaluation of Small Leather Samples Following Successive Analytical Steps. In *Postprints of the ICOM Committee for Conservation Leathercraft Group Interim Symposium, London*. ICOM.

Wouters, J. and Claeys, J. (1996) Analysis of Tannins, Sulphate, Fat, Moisture and Ash: Evolution and Significance of Parameters to be Followed for Measuring the Degradation of Vegetable Tanned Leather. In *ENVIRONMENT Leather Project* (R. Larsen, ed.), pp. 87–94. Copenhagen: Royal Danish Academy of Fine Arts.

Young, G.S. (1990) Microscopical Hydrothermal Stability Measurements of Skin and Semi-tanned Leather. In *Preprints of the ICOM Committee for Conservation 9th Triennial Meeting, Dresden 1990*, **II**, pp. 626–630. Los Angeles: ICOM.

# 7

# The manufacture of leather

*Roy Thomson*

## 7.1 Tanning in prehistoric and classical times

The hides and skins of animals killed for food appear to have been used for clothing, shelter and other purposes long before the evolution of modern man.

The earliest creature which appears to have made a systematic use of animal skins was *Australopithecus habilis* who roamed East Africa some two million years ago and seems to have developed a diet in which meat played an increasingly significant part. Unlike other carnivores, the *Australopithicines* were not equipped with specialized teeth or claws to penetrate the tough, fibrous, outer protective layer – the skin – but artefacts indicate that they discovered that the serrated edge of a chipped stone was capable of cutting through the thickest hide. This fundamental technological discovery led to a wider use of edged stone tools – an essential factor in the evolution of man. The presence of accumulations of specific bones at *Australopithecine* sites suggests that the more succulent joints of larger animals were butchered from the carcass and possibly dragged back to the living quarters on the skin of the animal itself. Once back at the living site the skins, relatively large sheets of a tough, flexible material, could have been used for a wide range of purposes.

The next stage in the development of the use of hides and skins is associated with the *Pithecanthropoid* group of hominids. These creatures were systematic and successful hunters. They had learned to control fire and cooked their meat. They had also developed a wider range of tools including the typical Acheulean *coup-de-point* which has been shown to have been used for both butchering and skinning. Some bands of these *Pithecanthropoids* lived in large, tent-like shelters constructed by spreading skins over wooden frameworks, supported by rows of upright poles. Skins also appear to have been used to make rough bundles running along the sides of the shelters. These dwellings were warmed by a fire in a central hearth. With such a structure, the curing effect produced by drying the skins out slowly would have been demonstrated. More importantly, the mild tanning action of wood smoke would also have manifested itself. The suggestion that *Pithecanthropus erectus* observed and exploited these changes is supported by the fact that they devised special stone implements for scraping away the unwanted fat and flesh. This scraping would also have distributed the fats throughout skins giving a more supple product. Flexing the pelt while it was drying would also have given a softer product. It is very possible that a crude form of leather was being produced by these hominids over a million years ago (Schwerz, 1958).

The next major stage in human development was the evolution of Neanderthal Man about one hundred thousand years ago. This close relative of modern man is typically associated with a wider range of specialized stone implements including a higher proportion specifically made for working the hides and skins. The Neanderthals were skilful hunters preying on animals such as deer, horses, bears, rhinoceros or even mammoth, and had access to large quantities of hides and skins. Neanderthal Man is associated with the last great ice age when they flourished despite the intense cold, living in skin-covered tents. The fact that they expanded their territory into the bleak tundra regions supposes the production of warm, protective clothing had been perfected. Neanderthal Man employed stone, horn, antler and bone for skin-working tools. The bone implements included lower leg bones sharpened along the concave edge to give fleshing knives. A similar range of stone and bone tools were used in an analogous manner by Native American peoples until the end of the nineteenth century. The earliest skin remains also date from this period.

Between thirty-five and forty thousand years ago, Cro Magnon Man, *Homo sapiens*, appeared in Europe. Their tools included many specialized skin-working implements including different stone and bone scrapers, burnishers, knives and awls each designed for a specific and slightly different purpose. By twenty thousand years ago, delicate skin-working implements including fine needles, awls and knives were being made from bone and flint. These were of such fragile construction that it can only be surmised that the art of preparing soft flexible leathers had been completely mastered. The works of the prehistoric artists have given us the first direct evidence of the use of leather for clothing. Rock paintings dating from thirty-five thousand years ago show males wearing leather loin cloths and a female wearing a skirt. Similar pictures ranging over the next twenty-five thousand years depict people wearing trousers and dresses presumably made from leather (*Figure 7.1*). In the colder, more northerly regions, warmer garments were necessary and stone statuettes have been excavated in Central Asia showing people wearing fur clothing closely resembling the anorak and trousers of modern Inuit. Cave paintings also show the use of leather for bags, belts and rope, and leather bindings were used to attach stone blades and axe heads to wooden, bone and antler shafts. It must therefore be concluded that by the time *Homo sapiens* evolved, the art of leathermaking had been fully developed and that tanning was probably man's first manufacturing process (Forbes, 1957).

Excavations in Egypt have given us the first pictures of tanning operations in the form of wall paintings from tombs, ranging from the fifth to the twenty-sixth dynasties. A scene from the tomb of Rekhmire, dating from about 1500 BC, depicts skins being fleshed, steeped in large jars containing an unknown liquor and staked or softened by pulling it over an implement similar to a lawn edging tool (*Figure 7.2*). The finest gloving leathers are still softened in this

**Figure 7.1** Upper paleolithic depiction of men and women wearing skin garments.

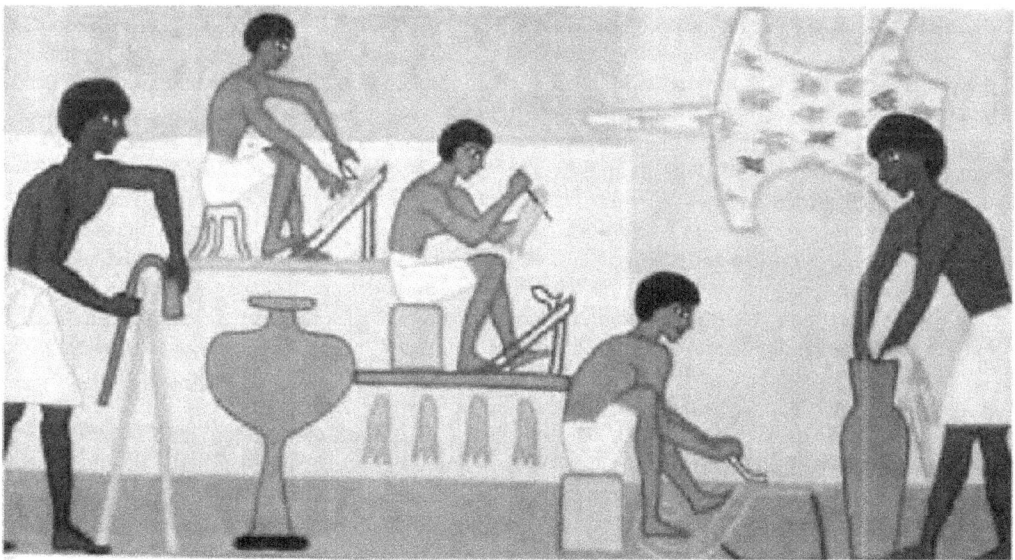

**Figure 7.2** Leather manufacturing processes from fifteenth century BC Egyptian tomb.

manner using an identical implement. The first recipes giving details of the tanning processes also date from about the same period. A series of Sumerian ritual magical texts describe the preparation of bullock and goatskin leathers to be used in the manufacture of drums employed in religious festivals (Levey, 1959). The texts themselves date from about 1000 BC but probably reflect techniques dating back to the third millennium or earlier. One reads:

> *You will steep the skin of a young goat with milk of a yellow goat and with flour. You grease it with fine oil, ordinary oil and the fat of a pure cow. You will soak alum in grape juice and cover the surface of the skin with gall nuts of the tree growers of the Hittite country.*

The rise of the city-state and the move towards urban living led to an increased demand for leather for civilian and, particularly, military purposes. Tanning developed from a small-scale craft activity to an important occupation organized on an industrial basis. Early Sumerian and Egyptian texts indicate that a complex trade in raw and cured hides and skins, tanning materials, finished leather and made-up goods flourished throughout the Near East over 5000 years ago. This trade continued and expanded until, by the period of Classical Greece, tanners were a prosperous and influential section of the community and a single factory at Piraeus, specializing in manufacturing leather shields, employed 120 slaves.

The development of leathermaking was not, however, restricted to the urban civilizations. Remains of high quality leathers have been found from Late Neolithic, Bronze and Iron Age sites as diverse as the Frozen Tombs of Central Asia, the Hallstatt salt mines of Upper Austria and the bog burials of Denmark. The Romans, too, recognized the vital importance of leather to their military forces. Hides and skins were demanded as a tribute by the legions and it has been suggested that cattle were bred for their hides rather than for their meat. Rather than rely on local supplies, the Roman Army organized its own tanneries and leather workshops.

Some idea of the tanning techniques used at the time of the Romans can be deduced from the remains of the civilian tannery excavated at Pompeii (Faber, 1938). It housed 15 tanning pits, each 1.5 m in diameter and about 1.2 m deep. There were also a series of smaller vats in which the tanning materials were prepared. A separate area was for the mechanical working of the hides. Here, two-handed scrapers, a currier's knife, sharpening steel and half-moon knives were found. The layout of the plant and the tools found suggest that a Roman tanner would have felt completely at home in the tanneries depicted by the eighteenth-century French encyclopaedists. Indeed, there appear to have been few fundamental advances in the technology of leather manufacture between the classical Greco-Roman period and the Industrial Revolution of the nineteenth century.

## 7.2 Tanning in the medieval and post-medieval periods

Leather and leathergoods manufacture were crafts widely practised throughout Europe and from the early medieval period until the mid-eighteenth century were second only in importance to the wool textile industries (Cameron, 1998).

The first job of the tanner was to wash the hides free from blood, dung and curing salts and to rehydrate them. This was often done by immersion in the local river or stream (*Figure 7.3*). The rehydration of dried hides could be accelerated by pounding them with hammers or by treading them underfoot.

It was then necessary to treat the hides in various ways to loosen the hair and enable it to be scraped off without damaging the grain surface. The most primitive method was to fold the hides and pile them until putrefaction set in just enough to loosen the hair roots. This putrefaction action was often speeded up by sprinkling the hair side with biologically active liquors prepared from such materials as stale beer, urine, dung, fermenting barley or mulberry and bryony leaves. An alternative method was to soak the skins in alkaline liquors prepared from wood ash or, more commonly, lime. A third method can be considered as a combination of the first two. In this, the hides were immersed in lime liquors which had been 'mellowed' by repeated use. These liquors would have contained large concentrations of organic breakdown products which would have accelerated the depilatory action. The hide was then spread over a wooden beam and both sides scraped with tanners' two handled knives (*Figure 7.4*). The hair side was scraped with a blunt, single-edged unhairing knife and the flesh with a sharper, two-edged fleshing knife. The hide was then put back into the liquors to open up the skin structure further, prior to another scraping with a blunt scudding knife.

The pelts were then washed and given a further cleansing and opening-up treatment using the alkaline bating, puering or mastering process or the acidic raising or drenching process.

In the alkaline processes, the pelts were immersed in a warm infusion of bird droppings or dog dung.

**Figure 7.3** Washing hides in stream.

**Figure 7.4** Removing hair with two-handled knife.

This removed excess lime and biochemically altered the hide structure to give a softer leather with a finer, more flexible grain. In the first decade of the twentieth century the active ingredients in the dung were found to be proteolytic enzymes secreted in the pancreas and activated by ammonium salts. The raising or drenching process involved treating the hides in liquors prepared by fermenting barley, rye and other vegetable matter (Thomson, 1981). The fermentation was often promoted by the addition of materials such as stale beer, urine or rotting pieces of hide. The action of the fermentation was to produce a complex mixture of organic acids and enzymes which again dissolved the non-fibrous protein of the skin and removed excess lime. Once the pelts were judged to be in the correct condition, the hides were again washed and worked over the beam to remove the slime that had been liberated.

The majority of hides were vegetable tanned and this technique will be considered first (Thomson and Beswick, 1983).

The preliminary stage of the tanning operation itself was to immerse the hides in weak, almost spent tanning liquors, moving them around almost continuously (*Figure 7.5*). Once the colour of the grain was judged to be satisfactory, the hides were transferred to a further set of pits. To prepare these, a layer of ground vegetable tanning material was tipped into the bottom of the pit and a hide laid flat

**Figure 7.5** Handling hides in weak tan liquors.

over it. The vegetable tanning material employed depended on what was available locally. Birch, willow, spruce and larch were used in northern Europe and Russia; various species of oak in Britain and central Europe; sumac, valonia, oak galls and various acacias around the eastern Mediterranean. A second layer of ground tanning material was strewn over the first hide and a second hide spread out. In this manner alternate layers of tanning material and hides were added until the pit was nearly full (*Figure 7.6*). A final layer of tanning material was piled on top and the whole pit filled either with water or an infusion prepared by extracting tanning materials with cold water. The hides were generally kept in these pits for at least a year. When the tanner had judged the hides to be fully tanned, they were rinsed off and smoothed out using a two-handled setting pin. The leather was then dried out slowly in a dark shed. This was often fitted with louvered panels to control the rate of drying.

The dried rough leather was sold to a currier whose first operation was to dampen it and then soften it by pummelling with heavy wooden mallets or by trampling under foot. This was followed by the scouring operation in which the leather surfaces were scrubbed clean using stone blocks or stiff brushes and smoothed out using stone or metal-bladed slickers.

The next operation was to pare the skin down to the required thickness using the currier's shaving knife (*Figure 7.7*). This had a rectangular, doubled-edged blade fitted with two handles, one in line with the blade and one at right angles to it. Both edges of the knife were sharpened and turned by pressing along them with a steel pin until the keen edge was bent almost perpendicularly to the face of the blade. The dampened leather was placed over a near vertical currier's beam and, holding the blade of the shaving knife almost at right angles to the leather, thin shavings were pared from the flesh surface until the desired thickness was achieved.

The shaved hides were worked on a bench with various stones, slickers and brushes to flatten the leather, remove loose tanning materials and stretch it. They were then partially dried and impregnated with a warm mixture of tallow and fish oils. After piling the skins to allow the fats to penetrate evenly, they were hung in a warm room. Once dry, the surplus grease was removed. On the grain side this was done using smooth burnishing stones or leather pads. The flesh side was often cleaned by removing the surface layer completely using a sharpened steel.

If the leather was to be used where a firm product was required, the hides were simply hung to dry before selling them to the various leather working

**Figure 7.6** Laying skins away in pits.

**Figure 7.7** Shaving leather to uniform thickness.

craftsmen. If a softer, finer product was required, further mechanical operations such as boarding or staking were carried out (*Figure 7.8*). In the boarding operation, the hide was folded grain side in, and rolled firmly up and down using boarding arms. Boarding emphasized the natural grain of the skin and a wide range of surface effects could be obtained. In the staking or perching operations the leather was softened by rubbing the flesh side over the convex edge of various blunt, curved blades in a manner similar to that shown in Egyptian tomb paintings.

If required, the leather was coloured, using natural dyestuffs or pigments. Finally, after swabbing the surface with such materials as weak animal glue, waxes, milk and blood, the leather was polished using a range of slickers, brushes and pads. It was only after all these operations had been carried out that the leather was ready to be sold to the various craftsmen who made it up into finished articles.

The above operations were employed by the tanner and the currier who generally worked on the larger cattle hides. Sheep, goat, deer and dog skins were processed by the fellmonger, the whittawyer, or the glover. Smaller, fur bearing pelts, on the other hand, were prepared by the skinners (Thomson, 1981–82).

The trade of fellmongering was to take sheepskins from the butcher or the farmer, remove the wool from the skin, sell the wool to the textile trade

72  Conservation of leather and related materials

**Figure 7.8** Working leather mechanically to obtain desired softness and appearance.

and the pelt to the whittawyer or the glover. The process employed was to hang the skin in a warm room and allow it to sweat until the wool root structure just started to rot. The wool was then pulled from the skin and sorted into various grades. The skin was washed thoroughly and salted to preserve it. An alternative method was to paint the flesh side with a paste of slaked lime and water. The alkali penetrated through the skin loosening the wool roots allowing the wool to be pulled out.

The pretanning processes used by the whittawyer were similar to those used by the tanner. Once the skins had been limed, unhaired, fleshed and given a thorough bating they were put into large wooden tubs. There they were kneaded with a mixture of materials such as alum, salt, egg yolks, butter, oatmeal, olive oil and flour. Traditionally the tawyer worked the mixture into the skins by trampling it in with his bare feet (*Figure 7.9*). Once the required amount of the tawing paste had been taken up by the skins, they were stretched out flat and piled overnight. The next day, they were mechanically worked again, often by twisting them into a rope and pulling this over an uneven surface such as a ram's horn set into a wooden upright. The skins were flattened and smoothed and hung to dry. The leather was then softened by staking or perching.

**Figure 7.9** Working the tawing mixture into the skins mechanically in the foot-tubbing process.

Widely used alternatives to the alum-based tawing pastes were the various oxidizable marine oils that were applied in the production of the chamois and buff leathers. The oils were trampled into the skins in a similar manner to that used in tawing. The skins were then hung in warm, airy stoves for the oxidation process to take place. After the oiling and stoving sequence had been repeated, three or four times, the leather was washed off in alkaline liquors to remove excess oil.

The whittawyer also used vegetable tanning materials to produce bazils, vegetable-tanned sheepskins, and as this trade increased, the more general term leatherdresser was introduced to describe the craftsmen who undertook these operations.

Originally glovers made up leather purchased from whittawyers. By medieval times, however, they not only produced their own leathers but also dealt in raw sheepskins and wool. They used the same methods of tanning as the whittawyers.

Skinners processed pelts from a wide variety of fur bearing animals ranging from rabbits and cats to wolves and bears (Veale, 1966). The skins were usually obtained in a dried condition and required careful wetting back before excess flesh could be removed and the pretanning operations undertaken. Great skill was required to ensure that the structure of the pelt was opened up sufficiently, while at the same time the hair remained firmly attached. Skinners employed the alum and oil tanning processes similar to those used by whittawyers. Alternatively, fur skins were treated in baths of fermenting grains similar to those described in the raising or drenching processes used by tanners. The organic acids and enzymes helped to remove the non-collagenous interfibrillary materials and give a softer product. These drenches did not tan the pelts but made them acidic enough to prevent bacterial attack.

## 7.3 Tanning in the nineteenth century

The nineteenth century in Europe was a period of rapid expansion and tanners were forced to keep pace with the general economic climate. The rapidly increasing population and the gradually improving standard of living during the nineteenth century added their own demands on the leather trade.

The tanning process was based on the application of the by-product of the forester, to the by-products of the farmer. It is not surprising, therefore, that the profound changes that took place in agricultural practices throughout the nineteenth century had their repercussions on the tanning trades. In particular they led to a shortage of both indigenous tanning materials and hides with a resultant increase in prices. The hides and skins themselves were also affected. Breeding experiments resulted in larger animals being produced and larger and thicker hides coming on the market. But larger size and artificial feeding generally leads to poorer quality skins. With sheepskins, the introduction of Merino strains to improve wool yields also led to a deterioration in skin quality. This resulted in shortages of both suitable quality hides and of tanning materials that could only be made up by importing, first, raw materials and then finished leathers from the expanding European empires and elsewhere.

The shortage of indigenous tanning materials also stimulated a search for alternatives. In addition, it encouraged attempts to use these materials more effectively. It was recognized that the traditional methods of layering and preparing oozes by leaching with cold water failed to extract all of the tanning materials. It was hoped that by employing more efficient extraction techniques, larger amounts of purer forms of tannin would be produced. These improved methods worked to some extent but despite this, there was just not enough European material available to tan all the leather required and the newly discovered imported materials had to be made to work. Tanners had always known that hides had to be fully penetrated with weak, almost spent, tanning materials before they could be subjected to the action of stronger, fresher extracts. The tanners learnt that this principle was even more important when dealing with these newer, more astringent products. They also learnt how to blend the new materials to obtain the required properties and to apply gentle mechanical action to get more uniform results (Thomson, 1991).

In parallel with the slow development of the vegetable tanning process, improvements were gradually introduced into the methods of pretanning. Traditionally, only lime was used as a depilating agent, although the beneficial effects of the repeated use of the same lime liquor had been known for centuries. To speed the process, alternative alkalis had been suggested, either as additives or as alternatives. The use of a mixture of orpiment and lime was introduced which unhaired in less than 36 hours instead of the usual three to four weeks and the modern sulphide unhairing system was born. Later, further improvements were introduced including the use of cheap, waste lime liquors from gas works, where lime was employed to scour hydrogen sulphide from

**Figure 7.10** Water-driven tanning drum.

the coal gas. Mixtures of lime and sodium sulphide are still used today in most tanneries to remove the hair prior to tanning.

In the 1830s Robert Warington patented the use of potassium dichromate and chromic acid for preserving the tanning skins but no use appears to have been made of the discovery (Thomson, 1985). Over the next 20 years, a number of chemists studied the tanning action of various chromium salts. It was not, however, until the 1870s, with the Heinzerling process, that the leather was tanned on a commercial scale using chromium chemicals. In the early 1880s Augustus Schultz was asked to devise a process for the production of leathers used to cover corset steels. Schultz used his experience in the textile industry and developed a system based on the processes employed for mordanting cotton. The skins were immersed in an acidified solution of sodium bichromate and then in a solution of sodium thiosulphate. This two-bath method of chrome tanning was examined by a number of tanners in the USA who found that excellent leather could be produced in a matter of hours rather than weeks. In 1893 Martin Dennis patented the use of trivalent chromium compounds for tanning and the so-called one-bath method of chrome tanning spread rapidly throughout the world. Today, 80% or more of leather is manufactured using variants of this process.

The first introduction of mechanization to the working of hides appears to have been the use of various types of fulling stocks used to assist in the chamois tanning process and to speed up the rehydration of the dried hides. A wide range of other methods for applying mechanical action to speed up the pretanning and tanning processes was also devised. Of these, the most widely accepted was the tanning drum which appears to have been developed in northern Italy in the late seventeenth century and is still the tanners' main wet processing vessel (*Figure 7.10*).

While leather manufacturers showed a reluctance in accepting many of the early mechanical aids, the one exception was the splitting machine. This split the hide into two layers, a top split equivalent to a shaved hide, and an extra, bottom split which had all the properties of leather except for the grain pattern. With the incentive of obtaining two hides for the price of one, the tanners were prepared to undertake development work and a number of different types of splitting machine were developed throughout the century.

A further development vital to the mechanization of the leather industries was the invention of spiral-bladed cylinder knives. These were originally developed for shearing furskins and were similar to those used today in the blades of a lawn mower. The use of two sets of these blades, opposing each other at an oblique angle, enable soft, irregularly shaped skins to be smoothed and stretched out, thus preventing creases, while the blade carried

**Figure 7.11** Spiral-bladed fleshing machine.

out its function (*Figure 7.11*). A considerable number of machines for unhairing, fleshing, scudding, setting, scouring, striking, staking and shaving were designed over the next hundred years based on this idea (Watt, 1855). Many leather working machines used today throughout the trade are based on late nineteenth century designs.

By the beginning of the twentieth century, the majority of tanneries were using improved tanning processes and were mechanized. A wide range of machines was employed, supplied by specialist tanning machinery manufacturers.

While many of the changes that took place in processing during the nineteenth century were improvements on previous techniques or produced a less expensive, more uniform, product, some had deleterious effects which only became apparent decades after they were introduced. This was particularly significant in the case of leathers for bookbinding, which, unlike those for other purposes, are expected to last a lifetime.

The first major change which was to lead to problems was the introduction of a wider range of tanning materials. As has been said, alternative products were imported including many species not previously employed in Europe. Although most of the new imported tannins were used for the production of heavy leathers, it is likely that some were tried for bookbinding leathers.

Vegetable tannins can be divided into two main classes depending on their chemical composition; the hydrolysable tannins, such as sumac and Turkish galls, and the condensed tannins, such as mimosa and mangrove. We now know that leathers prepared using condensed tannins decay more rapidly than those manufactured using hydrolysable tannins. It is fortuitous therefore that the new imported materials which gave the best quality light leathers such as divi divi, myrabolams, algarobilla, tara and bulbool are of the hydrolysable type. It is quite possible, however, that materials such as mimosa, mangrove or gambier were employed by some tanners, particularly towards the end of the nineteenth century. The gradually increasing introduction of these newer materials throughout the century resulted in more and more bookbinding leathers with poor ageing properties (Thomson, 2001).

A potentially more damaging improvement was the introduction of synthetic dyestuffs from the middle of the century onwards. Initially, these were developed for colouring textiles, but soon they were being used on a wide range of leathers. The replacement of the traditional methods with these new products adversely affected the ageing properties of the leathers in two ways. First, the natural dyestuffs previously employed were almost invariably applied in conjunction with alum. The beneficial effect of aluminium compounds on vegetable-tanned leathers

**Figure 7.12** Shaving machine.

is well known and is the basis of the British Standard for Archival Leathers. The new dyestuffs were either used alone or together with mordants such as potassium bichromate or iron sulphate which, unlike alum, actually have a damaging effect. Second, in order to 'fix' the dyestuff to the leather and 'clear' the dyebath it was necessary to increase its acidity. Sulphuric acid was generally used with extremely deleterious results. It is now known that sulphuric acid derived from atmospheric pollution is the major cause of the deterioration of bookbinding leathers and that the presence of strong acids is particularly damaging.

Just as the demand for leather outstripped the supply of vegetable tanning materials, as the century progressed, there was an increasingly serious shortage of skins for the tanners to process. This problem was overcome by the importation of crust EI leathers originally from East India, hence the name, but later from throughout the Middle East, the Indian subcontinent and Africa. These leathers were tanned in their country of origin either using local, traditional methods or local variations of 'European' systems. In all cases local indigenous tanning materials were employed. The skins were prepared on a very small scale and gathered together from over a wide area to a centralized trading centre. Here they were sorted, mainly into size and quality, and exported. Once in Europe, they were often sorted again before selling on to a tanner or leather dresser. Crust skins were sold by weight and often contained adulterants such as excess tannins, oils, earthy materials and soluble salts. In addition, the dresser was purchasing a very mixed lot of skins from a wide range of sources having very different processing histories. It was his job to minimize these differences and produce as uniform a batch of leather as possible.

In order to achieve this, he first washed the skins thoroughly to remove as much of the unwanted weighting material as possible. He often added alkalis to the wash water in what is known as the stripping process. This was designed to remove excess tannins but, unfortunately, it also removed nontans. These are various organic salts which are present in the leather but do not actually contribute to the tanning reaction. They are, however, effective buffers and have a protective action against the action of acidic atmospheric pollutants.

Next, the skins were shaved to the required thickness. By the last quarter of the nineteenth century various shaving machines had been developed which replaced the laborious skilled hand operations (*Figure 7.12*). With these machines it was possible to cut skins down to the required thickness, cheaply and accurately. One feature of these machines, though, was that in order to ensure a clean cut, their blades had to be sharpened continuously. This resulted in a shower of sparks and small specks of iron dropping onto the leather. These caused a pattern of blue–black iron stains which had to be removed in the clearing process. In this, the skin was immersed in a solution of sulphuric acid which not only dissolved the iron salts but changed the colour of the leather from a reddish brown to a pale yellowish buff. However, once again, dangerous sulphuric acid was introduced into the leather. In order to produce a more uniform substrate for the dyeing process, the leathers were then retanned either using sumac or one of the newly available tanning materials. They were then dyed, possibly with synthetic colours, acidified, again with sulphuric acid, dried out and finished.

As the century progressed, more and more machinery was introduced into the tanneries. This included a range of embossing presses or rolling machines which produced artificial grain patterns. In this way, for instance, it was possible to make sheepskins look like goat, pig or seal skins (Lamb, 1905). They were also employed to disguise scuffs, scratches and other damaged areas. These artificial patterns were formed using a combination of heat and pressure which, at best, distorted the natural fibre structure and, at worst, caused heat damage.

By the last decade of the century the effects of these changes had become only too apparent and work was carried out to determine the causes of the problems and eliminate them. Conservators are, however, working today on leathers manufactured during the period when the long-term effects of these 'improvements' had not been appreciated.

During the nineteenth century, then, leathermaking developed from a traditional craft into a science-based technological industry. The skills of the chemist and the engineer were being allied to the age old sense of craftsmanship of the practical tanner to produce an ever-widening range of materials, at the same time both utilitarian and aesthetically pleasing.

## 7.4 Tanning in modern times

A brief outline of the stages involved in converting raw skins into the most important types of leather today is indicated in the flow diagrams in Figures 7.13 and 7.14. These should, however, only be used as a rough guide to tannery practice as individual tanners may depart considerably from these schemes (O'Flaherty et al., 1956; Sharphouse, 1995; Thorstensen, 1992). The various tannery operations are still generally divided into three main groups:

(a) The beamhouse or pretanning operations in which the non-leathermaking constituents of the skin, such as hair, fat cells, non-fibrous proteins and flesh are removed and the collagen fibres converted into a chemical and physical state suitable for tanning.

(b) The actual tanning processes where the skin is turned into leather by the formation of chemical crosslinks and supramolecular matrices between adjacent collagen molecules.
(c) The post-tanning operations where the skins are treated to give the required properties such as handle, colour and appearance.

The various chemical operations generally take place in dilute aqueous solution. The hides or skins are usually placed, together with the appropriate liquor, into large wooden drums fitted with internal shelves. As the drums revolve, the skins are lifted up on the shelves and then allowed to fall back under their own weight, thus receiving a vigorous kneading and pummelling action which accelerates the penetration of the various chemicals into the skin.

### 7.4.1 Pretanning

Hides and skins are generally received by the tanner in the cured condition, that is, with the effective moisture content of the fibres reduced to such an extent that bacterial decay can no longer take place. This reduction is normally done at the point of slaughter by physically drying the skins under more or less carefully controlled conditions or by saturating the pelt with a concentrated brine solution containing a small amount of a bacteriostat. The cured skins are passed to the beamhouse where the following pretanning processes are carried out:

*Washing and soaking* in dilute solutions of alkalis and wetting agents in order to remove blood, dirt and excess salt.

*Liming* in a suspension of calcium hydroxide, usually containing sharpening agents such as sodium sulphide or sodium hydrosulphide which accelerate the unhairing process. In this treatment the hair and epidermal layer are attacked and loosened to such an extent that they are completely pulped away. The liming process also breaks down various non-collagenous constituents of the skins, opens up the skin structure and chemically modifies the collagen in preparation for tannage.

*Fleshing* in which any loose pieces of membrane, fat or meat still adhering to the underside of the skin are cut away.

*Splitting* by machine to give the desired thickness.

*Deliming* using various weak acids or buffer salts to reduce the alkalinity of the skins.

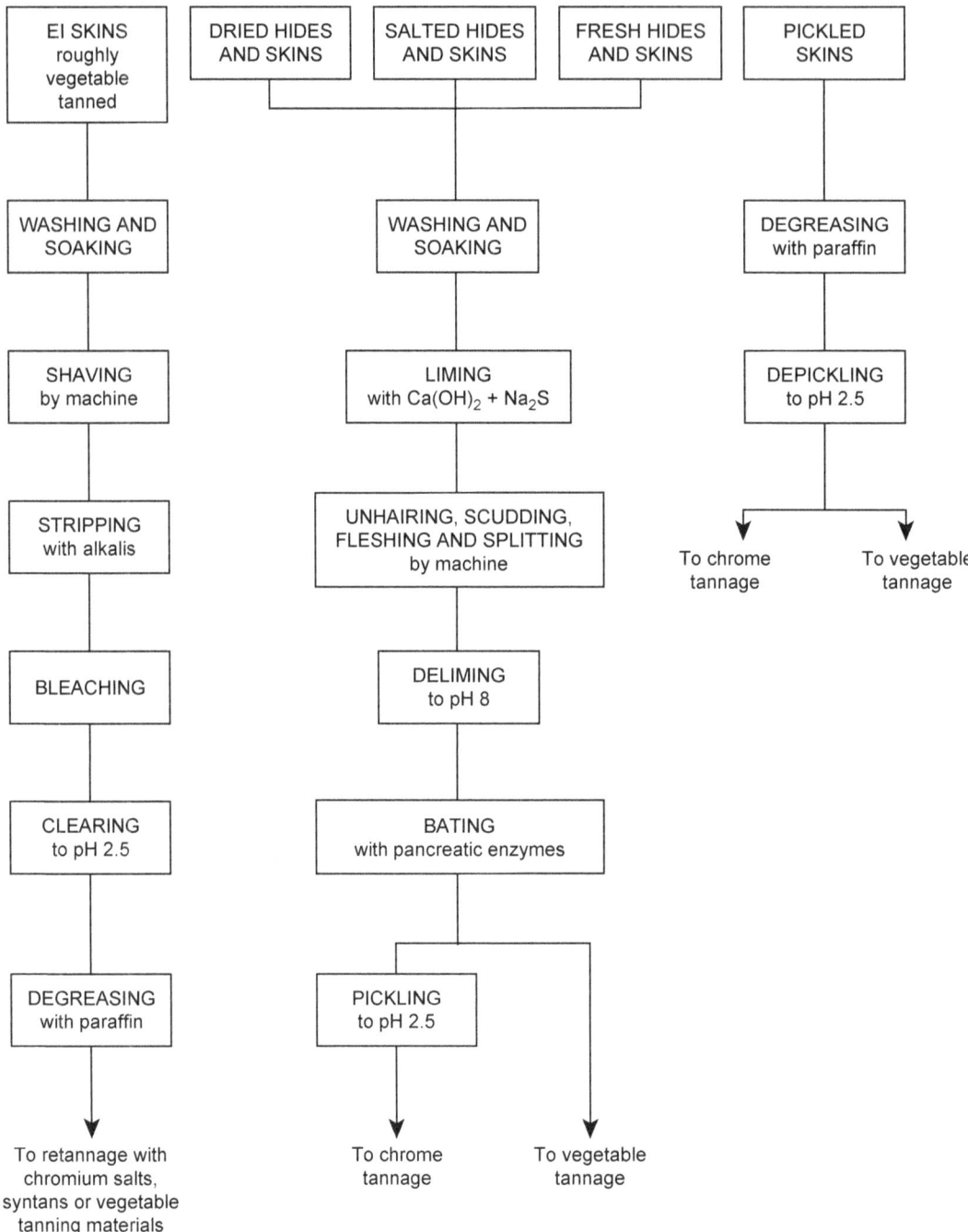

**Figure 7.13** Pre-tanning treatment of hides.

*The manufacture of leather* 79

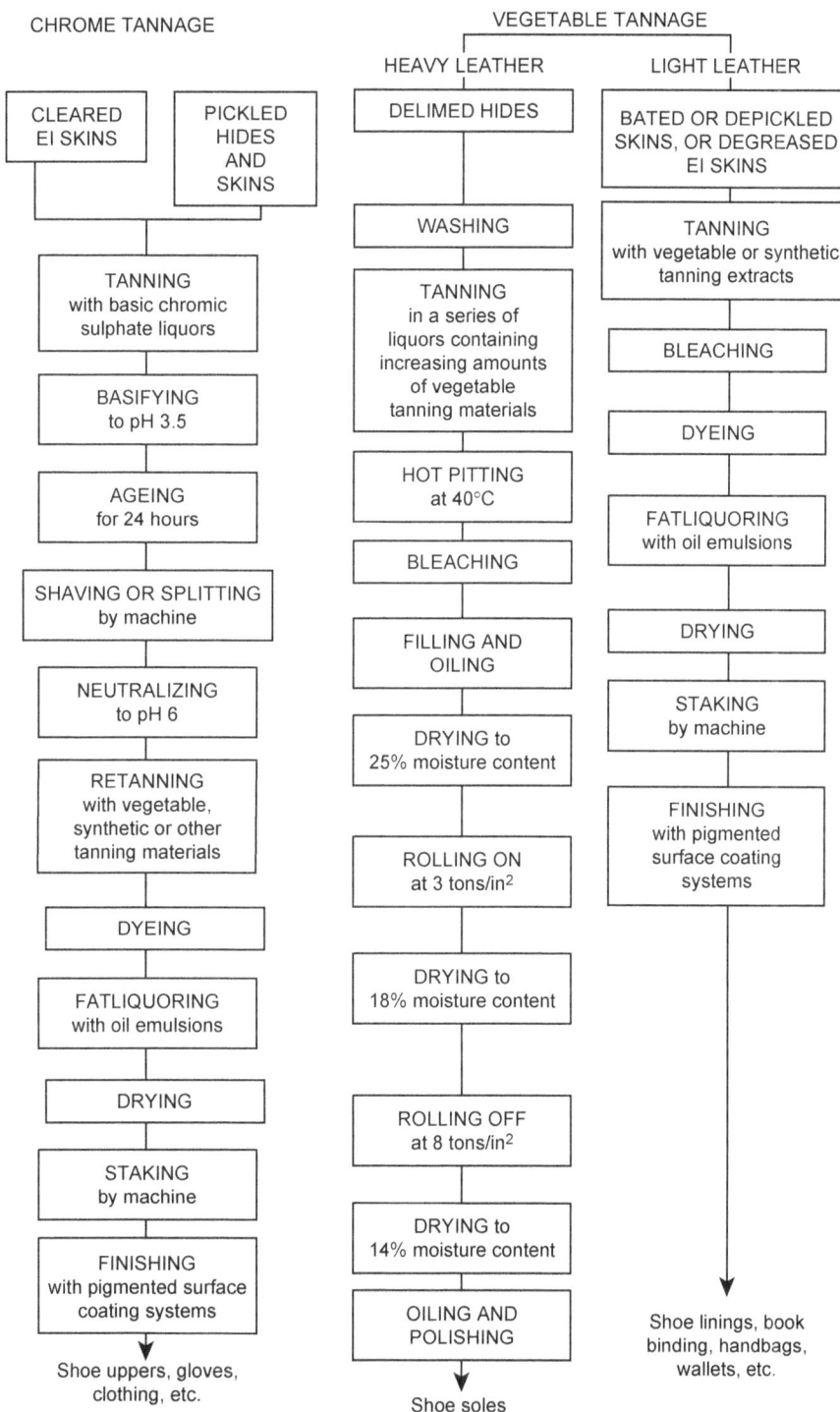

**Figure 7.14** Tanning and post-tanning operations.

*Bating* is a term given to the whole process of treating delimed pelts with enzymes to give the required flexibility and appearance to the finished leather. As has been said, dog and pigeon dung were once the source of the proteolytic enzymes which attack specific types of protein, but standardized products consisting of mixtures of deliming salts and enzymes have now completely replaced the traditional materials. The action of the enzymes is to clean up the skin structure even further by making soluble various non-collagenous proteins already partially degraded during the liming process.

*Pickling* in which the pH of the skins is reduced to between 2.0 and 2.5 by the action of sulphuric or hydrochloric acid. Salts such as sodium chloride or sodium sulphate are added to the pickle liquor to prevent osmotic swelling. Skins are often preserved in the pickled state and large quantities of sheepskins are transported around the world in this condition.

Another method widely used for preserving skins was to tan them lightly with vegetable tanning materials after the soaking, liming and unhairing processes: the dressing of such vegetable-tanned East Indian stock was an important section of the European leather industry. Large quantities of kips, small 'rough-tanned' cattle hides, were imported from the Madras area of India and goat and sheep skins were imported in the same crust condition from India, Pakistan, the Near East and West Africa. The skins were treated first by removing any excess vegetable tanning materials or adulterants and then by retanning them as required. The skins were 'wet back', shaved to the required thickness and the following processes carried out:

*Stripping* the excess vegetable tanning materials and adulterants from the grain by the action of dilute solutions of weak alkalis.

*Bleaching* using dilute solutions of oxalic acid or acidified solutions of sodium metabisulphite or sodium hydrosulphite.

*Clearing* in which the pH of the skins was reduced to about 2.5 at which point they have the palest colour and are in a suitable state for retannage with chromium chemicals.

### 7.4.2 Tanning

An increasing number of tanning materials became available to the tanner since the beginning of the twentieth century to enable him to produce leathers having exactly the required characteristics, but the three groups most widely used are the various vegetable tanning extracts, basic chromic sulphates and syntans.

Vegetable-tanned leathers are generally firm and dense and have a high dimensional stability. They are, therefore, widely used for the soles and linings of shoes, bags, belting, etc. They can also be embossed satisfactorily and are, therefore, chosen for bookbinding and for the manufacture of fancy leathergoods.

Chrome-tanned leathers on the other hand are relatively loose and soft and are more stretchy. The also tend to have a smoother, more supple grain which is resistant to scuffing and flexing. Leathers for shoe uppers, gloving, clothing, etc. are therefore usually chrome tanned.

The various syntans are normally used only in conjunction with other tanning materials, e.g. to improve the handle of chrome-tanned leathers or to speed up the vegetable tanning process for light leathers. However, some of these materials are used on their own when very pale, light-fast leathers are required. In general, such leathers resemble vegetable-tanned rather than chrome-tanned skins.

Most leathers are now produced using a combination of tanning materials chosen to give the blend of chemical, physical and aesthetic properties demanded by the customer.

Vegetable-tanned leathers to be used for shoe soles were normally tanned in a series of ten or more pits using the countercurrent principle. At one end of this system of pits, washed, delimed hide was immersed in an almost completely exhausted tanning liquor, while at the other end a strong, freshly prepared solution of tanning extract met the almost completely tanned leather. Thus over a period which may have extended up to ten weeks, the hide was immersed in a series of liquors containing an increasing concentration of tanning materials. When tanning was judged to be complete, the hides were rinsed to remove excess tanning material from the grain surface and drummed in a mixture of glucose, magnesium sulphate and china clay. The leather was dried out very slowly, during which time it was compressed under hydraulic rollers to give a compact, abrasion-resistant product. After conditioning, it was lightly oiled on the grain and given a final polishing treatment. Today drum tanning systems are more frequently found.

The vegetable tannage of light leathers, where solidity and firmness are less important, for example for the linings of shoes, is a more rapid process. The skins are drummed in dilute solutions of vegetable tanning materials, often containing additions of syntans. Once the grain surface is 'fixed', the concentration is gradually increased over a period of ten hours or more by the addition of more vegetable or synthetic tanning extract. The skins are then washed to remove excess tannin from the grain surface, drummed in a solution of oxalic acid to improve the colour and then dyed, fatliquored, dried, staked and finished.

The chrome tanning process usually consists of drumming the pickled skin in a solution of basic chromic sulphate until the tanning salt has penetrated through to the centre of the skin and then gradually increasing the alkalinity. This basification process renders the chrome tanning materials more astringent and accelerates the tanning reaction. The skins are then piled in a damp state for one or two days to complete the tanning reaction. In this state the tanned hides are termed 'wet blue'. Today many hides and skins are processed to this wet blue stage in the country of origin and traded throughout the world in this condition. Sales of wet blue hides and skins have almost completely replaced the traditional trade in EI crust stock.

### 7.4.3 Post-tanning

After tanning some or all of the following post-tanning operations take place:

*Shaving or splitting* by machine to give the required thickness.

*Neutralizing* using dilute solutions of mild alkalis.

*Retanning* with vegetable or synthetic tanning materials to give the required handle to the leather and to modify the dyeing characteristics as required.

*Dyeing* with selected acid and direct dyestuffs. A very wide range of dyeing auxiliaries, syntans or surface active agents, is used to give the required levelness and depth of shade.

*Fatliquoring* by drumming the skins in an emulsion of various animal, marine, vegetable or mineral oils. These oils were generally emulsified with the aid of the anionic sulphated oils, which were prepared by the action of sulphuric acid on oils of animal, marine or vegetable origin. A wide range of other cationic, nonionic or anionic wetting agents are, however, now widely used. These oils penetrate into the skin structure and prevent the fibres from sticking together, producing a hard, brittle material when the skins are dried out.

*Drying* under very carefully controlled conditions.

*Staking* in which the slightly damp skins are flexed and stretched.

*Finishing* using a pigmented surface coating system. Casein and nitrocellulose finishes were widely employed in the trade but the majority of leathers are now finished using coating systems similar to emulsion paints. These 'resin finishes' are now generally based on acrylic or polyurethane polymer systems.

## References

Cameron, E., ed. (1998) *Leather and Fur: Aspects of Early Medieval Trade and Technology*. London: Archetype.

Clarkson, L.A. (1960) The English Leather Industry in the Sixteenth and Seventeenth Centuries (1563–1700). Unpublished thesis, University of Nottingham.

Faber, G.A. (1938) Greek and Roman Tanners. *Ciba Review*, **9**, 303.

Forbes, R.J. (1957) Leather in Antiquity. In *Studies in Ancient Technology, 5*. Leiden: Brill.

Lamb, M.C. (1905) *Leather Dressing including Dyeing, Staining and Finishing*. London: Anglo American Technical Co.

Levey, M. (1959) *Chemistry and Chemical Technology in Ancient Mesopotamia*. Amsterdam: Elsevier.

O'Flaherty, F. *et al.*, eds (1956) *The Chemistry and Technology of Leather*, 4 volumes. New York: Reinhold (reprinted Krieger, 1995).

Schwerz, F. (1938) Leather Dressing in the Stone Age. *Ciba Review*, **8**, 259.

Sharphouse, J.H. (1995) *Leather Technicians' Handbook*, 75th Anniversary edition. Northampton: Leather Producers Assn.

Thomson, R.S. (1981) Leather Manufacture in the Post Medieval Period. *Post Medieval Archaeology*, **15**, 161.

Thomson, R.S. (1981–82) Tanning – Man's First Manufacturing Industry? *Trans. Newcomen Soc.*, **53**, 139.

Thomson, R.S. (1985) Chrome Tanning in the Nineteenth Century. *J. Soc. Leather Technologists and Chemists*, **69**, 93.

Thomson, R.S. (1991) The English Leather Industry 1790–1900: The Case of Bevingtons of Bermondsey. *J. Soc. Leather Technologists and Chemists*, **75**, 85.

Thomson, R.S. (2001) Bookbinding Leather: Yesterday, Today and Perhaps Tomorrow. *J. Soc. Leather Technologists and Chemists*, **85**, 66.

Thomson, R.S. and Beswick, J.A., eds (1983) *Leather Manufacture Through the Ages*, Proceedings of the 27th East Midlands Industrial Archaeology Society Conference, October 1983, Northampton.

Thorstensen, T.C. (1992) *Practical Leather Technology*, 4th ed. Melbourne: Krieger.

Veale, E.M. (1966) *The English Fur Trade in the Later Middle Ages*. Oxford: Oxford University Press.

Watt, A. (1885) *The Art of Leather Manufacture*. London: Crosby Lockwood & Co.

# 8

# The social position of leatherworkers

*Robert D. Higham*

With demand for leather by a wide range of manufacturing trades having been fairly consistent over many centuries, one would imagine that the social status of those who have worked in leather production to be high. However, this has not always been the case, despite the importance of the tanning industry to most national economies. For example, in 1783, leather was second to woollen textiles in value of production in Britain (MacPherson, 1805) and in 1830 leather and leather product manufacture accounted for 8% of the country's Gross National Product (Thomas, 1983).

Most towns in the Middle Ages had a tannery and in market towns there could be as many as five. In cities this could rise to 20 and in London and its suburbs there were nearly 200. Between 10 and 20% of adult males in many towns in Tudor times worked in the industry (Thomas, 1983). Tanning had also been a part-time craft pursued by farmers. With an abundance of native oak trees whose bark yielded the tannin and with availability of lime for unhairing and plumping hides and skins before tanning, farmers could lay away the hides and skins from their slaughtered beasts for the year or so necessary to complete the process in their own tan pits. In many parts of the countryside one can find such names as 'Tan Pits Lane' as the lingering evidence of rural tanning.

The problem for tanners ensued from the development of trade guilds. In reality these trade associations became a necessity. In medieval times there was no recognition of voluntary associations that protected trades. They began in feudal times as Christian brotherhoods formed for mutual protection before there was any state protection of individuals and their property. The Church was exempted from the feudal hierarchical system and as trade became increasingly important, businessmen managed to secure exemption from the feudal duty of military service for the local liege lords by paying fees to the Crown. These agreements were set out as borough charters and often included rights to control local markets by borough councils elected by 'freemen' who usually belonged to the craft guilds. They determined who could trade and the terms of trade so that oversupply did not cause price reductions. They also controlled raw material prices and thereby ensured reasonable profitability to the various trades (Thomson, 2002).

Several trades were involved in the production of leather; tanners, whittawyers, fellmongers, skinners (furriers) and curriers. These were formalized throughout Europe during medieval times under guild control into heavy and light leather manufacturing. Heavy leather production involved the tanning and currying of hides (currying is the transformation of rough-tanned leather into finished leathers for a variety of purposes through the use of oils and mechanical processes). Light leather production involved fellmongers (producers of pelts and wool from sheepskins), whittawyers (who applied alum and oil in making soft leathers from sheep, goat, deer, horse and dog skins) and leather and fur dressers. The trades which utilized leather were more numerous: cordwainers (shoemakers), girdlers, loriners (bit and stirrup manufacturers), saddlers, harness-makers, pouch-makers, cofferers, glovers, bottle-makers, male- makers (who made leather trunks) and leathersellers (traders in raw material and leathers).

Craft guilds ensured that one craft could never monopolize and control the whole leathermaking and leather using sequence of operations. The tanner was required to take rough dried leather to the local market for examination, certification and sale to a

**Figure 8.1** Sarkis are Nepal's traditional leatherworking caste. This hide, which has been tanned in a crude pit sunk in the ground, is being cut into sides before further simple processing. The stigma attached to this low-caste occupation caused a decline in rural tanning and leather product manufacture until the United Mission to Nepal established the Village Leather Training Association to encourage its revival. These simple methods reflect today what rural tanning and shoemaking would have been like in Europe in medieval times.

currier. A tanner could not be a currier or a butcher, a currier could not be a tanner and a shoemaker could not be a currier. In London, where there was a proliferation of trades, the demarcation between them was more strictly enforced than in the English towns where single guilds incorporated several crafts.

The severe disadvantage for tanners was their subjugation to suppliers of raw material and to the leather finishing and utilizing guilds. Indeed, despite the close association between tanners and curriers, only the curriers became a chartered body during the time when guilds had effective power. In 1410, the court of Aldermen of London issued an ordinance for assaying and proving tanned hides. The appointed examiners were from the cordwainers, girdlers, male-makers, bottle-makers and curriers guilds. The absence of tanners among the appointees is evidence of their subservience. Although they were listed in 1376 as being among 'the 47 misteries' (crafts) in London, they did not attain the dignity of a London livery company until guild authority had declined. While whittawyers were incorporated in 1346 (later to be absorbed in the Leathersellers Company in 1479), it was not until 1703 that Queen Anne granted a charter of incorporation to the 'Masters, Warden and Commonalty of the Art and Mistery of Tanners in the Parish of St Mary Magdalene, Bermondsey' (Waterer, 1944).

Because of the lengthy tanning process, tanners were often suspected of cheating by taking hides out of their pits too early. The Parliament of 1548 and 1562 attempted to enact measures to prevent such practices by stipulating various parameters for leather production. Unfortunately the legislators knew little about tanning and if a tanner abided by these rules he produced unsaleable leather. Consequently tanners broke the rules and were often fined.

In 1575 a compromise measure allowed a court official, Sir Edward Dyer, to licence tanners to produce unlawful leather for fees paid by tanners to Dyer's agents. Dyer's patent expired in 1596 and was not renewed because it had been seen as an extortionate abuse of the tanners (Thomson, 2002). The ludicrous situation was partially resolved in 1604 with the 'Great Statute of Leather' which defined leather and controlled its preparation and sale. All artificers were allowed to buy any kind of leather to

84  *Conservation of leather and related materials*

**Figure 8.2** Shoemakers' simple tools.

convert into manufactured goods. Several companies were still involved in quality inspection and seizure of faulty goods (Waterer, 1944), so tanners remained a suspect breed.

The companies had long since become monopolized in their memberships by capitalists and traders. In 1445 only one actual skinner was left in the skinners company. Consequently artisans were often oppressed in their wages and conditions and 'rich traders sold rubbish at high prices' (Waterer, 1944). Nevertheless the livery companies, despite the ills of self-interest and exclusivity of membership, administered a control upon undesirable trade practices.

The nature of certain parts of leather processing ensured that those involved could be at a disadvantage in the social order. Before the chemistry of leather production was understood, there was a reliance upon noxious processes. Unhairing was by partial putrefaction, often accelerated by soaking hides and skins in urine, dung and stale beer. Prior to tanning, skins would be puered or mastered in dog dung to render them very soft. An alternative process was bating, which employed pigeon dung for the same purpose. Tanneries were obliged to be outside any centre of population and downstream so that effluent could not pollute drinking water. In London's archives there is a reference to 'Tanners *without* Crepelgate' and 'Tanners *without* Neugate'. Bermondsey became the centre for tanning in the London area because it was across the Thames from the City. Chaucer made reference to the 'tan yeardes of Bermondsie' in the *Canterbury Tales*.

This separation of the tanning trade from the rest of society is reflected in the Jewish Mishnaic regulations that date from pre-Christian times. They required that deposits of dead bodies and carrion, burial grounds and tanneries be at least 50 cubits downwind of all centres of population (Mishnah Tractate Baba Bathra, ii.9). The separation even invaded personal relationships. All claims for divorce in Jewish marriages were the privilege of dissatisfied husbands, except where a husband had leprosy or polypus or was a tanner or dung collector. In these cases a wife could sue for divorce even though she had known she was marrying such a person. She could plead that she thought she could endure it, but now she could no longer do so (Mishnah Tractate Ketuboth, vii.11 270 and vii.10).

At Jerusalem tanneries were located outside the Dung Gate which borders on the Hinnom Valley where the city rubbish dump smouldered perpetually.

**Figure 8.3** The finished products. Village leatherworkers also produce simple leathergoods such as shoes. (*Figures 8.1–8.3* by courtesy of Anthony Tilly, TEAR Fund)

*Ge Hinnom* in Hebrew became *Gehenna* in Greek and became the source of the Biblical description of Hell. In crusader times, plans of Jerusalem show a cattle market and skinners' premises inside the city, near the Dung Gate. But by this date the gate had become known as the 'Postern of the Tannery', indicating that tanneries were outside the city's back gate (Higham, 1999).

The pattern is similar in many parts of the world. For example, the former tannery suburb of Istanbul, Kaslicesme, lies outside the city's ancient Byzantine walls. In India, tannery zones can be found near but outside the main centres of population such as Chennai (formerly Madras), Bombay, Kanpur and Calcutta. Today advances in process technology have reduced the noxious nature of the industry and in most societies tanners no longer suffer serious social disadvantage. The grouping nowadays of tanneries in specific industrial zones is for the purpose of sharing common effluent treatment facilities and has nothing to do with any social obloquy. It would be a mistake to assume from the disadvantages that tanners endured through the noxious nature of some of their processes and the suspicions of the craftsmen and traders in other guilds that they were social pariahs. Many tanners became wealthy and influential.

John Tatam, a tanner, was mayor of Leicester three times at the end of the sixteenth century. William House of Buckinghamshire was described in 1575 as 'a man both of great wealth, very wise and most skilful in his art and such a one as made the best leather in the land' (Clarkson, 1960). Many York dignitaries were tanners. However, economies of scale were essential to profitability. Leather production could be undertaken with little capital, but this would have to be combined with another occupation, such as farming, in order to attain a reasonable standard of living. Tanning required capital to be tied up in process for over a year. This required regular input to produce regular output and the pit capacity to wait until the best leather prices could be achieved. Less capital was required in light leather production, in currying and in shoemaking because of lower processing times and less space requirements, but their labour costs were higher than in hide tanning. Heavy leather crafts were generally located where hides, oak bark and water were available. In this respect Bermondsey was an ideal situation being next to England's largest centre of meat consumption and taking advantage of the many streams that flowed into the Thames and the oak woods of northern Kent. Shoemakers were scattered throughout London so that they were near their bespoke customers. In the early seventeenth century there was said to be about 3000 of them in the city and an equal number of light leather workers.

Curriers, the intermediate trade, were located near the shoemakers rather than the tanners simply because it was cheaper to transport rough leather from the leather market than it would be to deliver the heavier curried leather to their customers. While tanners had to be outside the city, curriers could work within the city provided that their windows did not open onto the street and that they ceased work at midday on Saturday so that the workmen could clean themselves and their homes before Sunday dawned (Clarkson, 1983).

Light leather crafts also tended to be nearer the sources of their inputs; but, with less bespoke work in gloves, belts, bags, purses, etc., manufacturers did not need to be near their customers so they stayed close to the leather dressers. Their products were also relatively light to transport. These crafts were also more suited to 'putting out' than the heavy leather crafts because of the ease of transport of leather and goods, and the availability of cheap labour. Light leather crafts were often integrated through one central merchant and producer who subcontracted out the leather dressing and goods manufacture.

Oak bark was the cornerstone of the tanning industry until the late nineteenth century. In early times the term tanner was synonymous with the term barker. A tanner collected his own bark and barkery was a term often used for a tan house. No other tanning material would be envisaged and the law demanded that hides lie in pits layered with oak bark for a year. The use of heat to extract tannin from oak bark was said to produce dark, cracky leather that was undertanned in the centre. The law of 1604 forbade them from setting their pits in old tan hills where fermentation of waste tanning materials could generate heat to warm the tanning liquors. Being proscribed by law from developing more efficient processing conditioned tanners to be suspicious of new theoretical technologies. They preferred the tried and tested methods.

Between 1780 and 1850 a boom in leather demand coincided with a boom in shipping, the former emanating from the spread of the industrial revolution and the latter with the growth in naval and merchant shipping that came with the expansion of the British Empire. Demand for oak bark for tannery and oak wood for ship building escalated but it was said that the tanyards were greater consumers of oak trees than the naval and merchant shipyards (Rackham, 1986).

Eventually the shortage of native oak bark forced the import of oak bark. In the 1850s tanners in England used 200,000 tons of native oak bark and imported 30,000 tons. In the 1890s the figures were respectively 100,000 tons and 700,000 tons. Chrome tanning technology had been discovered in the 1830s but it was not widely taken up until the 1890s. Such was the culture of resistance to change. Indeed there had been few radical changes in leather processing from the Greco-Roman period until the late nineteenth century, as can be witnessed from the layout and artefacts discovered in a tannery at Pompeii (Thomson, 1983).

The industrial revolution powered by coal and steam made greater economies of scale possible in many industries including the leather and its downstream industries. Workers were drawn from their rural and small town communities into the expanding towns where they were forced into very poor, unhealthy living conditions. Women and children worked alongside their men for long hours. Rural craft industries were turned into forms of drudgery and work became a necessary evil. Machines made economies of scale even more realizable and they were often better cared for than the factory workers. The former were more expensive and difficult to acquire whereas the latter could easily be replaced when worn out (Brown, 1954). However, this dark picture can be lightened to some degree by the enterprise of industrial entrepreneurs who had strong Christian beliefs and principles and practised a paternalistic benevolence towards their workforces.

In the leather industry the industrial revolution did not initially alleviate the noxious nature of some of the processing. Puering continued and boys, indeed anyone desperate for work, were sent out with buckets and shovels to collect dog excrement. In health terms, anthrax remained a minor but fatal health hazard and dangers from the use of chemicals, as their favourable properties in leather processing were discovered, were added. Respiratory problems from hydrogen sulphide and sulphur dioxide gases, lime burns, chrome ulcers and the discovery that some dyestuffs and finishing materials contained carcinogens ensured that the tannery environment was not the healthiest.

It is ironic that the old natural tanning process was a source of good health. Tannery workers were renowned for being the healthiest people in a community. Nobody knew why this was so. In the Great Plague in London (1665) many people crossed the Thames to Bermondsey because of the belief that the tan pit odours had medicinal virtue (Waterer, 1944). This was, in fact, true because surface crusts of mould formed on the pits during the long tanning process.

The mould was *penicillium glaucum* and, when this was disturbed, spores were released into the air thus ensuring that tannery workers had regular inhalations of penicillin.

The widespread perception of the tanning industry as offensive in Victorian times is amply illustrated in the novel *John Halifax Gentlemen* by Mrs Craik. Being a tanner is depicted as being a definite bar to his advancement in local polite society. By changing to being exclusively a flour miller he finds acceptance. Today such a situation would not arise as modern processing technologies have replaced the former noxious elements. But where concepts of personal defilement through contact with dead or decaying material are enshrined in theology, those who handle hides, skins and leather find themselves in situations of hereditary social disadvantage.

This is particularly the case in India, Nepal and Japan where there is sensitivity to imparted impurity that stems from Hindu and Buddhist religious beliefs. In India and Nepal tanners and shoemakers are drawn from the scheduled castes, the term that officially replaced 'the untouchables' in 1935. They live outside the caste system and often outside society altogether, being confined to 'no caste' villages and ghettoes. They are forbidden to enter temples because of their ritual impurity. Sensitivity to pollution by contagion is greatest among the Brahmins, the highest caste group. The sensitivity decreases with lower rank. Brahmin attitudes to contact with lower caste groups were evaluated in socio-anthropological research in Uttar Pradesh in the late 1950s. The shoemaker, tanner and sweeper were untouchable in all respects. Nevertheless, they had rank according to source of pollution. The shoemaker who utilized leather ranked lower than a washerman who was in contact with the grime of humans. The tanner was closer to dead organism than the shoemaker so he ranked lower still. While the sweeper who was in contact with animal waste ranked the lowest of the three.

In Japan, the Burakumin, meaning ghetto dwellers, are a people who derive from the stratified Japanese society of centuries ago with its warrior administrators at the peak (Samurai) followed by merchants and landowners. Beneath them were two despised classes, the '*hinin*' or non-people who included beggars, prostitutes, mediums, itinerant entertainers and criminals and the '*eta*' meaning filthy, who undertook polluting work such as animal slaughter and disposal of the dead. The despised groups are now the Burakumin and constitute 2% of Japan's population. The governments of India and Japan have striven to eradicate concepts of caste. As more and more people live in cities, they can become more anonymous and so caste awareness diminishes. But with sensitivities being rooted in theology it will be difficult to remove the old prejudices (Higham, 1999). In *A Suitable Boy*, a novel by Vikram Seth, an Indian mother seeks a suitable husband for her daughter. The young man who seems destined to marry her is a shoemaker in a factory that owns a tannery. A saving feature is that it processes only hides from fallen animals and not hides from animals slaughtered in Muslim abattoirs. The inferred loss of caste integrity by the young lady through associating with a person who works with leather is portrayed as an outdated prejudice (Seth, 1993).

## References

Brown, J.A.C. (1954) *The Social Psychology of Industry*, pp. 31–32. London: Pelican.

Clarkson, L.A. (1960) The Organisation of the English Leather Industry in the Late 16th and 17th Centuries. *Economic History Review*, **13**(2), 245–256.

Clarkson, L.A. (1983) Developments in Tanning Methods During the Post Medieval Period, 1500–1850. In *Leather Manufacture Through the Ages*. (R. Thomson and J.A. Berwick, eds) Proceedings of the 27th East Midlands Industrial Archaeology Conference, Northampton.

Higham, R.D. (1999) The Tanner's Privilege, a Historical Review of the Tanner's Status in Different Societies. *Journal of the Society of Leather Technologists and Chemists*, **83**, 25–31.

MacPherson, D. (1805) *Annals of Commerce*, **4**, 15.

Mishnah Tractate Baba Bathra, ii.9.

Mishnah Tractate Ketuboth, vii.11 270 and vii.10.

Rackham, O. (1986) *The History of the Countryside*, p. 92. London: J.M. Dent.

Seth, V. (1993) *A Suitable Boy*, pp. 623–626. London: Orion Books.

Thomas, S. (1983) Leathermaking in the Middle Ages. In *Leather Manufacture Through the Ages* (R. Thomson, and J.A. Beswick, eds). Proceedings of the 27th East Midlands Industrial Archaeology Conference, Northampton.

Thomson, R. (1983) The 19th Century Revolution in the Leather Industries. In *Leather Manufacture Through the Ages* (R. Thomson, and J.A. Beswick, eds). Proceedings of the 27th East Midlands Industrial Archaeology Conference, Northampton.

Thomson, R. (2002) Post Medieval Tanning and the Problems Caused by 16th and 17th Century Bureaucrats. In *Le Travail du Cuir de la Préhistoire à Nos Jours* (F. Audoin-Rouzeau and S. Beyriers, eds), pp. 465–472. Antibes: Editions APDCA.

Waterer, J.W. (1944) *Leather in Life, Art and Industry*. London: Faber and Faber.

# 9

# Gilt leather

*Roy Thomson*

Leather has been employed for a range of utilitarian purposes throughout time. Its properties also allow it to be decorated using a wide variety of techniques. Among these is the production of what has been called Spanish leather, cordovan, guadameci or, most correctly, gilt leather.

The production of gilt leather seems to have developed in Moorish Spain in medieval times and there is evidence of this material being made into wall hangings from the fourteenth and fifteenth centuries. Spain remained the centre for its manufacture until the late sixteenth century when a combination of factors including the persecution of the Moors and an incident of bubonic plague in Cordoba led to an exodus of skilled craftsmen either to Italy, particularly Venice, or to the Low Countries which were at the time under the rule of Spain. Two distinct schools of design and production methods developed in these new centres and it was from these that the techniques of gilt leather manufacture were disseminated throughout Europe in the seventeenth and eighteenth centuries, the decorative patterns

**Figure 9.1** Late seventeenth century Dutch gilt leather in dining room at Levens Hall, Cumbria.

changing to suit the varying fashions of place and time.

Briefly, gilt leather was produced as follows: uniform-sized pieces of leather, usually vegetable tanned, were cut out and covered with leaves of silver foil using a proteinacious adhesive such as egg white or parchment glue. After the silver had been cleaned and burnished, two or more coats of orange or yellow varnish were applied giving a rich, golden appearance. A three-dimensional design was then produced using hand stamps and/or embossing moulds. This was then coloured using a range of oil- or natural resin-based paints or glazes. Occasionally, with the Italian school, clear varnish was used as part of the design, enhancing areas of silver. In this way panels of leather, sumptuously decorated with three-dimensional motifs were produced which could be joined together to give sheets large enough to cover whole walls. One can imagine the effect of these with their bright, often fairground colours and golden stamp marks glittering in the candlelight, contrasting with the then more usual two-dimensional tapestries and other textile-based wall coverings.

While the main use for gilt leather throughout this period was for wall hangings, it was also employed widely for upholstering chairs, covering trunks, etc. Gilt leather also appears to have been a preferred material for covering screens. Sometimes newly produced leather was utilized but often leather from wall hangings which had gone out of fashion or parts of which had deteriorated were recycled for this purpose. Many country houses still have one or more gilt leather screens standing in corners of their main reception rooms.

Because of the importance of this material in the field of interior decoration and the specific problems associated with the combination of a deteriorated vegetable-tanned leather substrate and a decorative top surface, the conservation of gilt leather has been studied in some depth. These problems include damage associated with its use, either as part of a piece of furniture or as a wall covering in an often damp or otherwise unsuitable environment. They also include complications associated with past conservation treatments which may have included overoiling or the application of inappropriate surface coatings. A number of case histories giving practical examples of the treatment of gilt leather are included in Chapter 23 which will serve as an introduction to the conservation of this material. There are, in addition, a large number of publications dealing directly with all aspects of gilt leather: its history, its conservation,

**Figure 9.2** Levens Hall gilt leather showing nineteenth century remounting.

**Figure 9.3** Levens Hall gilt leather detail.

and the rehanging of large panels in historic interiors. The reader is directed to consult these for further information. Important papers include those which follow. It should be noted that the majority of these references have been taken from the bibliography on gilt leather prepared by Dr Eloy Koldeweij and distributed by him freely at the ICOM Committee for Conservation Leathercraft Group Interim Meeting in Brussels in 1998.

## 9.1 Production and art historical aspects

Bergmans, A. and Koldeweij, E.F. (1992) Inventaris van het 17de en 18de-eeuwse goudleder in Vlaanderen. *Monumenten en Landschappen*, **11**(6), 33–46.

Bonnot-Diconne, C., Couroal, C.N. and Koldeweij, E. (2002) Meeting between Solomon and the Queen of Sheba: History, Technology and Dating of Gilt Leather Wall-hangings. In *Preprints of ICOM Committee for Conservation 13th Triennial Meeting Rio de Janeiro*, **II**, pp. 764–769. London: James & James.

Clouzot, Henri (1925) *Cuirs decorés*. Paris.

Colomer Munmany, A., Ainaud de Lasarte, J. and Soler, A. (1992) *L'art en pell. Cordovans i guadamassils de la collecció ColomerMunmany*. Vic: Colomar Munmany.

Diderot, D. and Rond d'Alembert, J. le (1751–1772). *L'Encyclopédie*. Paris.

Doorman, G. (1953) *Techniek en octrooiwezen in hum aanvang. Geschiedkumdige aanvulingen*. Gravenhage.

Dossie, R. (1763) *The Handmaid to the Arts*, 2nd edition. London: J. Nourse.

Ferrandis Torres, J. (1955) *Cordobanes y Guadamecies. Catalogo ilustrado de la exposición*. Madrid.

Fougeroux de Bondaroy, A.D. (1762) *Art de Travailler les Cuirs Dorés ou Argentés*, Paris. (Déscription des Arts et Métiers).

Gall, G. (1965) *Leder im Europäischen Kunsthandwerk*. Braunschweig.

Glass, H. (1991) *Ledertapete*. Essen: Galerie Glass.

Glass, H. (1998) *Bedeutende Ledertapeten*. Essen: Kunsthandel Glass.

Huth, H. (1937) English Chinoiserie Gilt Leather. *Burlington Magazine*, **71** (July), 25–35.

Johannsen, O. (1941) *Peder Månssons Schriften über Chemie und Hüttenwesen, Eine Quelle zur Geschichte der Technik des Mittelalters*, **16**. Berlin.

Koldeweij, E. (1992) How Spanish is Spanish Leather? In *Conservation of the Iberian and Latin American Cultural Heritage. Preprints of the Contributions to the Madrid Congress 1992* (H.W.M. Hodges, J.S. Mills and P. Smith, eds), pp. 84–88. London: IIC.

Koldeweij, E. (1992) Het Gouden Leer. *Monumenten en Landschappen*, **11**(6), 8–32.

Koldeweij, E. (1996) The Marketing of Gilt Leather in Seventeenth Century Holland. *Print Quarterly*, **XIII**(2), 136–148.

Koldelweij, E. (1997) Leather. In *Encyclopaedia of Interior Design* (J. Banham, ed.), pp. 705–707. London/Chicago.

Koldeweij, E. (2000) Gilt Leather Hangings in Chinoiserie and Other Styles: An English Speciality. *Furniture History*, **36**, 61–101.

Leiss, J. (1969–70) Leder Tapeten. In *Tapeten, Ihre Geschichte his zur Gegenwart*. (H. Olligs, ed.) **I**, pp. 45–98, **II**, pp. 191–194. Braunschweig.

Madurell Marimon, J.-M. (1972) *Guadamacilers i Guadamaccils*. Barcelona.

Madurell Marimon, J.-M. (1973) *El Antiguo arte del Guadameci y sus artifices*. Vic: Colomar Munmany.

Mühlbacher, E. (1988) *Europäische Lederarbeiten vom 14. bis 19.jahrhundert aus dem Sammlungen des Berliner Kunstgewerbemuseums*. Berlin: Staatliche Museen zu Berlin.

Paz Aguilo, M. (1982) Cordobanes y Guadamecies. In *Historia de las artes aplicadas e industriales en España*, pp. 325–336. Madrid.

Scholten, F. and Koldeweij, E., eds (1989) *Goudleer Kinkarakawa. De geschiedenis van het Nederlands goudleer en zijn invloed in Japan*. Zwolle: Uitgeverij Waanders.

Soest, H.A.B. van (1989) The Manufacture of Gilded Leather-tapestry. In *Postprints of the ICOM Committee for Conservation Leathercraft Working Group Symposium*, p. 185. Offenbach am Main: ICOM.

Waterer, J.W. (1971) *Spanish Leather, A History of its Use from 800 to 1800 for Mural Hangings, Screens, Upholstery, Altar Frontals, Ecclesiastical Vestments, Footwear, Gloves, Pouches, and Caskets*. London: Faber.

Wormser, Jac. Ph. (1913) Het Goudleder. *Bouwkunst*, **5**, 45–87.

## 9.2 Conservation and restoration

Amoore, J., Bergeon, S. and Erlande-Brandenburg, A. (1980) La restauration des tentures en cuir peint du Musée d'Ecouen. *L'Estampille*, **127**, 39–46.

Bansa, H. and Mair, H. (1981) Bemerkungen und Betrachtungen zur Ledermalpflege. *Maltechnik-Restauro*, 111–114.

Berardi, M.C., Nimmo, M. and Paris, M. (1990) The Seventeenth-century Flock-leather Wall Hangings of the Chigi Chapel in Ariccia: A Case Study. In *Preprints of the ICOM Committee for Conservation 9th Triennial Meeting, Dresden*, **II**, pp. 611–615. Los Angeles: ICOM.

Berardi, M.C., Nimmo, M. and Paris, M. (1990) Parati in cuoio a cimatura. Cenni su una tecnica poco nota e problemi di conservazione. *Rivestimenti murali in carta e Cuoio: tecniche esecutive, conservazione e restauro*, pp. 5–24. Roma.

Berardi, M.C., Nimmo, M. and Paris, M. (1993) Il cuoio dorato e dipinto, richerche e conservazione. *Materiali e strutture. Problemi di conservazione*, **III**(3), 95–129.

Bergeon, S. (1990) France, XVIIème siècle–Tenture de Scipion. In *'Science et patience', ou la restauration des peintures*, pp. 236–239. Paris.

Bilson, T. and Bonnot, C. (1997) The Transfer and Restoration of an Eighteenth Century Gilt Leather Screen. In *Postprints of the ICOM Committee for Conservation Leathercraft Working Group interim meeting Amsterdam 1995* (P.B. Hallebeek and J.A. Mosk, eds), pp. 6–14. Amsterdam: ICOM.

Calnan, C.N. (1992) The Conservation of Spanish Gilt Leather – An Introduction. In *Conservation of the Iberian and Latin American Cultural Heritage. Preprints of the Contributions to the Madrid Congress 1992* (H.W.M. Hodges, J.S. Mills and P. Smith, eds), pp. 23–26. London: IIC.

Dudley, D. (1981) Conservation of Leather Wall Hangings in the James J. Hill Mansion. *Journal of American Journal of Conservation*, **20**, 147–149.

Erlande-Brandenburg, A. and Guichard, R. (1992) Une experience de restauration à la Bibliothèque Nationale. Historique et description: la tenture de cuir peint des héros romains d'après Goltzius. Techniques utilisées. *Revue de la Bibiotheque Nationale*, **2**, 33–39.

Hallebeek, P.B. and Soest, H.A.B. van (1978) Gilded Leather. In *Preprints of ICOM Committee for Conservation 5th triennial meeting*, pp. 1–8. Zagreb: ICOM.

Hallebeek, P.B. (1980) Restoration and Conservation of Gilt Leather, Part II. In *Conservation Within Historic Buildings*, pp. 164–165. Preprints of the Contributions to the Vienna Congress. London: IIC.

Hallebeek, P. and Keijzer, M. de (2002) Conservation Procedure Based on Methodical Research to Remove an Alkyd Resin from a Gilt Leather Wall-hanging. In *Preprints of ICOM Committee for Conservaton 13th triennial meeting Rio de Janeiro*, **II**, pp. 770–776. London: James & James.

Hejtmanek, R. and Lepeltier, C. (1988) *Restauration d'un fragment de tenture en cuir doré: étude des produits de nourriture et d'assouplissement pour le cuir ancien*. Paris.

Jagers, E. (1980) On the Deterioration after Treatment of Painted Leather Wall Hangings in Three Castles in Rhineland. In *Conservation Within Historic Buildings*, pp. 166–167. Preprints of the Contributions to the Vienna Congress. London: IIC.

Jagers, E. (1987) Die Konservierung von Ledertapeten – ein Unfallstudie. In *Conservation-Restoration of Leather and Wood. Training of Restorers (Sixth Internat. Restorer Seminar, Veszprem)*, pp. 225–232. Budapest.

Konigsfeld, G. (1989) Anmerkungen zur geschichte der Herstellung, zu Eigenschaften und Fehlern des Leders, Probleme der Lederpflege. In *Restaurierung van Kulturdenkmalern. Beispiele aus der Niedersachsischen Denkmalpfleg* (H.H. Moller, ed.), pp. 327–333. Hameln.

Lonborg Andersen, V. (1997) Remounting Gilt Leather Tapestries with Velcro. *Conservation of Leathercraft and Related Objects*. In *Postprints of the ICOM Committee for Conservation Leathercraft Working Group interim meeting Amsterdam 1995* (P.B. Hallebeek and J.A. Mosk, eds), pp. 121–122. Amsterdam: ICOM.

Moris, B. (1991) De restauratie van goudleer in het museum Plantin-Moretus. *Cultureel Jaarboek Van de stad Antwerpen*, **IX**, 41–42.

Moroz, R. (1996) Die Trockenreinigung von Lederobjekten. *Restauro*, **102**(4), 242–247.

Moroz, R. (1993) The Conservation of the 18th-century Leather Tapestries Covering a Travelling Chest. In *Preprints of the ICOM Committee for Conservation 10th triennial meeting Washington DC*, **II**, pp. 651–655. Paris: ICOM.

Moroz, R. (1997) The Formation of Cracks on Leather with Respect to Gilt Leather. In *Postprints of the ICOM Committee for Conservation Leathercraft Working Group interim meeting Amsterdam 1995* (P.B. Hallebeek and J.A. Mosk, eds), pp. 19–25. Amsterdam: ICOM.

Muller, D. (1986) The Degreasing of a Set of Gilt Leather Wall-hangings in the Rijksmuseum. In *On Ethnographic and Water-logged Leather*. Proceedings of ICOM Committee for Conservation Symposium, pp. 11–16. Amsterdam.

Muller, D. (1987) Das Entfetten von vergoldetem Leder. Die Ledertapeten im Raum 259 des Rijksmuseum. *Maltechnik*, **93**, 47–52.

Mors, B. (1991) De restauratie van goudleer in het museum Plantijn-Moretus. *Cultureel Jaarboek van de stad Antwerpen*, **IX**, 41–42.

Nimmo, M., Paris, M., Rissotto, L. et al. (1996) Tensioning Gilded and Painted Leather. In *Preprints of ICOM Committee for Conservation 11th triennial meeting Edinburgh*, **II**, pp. 751–758. London: James & James.

Nimmo, M., Paris, M. and Rissotto, L. (1999) Tensioning Gilded and Painted Leather. Part 2: First Verification of Supporting Structures with Elastic Tensionsing. In *Preprints of ICOM Committee for Conservation 12th triennial meeting Lyon*, **II**, pp. 697–701. ICOM.

Paris, M. (1997) Tensioning Gilded and Painted Leather: A Research Project. In *Postprints of the ICOM Committee for Conservation Leathercraft Working Group interim meeting Amsterdam 1995* (P.B. Hallebeek and J.A. Mosk, eds), pp. 15–18. Amsterdam: ICOM.

Peckstadt, A., Watteeuw, L. and Wouters, J. (1992) Conservering van het goudlederbehang van de dekenij van Zele. *Monumenten en Landschappen*, **11**(6), 55–61.

Philippe, M.M.J. (1952) Le traitement des cuirs de tenture et la façon de cordove conservés au Musée d'Ansembourg à Liege. *Bulletin des Musées de Belgique*, **1**, 96–100.

Pracher, P. (1974) Konservierung des Ledertapeten aus dem Schlösschen Veithöchstein. *Maltechnik*, **3**, 144–148.

Pracher, P. (1989) Die Ledertapeten aus Schloss Jever und aus das Kloster 'Zur Ehres Gottes' in Wolfenbuttel. In *Restaurierung von Kultulrdenkmalern. Beispiele aus der Niedersachsischen Denkmalpflege* (H.H. Moller, ed.), pp. 339–344. Hameln.

Reff, R. and Pavel, V. (1983) Restaurarea unor piese de piele din secolul al XVII lea. *Revista Muzeelor si Monumentelor*, 48–54.

Rosa, H. (1996) Gilt Leathers and the Problems of their Conservation. In *Postprints of the International Conference on Conservation and Restoration of Archive and Library Materials*, Erice (Italy).

Schulze, A. (1988) *Die Restaurierung von Ledertapeten – Literaturstudie*. Dresden.

Schulze, A. (1989) Leather Tapestry of Moritzburg: Its Actual Situation, and First Results of Restoration. In *Postprints of the ICOM Committee for Conservation Leathercraft Working Group Symposium*, pp. 171–180. Offenbach am Main: ICOM.

Schulze, A. (1997) Die Ledertapeten im Schloss Moritzburg. In *Denkmalpflege in Sachsen 1894–1994* (H. Magirius and A. Dulberg, eds), pp. 229–240. Weimar.

Soest, H.A.B. van (1972) Over goudleer en zijn conservering. *Vereniging Musea Restaurateurs en Technici*, 4–20.

Soest, H.A.B. van (1975) Goudleer. *Spieghel Historiael*, **10**, 36–46.

Soest, H.A.B. van (1975) Enkele historische aspecten van de goudleerfabricage. *Oud Nieuws*, **8**(3), 24–33.

Soest, H.A.B. van (1976) Het goudleer in Huize Voormeer. *Bulletin van de Koninklijke Nederlandse Oudheidkundige Bond*, **75**, 297–302.

Soest, H.A.B. van (1976) Beknopte beschrijving van de restauratie aan het goudleer van het Frans Hals Museum te Haarlem. *Oud-Nieuws*, **9**(1), 28–34.

Soest, H.A.B. van and Hallebeek, P.B. (1975) *Goudleeren behangsels*. Haarlem/Amsterdam.

Soest, H.A.B. van (1980) Restoration and Conservation of Gilt Leather, Part 1. In *Conservation Within Historic Buildings*, pp. 162–163. Preprints of the Contributions to the Vienna Congress. London: IIC.

Soest, H.A.B. van, Stambolov, T. and Hallebeek, P.B. (1985) Die Konservierung von Leder. *Maltechnik Restauro*, **91**, 49–54, 57–65.

Soest, H.A.B. van and Hallebeek, P.B. (1984) Chemical Solutions of Problems Encountered in the Conservation of Old Leather. In *Preprints of the ICOM Committee for Conservation 7th triennial meeting Copenhagen 1984*, **II**, pp. 16–18. ICOM.

Soest, H.A.B. van, Stambolov, T. and Hallebeek, P.B. (1984) Conservation of Leather. *Studies in Conservation*, **29**, 21–31.

Sturge, T. (1999) Gilt Leather Wall Coverings: Some Options for Repair. In *Postprints of the Conservation of Decorative Arts Conference*, London, pp. 65–69. London: UKIC.

Szalay, Z. (1980) Conservation and Restoration of Dyed and Embossed Leather Objects. In *Conservation Within Historic Buildings*, pp. 168–169. Preprints of the Contributions to the Vienna Congress. London: IIC.

Talland, V., Mangum, B., Tsu, C.M. and Fullick, D. (forthcoming) The Conservation of Leather Wallhangings at the Isabella Steward Gardner Museum. In *Postprints of the ICOM Committee for Conservation Leathercraft Working Group interim meeting Brussels* 1998.

Thomson, R.S. (1999) Choices in the Treatment of Gilt Leather Wallhangings. *J. Soc. Leather Technologists and Chemists*, **84**, 20.

Torre, S. de la (1970) Conservacion de cordobanes y guadameciles. *Informes y trabajos del instituto de conservacion y restauracion de obras de arte, argueologia y etnologia, Madrid*, pp. 91–96.

Tsu, C.M., Fullick, D., Talland, V. and Mangum, B. (1999) The Conservation of Leather Wallhangings at the Isabella Steward Gardner Museum. Part II. In *Preprints of ICOM Committee for Conservation 12th triennial meeting Lyon*, **II**, pp. 708–713. ICOM.

Winter, J. and West Fitzhugh, E. Some Technical Notes on Whistler's 'Peacock Room'. *Studies in Conservation*, **30**, 149–154.

Wouters, J. (1986) *De restauratie van het goudleder behang van het Museum het Erasmushuis te Anderlecht*. Brussels.

# 10

## *Cuir bouilli*

*Laura Davies*

### 10.1 The *cuir bouilli* technique

*Cuir bouilli* is a Norman French term that translates into 'boiled leather'. It describes the process used to change flexible leather into rigid, moulded and often intricately shaped objects like firemen's helmets, drinking vessels and boxes. The term directly suggests that boiling water was used to manipulate the properties of leather in a similar way to embossing, but the nature of the technique has never really been established.

Surviving objects and a number of historical accounts of methods for moulding leather give us many indications to the processes involved. However, the term *cuir bouilli* has been used to describe all manner of moulded leather objects, which exhibit the familiar, rigid character, but which vary greatly in style and function. There are distinct physical and visual differences in these objects, which indicate that methods for making *cuir bouilli* consisted of a number of different techniques which can be related to other leather moulding processes.

### 10.2 Leather moulding techniques

The familiar leather moulding techniques such as embossing and lasting rely on the use of mild heat and water to manipulate the properties of leather. These moulding and shaping techniques exploit the plastic character of vegetable-tanned leather when it is softened by water. When water is applied to leather it becomes malleable and easy to mould and it will retain this moulded shape or ornamentation when it dries. The ability of leather to do this derives from the action of warm water on the fibrous structure of its collagen chains and on the thermoplastic characteristic of vegetable-tanned leathers. However, once the process of moulding leather in this way is complete, and the object is dried, the structure of the leather is unaltered and it remains flexible unless coated with a stiffening medium or mounted on a backing material.

The nature of *cuir bouilli* leather after-treatment differs in the fact that the leather is rigid and water resistant. This indicates that when leather is moulded using water and higher temperatures, a chemical reaction occurs which permanently alters the structure of the leather. If one were to follow strictly the implied definition of the term, then leather would be boiled in water to achieve this alteration to its chemical and physical properties. This treatment though would result in the leather shrinking irretrievably and becoming brittle as it dries. Historical texts describe three main methods called the *cuir bouilli* technique which were used to harden and mould leather objects.

### 10.3 The origins of the *cuir bouilli* technique

John Waterer, who published many significant texts on the history of leather craft techniques, suggested the primitive method of heating water in skin bags over a fire, as the origin of the technique.

The early origins of the *cuir bouilli* technique are established by the finds of silver drinking vessels intact with mineralized organic deposits, thought to be the remains of leather cups.[1] Another important early evidence of the technique was the discovery of an Iron Age leather shield complete with its wooden former in a peat bog in Ireland.[2] It is claimed that these shields, called bucklers, were made entirely from wet

**Figure 10.1** *Cuir bouilli* leather porter's hat from Billingsgate fish market.

vegetable-tanned leather that had been pressed into moulds and dried in a warm place until they were rigid. However, the shield had survived two and a half thousand years, without losing its original rigid character and it is therefore most likely that some form of hardening technique had been used to provide this incredible durability. The shield displays the popular use of leather for making protective clothing and armour. It was ideally suited to the purpose because it could be moulded into close fitting, lightweight but rigid protective forms that allowed greater freedom of movement than metal.

Accounts of the methods employed for manufacturing such armour indicate that a variety of different moulding techniques were employed to achieve this rigidity. Leather horse cuirasses, sheaths and scabbards were made by laminating strips of leather with glue onto a textile support which may have been shaped over a mould (Cameron, 1998). Such leather is treated in no other way, but the resultant rigid material has often been misleadingly described as *cuir bouilli*. It is the surviving leather fire helmets and porters hats dating from the sixteenth to nineteenth century, however, which clearly define the use of the *cuir bouilli* technique. Moulded leather helmets were in use during the great fire of London in 1666 and continued to be used until the outbreak of the Second World War. Their use was popular until their replacement with heavy duty plastics, because they protected the head and did not conduct heat like metal versions. These helmets were often coated with pitch to provide added protection and make them more waterproof (*Figure 10.1*).

The manufacture of these helmets is paralleled by techniques used to make military helmets in nineteenth century America.[3] The technique used to make such helmets involved immersing oak bark vegetable-tanned cow hide into hot water (or raising the temperature of the water to this point with the leather already immersed). The leather would be removed as it began to shrink and it could be moulded immediately into the desired shape over a wooden former. The finished article could then be stitched while still wet to form a helmet or vessel-like structure, and held in place over the former to dry.[4]

A variation of this technique is associated with the manufacture of bottles called blackjacks, bombards and costrels (*Figure 10.2*). The earliest written reference to leather bottles is in the record of the Guild of Bottellars of 1373.[5] These vessels were made from oak bark-tanned cattle hide and were very strong and substantial. Their manufacture and use was a

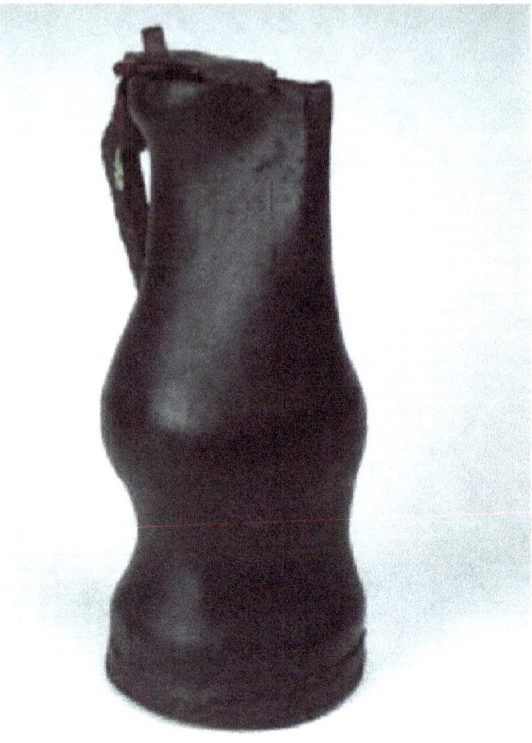

**Figure 10.2** *Cuir bouilli* leather. (Courtesy Museum of London)

**Figure 10.3** Postillion's boots with *cuir bouilli* leg protectors.

consequence of the scarcity of other materials for domestic utensils. Towards the end of the sixteenth century large pitchers were made of earthenware and the craft was eventually replaced by the expansion of the stoneware industry. Originally these objects were hand sewn but later examples are made with two seams riveted together by copper alloy studs.

The manufacturing technique used to make such bottles involved wetting the leather with cold water and then exposing it to a heat source. This was achieved either by the use of actively applied heat with the leather stretched over a wooden last, or by passively applied heat with the shaped leather heated in an oven.[6] The original technique would also have included a process where small vessels whose openings were too small to admit any type of former were shaped and sewn when wet, filled with sand to form a temporary firm support and then heated or baked to make the structure permanent. Once the vessel was dried the sand could be poured out.

Another method for hardening leather involves the use of materials such as waxes, resins, fats and flour pastes applied to bulk and harden leather. This impregnation technique has been explored by scientists, conservators and practising artists. The process involves the application of a hot resinous mixture onto a previously wetted leather surface. This technique has been called 'pitching' leather and has been documented during the manufacture of *cuir bouilli* boots popularly called postillions' boots which consisted of coating leather with such materials as heated colophony pine resin mixed with beeswax and lamp black pigments (*Figure 10.3*).[7]

The boots would be constructed with their grain surface inside so that the mixture could be easily absorbed into the exterior flesh side. They were extremely strong and warm and would be lined with quilted leather filled with wool for added insulation.

Each of these methods described were practised by highly skilled, master craftsmen who passed on their knowledge to apprentices orally. They have not been outlined specifically in any text. In practice they are very similar to other leather moulding techniques but share the characteristic that they are all based on processes which permanently alter the physical and chemical characteristics of leather. It is for this reason that much confusion has arisen in coming to a single simple definition of *cuir bouilli*.

## 10.4 Changes undergone by the leather in the *cuir bouilli* process

The physical and chemical changes of leather subjected to the *cuir bouilli* process are determined by the action of heat and water on the unique leather structure and the thermoplastic nature of excess vegetable tannin, particularly condensed tannins, within that structure.

It is necessary to understand these changes in the material in order to anticipate its process of deterioration and propose appropriate conservation treatments.

*Cuir bouilli* leather is made from vegetable-tanned leather whose manufacture and chemical structure has been well documented. During the tanning process chemical bonds are introduced between tanning agents and leather fibres which increase the chemical and physical stability of the collagen fibres. This network prevents the fibrous structure of leather from collapsing as water is lost. Vegetable tannins from oak bark consist of large phenolic molecules predominantly of the condensed tannin type. They are weakly acidic and through hydrogen bonding are able to bond efficiently with the collagen protein molecules. This action displaces water from the reactive groups and the hydrophobic areas of the tannin and protein are then drawn together by van der Waals forces. The collagen/tannin framework is therefore strongly networked, but at the same time flexible and relatively stable to hydrolysis by water.

The nature of the condensed tannins used in the manufacture of authentic *cuir bouilli* leather greatly influences the final properties of the material. For heavy duty purposes, leather is sometimes deliberately subjected to large quantities of the tanning material, with an excess left in place to make the leather more rigid. It is thought that much of the tannin may be physically held within the structure of the leather, but be chemically independent of it (Reed, 1966). The high molecular weight of these tannins dispersed within the fibre spaces of the leather could account for the restricted movement of the leather and an increase in stiffness.

When vegetable-tanned leather has been saturated with water and heated to a temperature around 75–90°C it begins to shrink (Bickley, 1991). The exact temperature will depend on a wide number of factors such as the type and amount of tanning material present, the rate of heating, etc. At this point of shrinkage temperature, the chains of the collagen molecule, which are normally held in an extended form by hydrogen bonding, become unstable as the energy input exceeds that of the hydrogen bonds holding the molecular shape together. Above this point, therefore, there is a sudden release from the extended form of the collagen molecule and the leather becomes incredibly elastic as the chains retract and shrink. As it cools the hydrogen bonds are able to reform between the realigned collagen molecules in a tighter formation. In this case if the leather has fully shrunk, when it dries it becomes inelastic. If the leather is heated to just below the shrinkage point, the bonding is weakened but not fully released allowing limited realignment to take place in a more controlled manner. In this case the new dense strong network makes *cuir bouilli* leather rigid and strong.

Alternatively, if the heat source application is limited to only the surface of the wet leather then it is possible that this shrinkage solely occurs in the outer layer of the leather, producing a surface hardening effect. The leather could therefore have an altered rigid outer layer and a less affected internal area, the combination of which produces a very strong structure.

The influence of heat on the condensed vegetable tannins within the leather structure is an important factor in the hardening of the leather, as these large molecules behave thermoplastically. If there is an excess of tannins present then they allow the structure of the leather to be moulded when hot, but as they cool they contribute to the hardening of the leather by supporting the fibres in a strong matrix in their new alignment. It is thought that the oils, resins, and waxes used in the impregnation technique of *cuir bouilli* manufacture would behave in a similar way.

The effects of the shrinkage temperature and the tanning process can therefore have a profound effect on leather, which is irreversibly altered during the process of making *cuir bouilli*. This has implications for some of the recommended treatments for conserving objects made from this material.

## 10.5 Conservation of *cuir bouilli*

### 10.5.1 Stability

The condition of *cuir bouilli* leather can be detrimentally affected by environmental damage factors and by restoration techniques. All leather in museum storage is susceptible to decay in the presence of atmospheric pollutants, light and fluctuating climatic conditions. But it is generally found to be the surface treatments applied to *cuir bouilli* leather objects during manufacture, maintenance treatments throughout their use and subsequent conservation processes which alter its original appearance.

*Cuir bouilli* leather is generally found to be resilient to attack by the atmospheric pollutants such as sulphur dioxide, which causes the common form of chemical decay in vegetable-tanned leather called red rot. This could be because the acidic pollutants cannot penetrate the dense structure. It remains relatively stable in a fluctuating environment but has been known to progressively harden and crack over time. This deterioration can be attributed to fluctuations in the environment but can also be caused by the routine application of leather dressings more appropriate to conventional leather conservation.

### 10.5.2 Damage caused by old treatments

A misconception regarding *cuir bouilli* is that leather dressings should be applied to lubricate the leather. This treatment is intended to impart flexibility and enhance the object's appearance by saturating the grain surface. However, this material was never meant to be flexible and most *cuir bouilli* leather objects have characteristic darkened surfaces. These dark surfaces are partly derived from the alteration in the colour of leather when it is heated but are also a result of the original application of surface finishes and coatings to *cuir bouilli* after manufacture. The presence of these materials also has implications for the conservation of *cuir bouilli* objects.

Many of the conservation dressings used on leather may have a long-term detrimental effect on the material they are meant to be preserving. There are an enormous number of oils, waxes, natural and synthetic fats and resins that are recommended in conservation literature for the care and preservation of leather. However, their use on museum objects has been shown in some cases to actually cause damage to both the body and surface of the leather (Soest *et al.*, 1984). Most leather dressings do not penetrate into the leather and as they are reapplied they accumulate as an excess on the surface. This alters the ability of the leather to regulate its moisture content relative to the environment, eventually resulting in it becoming brittle and prone to cracking.

These mixtures of materials eventually deteriorate, however, and alter the surface appearance of the leather. As they oxidize they can become brittle, discoloured and, as they lose their original properties, becoming sticky and gummy, they imbibe dust and dirt. For the preservation of *cuir bouilli* leather it is recommended these dressings are removed in order to preserve both the body and surface of the leather. This treatment also ensures that the original nature and surface quality of *cuir bouilli* leather can be appreciated.

The removal of non-defined layers of treatment from an original surface is very difficult. Historic references can provide information about the components of original coatings but there are an enormous variety of modern leather dressing/improving treatments, made up of complex mixtures of materials which have been used to conserve this material. They are therefore extremely difficult to identify and isolate during cleaning. However, with careful solubility testing and awareness of the quality and appearance of original coatings cleaning can be very effective.

### 10.5.3 Original treatments of *cuir bouilli* leather

*Cuir bouilli* objects were finished or dressed according to their intended purpose. If bottles or buckets were intended to hold liquids then *cuir bouilli* leather was lined on the internal flesh side with pitch, resins or waxes. The exterior of vessels was often coloured black with dye or pigmented wax, or varnished to make an enamelled or Japanned surface (Seeley, 1991). Linseed oil and bitumen coating is known to have been another alternative. Oils, waxes and resins were applied to objects like *cuir bouilli* fire helmets and coach horsemen's boots where water splashes would be a regular occurrence.

Some of these coatings may survive intact and are visible on objects, whereas others may be completely hidden by later conservation dressings and their presence may not be immediately apparent. The conservator must be aware of this when evaluating the possible removal of surface layers in a treatment proposal.

## 10.6 Case study of the conservation of *cuir bouilli* leather

The Museum of London have a comprehensive collection of *cuir bouilli* objects including fire helmets, porters' helmets, postillions' boots and bottles that provide a fantastic reference to the *cuir bouilli* technique. In particular they represent a complete range of the surface treatments applied to this material either as original manufacturing preparations or later conservation treatments which help enormously with the evaluation of treatments. The following case study of the conservation of a *cuir bouilli* fire helmet from this collection will outline a method for evaluating the surface finish of *cuir bouilli*. Visual and analytical methods were employed to investigate a possible technique for identifying and removing layers of leather dressing from original coating materials.

The fire helmet which was thought to be dated around the early eighteenth century was made from a cow or horse hide (Pyne, 1808) which had been hardened using the *cuir boulli* technique. The entire crown and brim of the hat was constructed from four pieces of leather which would have been saturated, heated and then stretched and stitched over a wooden block or former. The seams of the sections were exaggerated to give a deep, cockscomb raised crown. A strip of leather had also been added between each section to reinforce these areas. The stitches would have been completed and trimmed while the leather was still wet. This is apparent because the stitches have shrunk into the leather. These helmets were then lined with a quilted inner skull cap made from sheepskin padded with vegetable fibre or woollen wadding. A small space was allowed above the lining to provide an air gap, which would both insulate the helmet and ensure added protection from a blow to the crown of the helmet.

The helmet had an outer coating which was very black, shiny and evenly applied to the surface of the helmet. The leather had become very friable at the edges and in places the coated leather surface had become completely detached from the leather, lifting the grain layer with it to reveal the flesh layer below. After careful evaluation of the coating several factors indicated that this was not original:

(a) The nature and appearance of the surface was unusually black and shiny when compared with coatings on similar objects in the collection.
(b) The coating covered original physical damage, probably caused by a blow to the helmet, which should have disrupted the surface.
(c) There was no evidence of the shiny black coating near the tight corners of the construction of the helmet, around the underside of the brim or underneath the obvious museum restoration of the shoulder flap.
(d) Also there was evidence of raised lettering underneath the outer coating.

### 10.6.1 Analysis and use of non-invasive xeroradiographic imaging

In addition to these subjective methods of assessment, the use of analytical techniques was considered. The helmet was xeroradiographed and the results not only showed the detail of manufacturing techniques, such as stitching and quilting, but also revealed the presence of bold white lettering on all four sides of the helmet (*Figure 10.4*).

Cross-sections were prepared in the same way as for conventional pigment identification, from small samples taken from the surface of the leather helmet. These were examined in order to assess the nature of the coatings below the surface layer. These showed that the black outer coating, which fluoresced slightly under the UV microscope, was very thin, even, black and synthetic in appearance. This layer covered three further layers of a thick, resinous and waxy material which was pigmented with earth brown and lampblack particles.[8] This indicated that the layer below the black outer coating was a substantial surface treatment and highly likely to be the original surface of the helmet.

### 10.6.2 Removal of inappropriate surface coatings

The cross-sections proved that it was possible to distinguish between the layers of coatings on *cuir bouilli* objects. This information was used to aid the removal of these disfiguring surface coatings to reveal the original layers below. Because of the complex chemical nature of the coatings, cleaning tests were carried out using a variety of different solvents, solvent mixes and solvent gels. The composition of the outer coating demanded that the solvent(s) used in its removal had to have both polar and non-polar characteristics. Chlorinated hydrocarbon solvents, used in the past to remove old dressings and coatings on leather, are no longer an option as their use is proscribed by environmental regulations. After several cleaning tests it was found that benzyl alcohol was an effective alternative solvent, being both efficient and controllable when applied on cotton wool swabs. This solvent contains an aromatic benzene ring, appropriate for the removal of non-polar materials and a single alcohol side group

100  *Conservation of leather and related materials*

**Figure 10.4** Xeroradiograph of the helmet showing construction of helmet and lead white lettering on original surface of the helmet. (Courtesy Barry Knight, Science Conservation Department, English Heritage)

**Figure 10.5** *Cuir bouilli* leather fire helmet after cleaning.

that gives the solvent a sufficient level of polarity to dissolve some polar materials.[9] During cleaning, swabs used to remove the coating were rolled onto an absorbent paper and at the point where the blue/black outer coating changed to brown, the cleaning was stopped. The xeroradiographs were also used as a reference to assist with the cleaning of the lettering.

The final appearance of the surface of the helmet after cleaning was both pleasing and far more convincing as an original coating. The interpretation of the lettering revealed that the helmet had been used by a member of the voluntary fire brigade attached to a parish, in this case St Katherine Cree Church in the City of London (*Figure 10.5*).

A reassessment of the nature of *cuir bouilli* shows that it is necessary to re-evaluate traditional conservation treatments of this material. It is no longer acceptable to continue to apply leather dressings to a material that derives little or no benefits from its application. In most cases it is preferable to remove such coatings. It has been proved possible to cross-section and use UV microscopy to determine original surface finishes from later treatments, which can then be removed using appropriate solvents. This removal of these non-original surface coatings not only prevents the substrate of the leather from deterioration but reveals the true quality of these fascinating objects.

## Endnotes

1. 1848 discovery of a silver rim from a fourth century cup in Benty Grange. Waterer, J.W. (1946) *Leather in Life Art and Industry*. London: Faber and Faber.
2. Discovered in Clonbrin, Co. Longford, Eire, 1908. Now kept in the National Museum of Ireland, Dublin. Waterer, J.W. (1981) *Leather and the Warrior: An Account of the Importance of Leather to the Fighting Man from the Time of the Ancient Greeks to the Second World War*. Northampton: Museum of Leathercraft.
3. Storch, S.O.P. (1989) Military Leather Objects in South Carolina State Museum Collections: Manufacture. Condition. Treatment. In *International Leather and Parchment Symposium*. Offenbach am Mein, May 8–12.
4. The technique of hardening leather by saturating it in cold water and then exposing it to a heat source like a blow torch or oven is still used in the twenty-first century. Beaby, M. and Richardson T. (1997) *Hardened Leather Armour*. Leeds: Royal Armouries Yearbook.
5. Waterer, J.W. (1946) Elric colloquy, eleventh century Guild of Bottellars 1373. Dialogues of the allied group of craft guilds known as Archbishop Elfric's Colloquy dating from the eleventh century.
6. Baker, O. (1946) *Blackjacks and Bottles*. Unpublished document. Cheltenham Spa: W.J. Fieldhouse Esq and J. Burrows Co. Ltd.
7. The 'pitching' of carriage men's boots using an impregnation technique as described in 'Art Du Cordonnier' De Garsault, 1767, was successfully replicated using authentic methods and materials by Conservators and Curators at the Centre of Calceology and Historical Leather Works in Lausanne, Switzerland. De Garsault. (1767) Art Du Cordonnier. Facsimile Reprints. In *Arts du Cuir* (1984). Geneva: Slatkine.
8. The cross-sections were interpreted by Professor Jo Darrah, Senior Microscopist and Conservation Scientist at the Victoria and Albert Museum.
9. COSHH assessments should be carried out before undertaking cleaning work with this substance. Benzyl alcohol is harmful by inhalation and if swallowed. It has a toxicity level of LD50 1230 mg/kg oral, rat. There is no evidence of carcinogenic, mutagenic or teratogenic properties.

## References

Baker, O. (1946) *Blackjacks and Bottles*. Unpublished document. Cheltenham Spa: W.J. Fieldhouse Esq and J. Burrows Co. Ltd.

Beaby, M. and Richardson, T. (1997) *Hardened Leather Armour*. Leeds: Royal Armouries Yearbook.

Bickley, J.C. (1991) Vegetable Tannins. In *Leather, its Composition and Changes with time* (C. Calnan and B. Haines, eds), pp. 18–21. Northampton: Leather Conservation Centre.

De Garsault (1767) Art Du Cordonnier. Facsimile reprints. In *Arts du Cuir* (1984). Geneva: Slatkine.

Landman, A.W. (1991) Lubricants. In *Leather, its Composition and Changes with Time* (C. Calnan and B. Haines, eds), p. 29. Northampton: Leather Conservation Centre.

Pyne, G.B. (1808) *The Costume of Great Britain*, p. 38. London: William Miller.

Reed, R. (1966) *Science for Students of Leather Technology*, pp. 154–258. Oxford: Pergamon Press.

Seeley, N. (1991) Enamelled, Japanned and Patent Carriage Leather. In *The Conservation of Leather in Transport Collections*, p. 23. UKIC Conference Restoration 91. London: United Kingdom Institute for Conservation.

Soest, H.A.B. van, Stambolov, T. and Hallebeek, P.B. (1984) Conservation of Leather. *Studies in Conservation*, **29**(1), 21–31.

Storch, S.O.P. (1989) Military Leather Objects in South Carolina State Museum Collections: Manufacture,

Condition, Treatment. In *ICOM International Leather and Parchment Symposium*, p. 222. Offenbach am Mein, May 8–12.

Vincent, J. (1990) *Structural Biomaterials*, p. 64. Princeton New Jersey: Princeton University Press.

Volken, M. (1995) The Conservation of a Pair of Bucket Top Boots. Gentle-Craft Centre for Calcecology and Historical Leather Works. Lausanne, Swtizerland. *Leather Conservation News*, **11**(1, 2).

Waterer, J.W. (1946) *Leather in Life Art and Industry*. London: Faber and Faber.

Waterer, J.W. (1981) *Leather and the Warrior: An Account of the Importance of Leather to the Fighting Man from the Time of the Ancient Greeks to the Second World War*. Northampton: Museum of Leathercraft.

# 11

# The tools and techniques of leatherworking: correct tools + skills = quality

*Caroline Darke*

A full assessment of any object prior to conservation, whether it be made from leather or any other material, will require an appreciation of how that object was made. A specialist leather conservator will therefore need an understanding of the methods employed by leatherworkers. In addition, conservation and restoration treatments may require repair and replication of damaged and missing areas so knowledge and some experience with the techniques and tools of leatherworking are essential.

## 11.1 Leatherworking tools

This list of tools is a small part of a very large catalogue. For each type of tool there are many variations depending on the actual trade and individual craftsperson. Leathers vary in substance, texture and size, products and patterns vary in size, function also varies. Most tools are designed specifically for the intended job and many have not changed through the ages in their use, and skills are handed down from generation to generation. Leather after all is man's earliest natural material for clothing, shoes and containers. The qualities of leather are unique and when it is made up the product reflects this.

Tools of the trade should ideally be of good quality, as inferior steel tools will result in substandard work. It is important that these are cared for and well maintained. Individuals frequently 'customize' tools according to the task in hand and they will have a wide range to satisfy the needs and requirements of their own highly skilled and specialized trade. It is in the interest of the craftsperson that their tools are kept in good condition as most leatherwork has to be precise in taking measurements, cutting and making up.

### 11.1.1 The awl

This consists of a small wooden handle with an interchangeable pointed blade. There are two types of blade; a diamond-shaped blade used to penetrate the leather at a particular angle and for the preparation of hand stitching, and a round, pointed blade generally employed to mark positional points and to make small holes in the leather.

The diamond-shaped blade is often used in conjunction with 'pricking irons' or a stitch marker.

The stitching awl, as its name suggests, is designed specifically for hand stitching. The interchangeable blade is diamond shaped and will vary in size according to the stitch size and the substance of the leather. The awl is pushed through the leather at an angle following the pattern of the pricking iron (see below) making a long slanting shape. These blades are inserted into a small wooden handle that fits comfortably into the palm of the hand. The handle has indentations around the circumference in which the thread is wound and enable the worker to pull hard enough to tighten the stitches and form a firm line of stitching.

### 11.1.2 The knife

There are so many categories of knife it would be difficult to justify the space to describe them all. Ultimately knives have two uses, to cut the leather and reduce the thickness/substance when a fold or join is required. The name, size and shape of the knife depend upon the length of cut, the substance of the leather and the trade.

(a) *Head knife*. For a long cut the saddler will use a head knife. This half moon-shaped knife has, extraordinarily, survived from very early ages. There are different versions of the head knife, the single, with a quarter circle, hooked blade and the double, with a semi-circular blade, half the size of a round knife. The blades can also be used to pare down the thickness of leather.

(b) *Clicking knife*. This is traditionally employed by shoemakers, with the name derived from the clicking noise the blades make when the 'cutters or clickers' either change their blades or cut around the curves of leather and patterns. There is a choice of two types of blade, a hooked type, with a vertical blade used to cut easily around curves and contours and thicker leathers and a straight blade, mainly used on thinner leathers, paper and linings. This latter is often used in conjunction with a straight edge steel rule for a precise, straight cut. Sometimes a wooden handle with a mechanism enables the blades to be changed and screwed tightly into place for safety. In this case the sharp end of the blades may be inserted inside the handle when not in use, hence protecting the sharp edge. The clicking knife is not entirely confined to the shoe trade, many others use it. If a large production is required, then the patterns will be cut out using steel press knives and a press. This is clearly much quicker than cutting out by hand.

(c) *Paring or skiving knives*. These are made out of high quality steel, as the edge must be razor sharp. (A wide hacksaw blade sharpened at one end with a piece of leather wrapped around the handle is often made as a cheaper but not inferior substitute.) The blade is sharpened differently to the clicking blade either at a right-handed or left-handed angle. The angle varies from 45 degrees to 75 degrees according to the desires of the individual. The edges are sharpened on fine emery for a rough skive. For a smooth edge and skive a whetstone is used. In most cases the bevel is on one side only with the reverse side quite flat. There is a definite art to skiving and once mastered it can be invaluable in all areas of leatherwork and adds quality to the product. The main use is to reduce the substance/thickness of leather when joining two strips (this join is better skived at an angle) or reducing the edge prior to a turnover. It can be time consuming and unless the worker is skilled can be risky as the leather can be cut accidentally in the wrong place.

### 11.1.3 The strop

This is necessary whenever clicking and other knives are used. The blades have to be sharpened frequently for an efficient cut. Most clickers will wear their blades down to very different shapes and angles depending on their own individual techniques. The strop is approximately 35 cm × 5 cm × 1 cm in thickness with a strip of emery cloth on the one side and a strip of buff leather on the other. Alternatively, one surface of the leather is treated with an emery paste of the desired coarseness. The blade is sharpened with the desired bevel on both sides using the emery cloth first followed by the buff side. The coarseness of the emery cloth leaves a slightly serrated edge to the blade which has a saw-like action and will last longer than a razor edge.

### 11.1.4 The bone folder or crease (*Figure 11.1*)

This is an excellent, underrated tool, with endless uses. As the name suggests it is a folding tool made from pieces of bone, hardwood, horn or occasionally ivory. Nowadays it is often made from plastic or nylon but these lack the superior qualities of bone. Bookbinders originally used it for folding very thin leathers and papers. The typical folder used in the bookbinding industry is rounded at both ends. For making leathergoods the bone folder has a smooth pointed end with a wide bevel on one side and slightly flattened on the back. This smooth point prevents any damage to the leathers when either marking positional points on the grain side or when pushing out corners. With such a smooth point there is no fear of penetration through the leather. Likewise when lifting leather when turning it on to reinforcements.

A notch may be found on the side of a folder which is used when an edge is rubbed down to close the fibres for a smooth result.

### 11.1.5 The steel rule

This is usually about 90 cm long and marked with centimetres and millimetres for accurate measurements. It is also used as a straight cutting edge.

### 11.1.6 The dividers (compass)

A divider is an instrument with two adjustable hinged legs both ending in points, thus varying from a compass which has one point and the other designed to take a marker such as a pencil. The most common divider will have one point with a flat side which acts as a guide along the edge of the leather and the other with a rounded point ensuring accurate and repetitive measurements. It also enables positional points such as

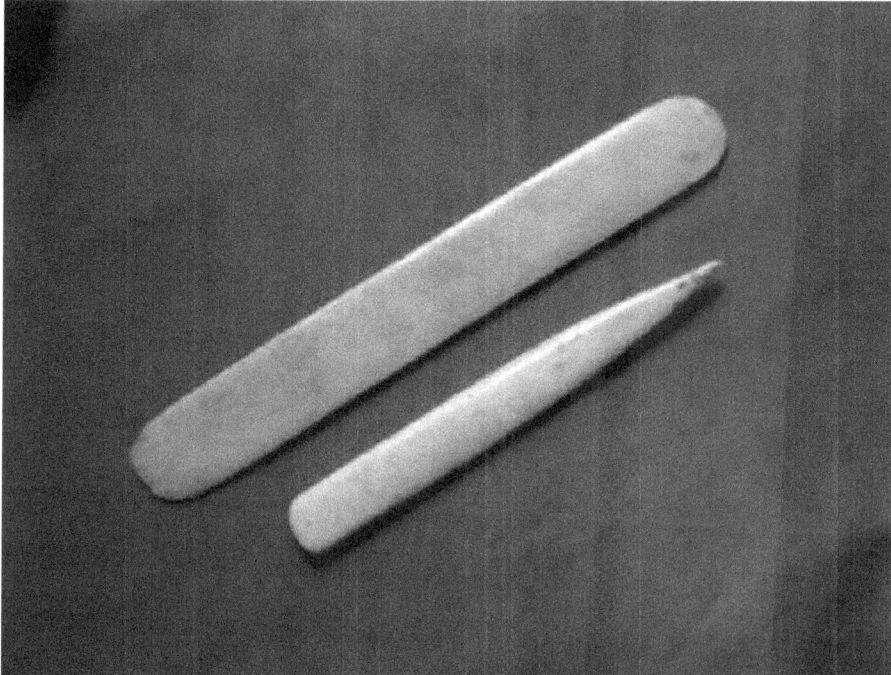

**Figure 11.1** Bone folders.

screw holes, stitching lines, the width of straps, etc., to be marked accurately.

The length of tool varies according to the span required. A bag maker will use a fairly long pair of dividers whereas a shoemaker will require a short span only.

### 11.1.7 The revolving hole punch

This implement is a simple tool owned by many households for the job impossible to do properly with any other tool. It has six differently sized punches, selected by revolving them around to the desired position. With a scissor action it will cut the hole in the material cleanly and efficiently.

### 11.1.8 The hammer

For leatherworking this tool preferably has a slightly domed face. It can be employed to reduce the substance of leather in preparation for stitching, although care must be taken not to bruise the work. Varieties of hammers are used across the trades but they are generally used for a similar purpose. To protect delicate work the head will often be covered with a piece of leather.

### 11.1.9 The race

This specialist tool is used to make a groove or a channel in a U or V shape when pulled over the surface of the flesh side of leather. It is used by a saddler on thick leather in preparation for making a sharp bend in the leather. It has the same effect as a score on paper when a neat fold is required.

The shoemaker will use a channel to prepare for a line of stitching which is to be sunk below the surface of the leather or for decoration.

### 11.1.10 The clam

This is a large, wooden, tweezer-like tool. Held between the knees it clamps on to the work allowing the craftsman to have two hands free for stitching. The holding ends are usually covered with leather to protect the piece being stitched.

### 11.1.11 The edge shave

These come in many sizes according to the thickness of leather, and the size and shape of bevel required. They consist of a slender, curved shank with a groove

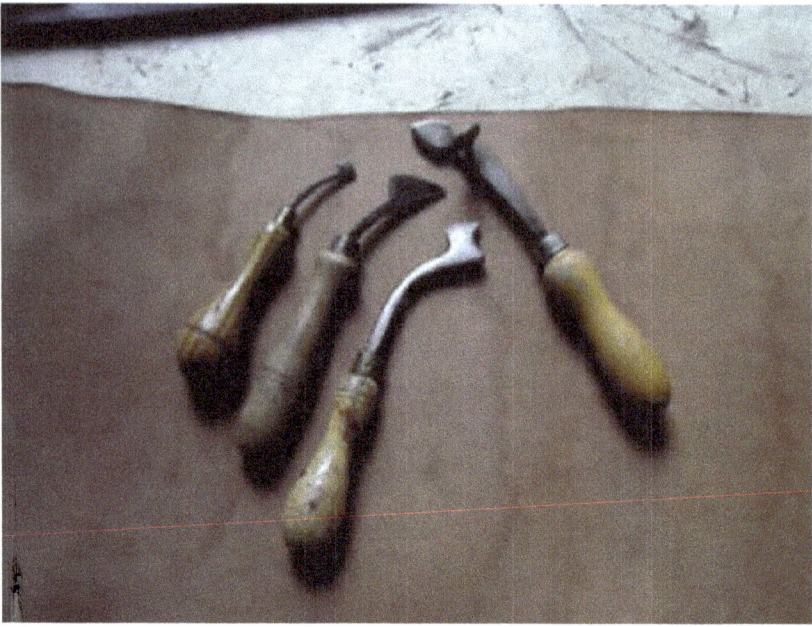

**Figure 11.2** Crease irons.

on the upper side forming a chisel-shaped cutting edge set into a wooden handle.

The hand cut edge tool is particularly vulnerable to wear and damage. If special care is taken to prevent this, the smooth bevel made along the top edge of a piece of work prior to polishing with an edge shave will guarantee a smooth aesthetically pleasing finish.

The harness maker, shoemaker and the belt maker all use this tool.

### 11.1.12 The crease iron (*Figure 11.2*)

This tool is used to burnish an edge or mark a line parallel to the edge of a piece of work. It is used by shoemakers, saddlers, bag makers and others. According to the task in hand, the metal head comes in various types and shapes. The two most common are the single-line creasers without a guide which are suitable for general creasing on turned edge articles and the double line irons which have a fixed or adjustable fence which is used as a guide along the edge of the leather. The screw crease is adjustable allowing the worker to choose the distance of the crease line from the edge. The different-shaped heads can be used for cut-edged work, box work and types of ornamentation.

Creasers can be made of steel, wood or bone. The steel ones have a curved shank set in a wooden handle. They are often heated and used on dampened leather to make a stronger impression. Vegetable-tanned leathers when wet will mark badly from a metal head so it is necessary to use a bone or wooden folder.

Although there are so many types of crease iron they all have the common element of a blunt blade, to ensure that the leather surface is not damaged. Lines produced by crease irons can be for decoration or as a guide or channel for stitching. They are generally used by craftspersons making leather goods from relatively thick leathers, i.e. harness makers, and shoemakers.

Crease irons must not be confused with the edge irons which are a different type of tool used by shoemakers. As with crease irons there are many variations of edge iron. Essentially they do a similar job but are a chunkier looking tool. The heads, usually made in seven widths, are rectangular in cross-section and set in short wooden handles. The longer-handled irons are known as jigger or waist irons. The purpose of the edge irons is to set the edge of the shoe to improve the appearance.

### 11.1.13 The stitch marker (*Figure 11.3*)

As the name suggests this is a tool used to mark a line of stitching. It is a disc with blunt points around

**Figure 11.3** Stitch marker.

the edge mounted into a wooden handle. It is often confused with a dressmaker's tracing wheel that has spiked points. Each point is equidistant allowing the marks on the leather to be at the correct stitch length. A leatherworker will have a range of different stitch markers according to the stitch length required. The rowels do not penetrate the leather. This tool is a marker only, and is followed by the use of an awl.

### 11.1.14  The pricking iron

The 'pricker' is a toothed punch and is employed to achieve uniformity of stitch size. It is available in different stitch lengths. There are two types, one with approximately ten stitch marks and the other with only three. The latter is for corners and curves. The tool is placed carefully on the leather and gently hammered to make an impression which can be pierced later with a stitch awl. It is important that the marker does not penetrate the leather as this weakens the areas between the stitches. The next time the tool is placed on to the leather it is positioned in the last two stitch impressions to give an accurate continuous mark. The impression the stitch marker makes is at a diagonal slant so when the awl is pushed through the leather at this angle it gives the stitch an oblique pattern.

### 11.1.15  The needle

Needles come in various lengths and sizes according to the use, the leather and thickness of the thread. There are two main types, a diamond shape with a sharp end enabling penetration of soft leather, and a rounded shape with a blunt end requiring preformed holes or slits. These are prepared with the aid of pricking irons or stitch markers to mark the leather, and awls to penetrate the leather. The shoemaker traditionally used a bristle in place of a needle, in conjunction with an awl. With awkward corners and curves a bristle is preferable to a needle as it is flexible and will bend. Today nylon bristles are sometimes used as a substitute for the natural product.

Long quilting needles are used for quilting saddle panels, girth straps, etc. and half round needles can be used, mainly for concealed stitching.

Collar needles are made from heavy gauge steel and have diamond-shaped ends. They are used primarily for edge stitching. These needles can take leather thongs as well as the heavy thread that are also for decorative edging.

The panel needle is similar to the collar needle but is bent. It is used for sewing in the panel of a saddle.

For hand sewing glovers use an eyed needle with a round butt and a three-sided point which is used for the penetration of the leather. The three-sided point will cut through the fibres of the leather.

Furriers use a round needle or a three-sided needle for heavier work.

### 11.1.16 Thread

For hand sewing threads are traditionally made from linen and are waxed every time the length of thread is used. The wax keeps the thread intact while stitching and grips the leather.

## 11.2 Adhesives

There are a number of categories of adhesives used by leatherworkers. Using the terms employed in the trades these include: animal glue, fish cement and starch pastes, rubber solution and latex, which are natural adhesives, and neoprene and PVA emulsions which are synthetic. The qualities they possess are tenacity and bond strength; elasticity, the pliability; mobility, the ease with which they can be applied; and compatibility, the power of entering the structure of the leather.

The choice of adhesive therefore will depend entirely on the type of work being undertaken and strength of adhesion required. The need may be for a total bond or a temporary hold prior to stitching.

It must be noted that natural glues and pastes can be susceptible to climatic conditions or insect damage.

(a) *Starch pastes* are prepared by boiling and simmering wheat, rye or other flours with water to form a thick smooth paste. They are easily applied by brush over large areas if correctly cooked. They are not considered very strong adhesives but are very penetrative and in the case of leather they soften the fibres allowing it to be shaped to a new form. They have thus been traditionally used for covering metal purse and handbag frames, as well as moulded articles. When dry they are fairly stiff without being brittle.

(b) *Animal glue* is made from bones, trimmings and other waste parts, boiled gently for a long time and then allowed to cool and solidify. It is supplied in block or a granular powder form. Animal glue is non-penetrative but quite strong. It is softened by warming and applied in a runny condition. Once applied it cools and dries fairly quickly when it becomes hard and brittle. Thus it is suitable only for stiffened work. Nowadays animal glue is rarely used because it is inconvenient to prepare and keep in a good condition.

(c) *Fish glue* or *cement* is made from gelatine extracted from fish skin and waste dissolved in acetic acid with an added preservative. It is not often used by leatherworkers but is extremely tenacious and inelastic. It is used mainly for attaching small metal corners and ornamental parts to leather, although it is not totally effective and small rivets or the like are generally used as an added precaution. It is a very penetrative glue making it difficult to use in the case of thin leathers. It can cause irreparable damage if accidentally brought into contact with the surface. A coating of shellac varnish was often applied to metal parts to assist adhesion and avoid rusting caused by the cement.

(d) *Rubber solution* gives a low strength, temporary adhesion. It is a contact glue and generally holds the work in place prior to machining.

(e) *Latex* is a thin liquid made from the milk of the rubber tree and preserved in ammonia. It is very strong and flexible when dry. Because of its penetration it is inadvisable for use on thinner leathers. Latex is often used as an alternative to rubber solution as it is water based, safer and more convenient to use.

(f) *Neoprene* is a very strong synthetic rubber solution, and if the leather is suitably prepared the bond can be permanent.

(g) *Polyvinyl acetate* (PVA) is a smooth, creamy white, liquid adhesive. It is very strong and flexible when dry. It is a one-way adhesive and allows for movement and adjustment of the components. It is also suitable for large areas.

(h) *Double-sided tape* is now used as an alternative to rubber solution as a temporary holding adhesive. It is quicker and easier to use but expensive, and if the wrong type is used will coat the machine needle with the sticky substance thus causing problems with the stitching.

*Editor's note*: The adhesives described above relate to those used both historically and currently by the various trades involved with the manufacture of leather goods. It is suggested that the reader should also see the section on adhesives used in conservation treatments, Chapter 13, and the section on animal glues and collagen products as described in Chapter 18.

## 11.3 Reinforcements

These are used to strengthen an article partially or all over, or as a support for softer leathers when a firmer look and feel are required. The area of an attachment such as a handle or closure will generally be reinforced and the area where a seam has been skived to reduce the thickness will also require a strong woven material to prevent weakness. There are many types of reinforcement depending on the trade and function of the product. The following list relate mainly to the bag, light-leather goods and similar trades.

(a) *Boards*. These are used for reinforcing heavy leather goods. These include tip, split cane, plywood, fibreboard, millboard, leather board, strawboard and wood pulp boards. These were used prior to the much lighter and stronger plastics becoming available. Many boards, both heavy and light weight, are stronger in one direction than the other and if used in the wrong direction will result in cracking. It is important therefore to discover the 'roll' or 'machine direction' of the board prior to cutting.

(b) *Papers*. Unglazed papers with absorbent features are often used which will make up for the deficiency in strength of the leather. These include kraft brown, stout brown, white cartridge, sugar, blotting and felt papers. The use of highly finished, non-absorbent paper will result in cockling and shrinking and cause a 'rattle' noise to the final product which is an unwanted characteristic.

(c) *Woven fabric materials*. These are used mainly for strengthening purposes. They have greater strength than leather and will not be weakened by stitching. These fabrics may also be used to strengthen a paper or board to increase its serviceability; for instance preventing cracking on boards in areas subject to constant bending as on a strap or handle. Thick leathers are skived at the seams to reduce their substance but as this weakens the leather, strength can be added with a woven fabric. The fabrics are manufactured mainly from cotton, linen and hemp. Most have been dressed or finished which aids in the cutting and reduces fraying. Other woven materials introduced in the mid-1900s were dressed with low melting point adhesives suitable for leathers. An iron-on backer will add body to the leather but care has to be taken not to use leather with a loose structure, as this will tighten the flesh side causing a wrinkled effect on the grain side.

(d) *Non-woven materials*. These are used for the same purpose as the woven fabrics but with the advent of man-made fibres provide additional properties similar to leather but with strength the leather lacks. Non-woven fabrics are also available with an adhesive coating.

(e) *Wadding*. This is usually made from waste cotton. It is available in sheets with a kind of skin on one side allowing easy separation of the layers. This enables the craftsperson to select the thickness of wadding required for any particular job. For example, where protection is required many layers can be used.

(f) *Foam*. This has generally replaced the use of cotton wadding. It will give a soft look and feel supplementing the properties of both leathers and woven materials. It also is used for padding. It is more effective in conjunction with soft leathers and is often used in combination with chipboard for butted bags.

## 11.4 Techniques

### 11.4.1 Skiving

This is a process of reducing the thickness or substance of the leather; this is attained by hand or with a skiving machine or a combination of both. Skiving is usually carried out on the edge of leather. If, however, an overall reduction of thickness is required this will be carried out on a splitting machine.

Skiving around the edges is generally used for the turnover method of construction, which is butted. If the leather is skived too thin there is a risk that this will weaken the leather and cause the machine line to split.

Similarly, skiving for a turned seam must not be too thin as the result will be floppy and mediocre and can cause the seam to tear when the bag is turned the right side out.

Skiving by hand is laborious but where a small or intricate piece of skiving such as a join on a strap is required the result is superior to a machine skive.

### 11.4.2 Preparation

As the name suggests, this is the preparation of the pieces once the components have been cut out and possibly marked. It is an essential part of the making up of a product and can often be overlooked resulting in poor inefficient work.

### 11.4.3 Sewing – stitch formation

(a) *Saddle stitch*. This technique is used on most hand-stitched products where two flat pieces of leather are sewn together. The method will vary slightly with different trades. The shoemaker, having marked the stitch, will attach a blunt needle or bristle to either end of the thread and holding one needle in each hand, with the work held between the knees or using a clam, push the awl through the leather. As the awl is withdrawn one of the needles is passed through, to be followed by the second needle from the other side. The thread is then pulled firmly and tightened to form an even stitch. It is important that the method is repeated in the same sequence for each stitch to ensure neatness and consistency. This form of stitching is considered to be very neat and strong and has been a method adopted by saddlers and other leatherworkers.

(b) *Box stitch*. This style is used on boxes and containers where the stitching is on the corner of the work and the two pieces of leather are at right angles to each other. It is sewn in the same way as the saddle stitch but at an angle. Firm leather is preferable for this particular style of stitching and a mould is often used to hold and support the work instead of a clam.

(c) *Butt stitch*. A butted seam is used on many traditional objects such as shoes, collar boxes, etc. where the edges are butted together and the needle is pushed through the middle of the leather forming a row of stitching either side of the join. The seam normally lies flat as opposed to the box stitch. The term 'split stitch' is often used for this particular join.

(d) *Back stitch*. This term is used where the stitch is longer on the underside, with the same appearance on the topside. The long stitch at the back of the work can strengthen the seam reducing the possibility of cutting through thin leather.

### 11.4.4 Decorative stitching

(a) *Thonging*. This is often used when there is no machinery available. The work is sewn together by means of strips of leather or braid passed through previously made holes or slits. The thong is a substitute for thread and is now available by the metre, but in the past the leatherworker would have cut the length of thong required from a circle of leather. This method enabled long strips to be obtained and eliminated joins. Some leathers have more strength than others due to the tightness of the fibre structure. Leather is probably more practical for making thongs than braid due to its strength. Thonging requires little training although it is a skill passed down from generations and can most typically be seen used by many civilizations such as in Native American clothing, saddles and containers. Used in conjunction with a pricking iron, awl and special thonging needle there are many fancy and complex styles.

(b) *Quilting*. This is the term used to describe the decorative stitching to hold padding in place.

(c) *Rounding* is the process of wrapping leather around a core and stitching it to form a cylindrical strap or handle.

### 11.4.5 Machine stitching

Since the mid-nineteenth century, the advent of the sewing machine has seen the demise of hand stitching in many trades. According to the type of material and its setting various stitching styles can be produced.

(a) *Corner stitch*. This is the machine version of the box stitch where the two edges of a box or case are butted at right angles and stitched through the middle, to form a stitch line either side of the join.

(b) *Chain stitch*. One thread only is used to form a chain on the underside of the work. This type of stitching is often found on work produced using early sewing machines.

(c) *Lock stitch*. This is a two-thread type of stitching where the top needle of the machine feeds the thread through the material where it links with the bottom thread and embeds the stitch in the centre of the material. Both tensions of the top and bottom must be equal to achieve an even looking stitch.

(d) *Top stitching*. This is a final stitch used on the topline of both a bag or shoe where the last row of machining completes the closing. This can often be used to stitch the lining in at the same time and in the case of handbags, the handle and the shoe, the straps as well. It can also be a term where the binding is stitched to an edge.

### 11.4.6 Decorative machine stitching

(a) *Fancy and mock*. These terms are used where the stitching is used for a non-functional effect. There are times when making a bag where functional stitching is not possible and where the components are glued into place. However, the design requires the visual appearance of stitching. This is when fancy or mock stitching is employed.

(b) *Cable*. This is a term used normally by shoemakers and bag makers where the stitching is longer and the thread is thicker than actually required for strength, thus producing a heavier looking line of stitching. This technique is known in the bag trade as a saddle stitch although the stitching is often alternate with a gap between one stitch and the next.

### 11.4.7 Seams and construction

Most seams are common to all leatherworking professions but differ in the way they are named which often derives from the French. There are many reasons for the type of seam chosen. These include the style required by the designer, the material, the skills of the maker and the machinery available.

(a) *Boot*. This is a term one would naturally associate with the shoemaker, it is therefore surprising that it is more often found being used by a handbag maker. If it is too bulky or thick, the joining of a seam will be butted edge to edge and sewn together by means of a strong piece of lining or thin cloth at the back of the work thus producing a very flat seam. Alternatively the two edges are skived, placed face to face, sewn together and then turned back. Because the skiving can weaken the leather a strip of reinforcement is placed over the back and machined either side of the seam. The term silk seam is the shoemakers' name for this technique.

(b) *Bound seam*. This term is used when a piece of leather is used to cover a raw edge, often where a style line or a definition to a seam is required. A butted seam will be bound, either in thin leather or with a contrast colour. Two versions can be found. In the first there is a special attachment on a machine which will stitch a tape or thin strip of leather over the edge of the work. Alternatively French binding, generally found on handmade goods, is a technique where the edge is bound in such a way that the stitching is not visible thus achieving a softer look to the binding.

(c) *Brooklyn seam*. This is yet another version of the silked seam but paper or tape is stuck over the seam.

(d) *Brosser or round seam*. This is where the leather is placed with the wrong sides together and the stitches cover the edge often with fancy stitches. This seam is used by glovers.

(e) *Closed*. This method is not dissimilar to a boot seam but the components are faced right sides together, machined with a small seam allowance, opened and flattened.

(f) *Faced or butted edge*. This is a term used to describe two edges that are sewn together face to face. On a structured bag the edges of the leather are skived and turned on to reinforcement on both sides, placed face to face and machined close to the edge. Otherwise it may be raw and can be machine or more generally hand stitched. This can be of any style of bag and is possibly the most used, although since the 1990s with the introduction of a specialist machine, a lapped raw seam can be seen on many quality bags.

(g) *French or invisible seam*. Here, two pieces of leather are stitched face to face along one edge. A reinforcement is laid parallel to the line of stitching and the seam is taken around the stiffening and glued down on the other side. The material on the unstuck side is able to hinge on the stitch line and opened out. This seam has no visible stitching and is generally used for side gussets or can be used as a styling or decorative seam.

(h) *Glove*. This is a generic term used in shoemaking and originates from one of the many stitches used on a glove. The machine has twin needles on the top but a single thread from the bottom thus making a slight ridge between the two lines of stitching. For a more defined ridge a cord can be run through between the two rows of stitching.

(i) *Inseam*. This term is used when the leather is stitched with the right sides facing and then turned. It is widely used by glovers.

(j) *Jean or roll seam*. This method is named from the similarity to the seams in jeans. It is used for joining one piece on top of another. It is necessary to use this method when a flat join is required, especially when joining curves. Reinforcement is usually needed. It is useful for decorative purposes and panelling. It is a stronger seam than the open seam and probably stronger than the boot seam.

(k) *Lapped/overlay*. Here the component pieces are placed one on top of the other and machined or stitched through either with a single or double seam. The underside is usually skived to reduce the bulk of the seam and the overlay skived when appropriate. A relatively new machine has been introduced which will stitch in this way and has quite dramatically changed the styles available to a handbag designer.

(l) *Moccasin*. This is a fancy but functional seam achieved when the shoe is constructed in a way that one side of the leather has to be gathered on to an apron top. It can either be stitched with an open seam or the top leather from the apron folded over to form a binding.

(m) *Open or flat seam*. As its name suggests this is the reverse of the closed seam. For shoemakers it shows on the outside of the upper whereas the handbag maker will have the open seam showing on the inside. It is a seam used to join soft materials of most types of tannage, vegetable, chrome, and semi-chrome, providing the leather is not liable to crack. Two pieces of leather are placed face to face and machined together, folded back and glued if required. It can then be top stitched and finally turned right side out thus hiding the seam.

(n) *Piped seam*. This is where a separate piping is inserted between the turned seams and will often be included in a style to add strength and a definition to the shape of the bag when particularly soft materials are used. They can also add to the visual appearance by using a contrast colour or material.

(o) *Pique*. This is similar to the lapped seam where one edge is lapped over the other with only one edge showing but making a flat seam. It is used by glovers.

(p) *Prix-seam*. This term is derived from the French, as are most of the glove seams where two edges of the leather are held edge to edge with the wrong sides facing and stitched together leaving a raw edge. This style is generally used on heavier gloves especially gentlemen's gloves.

(q) *Raw/cut edge or bound*. A craftsperson with limited equipment can use raw edge construction although it is now used by most levels of the market. The leather is not skived and the edge of the leather must be of hard or firm substance. A vegetable-tanned leather is ideal. Both edges are stained and polished and can be machined, hand stitched with a glove or harness stitch or thonged by punching the edges with holes and lacing together with a thick thread or narrow strip of leather. An alternative is to bind the edge with a strip of leather or material to neaten it.

(r) *Silked seam*. As a means of strengthening the seam the closed seam has a woven silk tape placed on the wrong side with two rows of stitching either side of the join to hold it in place. A row of zigzag machining can also be used. This is predominately found in boot manufacture. A gent's shoe will have it showing on the right side with leather instead of the tape for added strength. This is similar to a boot seam.

(s) *Sprung seam*. Sewing two different shapes together can give an extra dimension to the product. It is used for both footwear and bag making. It is not always necessary to use the same type of seam.

(t) *Turned*. The term describes the making and turning of the product from the wrong side of the work to the right side. Often used on soft materials of most types of tannage, vegetable, chrome, and semi-chrome, providing it is of a non-cracking type. Placing the wrong sides together using a flat seam to stitch the two pieces together joins the components. The seam can be stuck down with an adhesive if appropriate. Piping can be inserted for extra definition.

(u) *Turned over edge*. This construction is, as it describes, an edge turned over generally on to reinforcement; this can mainly be a board or a fabric with 'iron on' properties depending on the substance of the leather and the look of the product. This method requires a high level of skill, the maker has to skive the edge, prepare and machine. There are materials which do not lend themselves to this style simply because skiving is not always possible, this leads to bulky seams and on the corners an uneven turnover. Plastics and chrome-tanned leathers are typical of this group of materials.

## Bibliography

The following reference works may be useful to the reader requiring further information concerning techniques for leatherworking and craftsmanship.

Attwater, W.A. (1983) *The Technique of Leathercraft*. London: Batsford.
Cherry, R. (1979) *Leathercrafting: Procedures and Projects*. Texas: Tandy Leather Co.
De Recy, G. (1905) *The Decoration of Leather*. London: Constable.
Double, W.C. (1960) *Design and Construction of Handbags*. London: Oxford University Press.
Freidrich, R. (1986) *Cuir – Tradition, Création*. Dessain et Tolra.
Hamilton-Head, I. (1993) *Leatherwork*. Poole: Blandford.
Michael, V. (1993) *The Leatherworking Handbook*. London: Cassell.
Moseley, G.C. (1945) *Leather Goods Manufacture*. London: Pitman.
Salaman, R.A. (1985) *Dictionary of Leather-working Tools*. London: Unwin Hyman.
Waterer, J.W. (1946) *Leather and Craftsmanship*. London: Faber & Faber.

# 12

# General principles of care, storage and display

*Aline Angus, Marion Kite and Theodore Sturge*

## 12.1 Introduction

The overriding general principles, ethics and standards of practice for conservation are summarized in the 'Codes of Ethics' for the care of objects and collections drawn up and ratified by the International Council of Museums Committee for Conservation (ICOM-CC), and national institutions such as United Kingdom Institute for Conservation (UKIC) (now the Institute of Conservation) American Institute for Conservation (AIC) and Australian Institute for the Conservation of Cultural Material (AICCM). Conservation professionals will be aware of these and will have been trained to carry out their work with reference to the philosophies expressed in these standards.

The book, *Caring for the Past, Issues in Conservation for Archaeology and Museums* (Pye, 2001), focuses on the practical concerns and ethical dilemmas of conservators having to treat objects in a variety of circumstances. Some of these are governed by the necessity to make the object fit for continued use rather than static museum display. Conservation processes which change the object or its environment and lead to the long-term survival of the object are discussed in relation to the original purpose or function of the object and also its current situation in a museum or collection. This book is essential reading for all conservators, and also for those considering the issues and treatments discussed in this chapter.

Conservation treatments and principles of care applied in situations where leather objects may still be in use or where they are in private collections, historic houses or other places out of controlled museum environments can often cause dilemmas and conflicts of opinion. Standard museum practices and treatments may simply be unworkable in non-museum situations and compromise solutions may often have to be reached and adopted by the conservator in order to secure the best treatment or environment possible for the object concerned.

The following points will need to be considered and may help the conservator select the most appropriate treatments, evaluate general care requirements and make informed recommendations to the client for the long-term survival of the object within the context in which the object is kept or used.

## 12.2 Objects in use

The use to which the object is put is of prime importance as is the way it will be handled and cared for after treatment. There is no point in carrying out a delicate conservation repair on a piece of upholstery that is going to go into a family home and be used, at best, sat on regularly or, at worst, as a trampoline by small children.

Communication with the owner or curator responsible for the object is vital to find out what they expect from conservation, and to determine what conditions the object is returning to. An owner may have unrealistic expectations of what can be achieved considering the current condition of the object. Museums have very different conservation needs from private individuals who are still using and enjoying their objects. It also has to be established who will be caring for the object. Will it be looked after by professionals in a museum or a historic house? Is it in the care of an owner who will treat it in a way which is appropriate to its state? Will it be handled by untrained staff, who may lift the object inappropriately?

For example, if an object has a handle, but this is fragile, it is very difficult to ensure that untrained staff do not use it. It is easy to provide detailed advice on the future care of an object after treatment, but it is much more difficult to ensure that this is followed. As a result, a more pragmatic approach may be required, and the question 'What would happen if?' should be asked. If an object is likely to break when a very weak handle is used, the resulting damage could be extensive. This may lead to a decision to carry out a stronger but more invasive repair than might otherwise have been chosen. Alternatively it might be better to replace the original handle if it cannot be strengthened, thereby reducing the risk. However, such decisions should not be made without considerable thought and consultation. Such treatments should be fully documented and the original handle should be retained and kept carefully with the object should the situation change in the future and the original handle need to be put back.

## 12.3   Display or storage?

Is the object to be displayed or stored? If it is going to be stored for the foreseeable future, the prime need is for it to be stable and for the components to be secure so that they are not lost. The object needs to be in a fit state for handling, but an aesthetically attractive appearance is not a high priority.

It may be that the needs of objects which are to go on display will be the same as those to be stored, but they may differ in various aspects. Objects on display must be readily intelligible. For example, an object with a picture on it may be perfectly stable but if there is distracting damage to the image, it may be difficult to interpret. Inpainting, but not overpainting, to make an image readable may therefore be appropriate before it is displayed.

## 12.4   Levels of treatment

Good practice and theory require that all conservation treatments should be reversible, but to some degree all are irreversible and the most appropriate and least invasive treatment should be chosen.

Conservation of an object aims to ensure that the object is brought to a state where it is stable and as far as is possible safe from further deterioration. At its most simple, this could mean storing the object in a box in the correct environmental conditions.

The treatment of pests and mould in leather is essential. The stabilization of red-rot may also be appropriate, as may the securing of loose sections of leather or a flaking surface. However, the application of surface finishes or the inpainting of damage may not be considered fully compatible with best conservation practice.

Full restoration is at the other end of the scale. In the case of a chair, for example, this could involve refinishing the wood, a full rebuild of the underlying upholstery and a new leather covering. While this will give a very serviceable object, it will lose much of its feeling and appearance of age, and render it of less value as an historic object.

Usually, treatments fall between these two extremes. Some restoration work may be considered necessary to give the object structural integrity, to assist interpretation, or simply to make an object more visually attractive. Filling a hole in an object such as a leather screen may increase its strength and reduce the chances of further damage. It may also reduce the visual distraction caused by the area of loss, allowing the object to be viewed as a whole. The decision to inpaint the fill may well relate to the nature of the object. It is likely that the owner of a gilt leather screen will regard it as a work of art and will want nothing which will distract from the overall appearance of the object. However, an assumption that this will be required should not be made without consulting the owner or curator, who may prefer a simple background colour.

The context in which an object is to be displayed is significant and may affect the level of work carried out. For example, if a leather bag is being displayed in a museum in a leatherworker's workshop, as part of a social history display, it does not need to appear pristine. However, if the display is of a leathergoods shop, it needs to look plausible as shop stock. Any surface treatments applied to achieve this may not be actively conserving the object, instead they would serve an aesthetic function.

## 12.5   Handling by the public

Repeated handling and direct access by the public may cause excessive damage and this may influence decisions on the level of treatment to be undertaken. For example, boxes, cases or trunks which open and are accessible on open display may prove irresistible to visitors. If a box has leather hinges or straps, or perhaps an overhang around the edge of the lid, casual opening and closing by people passing by can rapidly lead to damage. The object either needs to be moved from open display, or those responsible must realize that in time the leather may be damaged beyond repair with replacement the only remedy. This does not sit comfortably with the concept that museums should preserve objects, as far as possible, for ever. Even if the harm caused in this way can be repaired this will never remove the underlying damage. A tear can be repaired but the object will never be as good as it was before it was torn. Accessibility is desirable but repeated handling should be avoided.

## 12.6 The 'finish'

The final finish or dressing given to an object as part of the conservation process is an issue that requires consideration. A finish or dressing is often applied for aesthetic reasons rather than on conservation grounds. A leather dressing may brighten the object and give it a 'finished' look, but may not actually contribute to its conservation. Finishes may lead to problems in the long run, so care should be taken. Dressings may absorb dirt and may create a tacky surface. If they contain inappropriate ingredients or if too much is applied these may spew onto the surface. Spew is a white surface deposit of free fatty acids which may be confused with mould.

Dressings should never be applied to the front or back of painted leather, be it gilt or otherwise. The oils and fats in them can migrate into the paint and varnish layers, causing irreversible discolouration and softening of the varnish. This may not happen immediately, so testing at the time of application cannot be relied on.

There is a long tradition of applying varnishes based on oils and resins, both natural and synthetic, to leather objects. Many of these are now giving problems because they have discoloured. Their removal from an object can be very difficult or even impossible, particularly from gilt leather. Gilt leather is problematic because the original varnishes applied over the silver to give the golden colour are often soluble in the same solvents needed for the removal of the over-varnish. The long-term consequences of any non-essential finish should be carefully considered.

It is possible to refinish leather completely using modern materials, and there are firms which specialize in this. They very often strip the original surface coatings completely in order to provide a uniform key for the replacement pigmented finish layers. It is very unlikely that this would be appropriate for historic material, but items such as modern sofas may well benefit and be given an extended life. It should be remembered, however, that many rare treasures on display in museums today were once in day-to-day use and only considered ordinary.

## 12.7 Preventive conservation

### 12.7.1 Environment

Leather does not differ greatly in its preventive conservation needs from other organic materials. However, it responds quickly to changes in the environment, so additional care is required.

The relative humidity (RH) of the room where the leather is located will affect it. If it is too damp, above 65%, mould may form and stain the leather. If the room is too dry, below 40%, the leather may lose some of its flexibility and become more prone to cracking and tearing. Changes in humidity also lead to significant changes in area. For example, leather mounted on a screen can become very tight in dry conditions, leading to splitting and pulling away from the edges. This also makes it more vulnerable to knocks. On the other hand, if the atmosphere becomes too damp, the leather on the screen will expand and it may sag. Leather accommodates the slow swing of seasonal RH change, but extremes of RH and sudden changes cause damage. A further consideration is that many leather objects are composite, and may include several differing materials in their construction. These materials may have very different RH requirements. In this instance environmental conditions will need to be maintained which will cause as little damage as possible to all materials included.

Room temperature does not in itself make a great difference. However, it is so closely linked to relative humidity that it has to be considered. The problem can be particularly acute when large rooms, used perhaps for functions, are unheated most of the time and then boosted to a high temperature when the room is being used.

All light is harmful and it can lead to a change in the colour of leather, particularly that which has been dyed, and may also affect decorative finishes such as paint. Ideally, the lux level should be kept to 50 or less but a higher lux level is sometimes suitable for a limited display time. The most damaging source of light for leather is direct sunlight. In addition to the lux level being very high, the leather will absorb heat from the sunlight very rapidly. A gilt leather screen which is exposed to direct sunlight will rapidly become noticeably warm, and loosely fixed leather can become drum tight in less than an hour and then the risk of damage to the object becomes very high. Prolonged exposure to sunlight can lead to vulnerable leathers becoming literally burnt. Spotlights should be avoided as they can cause similar problems and localized hot spots.

Dirt is difficult to remove from the grain crevices and cracks in leather, and is also problematic on the flesh side. It can come from a variety of sources, common ones being open fires, people's hands and general airborne dust. This last can be exacerbated by poorly filtered air conditioning. Air dragged into a museum in the middle of a city is likely to be dirty. If the air conditioning filters are only working at 90% efficiency, the 10% of dirty air that gets through into the building can rapidly build up in the galleries. The more that can be done to protect objects the better, be they on display or in store. It is preferable and cheaper to prevent contamination, rather than to have to undertake frequent cleaning programmes.

### 12.7.2 Pests

Regular inspection and insect pest monitoring is desirable so any infestation may be noticed quickly and early remedial treatment undertaken. Protection from damage by other pests such as vermin, or cats or dogs if the object is in a domestic environment, should also be ensured.

It should be noted that careless handling and clumsy usage of an object by curators, owners and other people can also cause serious damage to a fragile object.

### 12.7.3 Storage and display

Good storage and display can do much to protect objects from damage. The essential requirements are suitable environmental conditions including providing protection from light, from atmospheric pollution and from general dirt and dust. Providing sufficient space and support and thus preventing crush damage is also essential. Boxing an object carefully is often a storage solution that will go a long way to addressing several of the above requirements.

One issue which is sometimes raised in relation to the storage of leather is the use of acid-free tissue and card. Leather by its nature is an acidic material and it is likely that old vegetable-tanned leathers will have a pH in the range of 2.5–4. As a result, the use of acid-free packing materials is likely to be of limited benefit. However, if the object also contains acid-sensitive materials such as paper or textile, it may still be advantageous to use acid-free packing materials, but the benefit is likely to be limited by the acidic nature of the leather itself.

The ideal form of storage provides support to the whole shape of the object. Any interior voids should be padded which will give protection from crushing. However, the object also needs to be accessible for study and to monitor its condition. If the packaging is too complex and time consuming to open and close, damage may be caused during the process of unpacking. In addition, the packaging components may not be properly reinstated after inspection. All storage solutions must be clearly labelled and indexed. A great deal of time can be saved and damage prevented if the desired object can be located quickly and easily. If it is necessary to disassemble an object for storage, a reassembly diagram should be packed in the box with it as well as included in the notes. For smaller items, a sturdy box will usually give the best protection, provided it is not overfilled and the contents are not inappropriately mixed, with a heavy object overlaying a fragile one. As well as giving good physical protection against knocks and dust, a box provides its own microclimate and gives some buffering against changes in temperature and humidity. It is also easy to transport and avoids the necessity of handling the object.

Where boxed storage is not practical, open shelves, racks, cupboards and open storage have to be considered. Shelves in a dustproof cupboard or a roller rack with doors provide the best protection. Roller racking takes up less space and is cheaper in the long run. However, initial cost often leads to open shelving being used instead. Provided that the shelves are not overfilled, they offer reasonable protection from knocks. Unfortunately, they only give limited protection from dust and covering the objects adequately can be difficult. Sheets of tissue paper have a limited life, and cotton or Tyvek dust sheets can be difficult to manoeuvre in a tight space without damaging the objects under them. Items should never be stored directly on the floor if it can possibly be avoided because of the danger of flooding and accidental mechanical damage.

To sum up, the display needs of leather are not significantly different from those of objects made from other organic materials. Where appropriate the object needs to be supported to prevent collapse or damage, cushioned from environmental extremes and be protected from soiling. Appropriate handling while the object is being moved is of course also essential.

## 12.8 Shoes

Shoes are easily crushed and whether they are in storage or on display, they should be lightly padded out to hold their shape. The padding should hold them in their original shape, if unworn, or to the shape of the last wearer, and the ankles and legs of boots should be supported. If the leather is stiff it may be necessary to humidify it first. Acid-free tissue paper is the most easily moulded to the shape of the shoe or boot. Care needs to be taken to avoid overpacking as this can lead to distortion and loss of the individual shape worn into the shoe by the original owner. Shops and warehouses have long known that the most efficient storage is individual boxes for shoes and boots (ankle and knee sized). If bigger boxes holding more than one pair are used, it should be ensured that they are shallow so that the shoes cannot be stacked on top of one another. Shoes often have non-leather components, and if they are on display this needs to be taken into account. For example, textiles may be more sensitive to light than leather, and if these are present particular care should be taken to keep the lighting down to 50 lux. Some shoes are finished with a bright, shiny, patent finish. This can soften and become sticky over time so care should be taken to ensure other components do not rest on these finished surfaces, nor should shoes of this nature be wrapped in tissue paper. Silicone release paper should be used if such dust covers are to be employed.

## 12.9 Gloves

As a general principle gloves may be padded or lightly stuffed using rolls or 'sausages' of acid-free tissue paper. The tissue rolls should never be forced into the fingers but rather slid gently into place. Gloves of some age may have hard creases or fold lines formed at the sides so great care should be taken not to introduce too much support into the gloves causing strain or cracking along these lines. It is inevitable that gloves will have creases formed at the sides; the purpose of the padding is to lessen the damage caused to the leather or fabric. If the gloves are brittle no attempt should be made to introduce padding. This danger may often be reduced by humidifying the leather before padding.

## 12.10 Leather garments

The care needs of leather clothing, whether on display or in store, do not vary dramatically from that of other types of costume. The most significant factor to take into account is weight. An item such as a leather coat may be very heavy. If it is hung on a plain wooden hanger for any length of time, the weight can quickly start to pull the shoulders out of shape. The ends of the hanger will push up into the shoulders, leaving a visible area of distortion which in time will wear through. If hangers are used they should be carefully padded with inert materials and shaped to fit the coat. Buff leather coats and very heavy items of leather clothing should be stored on a customized specialist mannequin to provide adequate support. Alternatively, it may be more appropriate to store a garment flat in a box, provided it does not need to be folded, as this, too, can cause creases and distortion. Any necessary folds should be softened by packing them with rolls of acid-free tissue. Hats and helmets can be stored in boxes, but with soft leathers or headwear with vulnerable parts or trimmings it is preferable to store them on head-shaped or purpose-made supports.

## 12.11 Luggage

Luggage tends to have had a chequered history and may have spent long periods in attics when not in use, in less than ideal environmental conditions, but generally away from the more aggressive and catastrophic elements of deterioration. The principle risk with luggage is that it may crush if it is stacked. Leather suitcases are heavy and if they are stacked, be it for storage or display, they will scratch one another and the corners of the smaller ones will press into the surfaces of the larger ones and distort them. Although it is tempting, it is not a good idea to use historic luggage as storage boxes. Soft leather bags should be padded lightly to keep them in shape. It should be ensured that the handles are not used to move the luggage around, and luggage should also be kept off the floor out of the way of mops and vacuums. Ideally, luggage should be stored on shelves without stacking and with protection from dust and light.

## 12.12 Saddles

Poor storage of saddles can lead to severe damage (fig. 12.1). The nineteenth century racing saddle shown in *Figure 12.1* had not been supported or protected for many decades prior to acquisition by a museum. The damage shown is typical and

118  *Conservation of leather and related materials*

**Figure 12.1** Nineteenth century racing saddle as acquired by the National Horseracing Museum, Newmarket.

demonstrates what can happen when a saddle is stored unsupported. It had rested on the floor with the flaps folded under, and was heavily soiled. Layers of degraded newspaper probably originally intended as a protective wrapping had not improved matters. Saddles are composite objects and in addition to the leather components there is often wood, fabric, padding and metalwork present. The organic materials in the saddle are susceptible to insect attack and mould.

Care needs to be taken, when bringing all new items into a museum collection or conservation studio, that fresh infestations are not introduced. Items such as saddles are often stored, prior to their acquisition, in areas such as barns and outhouses and so are very vulnerable to infestation. *Figure 12.2* shows the same saddle with associated racing boots and whip after conservation.

The requirements for the storage and display of saddles are very similar. They need to be supported underneath so that the flaps and straps are free to hang down without distortion. There are two basic designs of support available on the market at present. One is free standing while the other screws onto the wall like a very large coat hook. Both are shaped roughly to resemble a horse's back and they are usually made of tubular metal. This creates two problems in that individual saddles do not always fit very snugly onto the support and may slip off, and the narrow rods do not spread the load equally. The pressure points may, over a period of time, distort the padding on the underside of the saddle. Therefore the supports should be padded to fit individual saddles. This will help them to sit more securely on the support and lessen the chance of pressure damage. A custom-made saddle support gives the best possible fit, and provides more scope for its design if it is to go on display. The simplest way to make one is to build up layers of Plastazote, an inert closed cell polyethylene foam, or a similar material which can then be carved into shape. If wished, this can then be covered with a fabric such as unbleached calico.

Saddles in store need to be protected from dust. A simple dust sheet can be used, but a made-to-measure loose cover is preferable. These can be made from a fabric such as unbleached calico or Tyvek with Velcro closures.

## 12.13 Harness

Harness is awkward to store. It is designed to go around the curves of the horse's body, and does not

*General principles of care, storage and display* 119

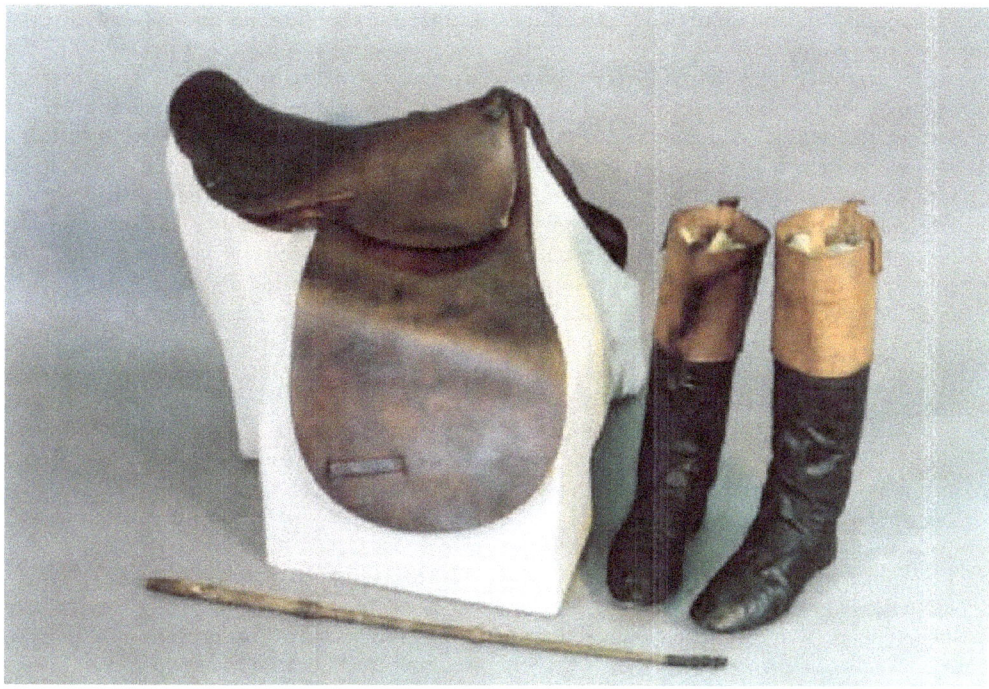

**Figure 12.2** Saddle after conservation, together with associated racing boots and whip.

readily lie flat. If stored on open shelves it tends to become entangled with itself and other items. Probably the least potentially damaging solution for the harness is to keep it in a large, shallow, flat box or tray. However, these can be difficult to move about. The alternative to storing it flat is to hang it up over curved blocks with a radius of at least 8–12 cm to spread the load. Some harnesses are finished with a linseed oil-based patent finish similar to that found on shoes. The same precautions should be taken.

## 12.14 Screens, wall hangings and sedan chairs

The storage of screens can be difficult. There is a strong temptation to lean them up against a wall, but this puts all the weight on one fold and, unless they are protected with boards, they are vulnerable to knocks. Vertical racks are a better solution, as they provide all-round protection, but sliding them in and out of these racks can be awkward. If space is available the best solution is to store them flat on long span shelving. The protection provided is good and it is relatively easy for two people to lift them out and stand them up.

The leather on screens is very vulnerable, particularly that on the two outer folds. If it gets too dry it will become very tight and may split. When tight, it is also more vulnerable to knocks than when it is relaxed. 'Too dry' here refers to the moisture content of the leather, which is governed by the relative humidity in the room, rather than the absence or otherwise of dressings. It is also possible that the tension will distort the frame, leading to joints breaking. Screens may be as high as 3 metres, and this makes them difficult to move around. They are best carried, folded flat, by two or more people. Screens should be handled as little as possible as fingernails can scratch or even go through the leather.

The display and use of screens needs some care. Sources of heat, including lights, radiators, fires and sunlight need to be avoided as these can rapidly dry out the leather and cause excessive tension. Sometimes screens are displayed flat against a wall. They can be held in place with cords or some other device. However, if something goes amiss and the screen starts to fall away from the wall, it can become dangerous and unwieldy, especially if it is very large.

Screens are designed to stand on the floor in a zigzag position, and this is probably the safest for them. Damage to hinges and frames should be repaired to maintain a stable configuration.

Members of the public may present an added hazard. Sometimes their reaction to being told that 'this painted screen is made of leather' is to poke it in disbelief. Wall hangings are also at risk from the temptation to touch and prod the material. It may be advisable to ensure that the leather is displayed out of reach, or to cover accessible areas with glass or acrylic sheet. Care should be taken that this is not in contact with the leather and can be removed for cleaning. Wall hangings and panels are susceptible to damage from moisture. This can be as insidious as a slightly higher RH between the back of the leather and an outside wall leading to secret mould growth, or as disastrous as a leak running down the wall. Loose leather wall hangings are usually stored flat as separate sheets of leather, often in plan chests. They should be interleaved with acid-free tissue paper and clearly labelled. Care should be taken that too many sheets are not piled one on top of another.

Sedan chairs are generally made using patent leathers. In addition to the difficulties associated with this material they present many similar problems to screens and wall hangings. The leather panels are under tension and are very vulnerable to accidental damage. Sedan chairs should preferably be stored and displayed slightly off the floor to protect them from mops and vacuum cleaners, and away from a thoroughfare.

## 12.15 Carriages and cars

Barns and garages do not provide ideal storage conditions, being susceptible to extremes of temperature and humidity through the year, and wide fluctuations throughout each day. The objects are also more accessible to a variety of pests such as mice nesting in upholstery, bird droppings, etc. and the associated damage they cause. The size of carriages means that this type of storage is often the only option, so the leather should be monitored carefully for deterioration and infestation, and protected from falling or windblown debris and leaks. A regular cleaning schedule should be built into the monitoring system so that debris is removed before it can attract further pests. Cars present similar problems, and in addition enclosed cars can set up their own microclimates. The leather hoods on cars and carriages are especially problematic. If they are left up they are under a great deal of tension and tend to split, and if they are folded down they get brittle along the folds and split. It is advisable to decide whether to have the hood up or down, and leave it there as moving it around just makes the situation worse. If it is necessary to fold the hood up and down, or actually to use the vehicle, it may be better to think of replication with the original stored appropriately.

## 12.16 Conclusion

While it is accepted that preventing damage is more desirable and cheaper than expensive conservation, objects will inevitably suffer some damage caused by use and wear. It is better for the long-term survival of the object to encourage careful and limited use and to educate the owner how to limit the damage while still continuing to enjoy the object, fulfilling its intended function.

The ideal environment will usually be 50–55% RH, 50 lux or less, 18°C, combined with protection from dust, handling and pests. Although this may be unachievable in domestic, non-museum or alas many museum contexts, a clean and stable environment where the object may be regularly inspected and an owner or curator who understands the ideal to aim for and what to do when something is going wrong, will go a long way to prolong the survival of the object.

## Reference

Pye, E. (2001) *Caring for the Past, Issues in Conservation for Archaeology and Museums.* London: James and James.

# 13

# Materials and techniques: past and present

*Marion Kite, Roy Thomson and Aline Angus*

## 13.1 Past conservation treatments

### 13.1.1 Introduction

When examining an object it is frequently detectable that previous conservation treatments have been carried out. Sometimes these treatments may have been undertaken at a time before precise records were kept detailing which chemicals, adhesives and methodology were used. Identifying which chemicals and methods have previously been used is essential before further conservation is carried out but defining past treatments on leather can sometimes be problematic. Included here are the results of surveys that identify many of the treatments, chemicals and adhesives which have been used at various times in the conservation of objects made from leather and skins. The results of these surveys provide an overview of procedures that may have been carried out to a leather object in the past and a starting point for the conservator who has to deal with an object that has obviously been treated but for which there are no existing records.

Some of the earliest treatments for leather were devised from those developed for the care and preservation of bookbindings and include the application of fungicides and insecticides and the use of buffer salts to impart resistance to acidic atmospheric attack (Plenderleith and Werner, 1971). Many of these treatments, such as the use of Lindane, pentachlorophenol, p-nitrophenol and orthophenyl phenol would now be considered unacceptable for health and safety reasons.

The first comprehensive work devoted solely to the care of leather was *A Guide to the Conservation and Restoration of Objects Made Wholly or in Part of Leather* by John Waterer (1972) and for a generation this became the standard reference work.

### 13.1.2 1982 Jamieson survey

In 1981, immediately after the formation of the Leather Conservation Centre Trust, a worldwide survey was carried out by F. Jamieson, a recently retired leather chemist with an interest in historical matters, to look at the current methods being used for the conservation of leather objects, to identify areas for future research and to determine whether there was a requirement for a specialist institution where research training and practical work could be undertaken. Sixty-nine museums in the UK, USA, Canada, Spain, Australia, the Netherlands and Germany took part in this survey and a detailed report was presented to the trustees. The major results of this report were summarized by Jackman (1982). The materials used at the time are listed below.

*Bavon ASAK APB.* Leather lubricating compound. Alkenyl succinic acid derivative. Soluble in white spirit and Genklene (1,1,1-trichloroethane).

*British Museum Leather Dressing.* Standard recipe:

20 g anhydrous lanolin.
15 g beeswax.
30 g cedarwood oil.
330 ml hexane.

*Connolly's Leather Food.* A dressing for hide upholstery, leather goods and clothing. Preparation containing lanolin, beeswax, water, morpholine and white spirit.

*Disinfectant 1473.* Used as a fungicide. Constitution not known.

*Dowicide A.* Water soluble fungicide. Sodium salt of orth-phenyl phenol.

*Facteka A.* Granular cleaner in fine and coarse grades suitable for cleaning suedes or leather with abraded surface. Rubber-like material made from vulcanized rape oil.

*Genklene.* Non-flammable solvent. 1,1,1-trichloroethane.

*Hydrophane.* Liquid leather dressing based on linseed oil.

*Invasol S.* Water soluble leather lubricant. Synthetic anionic oil.

*Lipoderm Liquor SA*

*Lipoderm Liquor LPK.* Water soluble leather lubricant. Synthetic anionic oil free of natural fat.

*Lissapol N.* Non-ionic detergent.

*Neutralfat SSS.* Water soluble leather lubricant. Stabilized olein soap which on drying loses its emulsifying property so that it no longer promotes absorption of water.

*Plexisol.* Consolidant for leather affected by red rot. Polyacrylate resin preparation containing 25% solids. Must be diluted with Genklene before use.

*Pliancreme.* British Museum Leather Dressing in the form of a cream, emulsified with water, containing a fungicide.

*Pliantex.* Consolidant for leather affected with red rot. Polyacrylate resin preparation containing 25% solids. Must be diluted with Genklene before use.

*Pliantine.* Traditional British Museum Leather Dressing.

*Polydiol 400 and 1000.* Impregnating material. Polyethylene glycol waxes.

*Preventol L.* Fungicide. Sodium salt of chlorinated phenol.

*Santobrite.* Fungicide for leather. Derivative of pentachlorphenol.

*Vulpex.* Material for cleaning leather articles. Potassium oleate soap. Soluble in water or white spirit.

### 13.1.3 1995 survey

In 1995 members of the International Council of Museums Committee for Conservation (ICOM-CC) Leather and Related Materials Working Group participated in a survey to find out and record the treatments which had been used on leather, prior to this date, by members of the conservation profession. The results were published as part of the post-prints for the interim meeting of this working group (Kite, 1995). The methods of application of these past treatments were also recorded if known. The aim was to assist conservators when faced with problems which may have been caused by these treatments. The possibility of problems with past treatments had already become a reality and was demonstrated on collections of shoes which were frequently found after prolonged storage to have spews and sticky, oily deposits on them. These problems were also found on many other leather objects which were known to have been treated and conserved within the previous 20 years or so.

The results of this survey are reproduced here, with comments as to efficacy of the treatment which were made at the time. This serves as a record for the current reader and as background before further discussing current methods of conservation.

*Aluminium alkoxide*, 1% in mineral spirits: used as a chemical stabilizing retanning agent for red-rotted vegetable-tanned leather, having no consolidative effect. In addition to having a buffering effect the aluminium ions crosslink with the collagen and vegetable tannins increasing the shrinkage temperature.

*Bavon* leather lubricants based on alkylsuccinic acid.

*Bavon ASAK ABP*: leather lubricant applied from a solvent solution. Provides good flexibility at low levels of use. Based on synthetic paraffin long chain polymers with non-ionic water in oil emulsifying agents. Soluble in white spirit, petroleum spirit or 1,1,1-trichloroethane, used in 2–25% solutions (10% is normal).

Painted on leather with 15 minute intervals between coats (ten coats normally sufficient to allow manipulation). Very brittle leather may be immersed and soaked.

*Bavon ASAK 520S*: highly polar leather lubricant based on an alkylsuccunic acid modified to give complete water solubility. Used on intestines, bladders and other fine membranes. Very effective on drumskins and can be used on objects particularly when reshaping; water content being used to advantage. 5–20% solutions. Often useful to start at 5% working up to 20%. Can produce a spew over time, which can be removed with white spirit.

*Bedacryl 122X*: polymethacrylate ester: a consolidant for wood and some types of leather. Supplied in a mixture of xylene and n-butanol, or xylene and cellosolve acetate, or a petroleum solvent. Cellosolve = 2 ethoxy ethyl acetate.

*Beva 371*: a heat seal adhesive dissolved in a petroleum fraction. A blend of ethylene-vinyl acetate copolymers, ketone resin and paraffin.

*DDT (dichloro-diphenyl-trichloroethane)*: now banned but older collections may have been treated with it as an insecticide. Used on skins, leather and wool.

*p-dichlorobenzene (PDB, 1,4-dichlorobenzene, one of three isomers)*: mothballs.

*Draftclean*: ground rubber.

*Ethylene glycol*: solvent, substitute for glycerol in conjunction with olein soap for softening ethnographic, semi-tanned leather.

*French chalk*: used to clean feathers, fur and chamois leather.

*Fuller's earth*: used as a powder cleaner for feathers, fur and chamois leather, often mixed with magnesium carbonate. Sprinkle over feather, leave overnight and then brush off. Also used mixed to a stiff dryish paste and brushed onto surface, left to dry then brushed or vacuumed off. Only suitable for strong good condition furs.

*Invasol S*: used as a leather lubricant. Miscible in water (up to 20%); apply in several coats with swabs. Softens leather well but it has been recorded that light-coloured leathers darken.

*Isopropanol or isopropyl alcohol (2-propanol)*: solvent used to soften and swell leather: employed in aqueous mixtures to soften hard and brittle parchment. Properties intermediate between ethanol and acetone.

*Lanoline, anhydrous*: used as a leather lubricant.

*Magnesium carbonate*: an absorbtive cleaner, particularly for feathers, fur, chamois leather.

*Opodeldoc*: soap liniment of the following composition:

Camphor 40 g
Oleic acid 40 g
Alcohol (90%) 7000 ml
Potassium hydroxide solution 140 ml
Rosemary oil 15 ml
Purified water to 1000 ml.

*PEG 400, polyethylene glycol 400 (25–35% in tap water)*: impregnation period between one and five weeks. Used for treatment of waterlogged archaeological leather.

*Pliantine (British Museum Leather Dressing, standard and special G)*: thick brownish liquid: lanolin, cedarwood oil, beeswax, 1,1,1-trichloroethane. Special G omits the beeswax. Used as a dressing for hard, brittle leather.

*Pliantex*: flexible polyacrylic resin based on ethyl acrylate. Supplied as a 30% solution in ethyl acetate. Used for the consolidation of fragile leather, particularly where 'red rot' is present. The polymer is stable in light and is flexible. It is not swelled by water and is non-tacky. The material will not harden because no C=C double bonds remain in the polymer molecule. Because of this, subsequent polymerization of the dried film cannot occur. Ageing does not produce crosslinking with its lack of solubility. It is diluted 1:4 for use. Diluents are esters, ketones and aromatic hydrocarbons. Produces a very soft flexible film after the evaporation of the solvent (Waterer, 1972).

*Renaissance wax*: microcrystalline wax used for cleaning and sealing leather, wood, ivory and metals. High shine can be achieved, useful for patent leather.

*Saddle soap (Proparts)*: commercial leather cleaner. Alkaline pH 9–10. Based on neatsfoot, cod or sperm oil, emulsified with soap in water to produce an emulsion fatliquor. Considered obsolete by 1995 as extremely alkaline when pH 4–6 is most favourable for leather. Spew formation can occur and in time stiffen the skin.

*Silicone leather wax*: used as a leather cleaner and lubricant.

*Tannic acid, gallotannin, gallotannic acid*: sometimes used as a retanning agent to treat archaeological leather. Dissolved in alcohol or acetone.

*Thymol*: fungicide for leather, furs, paper, parchment.

*1,1,1-trichloroethane*: solvent, component of most leather dressings.

*Vulpex (potassium methyl-cyclohexyl-oleate)*: soap; for leather, featherwork, etc. where use of water is impractical. pH 10.5–11.5; soluble in white spirit, trichloroethane or water.

*White spirit BS245 (Stoddard solvent)*: mixture mainly of alkanes of boiling range 150–200°C. Miscible with acetone. Used as a solvent, for dry cleaning and leather treatment.

### 13.1.4   2000 list

In 2000 Sturge presented a list of materials with a summary of their properties and methods of application. These had all been used in the conservation of a number of leather objects featured in a collection of case studies (Sturge, 2000). All the objects had been conserved within five years of the publication. This list covers a wide range of materials used at the Leather Conservation Centre at this period but was not intended to be a comprehensive survey. However, it gives a wide ranging overview.

### 13.1.5   2003 Canadian Conservation Institute (CCI) survey

In 2003 the results of further research were presented by Jane Down at the Adhesive and Leather Symposium prepared for the Library of Congress Preservation Directorate in partnership with the Folger Shakespere Library (Down, 2003). From this the following treatments may be added to the above list as having been used in the conservation of leather and skins.

*Glue*. Animal glue has been of importance to the bookbinder for many years. Glue-based recipes contain plasticizers, preservatives, perfume, thickeners and in some cases extenders. An early recipe for bookbinders' glue has the following ingredients: gelatin, water, yellow rosin, methylated spirit and glycerine.

*Gelatine*. This has been used to repair vellum. In order to increase the flexibility of the adhesive, two drops of a 2% aqueous sorbitol solution are added to each 100 ml of a 9% edible gelatine solution. The mixture is warmed in a water jacket for use, because if gelatin is boiled, it will darken and lose adhering power.

*Rubber cement*. Reported by Julia Fenn (1984) that the Royal Ontario Museum has several examples of skin and leather artefacts with hardened black stains – the results of repairs using rubber cement. Its advantages were that the adhesive was flexible, easy to apply and had good adhesion. The use of this material is no longer recommended.

*Soluble nylon* (N-meythoxymethyl nylon). This was first recommended in 1958 by Werner as a consolidant for objects with powdery surfaces that require soaking to remove soluble salts. This material is no longer recommended.

## 13.2   Notes on treatments in use in 2004 – additional information

### 13.2.1   Introduction

Much of the information in this section is based on the work published by Sturge (2000).

It is accepted that the first priority before any treatment is undertaken is to assess the condition and identify the tannage of the leather. Second, to determine if the object has been treated on a previous occasion and identify what has been used in this treatment if possible. The condition of the surface of the leather and the ongoing effects of any previous treatments, especially if oiling has been used, will have an effect on the current method chosen. If the grain surface is lifting and scuffed, it can snag the cotton wool of the swabs used in cleaning with solvents. If the leather is very deteriorated, it may be possible to consolidate first and clean once the surface is a little more sturdy. It is important to ensure that a consolidant has been chosen that can be cleaned through, and that the cleaning method is compatible with the consolidant. It may be necessary to carry out tests on a small area before proceeding generally with the treatment. Comments on methodology of use have been included in the following section as an aid for the conservator and to provide a record.

### 13.2.2   Dry cleaning

*A good brush vacuuming* can be very effective and may be all that is required.

*Smoke sponges* (dry cleaning sponges of vulcanized natural rubber) are very useful for surface cleaning and the dirty surface of the sponge may be shaved or snipped off as work progresses.

*Wishab erasers* (sulphur and chlorine free) have proved useful. They are available in three hardnesses.

*Groomstick* (molecular trap, made from natural rubber, neutral pH). This must be used with caution and the surface of the leather must be in good sound condition and firm. Groomstick is tacky and its use can result in damage and the lifting of any loose fragments from a friable leather surface.

*Glass bristle brushes*. These are used on very rare occasions and should be used with extreme caution as they are abrasive to the leather surface. In certain circumstances they may be the only treatment that works to remove stubborn surface deposits and soils. They are used particularly for thick suedes and should not be used on fine leathers or fine suede surfaces. Glass bristle brushes should be employed with extreme care and the correct personal protection worn. Gloves and eye protection must be worn as well as a mask. A thorough cleaning of the object and the work area after use is essential.

### 13.2.3 Wet cleaning and solvent cleaning

*Aqueous cleaning.* Cotton wool swabs moistened with deionized water have been used to remove surface soiling but care must be taken not to wet the leather if water is being used in this way. The aim is to soften and remove the dirt, not to soak the object. Overwetting the leather may cause all manner of problems including distortion, discolouration, hardening, movement of salts and tannins, tide marks, etc.

*Water and ammonia.* Water with the addition of a very small amount of ammonia has also been used. The ammonia is more aggressive to the dirt than water alone but is harmless to the leather in low concentrations. When used extremely diluted, to almost a homeopathic level, good results may be obtained. Tests should be carried out to determine the appropriate strength which should be as low as possible while still effective. The ammonia should be added to a measured amount of water a drop at a time. All water-based treatments should be used with caution, but water with a little ammonia may also in some circumstances be used as a cleaner for varnished and painted leathers. However, ammonia in too high a concentration can adversely harm some painted surfaces.

*Water and detergent* with a little non-ionic detergent or wetting agent have been used. The detergent helps the water to wet, soften and break down non-polar soiling more easily. It is essential to remove the residual detergent with a swab moistened with deionized water.

*Water and alcohol.* A 50/50 mix of water and alcohol is favoured by ethnographic conservators in some instances where this has been found to be more effective than water or alcohol alone.

*Polar organic solvents.* The use of materials such as ethanol, isopropyl alcohol, acetone and other polar organic solvents is sometimes necessary to remove soils. Such materials must be employed with care as these products can dissolve oils and tannins in the leather and harm finishes.

*White spirit (Stoddard's solvent)* being non-polar and slow evaporating is the mildest organic solvent for use with leather. *Mineral spirit* has a similar solubility profile, and does not smell as much. It contains less aromatic hydrocarbons and is therefore sometimes less effective. White spirit is generally the starting point for solvent tests, and is used for the majority of solvent cleaning of leather. Small objects which have been overdressed may be immersed in a bath of white spirit (or mineral spirits) to try to remove the excess oil. This can be necessary as an oily surface will resist all attempts at consolidation and repair. Excessive dressings also attract dust, can be unsightly and may cause damage to the object by accelerating the degradation of sewing threads used in the construction of the object.

*White spirit and water.* It has been found that a 50/50 mix of water and white spirit, with a small amount of detergent to form an emulsion, is an effective cleaner for gilt leather and other leathers with painted or varnished surfaces. The mixture may be applied with swabs or soft cloths, and then carefully rinsed off with a little distilled water, ensuring that the leather at no time becomes wet.

*Toluene* is an aromatic non-polar solvent which is effective for cleaning leather. It has particular health and safety implications and appropriate safeguards should be taken, as is the case with many organic solvents.

### 13.2.4 Proprietary leather cleaners

Proprietary leather cleaners must be used with caution as it is sometimes difficult to confirm the ingredients.

*Leather Groom* (oleaic acid soap with small quantities of neatsfoot oil and silicone oil). This is an effective mild cleaner for smooth leather surfaces. It is used as a mousse so is difficult to overapply. It was originally developed to clean and condition motorcycle leathers. In some instances it may disturb or remove surface finishes so testing of the product before use is essential.

*Microcrystalline wax* has some light cleaning properties attributed to the white spirit in its formulation.

### 13.2.5 Humidification

Leather becomes more malleable when humidified so raising the humidity renders it more receptive to reshaping. As it dries after reshaping the leather may set in its new shape naturally or it may need to be weighted or otherwise restrained and prevented from returning to the previous deformed state. A way must be devised to hold the leather in place as it is dried out, often with the aid of dry blotting paper. If using non-permeable weights, like glass, it is necessary to keep changing the blotting paper layers or the moisture will be trapped and unable to evaporate.

The aim of the process is to raise the general humidity without soaking the leather. While leather reacts well, in general, to a rise in humidity caused by the presence of water vapour, it will react very badly to getting wet through the action of liquid water. Degraded leather will have a lower shrinkage temperature than new leather so this should be borne in mind. Heat treatments should never be used at high humidity as leather is likely to shrink if heat is applied when it is wet.

Large commercially supplied humidification chambers built to a high specification, which can provide precisely controlled levels of humidity and temperature, are excellent but may not be available to all conservators. Good results can be obtained using a small homemade humidity chamber, constructed using racking, retort stands, polyethylene, Perspex or other materials available in most conservation studios. A relatively inexpensive ultrasonic humidifier may be used to provide the humidity source. For smaller objects, a plastic box with melinex or polyethylene film over the top and the ultrasonic humidifier feeding in may be adapted. Another solution would be to introduce a bowl of wet cotton wool into a simply made polyethylene chamber. In other circumstances areas of an object may be locally humidified using an ultrasonic humidifier. Flat objects can be humidified using a semi-permeable membrane such as Goretex or Sympatex, or using damp blotting paper with a separating layer of soft cotton fabric or Reemay. If damp blotting paper is used care must be taken to ensure that the leather is only subjected to the humidity and that it is not in direct contact with the blotting paper, nor that the blotting paper is too wet resulting in the separating layer or cloth becoming soaked through.

There is always a danger of mould growth being set off by a rise in humidity. Awareness of this and regular inspection of the object during the humidity treatment is essential. Humidity treatments should be carried out for as short a period as possible. If necessary, a series of treatments may be undertaken involving gradual reshaping carried out in stages.

Humidification should not be hurried nor taken to excess resulting in wet leather. The leather may want to return to an even earlier state than that which it is intended to achieve.

One should always be aware that the shrinkage temperature of badly degraded and old leather can be extremely low. It should not even be left near a window while it is humidified where it is in danger of being warmed by the sun.

## 13.3 Repair materials

A wide range of materials have been used for repair patches, strip lining or full supports.

*New leather* of similar tannage to that used on an object may have compatible properties, but generally will have a completely different appearance. It is often impossible to match one or two hundred years of wear with a new piece of leather. Modern tanning methods produce a very different looking product, and sourcing suitable compatible leather is difficult. However, a compatible leather does give a good robust repair that will react to changing conditions in a sympathetic way.

*Polyester sailcloth* has been used to support leather in the past, for instance in the relining of screens. It is very heavy and does not move at all with the leather, for instance as humidity changes take place.

*Spun-bonded polyester fabrics*, e.g. Reemay, Ceerex, Vilene. This is a modern alternative and has frequently been used in recent times. It is available in a variety of weights and because it is spun bonded has no nap or direction of weave. These materials are light in weight and easy to use, with just enough 'give' to move with the leather rather than imposing its own priorities on the leather. These materials have been used for a complete reline or for small repairs.

*Japanese tissue paper* is ideal where a less robust repair is suitable. A good thick handmade paper is easy to use, and has a pleasant 'organic' appearance which often resembles old partially degraded leather. It is easy to colour and will stick with most adhesives. It is useful where a sacrificial repair is required as it is generally weaker than the leather. If a thin repair is needed, it is in many cases stronger than leather pared down to a similar thickness.

*A woven textile* can be used. There is a danger though that over time the weave will evidence through the leather or impose its own tensions on the object.

## 13.4 Adhesives

There is a range of reversible adhesives which may be used depending on the circumstances, e.g. the condition of the leather, working conditions and the conditions which the object will be returning to.

*Beva 371* is supplied as a paste or as a pre-cast film. It is a mixture of ethylene vinyl acetate resin, polycyclohexanone and paraffin wax. It was developed for use in the conservation of paintings. The paste is dissolved in toluene, so needs to be handled with care and fume extraction. Beva is frequently used in the dry form using heat to set it which makes a good strong bond. It can be bonded to the backing fabric first which is then cut to size and shape. When bonding leather to leather it can be difficult to get sufficient heat in between the layers to melt the Beva, but it is good if the repair material is thin.

*Poly vinyl acetate and ethylene vinyl acetate emulsions.* Any general-purpose PVA or EVA used in the conservation world would most likely be suitable although there are specific ones supplied for leather conservation. PVA M155 is a general-purpose dispersion with good tackiness when wet. A water reversible PVA, M218, is very similar to M155, but is considerably easier to clean up after use. Vinamul 3254 is an EVA copolymer dispersion, with good flexibility and water resistance. It is not as easy to reverse as other adhesives, but in many cases can be peeled away carefully if necessary.

*Acrylics.* Lascaux 498 HV and 360 HV (butyl methacrylate copolymers, thickened with acrylic butyl-ester). These are supplied as a creamy emulsion, and can be solvent reactivated if necessary. These do not have the tackiness when wet as the above-mentioned PVA adhesives which makes repositioning easier. Acrylics have been found to make a good bond for flat objects. The 498 grade dries fairly hard, while the 360 is softer and slightly tacky when dry. The two grades are used as a mix to get the flexibility of the 360 with the strength of the 498. These acrylics are easy to remove, either with solvent or by peeling, as the adhesive does not readily penetrate the leather. The Lascaux acrylics are water resistant when dry, but can be solvent reactivated with acetone or toluene. The solvent reactivation is useful to form a fast stick without involving further moisture. Water can lead to problems with the expansion of the leather. Generally this is not a problem, and the moisture content of emulsion adhesives can be beneficial in helping to relax the leather, but expansion caused by water can throw joins and splits out of alignment. The back of the leather and the repair fabric are coated with the resin mix and left to dry thoroughly before solvent reactivating. Various solvents can be used, depending on what is needed and what kind of extraction is available for use. Acetone gives a quick, but not sticky bond, although it should be used with caution. Toluene gives more tack and a longer working time.

*Paraloid B72* is rarely used as an adhesive for leather as it lacks the required degree of flexibility. It has, however, been used as a consolidant.

*Wheat starch paste.* This may be bought ready made, although it is preferable to make it as required. It will not bond well with synthetic fabrics. As an adhesive it has very little bulk but a good contact between the two surfaces being stuck is essential. Its high water content can cause problems, but it has good reversibility and long-term stability.

*Sodium carboxy methyl cellulose (CMC)* is supplied as a powder and is prepared as a 1 to 5% dispersion in distilled water. This material is useful when facing objects such as gilt leather screens to enable repairs to be carried out from the reverse side. The moisture content of any adhesive must, however, be borne in mind when treating degraded leather. CMC is also used as a thickening agent for dispersion adhesives like Vinamul 3254. It increases the viscosity and working time by retarding the drying point.

*Animal glues.* These can be difficult to use for the repair of leather as generally they have to be applied hot, and most have a high moisture content. The shrinkage temperature of the leather concerned is an important factor if animal glues are to be considered as an adhesive option. Leather that has been stuck with animal glue in the past is often found to have become embrittled, and the bond has failed.

## 13.5 Surface infilling materials and replacement techniques

*Leather* is often chosen as it can give a pleasing finish but is frequently hard to match the original texture. Modern tanned leathers have a very different handle and appearance to historic leathers particularly if they are degraded and it can be difficult to source compatibly tanned skins.

*Japanese tissue* is often used as an alternative to leather where a thin fill is required. It can be coloured to tone.

*Beva 371* can be used as an infilling material. It bonds well, is plastic and flexible and moves with the leather. It can be painted, moulded and matted to blend with the leather surface. Beva may be prepared to be used as a hot melt filler as described by Sturge (2000).

## 13.6 Moulding and casting materials and techniques

Conservators will each have their own preferred moulding materials subject to availability and individual experience. One product is Steramould, a silicone rubber moulding material developed for hearing aids. It is considerably less expensive than the dental materials often recommended in the literature. It can leave an oily mark on unfinished leathers so tests should be carried out before use.

Fills can be made from solid Beva 371 to match original printing or tooling of the leather. The original is first moulded with Steramould with a piece of glass pressed down onto the silicone to give it a flat surface for working on. When the silicone rubber has hardened and been removed, solid Beva can be melted into the mould under silicone release paper. The moulded form can then be cut to fit and adhered into the gap in the object. The Beva fill can be inpainted to tone with the original.

## 13.7 Consolidation techniques

Consolidation is an option for leather which is badly degraded. Sometimes consolidation will need to be carried out to give a firm surface for repairs to attach to or for cleaning to be undertaken. It may be that it is not the leather itself which requires consolidation but the painted surface decoration. A variety of consolidants are available to the conservator.

*Paraloid B67 in white spirit*. This is occasionally used as a consolidant for painted surfaces on leather. It is generally only used as a final coating to give protection to the surface or to reform a disrupted varnish.

*Klucel G* can be dispersed in ethanol. It is sold as Cellugel ready dissolved in isopropyl alcohol. It has been used for the consolidation of friable and flaky surfaces on leather. It holds a degraded surface together but does not penetrate to any extent into the leather structure.

*Pliantex (flexible ethyl acrylate resin)*. Used until 2004 when it ceased to be commercially available.

*SC6000* (an acrylic wax blend) is used primarily as a surface coating material. It does, however, have a consolidative effect on the surface. Blends of SC6000 with Klucel G and isopropyl alcohol have been used widely in the United States to consolidate the surfaces of red-rotted bookbinding leathers. These appear to penetrate the fibrous structure to some extent and give a semi-matt natural appearance (see Chapter 21).

*Lascaux wax-resin adhesive (443–95)* has been used as a consolidant for flaky paint on leather.

*Gelatine*. This has been used particularly for consolidating bookbindings or by conservators trained in that speciality. Gelatine films are not particularly flexible and are probably most suitable for objects such as the coverings of book boards which are fixed and not intended to flex.

## 13.8 Dressings and finishes

The overapplication of dressings to leather can often cause irreversible damage so they should not be applied as a routine treatment and in many circumstances should be avoided entirely.

Traditionally dressings were applied in an attempt to slow down deterioration, improve the appearance and to restore some strength and flexibility. The long-term effects of overoiling leather are that oils and fats can encourage bio-deterioration, spew, oxidize and stiffen the leather, discolour and stain it, leave a sticky surface and wick onto nearby material, soften the original finishes and decoration, attract dust and impede future conservation treatments. The solvents present in many dressings can affect surface finishes, wet, swell and deform the leather, dissolve or dislocate the original components. The use of dressings therefore may lead to many disastrous side effects. Each object is different, so there can be no cure-all routine treatment. Unfortunately dressings have in the past being used as standard treatments over the whole of a collection resulting in some of the problems we have today.

*Saddle soap* contains soaps but is mostly emulsified oils. It is very alkaline with a pH 9 to 10, which can lead to a lowering of the shrinkage temperature of the surface of the leather. It is intended to be used on modern equestrian leathers as a lather or foam and to be washed off thoroughly. This technique is not ideal for historic objects. It can result in darkening and its application involves excessive wetting. Although it appears to clean leather it can be pushing the dirt further into the surface. It may even be the cause of the fine cracking of the surface sometimes seen on old leathers.

*Pliantine (British Museum Leather Dressing)* contains beeswax, lanolin and a small amount of cedar wood oil. It has a bad reputation for being overapplied and becoming a darkened and sticky dust trap, but if applied sparingly to sound leather and worked in well, it can be satisfactory.

*National Trust Furniture Polish (natural colour)*. This is not strictly a dressing but a wax-based surface coating. In most cases, leather does not need a heavy waxing and a cloth which has previously been used with a wax polish on furniture will generally have enough residual polish remaining on it to give a sheen without the danger of overapplication. Any good quality uncoloured polish of this type will give a similar result. If the leather is firm enough, a hint of wax followed by a good buffing can make a huge difference to the appearance.

*Microcrystalline wax* also gives a pleasant sheen, but must used sparingly. If overapplied, it can stay slightly tacky and can leave residues in the cracks. These can be removed with white spirit or mineral spirits.

*SC6000*. A commercial surface finish containing wax, acrylic resin and isopropyl alcohol, supplied in the form of a cream. It dries to a shiny and hard surface.

## References

Down, J. (2003) *Adhesive Research and Some Aspects of Skin and Leather Bonding*. Unpublished report presented to Adhesive and Leather Symposium at the Library of Congress. Canadian Conservation Institute.

Jackman, J. (1982) *Leather Conservation, A Current Survey*. London: The Leather Conservation Centre.

Kite, M. (1995) Some Conservation Problems Encountered When Treating Shoes. *ICOM Committee for Conservation Leathercraft Working Group interim meeting Amsterdam* pp. 91–95 ICOM.

Plenderleith, H.J. (1946) *The Preservation of Leather Bookbindings*. London: The British Museum.

Sturge, T. (2000) *The Conservation of Leather Artefacts. Case Studies from the Leather Conservation Centre*. Northampton: The Leather Conservation Centre.

Waterer, J.W. (1972) *A Guide to the Conservation and Restoration of Objects Made Wholly or in Part of Leather*. London: Bell.

# 14

# Taxidermy

*J.A. Dickinson*

## 14.1 A brief history

Taxidermy as we know it today, that is the dried or preserved skin of a vertebrate modelled to represent the living animal, dates from the latter half of the eighteenth century. Prior to that attempts had been made with varying degrees of success to preserve the whole creature by some form of evisceration and drying. While these were sometimes successful in the short term, most succumbed before long to the ravages of insects and damp, leaving dotted across Europe, a handful of specimens from the seventeenth century. The only example of this method to survive in the UK and indeed probably the oldest preserved bird is the Duchess of Richmond's African Grey Parrot in Westminster Abbey, which died in 1702. However, by the mid-eighteenth century Bécoeur had discovered arsenic as a preservative and this, coupled with widespread agreement that specimens should be skinned completely, meant that the many examples which survive from the late eighteenth century/early nineteenth century are prepared in a manner which differs little from those used today for birds and smaller mammals.

As the nineteenth century progressed, so too did taxidermy methods and by 1851 many taxidermists were able to exhibit work of a high standard at the Great Exhibition in London. Whereas much early work had been on birds and small mammals, now a full range, from fish to large mammals, was on show. This necessitated more varied preservation and preparation techniques. The publicity achieved at the Great Exhibition led to taxidermy's heyday from about 1870 through to the First World War. This period saw the opening of many public museums, a vogue for the natural sciences and a craze for birds (or parts of) as a fashion item.

As the world opened up, colonists expanded into lands from which a stream of big game trophies started to flow. So popular was taxidermy by the 1870s that General George A. Custer was collecting and mounting his own trophies while campaigning in Dakota. In the UK, shooting a tiger while visiting or working in India seemed to become an essential part of life for the upper classes. These, usually prepared as rugs, often with the head mounted with the mouth open, were eventually sent back home to serve as domestic obstacles for future generations. Museums and wealthy collectors sent expeditions to search far flung corners for new rarities and to this day there are species which we only know from a few preserved skins collected at this time. Indeed, famously, the last two specimens of the now extinct great auk (*Alca impennis*) were killed to be mounted and sold to collectors. The extinction of the passenger pigeon (*Ectopistes migratorius*) in 1914 was not due to overcollecting but taxidermy had enabled the preservation of several hundred specimens from a population, which numbered tens of millions only 50 years before. We can at least see these today, unlike the dodo, which became extinct too early for taxidermy, leaving only one poor specimen in Oxford. This was destroyed in 1755 leaving as its legacy a continuing debate as to the true appearance of the species.

Changing tastes and the Great War finally brought a halt to the boom, which in the UK had seen almost every town have at least one taxidermist. The larger and more reputable firms survived into gradual decline in the interwar period until the *coup de grâce* of the Second World War, which left only the two large London firms of Rowland Ward and Gerrard & Sons, plus a handful of small operators. North America with its huge wilderness areas and 'frontier' culture did not suffer in the same way and hunting and fishing trophies remained popular with many companies supplying a demand which is today probably larger than ever. The latter decades of the twentieth century have seen a revival in Europe with a new generation discovering taxidermy in an age when there is the threat for many species that only the preserved specimen may be left for posterity.

## 14.2 Taxidermy terms

The language of taxidermy uses the following terms.

*GRP.* Glass-reinforced polyester, what is commonly called fibreglass.

*Hide paste.* Usually based on Dextrin plus, possibly, whiting, talc and fine papier mâché.

*Base.* That to which the mount is fixed, often made to represent the ground but can also be a branch, artificial rock, or simply polished wood.

*Mount.* Taxidermists call the finished specimen a mount. They also refer to 'mounting' a specimen.

*Form.* Usually for mammals, this is what the skin is fitted onto.

*Pickling.* The method of preserving a skin by soaking it in a solution such as salt/alum or carbolic acid. This will not endow any of the qualities of a dressed skin.

*Tanning.* Taxidermists commonly use this to mean dressing.

*Prepare.* To mount a specimen.

## 14.3 Birds

### 14.3.1 Methods

X-rays have revealed that the early taxidermists employed a variety of methods as each experimented with this new subject. It was only with the publication of manuals that methods became standardized. Whatever method chosen, the bird would be skinned, usually by ventral incision, and the skull, legs and wing bones retained in the skin but scraped clean of meat. The inside may have had some of the excess fat and tissue scraped away, if time permitted, but the skin was often not washed as drying plumage was time consuming and difficult. At this stage a preservative, usually either a dry powder or an arsenic-based soap might be applied to the inside and sometimes the feathers would be treated to a wash containing either arsenic or 'corrosive sublimate' (mercury (II) chloride). Next, one of two methods would normally be used to mount the bird:

1. *Soft stuff.* Wires would be inserted through the foot, up the sinew tract at the back of the leg and probably bound onto the femur and tibia. Wires might be fixed into the wings in a similar manner. One or sometimes two or more wires would now be inserted through the tail, on up the neck and out the top of the skull. All these wires would now be twisted together in the body area and the skin would be filled out with chopped tow (jute strand), wool, cotton, grass or anything else considered suitable. After stitching up, the bird would be fixed to a base by its leg wires, modelled into shape and left to dry.
2. *Bind up.* Wires would be similarly placed in the legs but then a body, sometimes approximating in size to the bird's own but often ovoid, would be made by binding hay, wool, wood wool, tow or carved peat. Through this would be put a wire, which was wrapped with cotton, tow, wool, or similar, to make a neck. This was then inserted into the skin with neck wire coming out of the top of the skull and the leg wires clenched through the body. Again, after stitching of the skin and fixing to a base, it would be modelled into shape.

At this stage, with either method, glass eyes would have been fitted. Sometimes they would be simply on a bed of cotton wool held in by the skin. Others would be modelled in with clay and sometimes a rock hard mix of plaster and animal glue secured them.

When dealing with very old taxidermy, however, one should always be prepared for the unusual. In the mid-nineteenth century the English naturalist Charles Waterton prepared a large collection of birds, which are hollow!

Most birds are now mounted using a hard body, often still made of bound wood wool but increasingly made from balsa wood, Styrofoam or moulded polyurethane (PU) foam. Tow is still widely used to make necks and leg muscles and, while clay is the main method of setting eyes, epoxy putty is increasingly used. Some eyes, traditionally glass, are now made from acrylic. Skins themselves, if properly prepared using mechanical wire brushes and washed in a scouring agent such as Pastozol AZ or biological detergent, are virtually fat free, though nothing seems to be known of the long-term effects of such treatments.

### 14.3.2 Problems

Due to the use of arsenic and mercury (II) chloride as feather washes on some specimens, all old birds should be handled with caution. Measures such as wearing latex gloves and a dust mask should be taken. Museums may wish to use X-ray fluorescence to determine the extent of arsenic in larger collections. Provided common sense is used there should be no great threat. Indeed Morris (1982) found that the life span of a random sample of taxidermists born between 1782 and 1908 averaged 76.4 years with 50% passing the age of 80.

It is only in the last 50 years that most taxidermists have had access to deep freezers and modern equipment and chemicals for cleaning and drying. Prior to that, the taxidermist had to work fast to preserve the specimen and usually the best way to do this was to mount it as quickly as possible. Once this was done, it would dry out before it could decompose. Even though they might have been treated with a preservative on the inside, the skins of many old specimens are little more than raw dried. Inevitably, the speed of this process meant that many corners were cut, the worst being the failure to remove fat and properly degrease skins. In species such as waterfowl, the fat layer within the feather tracts can be at least 5–10 mm thick. This fat appears to oxidize with age and creates what is commonly called 'fat burn'. This can cause damage within only a few years of preservation and causes the skin to become very brittle and will either split or break into pieces if the specimen is touched or vibrated. Individual feathers will come away with a crusty lump of usually rusty-coloured skin attached. This problem does not seem to be reversible and is difficult to treat successfully. If the bird has been mounted with a firm form inside it is possible to fix small areas of damage with a suitable adhesive such as PVA. In some cases it may be possible to strengthen the whole of the inside of the skin by injecting small amounts of a consolidant such as Paraloid B72.

Old bird mounts which have remained in their original cases or glass domes are usually fairly clean but collections of uncased mounts like those in many museums can often be quite dirty. Most fresh bird skins can be washed by immersion and the plumage dried and fluffed to leave it in perfect condition because, while the skin is flexible, the feathers can all be separated and each dried individually. In dry, immovable skin, it is not possible to achieve the necessary movement of each feather that allows it to be fluffed up as it dries. Given these conditions, little more than surface cleaning by wiping should usually be attempted, particularly by the inexperienced. If washing is to be undertaken it is usually best done using a neutral detergent in small patches, drying each before starting the next. Even then, a small area should always be tested first as some feathers react badly to water. The fringes of feather tips, particularly if they are abraded or light damaged, of some species of *Accipitridae*, *Falconidae*, *Glareolidae*, *Columbidae* and *Upupidae* can curl when washed, though circumstantial evidence suggests that this may be partly due to using a too alkali detergent. Should this happen it is sometimes possible to reduce the curl by carefully wrapping the affected area in wet tissue which when dry can be left in place for up to several months. In many cases, wiping with a solvent merely forces the dirt into the feather. In these cases, dirt will only be removed by flushing in a suitable extraction bath. However, it is probably best, in most instances, to avoid this unless absolutely necessary.

Besides dirt, the problem found most obviously in bird mounts is that of leg wires (usually ungalvanized) rusting, erupting and splitting through the back of the legs. Often these wires will break off at the foot if the specimen is removed for repair. In all but the largest birds, it is usually possible to discard this wire and replace it with a new one fitted into the metatarsal by drilling through the base of the foot. The rust-damaged area can then be repaired and remodelled.

## 14.4 Mammals

### 14.4.1 Methods

Initially, small mammals were prepared in a similar way to that described for birds with the addition that the nose, lips, ears and toes had to be skinned out and modelled to avoid shrinkage and distortion. It was quickly found that anything larger than a fox needed a much stronger, more rigid framework to prevent them sagging under the weight of materials used. Larger mammals also had skin which needed immersing in a pickle to fix it and also needed shaving down to allow the pickle to penetrate and to help prevent shrinkage as it dried.

Many of the early, large mammal mounts were fairly crude but from the mid-nineteenth century standards improved steadily and by the 1880s competent mounts were being produced. These commonly used a method of a centre board shaped as a silhouette of the neck and body to which metal rods bent to the shape of the legs were fastened. The skull and sometimes the leg bones were attached, and leg muscles and body were created by binding on straw, hay, or wood wool. Often, this was covered with a layer of clay to smooth it out and allow some detail to be shown. The better taxidermists improved on this by using plaster and scrim instead of clay to strengthen and provide a firm base to fix the skin to. Sometimes a layer of papier mâché–glue–plaster composition would be applied under the skin at mounting. Folds, wrinkles, etc. could be modelled into it before it set helping to hold the details in place as the skin dried.

In the early twentieth century the great American Carl Akely (1864–1926) introduced modern taxidermy by producing highly detailed plaster or glued paper forms cast from an anatomically accurate clay sculpture. The 'Akely method' meant that more than ever skins had to be dressed or tanned, not merely pickled, and some taxidermists turned to the methods used in the fur trade. Thin, supple skins which hardly shrink on drying allow detail to be modelled in by gluing them on to hard forms without the risk of them splitting. Trophy heads of big game were prepared in the same way as large mammals, with similar variations in technique. In North America the 'Akely method' meant that forms for both full mounts and trophy heads could be mass produced and firms very quickly became established supplying taxidermists with a range of forms for species both large and small. For many years these were hollow and made of glued paper up to 5 or 6 mm thick, but from the 1970s they have slowly been superseded by high density PU foam. This trend has spread to Europe and most countries with commercial taxidermists. The market for these forms is large – firms can supply almost any species from a squirrel to a polar bear. There remains, however, some doubt as to the longevity of PU which may present interesting problems in the future. Individual taxidermists also create 'one-off' forms from GRP, others carve forms from Styrofoam and some still resort to the bind up method. Many ears are now filled with GRP, epoxy putty or custom-made plastic liners. The preparation of skins is becoming much more uniform with most species above squirrel size being tanned with Lutan FN, a low basicity aluminium tanning complex.

**Figure 14.1** Trophy heads of Argali sheep showing badly soiled condition (left) and after cleaning (right).

### 14.4.2 Problems

Many old mounts and especially trophy heads become very dirty due to years of open display. After ascertaining the condition of the skin and hair and the type of form used, it may be possible to remove much loose dust by gently brushing and blowing followed by further cleaning as for furs and skins. However, certain ungulates such as sheep of the genus *Ovis* and some deer of the genus *Cervus* have hair which becomes very brittle with age and these should on no account be treated by brushing. With most mammals, and particularly trophy heads, it is possible to wash the whole mount by spraying with warmish water containing a suitable detergent and gently working the hair with a sponge to loosen the dirt. The whole mount can then be rinsed thoroughly with running water. Brittle-haired species in particular can be cleaned in this way as the hair becomes much more pliable while wet. When rinsed, the hair can be 'blotted' with towels and then dried fairly quickly with warm air. It is important to dry as soon as possible to prevent the skin and any modelling materials such as clay from absorbing water. For other methods see the section on cleaning in Chapter 15.

One of the commonest problems in taxidermy, particularly in larger species, is skin splitting. This generally occurs for one of two reasons:

1. The whole mount is unstable or has been removed from its base allowing it to wobble whenever it is moved. Eventually this leads to the skin splitting in areas of movement. This type of problem can only be addressed by first stiffening the mount and preventing further movement. If the mount is not on a base, simply putting it on one may cure the problem. If, however, the form inside the skin is not rigid – perhaps a too soft 'bind up' or leg rods too thin – the solution may be achieved in a variety of ways. With small specimens, it may be possible to identify points of movement and to inject Polyfilla, or similar, into and around these areas using a syringe and large bore needle. In larger species, such as bear and tiger, I have successfully stiffened soft, straw bodies by driving lengths of plastic piping into areas of movement and pouring in a two part PU foam mix. Using a low rise foam in small pours it is possible to impregnate a large area of a body or leg until it becomes rigid. The same method will also stiffen sagging necks.
2. The skin has not been prepared properly and shrinks because it is too thick; has been incorrectly dressed or only pickled; has been stretched to fit a form that is too large. Any one or combination of the above can cause splits to appear almost anywhere on a specimen. Where the skin has not been glued to a solid form, shrinkage usually first shows itself by the bursting of stitched seams. Even if the skin is glued to a form, this will not prevent splitting if the skin is incorrectly prepared and/or is overstretched. Splits and bursts are also likely to happen due to large fluctuations in relative humidity (RH) and high temperatures, and even a well-prepared skin can be damaged by the expansions and contractions of a large bound body covered with a thick layer of clay.

Whatever the cause of splitting, an assessment must be made of the skin's condition in order to determine what method will be used to effect a repair. If the skin is in good condition and appears to have been well prepared, it may be possible to soften it down using damp cloths and stitch it together. It may, of course, be necessary to soften a quite large area in order to get enough stretch. At the same time, if the mount has a solid form a suitable 'hide paste' such as Dextrin may be used to prevent the skin from shrinking back. Crucially, any such repair must be allowed to dry out slowly to reduce further shrinkage. When softening any mammal skins, it is important to remember that old skins are often acidic due to being pickled with salt and alum and care must be taken to avoid 'acid swell', which causes the skin to become very rubbery and weak. To reduce the risk of this, salt should be added to softening water at a ratio of 50 g per litre.

Where the condition of the skin will not allow it to be stitched together, then it may be possible to disguise splits by patching. This can be achieved with bits of skin from the same species, if available, and it is always useful to have a collection of aged skin from unwanted specimens. Unlike furriers, taxidermists have never dyed skins so most old mounts have faded in varying degrees, often unevenly, which makes matching in far from easy. If no skin or hair is available, then an alternative may be found; natural or synthetic, it does not matter as long as it can be made to match.

This may be achieved on hard forms by simply gluing in place, having first made sure that the surrounding skin is similarly held. On soft mounts, Polyfilla can be injected into and around the split in sufficient quantity to create a firm base to which adhesive will adhere. The split can then be patched as above or can be 'haired' by painting with a suitable adhesive (PVA) in small amounts and laying in tufts of hair. Sparsely haired species such as apes can be 'haired' using a two part unfilled PU tinted to the skin colour as the adhesive. This will grip the hair

**Figure 14.2** Ear of antelope from rear showing traces of iron oxide in line along the centre leading to a damaged area where the rusted wire has broken through the weakened skin.

easily but then set quickly. Sometimes it is possible to use a longer hair, which can be trimmed to length when dry. If it is necessary to dye or colour hair completely to match, this is best done before patching takes place. As few mammal hairs are single coloured, dyeing can only be used for base colouring after which airbrushing can be a useful technique. Patched areas can have markings and other detail added in situ by this method. On species with sparse coarse hair the airbrush with acrylic paint can be very effective and easily reversed.

There may arise instances when the mount is in such poor condition that simple 'repairs' are not sufficient. Very occasionally even historic specimens can be completely taken apart and remounted. However, the costs of such an undertaking mean it is only likely to be considered for very important specimens. An excellent account of the renovating of an extinct quagga (*Equus quagga*) is given by Rau (1993). Suffice to say this type of project should only be attempted by those with great experience.

Initially, mammal ears were left untouched because, being composed of skin and cartilage, they might dry without rotting. The drawback is that any ear larger than a squirrel's quickly distorts badly as it dries. To avoid this, the ear must be 'skinned out'.

This involves either separating, from inside, the skin of the back of the ear from the cartilage or, better still, removing the cartilage completely. Either way, a stiffening and shaping substance can then be introduced into the ear. Older mounts often have just clay and the retained cartilage, but many from the twentieth century have a liner made of lead or paper often accurately shaped or, at least, a non-hygroscopic 'compo' mix, together with a larger wire rod to provide rigidity while setting. Often these rods were driven through the tip of the ear, down the length and into the form. Their legacy is a wealth of heads with rusty damage patches to the back of each ear tip. Ears which have not had the cartilage removed can become distorted over time. These can be reshaped to some degree by softening and clamping between shaped wooden formers while they dry. If kept at a steady RH, they may retain this shape.

## 14.5 Fish

### 14.5.1 Methods

It is undoubtedly the case that attempts were made to preserve fish in the eighteenth century but it is

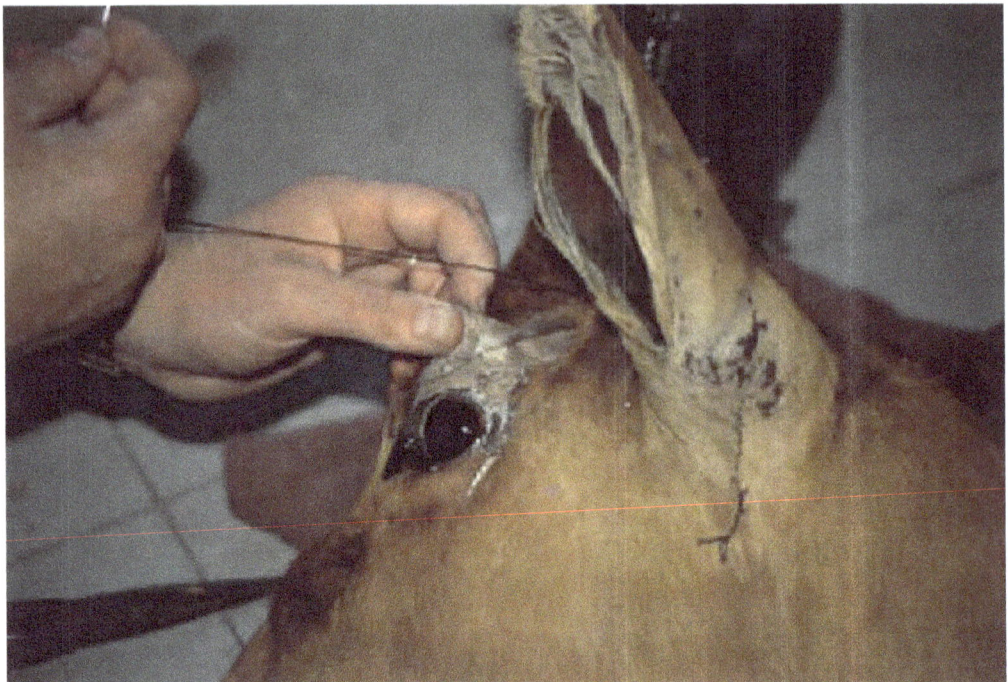

**Figure 14.3** Stitching a split in relaxed skin behind the eye of an antelope.

only from the mid-nineteenth century that fish of any quality and in quantity are to be found. Traditionally, until the late twentieth century, fish were prepared by skinning along the lateral line on the non-show side and removing the body. As much flesh as possible was taken from the head without skinning (as this is nearly impossible), the inside of the skin was scraped clean of fat and tissue and an arsenic mix might be put on the inside. The commonest method of mounting was then to fill the skin with sufficient sand or damp sawdust to recreate the body, stitch up, model into shape and leave to dry. Larger species such as tunny (*Thunnus sp.*) would require a body form and be mounted in much the same way as mammals. At this stage, it might be given a carbolic wash over the outside and a coat of shellac would help to hold the scales in place. This, no doubt, helped to mask the smell and keep off flies as it dried and fleshy areas of the head shrank. When dry, these might be modelled back with a glue/plaster mix, sometimes with papier mâché.

After drying, the sand, if used, was emptied out and the now discoloured, faded specimen might be painted to resemble itself in life. Often this was a cursory effort and more usually it would receive another coat of shellac to enhance any markings in the skin and impart a 'fishy' gloss.

The use of silicone rubbers has enabled fish to be reproduced in GRP in minute detail and all museums and many taxidermists now use this method. A few die-hard anglers, however, still want their actual fish and today many are prepared by replacing the real head with a cast copy fixed to the skin. This itself may be mounted on an accurate PU form that requires only some adhesive to attach the skin. Most specimens are now completely painted with several layers of paint to replicate the living fish.

### 14.5.2 Problems

Fish skin and bone are inherently oily and the preservation method described leaves relatively large amounts of fatty tissue in the fins and, of course, the whole skull with much untouched skin. Within decades, discolouration and eruptions through the skin can appear. These take the form of dark brown sticky globules of oxidized oil or fat, sometimes forming little runs. In smaller specimens, these may dry out and can be removed with a blade but in larger ones they may remain very sticky and are difficult to

**Figure 14.4** Soiled study skin of a garganey, showing test cleaned area on breast.

remove even with solvents. In larger, oily species, the whole specimen may be subject to 'fat burn' leading eventually to its total disintegration. It is probably not possible to maintain the original appearance of such a specimen and, if it is of historical or scientific value, consideration should be given to moulding it to make a GRP replica before it deteriorates too far.

## 14.6 Care

As most museums are aware of the environmental requirements of their collections and have in many cases addressed these needs the following points are intended for the guidance of those who care for taxidermy in the domestic environment. Museum curators should refer to the *Manual of Natural History Curatorship* (Stansfield et al., 1994).

### 14.6.1 Light

(a) Ultraviolet light (UV) is the commonest cause of irreversible damage.
(b) For display, try to use areas that have as little natural light as possible.

(c) Never display by using picture lights, spots, etc., which are used as main room lighting.

### 14.6.2 Temperature

(a) Ideal ambient temperature is 18–22°C.
(b) Avoid high temperatures.
(c) Avoid hot spots, i.e. above radiators.
(d) Lower temperatures (13–15°C) are acceptable if the RH is low.

### 14.6.3 Relative humidity

(a) Ideal RH is 50% ± 5%.
(b) Avoid fluctuations in RH. It is better to have a slightly high or low level that is steady.
(c) Lower RH is acceptable at lower temperatures.
(d) High RH is particularly bad as it promotes bacterial and mould growth, which can destroy specimens. This process is speeded up when combined with high temperatures.

### 14.6.4 Storage

(a) Always store in darkness, by covering if necessary.
(b) Taxidermy in cases and domes is always easier to maintain. These can be put in a black bag if a

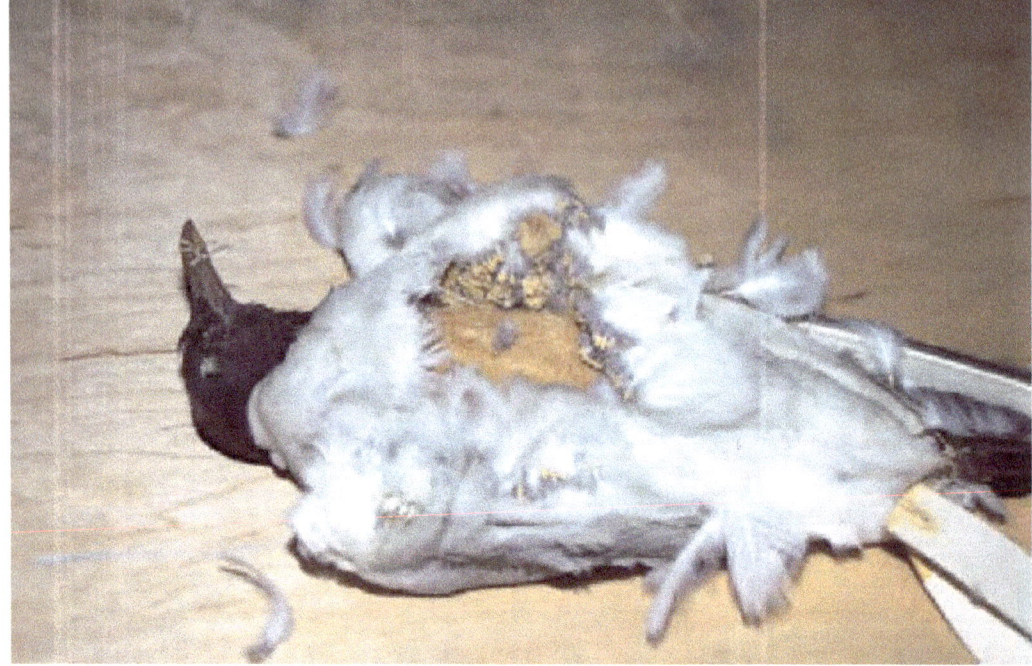

**Figure 14.5** Study skin of a blackheaded gull suffering from extreme 'fat burn' which has so damaged the skin that the specimen is falling apart.

**Figure 14.6** A mounted tuna showing on the lower half areas of oxidized oil/fat leaching through the paint.

Taxidermy 139

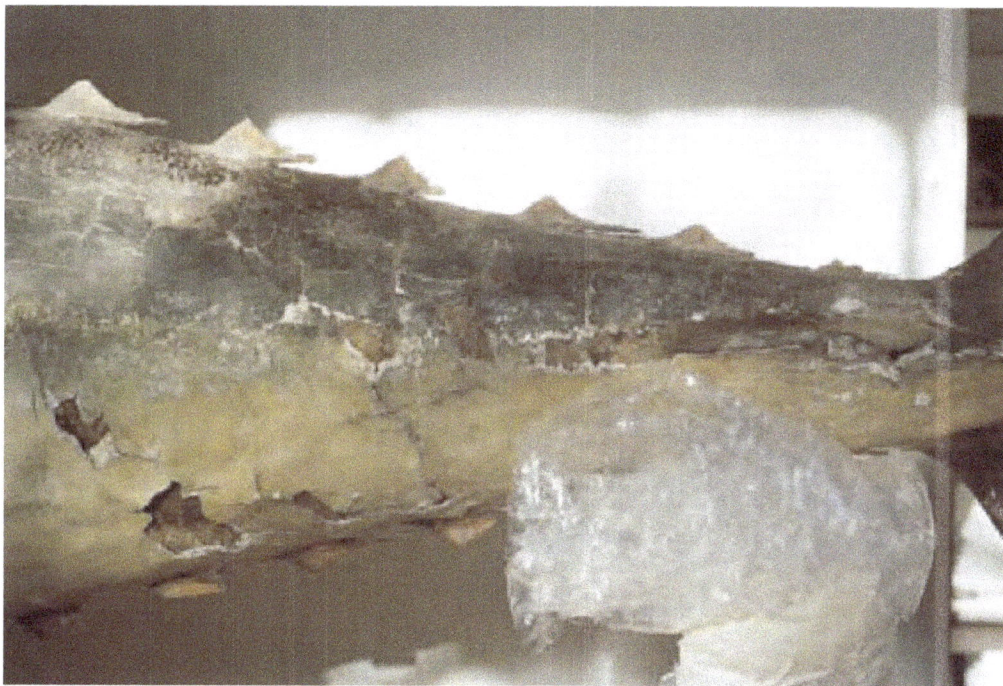

**Figure 14.7** Tail end of tuna, showing areas of filler and paint being cracked and lifted by oxidizing fats underneath.

**Figure 14.8** Detail of above.

dark cupboard is not available. Ensure cases are well sealed.

(c) Larger uncased specimens should be bagged or covered to keep off dust. This should be done simply so they can be checked for insects.

## 14.7 Preservatives

Bowditch (c.1821)
'Receipt' for making arsenic soap, invented by Bécoeur, apothecary at Metz.

| Camphor | 5 ounces |
| --- | --- |
| Arsenic in powder | 2 pounds |
| White soap | 2 pounds |
| Salts of tartar | 12 ounces |
| Lime in powder | 4 ounces |

Wood (1877)
Take half an ounce of corrosive sublimate
Take half an ounce of white arsenic
Take four drachms of spirits of wine
Take half an ounce of camphor
Take six ounces of white Windsor soap

Davie (1894)
Arsenical soap

| White soap | 2 pounds |
| --- | --- |
| Powdered arsenic | 2 pounds |
| Camphor | 5 ounces |
| Subcarbonate of potash | 6 ounces |
| Alcohol | 8 ounces |

Browne (1896)
Alcoholic solution of mercury, one part in 1000. For preventing and arresting mildew, and for external use upon skins

| Methylalcohol, 90–95% | 1 pint |
| --- | --- |
| Bichloride of mercury | 10 grains |

Non-poisonous preservative powder (M.B)

| Pure tannin | 1 oz |
| --- | --- |
| Red pepper | 1 oz |
| Camphor | 1 oz |
| Alum (burnt) | 8 oz |

Preservative for mammals

| Alum (burnt) | 4 parts |
| --- | --- |
| Saltpetre | 1 part |

A mixture made of the above proportions, with the addition of hot water and a little bichromate of potassa, makes a very fine bath in which to plunge a thick or slightly tainted skin.

Davis (1907)
Preservative powder for small skins.

| Saltpetre | 1 oz |
| --- | --- |
| Burnt alum | 5 oz |
| Plaster of Paris | 4 oz |
| Naphthalene | 1½ oz |

Rowley (1925)
Dry arsenical mixture for poisoning the interior of small fresh bird and mammal skins.

| White arsenic (arsenious oxide) | 3 parts |
| --- | --- |
| Powdered alum (aluminium sulphate) | 1 part |

Tan liquor for fur dressing.

| Water | 1 quart |
| --- | --- |
| Salt | 1 pound |
| Alum (aluminium sulphate) | 1 pound |
| Dissolve by heating, cool and add, Sulphuric acid (commercial) | ½ fluid oz |
| Formic acid | 4 fluid oz |
| Hyposulphite soda | 4 oz |
| Flour (dry) | ½ pint |

Fat liquor.

| Water | 1 quart |
| --- | --- |
| Laundry soap, sliced | ½ cake |
| Stearic acid, shaved | the same bulk as soap |
| Carbonate soda | 1 oz |
| Neats foot oil | 4 fluid oz |

(castor oil gives fine results and is easily obtainable)
Boil and stir until dissolved   1 tablespoonful
then add, Ammonia
(commercial)

## References

Bowditch (1821) *Taxidermy: or, The Art of Collecting, Preparing and Mounting Objects of Natural History*. London: Longman, Rees, Orme, Brown and Green.

Browne, M. (1896) *Artistic & Scientific Taxidermy and Modelling*. London: Adam and Charles Black.

Davie, O. (1894) *Methods in the Art of Taxidermy*. Columbus: Hann & Adair and H.T. Booth.

Davis, W.J. (1907) *Bird and Animal Preserving and Mounting*. Hythe and London: J. & W. Davis and Elliot Stock.

Morris, P. (1982) On the Supposed Occupational Hazards of Being a Taxidermist. Guild of Taxidermists. *Newsletter*, **9**, 36–37.

Rau, R.E. (1993) Remounting the Quagga. *Breakthrough Magazine*, **34**, 54–57.

Rowley, J. (1925) *Taxidermy and Museum Exhibition*. New York: D. Appleton and Co. Ltd.

Stansfield, G., Mathias, J. and Reid, G. (1994) *Manual of Natural Curatorship*. London: HMSO.

# 15

# Furs and furriery: history, techniques and conservation

*Marion Kite*

This chapter deals with furs and fashionable dress made from the skins of fur-bearing mammals. Additional information will be found in the chapters dealing with ethnographic materials and taxidermy as the methods of skin preparation used for objects which fall within these disciplines will differ considerably from those used for furs in fashionable dress. Basic information on structure and morphology of hair is, of course, common to all. Methods of cleaning, conservation and general care may also be applicable across the disciplines.

This chapter is divided into three sections. Section 15.1 describes the history of fur use in the context of species availability at certain times and its use in fashionable dress. Section 15.2 describes hair structure and morphology, fur-skin dressing and dyeing methods, furriery techniques including the 'dropping' or 'stranding' of skins and the method of making up furs into finished garments. Section 15.3 deals with practical conservation issues and describes briefly the cleaning and repair of furs as it is carried out by furriers to make clear how and why this differs from the conservation approach. The damage to furs that conservators may encounter is discussed, and the ways in which conservators repair furs and solve some of the care and conservation problems are described. Case histories of treatments are discussed as examples to illustrate further the conservation of fur within museum collections.

Personal feelings concerning the morality or desirability of wearing fur have no place here, as it is an indisputable fact that fur is an important part of the history of dress across many cultures. Conservators will therefore need to be as knowledgeable about fur as any other skin material, animal product, or any other organic material they may come across during the course of their work.

## 15.1 History of fur use

### 15.1.1 Introduction

In the latter years of the twentieth century and early years of the twenty-first, the wearing of fur as an item of fashionable dress is an emotive subject. Many museums do not have a collections policy which positively includes fur, and those which have it among their collections rarely show it or actively encourage its acquisition.[1] The reason is often given that it is impossible to store but this is not the case. Although not readily acknowledged there is often a quantity of fur in most textile collections which include dress and accessories and this should not be dismissed lightly. There may be hats, muffs, tippets, collars, or whole garments such as coats, jackets and stoles. Items may be edged or trimmed with fur. They may also be lined with fur, either whole or part garments, or accessories such as gloves, boots, etc. Fur has often been used on ceremonial robes and as part of military uniforms and will thus be found in most, if not all, collections dealing with this type of dress. For example, coronation mantles and ceremonial robes were trimmed with ermine and other furs. Guards' 'bearskins' or shakos are made from the skins of American brown bears (*Euarctos americanus*), busbies of the hussars troopers are made of hair-seal (*Phocidae* sp.) as are those of the rank and file of the fusilier regiments (Sachs, 1922) and flying helmets from the period of the First World War and the early years of aviation are lined with rabbit fur (*Oryctolagus cuniculus* sp.).

It is not only the textile conservator who might have to treat fur. Conservators working with ethnographic materials will certainly have to treat fur, so will conservators working with mixed collections of organic materials, possibly furniture conservators, those working with archaeological materials, natural sciences and taxidermy collections and possibly other disciplines too.

In ethnographic objects from diverse cultural origins fur is abundant, as it is in taxidermy specimens. It has been used in furniture; in the late eighteenth and early nineteenth centuries travelling trunks were covered with and lined using the pelts of fur seals (*Otariidae* sp.) (Ewing, 1981). Occasionally archaeological fragments may also be found and several fragments have been discovered with various Iron Age finds – bog bodies (Glob, 1971).

### 15.1.2 Background and history

*Unto Adam also and to his wife did the Lord God make coats of skins, and clothed them.*
*Genesis. Chapter 3 verse 21*

The practice of wearing furs is as old as humanity. Some method of dressing skins was discovered early in history as a skin flayed from an animal and used in its rough state would quickly go hard or if dried slowly without further treatment to preserve it, would putrefy. In neither case would it be of any use.

The discovery of a 5000-year-old body (now known as the Ice Man) preserved in a glacier at Hauslabjoch across the border of the Austrian/Italian South Tyrol gave evidence of fur being worn. The man was found fully clothed, and although the clothing was not completely preserved, his upper garment was described as of leather – and as fur which had lost its hair and was probably similar to a cloak or cape. Photographic evidence taken prior to disturbing the body showed that it was made of fur. In the course of recovery and stabilization the hair of the fur had come out and is now present as separate clumps (Spindler, 1994).

Found with the body of a woman discovered in a bog in Huldre Fen at Rainten, Djursland, in Denmark was the most complete woman's costume of Iron Age date yet discovered. She was discovered wearing a lambskin cape and another cape as an outer garment (Glob, 1971).

Another body of an Iron Age man, dating from some 2200 years ago and found in a peat bog at Tollund in Jutland in 1938, was discovered wrapped in animal skin. This was identified as some kind of sheepskin and had pieces of cowhide sewn onto it (Glob, 1971).

Arctic peoples and others from cold climates have of course chosen fur for warmth and protection but throughout history the wearing of furs in fashionable society has been synonymous with wealth, luxury and status.

In classical antiquity the Phoenicians, Assyrians, Greeks and Romans all used furs and the British fur trade began at least 2000 years ago. When the Romans arrived in Britain they traded furs for British wool and metals. Later, Norse traders brought exotic luxuries including furs from Russia and Armenia which they traded through London and were imported through Queenhythe docks in the city. The Anglo-Saxons and Normans wore furs and as London developed it became the centre of the world fur trade. By the thirteenth century furs from northern Europe, Scandinavia, Russia and the Baltic lands were traded through London and a business community of merchants and furriers developed. The Peltry in West Cheap is mentioned in 1274 and the shops and homes of skinners extended south from Cheap near the banks of the river Walbrook to the Thames (Veale, 1966).

From the fourteenth century onwards the Hanseatic League, founded by wealthy merchants, formed one of the great trading organizations of the world and furs were one of their principal commodities. Squirrel skins, sables and others from Russia and the Baltic lands were marketed through the City of London and the Worshipful Company of Skinners of London were one of the earliest medieval guilds drawing up regulations for fur trading and manufacture of skins into furs. The Craft Guilds had enormous power and ensured fair trading and quality control. The Skinners originated as a guild of furriers and were granted their charter by Edward III in 1327. Their coat of arms depicts three ermine caps tasselled and enfilled with gold crowns supported by a lynx and a marten. A lynx also forms the crest. Ermine was the most esteemed fur in which the skinners dealt at the time, and lynx and marten were also highly valued as the wearing of them indicated social prominence, success and wealth. Edward III ordered that no person of an income under £100 per year should wear fur except lamb, coney, cat and fox under penalty of forfeiture and that ermine was to be worn by royalty only (Baldwin, 1926).

From early on the upper classes had tried to limit the wearing of furs by the lower classes in order to maintain and protect their own status and the exclusivity of certain furs. The wearing of furs was strictly forbidden to the church, however in 1127 the Council of London allowed abbesses and nuns to wear the furs of lambs and cats only (Sachs, 1922). These must have been the lowest class of furs at the time.

In 1555 the British owned Muscovy Company was formed and opened new sources of trading in Russia. Beaver, hunted to extinction in the British Isles, but long valued as one of the principal furs and also used for making fur-felt hats, was available in quantity (Spriggs, 1998).

In Tudor England furs were thought to benefit the health. Skins of wildcat were supposed to cure rheumatism and gout, and mouse skins were thought to cure chilblains (Veale, 1966).

Furs were used as political and royal gifts with Russian sables being the most valuable. In 1587 Czar Ivan the Terrible gave to Elizabeth I four timbres (bales of 40 skins) of very rich black sables, six well-grown white spotted 'Luzerance' (lynx) and two gowns of ermines (Veale, 1966).

In the sixteenth and seventeenth centuries the wearing of furs in Europe was strictly controlled and Laws of Sumptuary dictated what estate of man could wear which fur; with the highest ranking having permission to wear furs of the greatest rarity and value.

Squirrel skins were much in demand by the upper classes and the skins of the grey squirrels with white bellies (not an indigenous species of the British Isles but imported from northern Europe) were versatile giving a variety of different decorative effects. They were known as *vair* if used whole, *gris* if only the backs of winter skins were used, *poppel* for the light skins of early summer, *rovair* for autumn skins where streaks of red appeared in the grey backs, *minever* if only the bellies were used (white with a little grey surrounding) and *pured minever* if the white bellies were used with all the grey trimmed off (Veale, 1966).

It is from a misunderstanding when the story of Cinderella was translated into French by Charles Perrault in 1697 that her slippers became lined with *verre* and not *vair*. Henceforth her slippers were glass and not made with the fur that would befit a young woman of a suitable social status to marry a prince.

In the late sixteenth century, with a view to seeking new territories and to fish the North American fishing grounds which were already known as exceptional, expeditions were sent to North America and Canada. It was by chance trading with the native peoples that the inexhaustible wealth of fur resources was discovered. In 1608 De Champlain, the leader of a French expedition, established a fur trading post at Montreal and trade was initially dominated by the French until the final conquest of Quebec in 1659. In 1670 Charles II granted a charter to the Hudson's Bay Company giving it a monopoly of trade and commerce in the lands within Hudson Straits. The first auction of Hudson's Bay pelts took place in 1672 in Garraways Coffee House in Change Alley, Cornhill, in the City of London.

Today the main auction houses are based in Scandinavia, Russia and North America but London commodity brokers are still responsible for more than 50% of the world trade in fur at a primary or wholesale level.

Throughout the earlier part of seventeenth century, before the Hudson's Bay Company came into being, most furs continued to be supplied from northern Europe. Both men and women wore furs and throughout seventeenth and eighteenth centuries sables continued to be the most valuable. By the eighteenth sumptuary legislation had been abandoned but sable (*Martes zibellina*), ermine (*Mustela erminea*) and squirrel (*Sciurus vulgaris*) were only affordable by the most wealthy. Furs continued to be used as linings. Fur muffs of varying styles and sizes were in fashion for men and women from the end of the seventeenth century throughout the eighteenth century. For men they were usually worn tied round the waist with a sash and for a time they became very large, almost pillowcase size. This was particularly an Italian fashion. In the eighteenth century the most popular furs worn by men were ermine and squirrel, but fox (*vulpes* sp.), lynx (*lynx* sp.) and others were also worn. For women, the contrast of textiles and furs became a fashion focus with luxurious velvets and rich silks being complemented by the choice of furs. In 1771 a sack back dress of pastel velvet trimmed with sable was admired by Lady Mary Coke at the court in Vienna (Ribeiro, 1979). A dress described by Mrs Delany in 1742 consisted of 'dark green velvet trimmed with ermine, and an ermine petticoat' (Llanover, 1861). Towards the end of the century when the silhouette changed, bulky fur linings of cloaks for outdoor wear were replaced by edgings only.

In the eighteenth century fur muffs were often perfumed. Sable and ermine remained the most exclusive and light white furs such as ermine and arctic fox were particularly fashionable for the wealthy, and appear in many eighteenth century portraits. Squirrel and lynx were also popular in the first quarter of the eighteenth century (Ribeiro,

**Figure 15.1** Detail from eighteenth century silk dress, V&A collection.

1997). In the early eighteenth century rabbit was worn by those of the poorer classes who were able to afford it but it became a fashionable choice by higher social classes by the end of the century.

Such was the fascination with fur that some silks were woven with patterns imitating fur worked into the design (*Figure 15.1*).

Until the mid-nineteenth century the primary use of furs in fashionable dress was as linings and for making accessories and for trimming gowns. Rarely were complete garments fashioned from furs with fur being used as the outside material except in certain European countries including Hungary where the *suba* (a type of cloak) was worn (Ewing, 1981). The dictates of the sumptuary laws had long ceased to apply and furs were an indispensable item of luxury and a way of displaying wealth now being accumulated by the newly prosperous middle classes. However, it was not until around the mid-nineteenth century that fur coats or jackets came into being in their own right as an item of fashionable dress for women.

Seal skin from the fur seal (*Otariidae* sp.) was one of the first furs to be used for coats as a method had been perfected for removing the harsh top hairs and refining the processing to give a light pliable skin. Sealskin was plentiful and had a beautiful rich velvety appearance when dyed black.

By the late nineteenth and early twentieth centuries a huge variety of furs were available and in demand by fashionable society. By 1880 all sorts of oddities were worn including cats heads, stuffed squirrels, tiny cubs or bears, small stuffed monkeys and also mice as trimmings. These were particularly used for hats and worn together with any number of stuffed birds, wings and feathers already applied to hats. Large animal skins had become fashionable for home decoration; whole specimens of stuffed bears and lions plus the skins of lions, bison, tigers, leopard and polar bears had become fashionable accessories in the domestic environment.

J.G. Sachs writing around 1922 states that over 100 species of fur-bearing mammals were used in the fur trade from the following classes – rodents, felines, canines, weasels, bear/racoon group, marsupials, ungulates and sundry, which correspond to none of the above but includes fur seals (aquatic mammals), moles (underground mammals) and monkeys (primates but usually only the colobus).[2]

Now comparatively few species are available to the fur trade as they are protected and come under the control of CITES (Convention on International Trade in Endangered Species) which regulates world trade in threatened and endangered species. The world populations of endangered and at-risk species are monitored and the Red List of most vulnerable species facing extinction is constantly being updated.

A leopard skin coat or tiger skin rug seen today may give rise to a feeling of horror and sadness knowing how few individuals of the species are left in the wild, but the wearing and owning of a leopard skin or ocelot coat was an expression of superior taste and elegance at the time they were fashionable. In 1956 J.G. Links, head of the fashion house of Calman Links, a director of the Hudson's Bay Company and President of the British Fur Trade Alliance, wrote:

*The buyer of such a coat generally has considerable experience and sophistication in the world of fashion as well as a number of other fur coats. She therefore wants the best and nothing but the best will do. The price of a fine leopard coat can easily exceed that of mink. (Links, 1956)*

Today, this subtle message has been lost to the casual viewer and we only see in each coat the wanton killing of several specimens of what is now an endangered species. Conservators treating furs, and those treating bird skins and feathered objects too, are likely to come upon examples of many endangered or protected species and should be aware that it is important to identify and safeguard the long-term survival of these examples. A leopard skin coat is much like the dodo. We will not see its like again.

In the 1960s designers targeted the youth market with less costly furs such as rabbit. Particular designs and colours were also tailored to the young and these became known as 'fun furs'. They were accessible, fashionable and relatively inexpensive.

In the last quarter of the twentieth century there was a marked decline in the wearing of fur in the UK but since the mid-1990s a resurgence has occurred with 150 of the world's top fashion designers showing fur in their 1997 collections, Dior, John Galliano, Alexander McQueen, Gucci, Givenchy, Valentino to name a few. In 1999 240 leading designers included fur in their collections.

### 15.1.3 Husbandry and harvesting

The skins of fur-bearing mammals are taken when they are in their prime. In the wild they are usually in their best condition during the winter months when it is coldest. The fur covering on the animal is then at its most abundant and best colour, and the skin is at its thinnest and most supple. In spring the animal will moult and have a thinner covering of hair for the summer months. The skin will also undergo changes and begin to take on a reserve of fatty tissue. In autumn the animal will grow a new winter coat and the skin will again change and the fatty tissues in the skin dissolve. These seasonal changes affect the quality, colour and grade of pelt which in turn affect the value. Prime skins are the best and therefore of the highest value.

It was not until the latter part of the nineteenth century that commercial raising and breeding of fur-bearing animals was introduced. In America the first experiments in mink (*Mustela vison*) farming were undertaken in 1866 followed in 1887 when silver fox (*Vulpes vulpes*) farming was begun; however, in the UK fur farming was not undertaken until the early twentieth century.

Commercial farming began slowly and tentatively and was not always successful. Many species breed only once a year and only certain species do well in captivity in pens. Animals that are confined are also more prone to disease, parasites and bacterial infections. Good husbandry, sanitation, feeding and management are essential and the early fur farmers had a great deal to learn in understanding the animals' habits and general characteristics. They also needed an understanding of genetics and a planned and well-recorded breeding programme in order to have the likelihood of breeding good stock and not to in-breed too closely and thus have unplanned mutations and weaknesses occur in successive generations. Today with all the benefits of modern science, zoos are still struggling with breeding certain animals in captivity and are not always successful.

'Pelting', the taking of the skin, is only done when the fur is at its best quality and colour and a diseased and stressed animal will not be of any use to the furrier as the animal's health and condition is shown in the quality of its fur.

In the wild furbearers will produce indefinitely if their habitat is viable. For species that are still hunted, population and habitat management ensures this viability. This is achieved by scientific monitoring by professional wildlife biologists and governmental regulations which allow for a certain number of a species to be harvested each year without causing threat to the survival of that species.

Today the most common farmed furs are mink (*Mustela vison*) and fox (*Canidae* sp.). Other species farmed on a smaller scale include nutria (coypu, *Myocaster coypus*), chinchilla (*Chinchilla lanigera*), fitch (polecat also known as foul marten, *Mustela putorius*), sable (*Mustela zibellina*), and finnraccoon (Tanuki *Nyctereutes procyonides*). From a scientific point of view, fur animals that have been domesticated (farmed) for more than ten generations are considered to be so far genetically removed from their ancestors that they must be treated as a fully domesticated subspecies.

Fur farming is regulated and in the European Union, Council Directive 98/58 sets down rules covering the welfare of farmed animals, including fur-farmed animals. Directive 93/119 deals with the slaughter and killing of fur and other farmed animals. In North America, fur farmers also follow strict codes of practice and conform to provincial, state or national animal welfare and other regulations.

The most significant change in the long history of the fur trade is that by the end of the twentieth century 85% of the world's commercial furs were produced on farms.

In November 2000 a bill to ban fur farming in England was passed by Parliament and came into force at the end of 2002 forcing the 13 mink farms to close down.

It is useful to the conservator to know what are the most likely species which may be found in fashionable dress and when they were used. It does not, however, preclude any of the 100 or so species appearing in an object but it may be relevant to know when a particular species was fashionable and available in order to help pinpoint the likely date of the object concerned, and how the fur may have been processed.

### 15.1.4 Some fashionable furs and dates

#### 15.1.4.1 Later middle ages and into sixteenth century

Lynx (*Lynx* sp.)
Genette (*Viverra genetta*)
Ermine (stoat; *Mustela erminea*)
Mink (*Mustela vison*)
Lettice (skins of the snow weasel; *Mustela* sp.)
Sable (*Martes zibellina*)
Squirrel (English, European, Russian, *Sciurus* sp.)
Coney (rabbit; *Oryctolagus cuniculus*)
Lamb (*Ovinae* sp.)
Wolf (*Canis lupus*)
Hare (*Lepus europaeus* and *Lepus timidus*)
Fitch (polecat; *Mustela putorius*)
Shankes or budge (kid; *Caprinae* sp.)
Leopard (*Panthera pardus*)
Otter (*Lutrinae* sp.)
Swan (*Cygnus* sp.) is also listed in the Wardrobe Accounts of Queen Elizabeth I, 1561 and 1590, and was prepared by her skinner William Jurden (Arnold, 1988).

#### 15.1.4.2 Sixteenth and seventeenth centuries

Squirrel (English, European, Russian; *Sciurus* sp.)
Ermine (stoat; *Mustela erminea*)
Beaver (*Castor canadensis*)
Lettice (skins of the snow weasel; *Mustela* sp.)
Mink (*Mustela vison*)
Sable (*Martes zibellina*)
Marten (*Martes americana*)
Foynes (stone marten; *Martes foina*)
Budge (lamb skin; *Ovinae* sp.)
Fitch (polecat; *Mustela putorius*)
Lamb (*Ovinae* sp.)
Otter (*Lutrinae* sp.)
Fox (*Canidae* sp.)
Genette (*Viverra genetta*)
Coney (rabbit; *Orctolagus cuniculus*)
Cat (*Felidae* sp.)

Lynx (*Lynx* sp.)
Hare (*Lepus europaeus* and *Lepus timidus*)

#### 15.1.4.3 Nineteenth century

In 1851 the Hudson's Bay Company exhibited at the Great Exhibition and showed examples of the following furs from Canada:

Racoon (*Procyonidae* sp.)
Beaver (*Castor canadensis*)
Chinchilla (*Chinchilla lanigera*)
Bear (*Ursidae* sp.)
Fisher (*Martes pennanti*)
Fox (*Canidae* sp.)
　Red (*Vulpes vulpes*)
　Cross (*Vulpes vulpes*)
　Silver (*Vulpes vulpes*)
　White (*Alopex lagopus*)
　Grey (*Vulpes cinereoargenteus*)
Lynx (*Lynx* sp.)
Marten (*Martes americana*)
Mink (*Mustela vison*)
Musquash (muskrat; *Ondatra zibethica*)
Otter (*Lutrinae* sp.)
Fur seal (*Oteriidae* sp.)
Wolf (*Canis lupus*)

Furs from the rest of the world were also shown and included:

European marten
　Stone (*Mustela foina*)
　Baum (*Mustela martes*)
Squirrel (*Sciurus* sp.)
Fitch (polecat; *Mustela putorius*)
Kolinsky (*Mustela sibirica*)
Ermine (stoat; *Mustela erminea*)
Blue fox (*Alopex lagopus*)
Astrakhan (karakul lambs; *Ovis aries*)
Leopard (*Panthera pardus*)
Coney (rabbit; *Oryctolagus cuniculus*)
Genet (*Genetta ginetta*)
Sable (*Martes zibellina*)

#### 15.1.4.4 Twentieth century

In 1922 the skins of the following animals were available and regularly used (Sachs, 1922):

Badger
　Canadian (*Taxidea taxus*)
　Japanese (*Meles ankuma*)
　Russian (*Meles meles*)
Bear
　Black (*Ursus americanus*)
　Brown (*Ursus arctos*)

Grizzly (*Ursus arctos*)
White (*Ursus mairtimus*)
Beaver (*Castor canadensis*)
Cat
  Wild
    European (*Felis sylvestris*)
    North African (*Felis sylvestris lybica*)
    North American (*Felis rufus*)
  House (*Felis catus*)
Civet (true civet cat; *Viverra zibetha*)
Chinchilla (*Chinchilla lanigera*)
Dog
  Chinese (*Canis familiaris*)
Ermine (Stoat; *Mustela erminea*)
  Canadian (*Mustela nivalis*)
  Russian (*Mustela nivalis*)
Fisher (*Martes pennanti*)
Fitch (polecat; *Mustela putorius*)
Fox
  Blue (*Alopex lagopus*)
  Cross (*Vulpes vulpes*)
  Grey (*Vulpes cinereoargenteus*)
  Kit (*Vulpes velox*)
  South American (*Dusicyon* sp.)
  Red Australian (*Vulpes vulpes*)
  Russian (*Vulpes* sp.)
  Japanese (*Vulpes* sp.)
Japanese fox (a species of racoon dog; *Nyctereutes procyonides*)
Goat
  Grey (*Caprinae* sp.)
  Black (*Caprinae* sp.)
Guanaco (*Lama guanicoe*)
Hamster (*Cricetus cricetus*)
Hare (*Lepus europaeus* and *Lepus timidus*)
Kid (*Capra hircus*)
Kolinsky (*Mustela sibirica*)
Lamb
  Persian (*Ovis aries*)
  Broadtail (*Ovis aries*)
Leopard (*Panthera pardus*)
Lynx, Canadian (*Felis lynx*)
Marmot (*Marmota* sp.)
Marten (*Mustela martes*)
Mink (*Mustela vison*)
Mole (*Talpidae* sp.)
Monkey (*Colobus polykomos*, *C. angolensis*, *C. satanus* and *C. guereza*)
Musquash
  Southern (*Ondatra* sp.)
Nutria (*Myocaster coypus*)
Opposum
  American (*Didelphis virginiana*)
  Australian (brushtail possum; *Trichosurus vulpecula*)

Otter, Canadian (*Oryctalagus canadensis*)
Rabbit
  White (*Oryctalagus cuniculus*)
  Chinese (*Oryctalagus cuniculus*)
Raccoon (*Procyonidae* sp.)
Sable (*Martes zibellina*)
Sea otter (*Enhydra lutris*)
Seal
  Hair (*Phocidae* sp.)
  Fur (*Oteriidae* sp.)
Skunk (*Mephitis* sp.)
Squirrel (*Sciurus* sp.)
Thibet (*Caprinae* sp.)
Tiger (*Panther tigris*)
Wallaby (*Macropodidae* sp.)
Wolf (*Canis lupus*)
Wolverine (*Gulo gulo*)

#### 15.1.4.5 Late twentieth and early twenty-first centuries

At the close of the twentieth century and the beginning of the twenty-first century the following animals were readily available to the fur trade. Many previously used species are now protected, i.e. monkeys, civet, large cats including ocelot, jaguar, and sadly several appear on the CITES Red List:

American marten (Canadian sable; *Martes zibellina*)
Australian opposum (*Trichosurus vulpecula*)
Beaver (*Castor canadensis*)
Chinchilla (*Chinchilla lanigera*)
Coyote (*Canis latrans*)
Fisher (*Martes pennanti*)
Fitch (polecat; *Mustela putorius*)
Fox (*Vulpes* sp.)
Lamb (*Ovinae* sp.)
Lynx, Canadian (*Lynx lynx*)
Marmot (*Marmota* sp.)
Mink (*Mustela vison*)
Mole (*Talpidae* sp.)
Musquash (muskrat; *Ondatra zibethicus*)
Nutria (coypu; *Myocaster coypus*)
Rabbit (coney; *Oryctolagus cuniculus*)
Raccoon (*Procyonidae* sp.)
Sable (*Martes zibellina*)
Squirrel (*Sciurus* sp.)
Ermine (Stoat; *Mustela erminea*)
Weasel (Tanuki, *Nyctereutes procyonides*)

Others may be available as part of licensed animal population management programmes – culls:

Seals (*Phocidae* sp., *Oteriidae* sp.)
Bears (*Ursus arctos* and *Ursus americanus*)

## 15.2 Structure, morphology, dressing and making

### 15.2.1 Definitions and terminology

Understanding the differences in terminology used in describing fur and its preparation as opposed to the preparation of leather is an essential introduction to the art and science of fur dressing.

In common usage the word 'fur' is used to describe the hairy outer coat of certain mammals.

To the fur trade the term 'fur-skin' is most usually used and applies to the hairy outer coat of a mammal which is inseparably attached to the skin proper. To a fur-skin processor (or fur-dresser) and to the furrier concerned with the manufacture of fur garments, etc. 'fur' applies to the hairy outer coat of mammals used in the fur trade. Included in these are sheep and lambs which by convention are commonly described as possessing wool.

The term 'pelt' is generally used to refer to the leather side of the 'fur-skin', which is also referred to as 'skin'.

Fur-skin processing or 'dressing' refers to the preparation (dressing) and dyeing of raw fur-skins for the purpose of making them non-putrescent and suitable for commercial use.

The term 'fur' also refers to the 'down' or 'undercoat' of a skin while 'hair' refers to the longer stiffer guard hairs which are the outer covering or top hairs. During fur processing sometimes these guard hairs are removed to expose the soft underfur. Some animals only have hair, e.g. horses and cows have no fur, only hair.

To the leather trade 'pelt' is commonly used to designate the hide or skin prepared for tanning by removal of the hair or wool, epidermis and flesh. 'Skin' is used to describe the outer covering of an animal of the smaller kind, e.g. sheep, goats or of immature animals of larger species such as calf and pony; also the skin of a fur-bearing animal dressed and finished with the hair on, e.g. sheepskin.

'Hide' is the outer covering of a fully grown animal of the larger kind, e.g. cattle, horse, camel, elephant.

In the leather industry a 'dresser' is distinguished from a 'tanner' in that the dresser carries out the finishing of tanned or semi-tanned leather and this includes dyeing, glazing and the final finishing processes as required before the skin is marketed (sheep, horses and cattle are treated both by skinners and leather dressers for different purposes).

There are few differences across the continents in the manner in which raw skins were and are commercially prepared. The differences lie primarily in the recipes used for the dressing.

### 15.2.2 Brief history of fur-skin processing and dyeing

The discovery of alum as a preservative was probably made by the Chinese who claim to have used furs for 3000 years. Alum is an important substance in the history of leather and fur-skin dressing.

Sylvia Matheson in her publication *Leather in the Lands of Ancient Persia* gives a first millennium BC Babylonian recipe for processing very special leather (Matheson, 1978). It reads:

*The skin of the kid thou shalt feed with the milk of a yellow goat, and with flour; thou shalt anoint [it] with pure oil, ordinary oil, and the fat of a pure cow. Thou shalt dilute alum in pressed grape juice, then fill the surface of the skin with gall nuts of the tree-cultivators of the Hittites.*

It is logical to assume that as the Babylonians used and traded furs, a recipe similar to this would have been used for their preparation and dressing.

By Roman times a distinction was made between tanners (*coriarii*) and fur-skin dressers (*pellioni*) and it was recognized that each was a separate distinct skill and process. In addition to dressing, pellioni included in their activities the making up of certain items and a general trading in fur-skins.

Tanners concentrated on bark tanning and tawyers concentrated on those skins for which the use of alum and oil was more suitable. Among these were fur-skins. (Tawe comes from the Anglo-Saxon meaning to prepare.) The word 'tawyer' has long been associated with alum dressing (Kaplan, 1971).

By medieval times the furrier combined the activities of fur-skin dresser with that of manufacture of garments and trimmings. The guilds covered many branches of a particular trade and it was considered fraudulent practice to dye fur-skins. However this practice did not include trimmings matched to a garment and dyed to red, blue or green, only the practice of dyeing to cover up natural faults in a skin, light spots or patches with the intent to deceive. Black dyeing was especially forbidden and remained so until the seventeenth century.

According to the records of the Skinners Company alum and oil were used for fur dressing but by 1593 alum dressing had been partially replaced by oil. The method used for preparing fur-skin was to stretch the skin on a frame and sponge the flesh side several times with a solution of alum and salt, or oil.

Lightweight fur-skins such as squirrel, beaver and fox were first greased with oil, butter or other fatty substances and trampled with bare feet in a barrel until the skin was rendered pliable. Fleshing with a scraper then followed.

In the seventeenth, eighteenth and nineteenth centuries dyeing was less restricted although still controlled. Black and brown were dyed onto fur-skins using plant products such as logwood, sumac and gallnuts together with mineral substances such as verdigris, iron filings, alum, or copper scale. The processing method of 'foot tubbing', trampling greased fur-skins in a barrel, to make them pliable was carried on until into the twentieth century.

Very little changed for nearly 300 years. By the mid-nineteenth century demand for furs had increased to the extent that large factories employing hundreds of workers were set up in Europe to process furs rather than small workshops with only a small number of craftsmen. Methods of dressing were still dependent on alum, salt and natural fats, and for dyeing upon vegetable products and metal salts. The fur seal which became popular around 1847 was dyed by repeated brush applications of dye in order to preserve the leather from damage by the strong dye, the composition of which was based on that used in the French silk industry (Kaplan, 1971). By the end of the nineteenth century mechanization of some processes had taken place. The 'foot tub' was replaced by wooden 'kickers' derived from the textile trade's 'fulling' mills and cleaning drums were rotated by power-driven belts. The discovery of synthetic coal tar dyes by Perkin in 1856 had no immediate application in fur dyeing and it was not until the end of the century when oxidation dyes were patented that coal tar derivatives could be applied to furs. The use of para-phenylene diamine enabled a greater range of colours to be obtained on fur.

### 15.2.3 Hair and fur fibres

The structure of mammalian skin has been described in detail in Chapter 1 but for the purposes of this chapter, I shall briefly revisit it in the context of fur/hair.

The mammalian skin is constructed mainly from a series of interlaced protein fibres of which the four main types are described as follows:

1. The hair fibres rooted in the follicles of the epidermis. These are frequently described as 'fur'.
2. Muscle fibres fixed between the grain and the base of the hair follicles.
3. Collagen fibres forming the bulk of the grain and corium, and constitute what is generally known as the pelt or leather.
4. Elastin fibres lying mainly on the grain and flesh layers.

The fibre, described variously as hair, wool or fur depending on the animal type, is composed of the protein *keratin*.

The fur-skin processor deals with the keratin fibres and the three other types of fibrous proteins in an inseparable form, his aim being to maintain and enhance not only the properties of the fur fibre but equally the skin fibres.

The function of these keratin fibres is to serve as a protection for the skin and to insulate against weather and other hazards of the environment. This is true of warm climates as much as cold; fur-covered tropical animals are insulated against strong sun and heat.

Keratin fibres are themselves poor conductors of heat. The air trapped between the fibres helps to form an efficient insulating layer, preventing rapid loss of heat which is why animals from cold regions are more fully furred than those from warm climates. Certain species of mammals, such as pigs, have very little hair and in this case a layer of fat under the skin plays the role of temperature regulator. Some aquatic mammals such as seals possess both a well-developed fur fibre and a considerable layer of fat.

The hydrophobic nature of keratin preserves the skin from direct contact with water.

### 15.2.4 Keratin

Keratin is the generic term applied to chemically related resilient protein materials including hair, scale, hoof, horn, nail, claw, beak and feathers.

Proteins are materials formed by amino acids linking together to form complex high molecular weight natural polymers.

A typical protein chain may be represented by

$$-NH-CHR^1-CO.NH.CHR^2.CO.NH.CHR^3.CO-$$

where R may be the same or different organic radical. It is the type and distribution of these different organic radicals which gives each individual protein or group of proteins their characteristic properties.

The amino acid composition of the keratins is complex and has been described by Vincent (1978).

All keratins contain significant amounts of the diamino acid cystine, which enables the formation of particularly stable covalent chemical linkages between adjacent protein chains of the material.

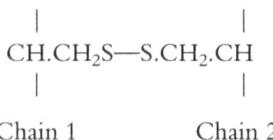

Keratin can therefore be considered a long chain protein that owes much of its properties to the disulphide bond of cystine. A typical molecular weight of the individual keratin chains varies from 2000 to 12 000.

The disulphide link S—S between adjacent polypeptide chains also accounts for the complete insolubility of keratin in polar solvents, and its limited lateral swelling and high wet strength.

In addition to the strong covalent disulphide links there are a large number of relatively weak hydrogen bonds between adjacent protein chains. Most hydrogen bonds are disrupted at temperatures between 40 and 60°C, which is why heat or steam can successfully be used to relax keratin enabling it or objects made from it to be shaped or moulded. Keratin is remarkably chemically stable to changes in relative humidity (RH) and not easily damaged by chemical attack. It is, however, damaged by strong acids and alkalis and reducing agents and is particularly vulnerable to poor handling practices and insect attack.

The keratins are subdivided into two groups: soft keratins (epidermis or skin layer) containing low levels of sulphur and hard keratins (hair, horn, etc.) containing more than 3% sulphur. The roots of the hair are more susceptible to chemical attack than the mature fibres because some of the cystine has not yet formed crosslinks and still exists as side chains.

### 15.2.5 Morphology of hair

Each hair is composed of three sections, the shaft, the root and the bulb. Hairs are formed from the epidermis. In the place where the hair is to appear, the epidermis becomes thicker and forms an indentation. This deepens into the derma to constitute a follicle. Hairs are always fixed obliquely in the surface of the skin and can be straightened, in life, by the contraction of the small muscles (erector pili) which are attached to the follicle. These come into effect when the animal is attacked or afraid. When the contraction ends, elastic fibres (elastin) surrounding the sheath of the hair return it to its original position. Cold operates in the opposite direction, realigning hairs so as to increase the volume of air trapped

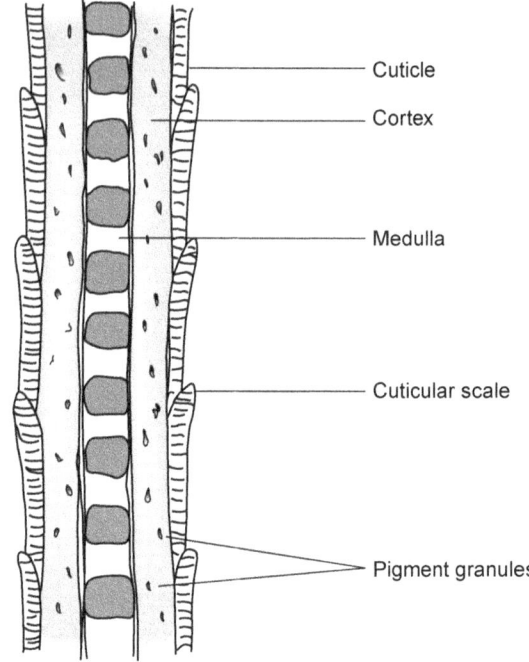

**Figure 15.2** Diagram showing morphology of hair.

between them and thus diminish heat loss from the body.

The character and appearance of hair fibres differ from species to species and also within the same animal.

Hair, examined under the microscope, is formed of three distinct layers: cuticle or exterior, corticle or cortex and medulla (*Figure 15.2*).

#### 15.2.5.1 Cuticle

The cuticle is formed in small scales and is composed of horny flat cells. The form of the cells varies according to the species and also according to the type of hair, e.g. guard hairs or fur. The disposition of scales may differ from root to tip.

(a) Scales may surround the shaft completely in ring form, the upper part of the scale covering the lower part of their scale directly above it. This type of structure is more often encountered in soft underfur.
(b) Scales do not entirely surround the shaft but have the same form covering one another. This usually occurs in guard hairs, which are generally more lustrous than those of the first category (a).
(c) Scales may be placed side by side without overlapping. This occurs usually in 'feeler' hairs.

The cuticle preserves the interior of the hair from destructive action. The lustre of the hair depends upon scale structure. The system of cuticle cells forming the imbricate (overlapping) surface of animal hair is responsible for the greater resistance to motion in tip to root direction. Gloss/lustre is dependent on the degree of light reflectance from the surface of the fibres and is highly prized by the furrier as a measure of quality. Two factors are involved, (a) the structure of the scales of the individual fibres, and (b) the parallelism of the fibres as a whole. If the fur fibres are untidy and run in random directions, light is scattered upon the surface and the fur looks dull, while if the fibres lie straight and parallel a good gloss is obtained.

### 15.2.5.2 *Cortex*

Below the cuticle lies the corticle layer, which consists of cells composed of fibrils and microfibrils in an amorphous matrix. The cells are positioned along the axis of the fibre and are surrounded by cementing material composed of amorphous keratin. The cells themselves are crystalline in form and constitute about a quarter of the total material. The thickness of the cortex varies with the species. In the hair seal 96% of the total diameter of the fibre is cortex, in the squirrel 34% and in reindeer hair it is absent entirely. The resistance of hairs depends on the development of the cortex. The outstanding feature of poor wear is a high percentage of medulla, with very little or no cortex and little cuticular layer. Reindeer hair is therefore very brittle while hair seal is very resistant.

### 15.2.5.3 *Medulla*

In the interior of the cortex there is usually the medulla composed of polyhedral cells, which may also enclose pigment cells and bubbles of air. The diameter of the medulla varies according to the dimensions of the hair shaft and often the medulla is absent from the tip. In underfur the medulla is less well developed than in guard hair. Some hairs have no medulla at all.

The results of chemical treatments and dye absorption will to a large part be determined by the properties of these histological phases (Kaplan, 1971).

Pigment granules are distributed in the cortex and sometimes in the medulla of most animal fibres, and are responsible for the wide and varied range of colours of fur-bearing mammals. These granules are protein and known as melanin. The colour of the pigment may appear as black, brown, reddish-brown, yellow or red. The pigment is insoluble in most solvents and is resistant to concentrated acids but may be dissolved in alkalis. It is the distribution and quantity of the pigment together with the colourless air bubbles in the hair fibre, which account for the wide variation of hair colour in animals. The colour is partly due to pigmentation and partly light interference.

## 15.2.6 Fur-skin dressing

### 15.2.6.1 *Introduction*

Very little has been written on the subject of fur preparation and finding detailed information about the practical aspects of fur dressing and furriery requires a long literature search for the conservator who may have to treat fur in the course of their professional activities. The fur trade has been in decline for a quarter of a century and although it is now on the increase due to an upturn in the fashionable status of fur, the militant stance taken by many anti-fur protesters has made it is almost impossible to find a furrier to talk with in person about the subject.

The author has found only five books published in English which describe in detail the methods involved with the processing of furs (Austin, 1922; Kaplan, 1971; Rosenberg, c.1920; Bacharach, 1930; Samet 1950). There are also a few publications in Russian and German (Pense, 1955; Hahn and Weigelt, 1967). All of these were published prior to 1971.

### 15.2.6.2 *Dressing*

For in-depth descriptions of the processes of fur dressing I refer the reader to Kaplan (1971); Austin (1922) and Bacharach (1930). The following, however, is an introduction for the conservator who will need to know something of the methods used for fur preparation and dyeing, especially during the last two centuries: this being the most likely era from which furs will date that the conservator has to treat. Awareness of these processes will enable the conservator to understand the likely degradation processes to expect and also help them work out a suitable and compatible method of cleaning and further treatment.

The priority in fur-skin dressing is the preservation of the fur properties but as the skin and fur are inseparable all processes must be compatible to both and not cause damage to either.

The dressing must be sufficiently permanent to withstand the wetting and drying processes used in the manufacture of fur garments and accessories and the end result should be a supple, pliable skin, of good tensile strength to hold the stitching. The dressing should not be too heavy, nor should it be elastic so the skin holds its shape after stretching, and the natural gloss, colour and quality of the fur should be maintained. The fur and skin should be able to withstand normal wear and storage and subsequent cleaning by accepted fur trade methods (Kaplan, 1971).

### 15.2.6.3  Soaking

Fur-skins arrive at the skin dresser in a raw dry state, either opened flat, an incision having been made down the centre of the flank, or cased, the skin having been removed by cutting across the rump and hind paws then peeling the skin off. Some may have been salted. Most of the adhering fatty tissue has been scraped away before drying. The first operation is soaking to restore to the dried collagen approximately the amount of water it had in life and ready the skins for the mechanical and chemical treatments to follow. Salt is added to the soak. Some longhaired fur-skins are not immersed but are wetted by applying water or salt solution to the raw pelt either by brushing or by means of a poultice of wet sawdust. Skins may need to be 'drummed' (treated in a revolving drum) after soaking in order to remove excess water prior to fleshing. They may also need 'caging' (treated in a revolving cage to free them from the sawdust).

### 15.2.6.4  Fleshing

The next stage is fleshing: the removal of the thin protein membrane layer which separates the skin proper from the fat layers and organs of the body. In the past this was done by hand working the skins over a curved knife for small skins or over a beam for larger skins. Today this process is done by mechanical blades but still requires a skilled operator. It is important to remove this layer as it is not readily permeable and could inhibit the following chemical treatments. Great care has to be given to this process as cutting too deep will expose the roots of the hair and this will lead to hair loss at a later stage.

### 15.2.6.5  Unhairing

The process of mechanical unhairing is resorted to in the case of skins possessing longer top hairs purely for the purpose of improving the general appearance of such skins. The types of skins which undergo this process are beavers, seals, otters and nutria. These have a growth of hair which is pulled leaving the underfur. In such cases the pile becomes shorter and closer.

### 15.2.6.6  Pickling

The fleshed fur-skin is now ready for the chemical treatment, which is the basis of skin dressing. Pickling prevents bacterial attack and contributes to the hydrolytic breakdown of the non-collagen interfibrillary material within the skin structure. The oldest and best-known method of pickling is the use of 'Leipzig dressing'. This is a mixture of sulphuric acid and salt, and replaced the earlier method of treating with a mixture of organic acids derived from fermenting bran. This pickle may be applied to fur-skin either by immersion in vats or by brush. Immersion has the disadvantage of leaving the hair in an acid state and if left with an excess of acid remaining in the pelt, there is a danger that over time it will be damaged. It is customary to add ammonia to the oil at the oiling stage to neutralize the excess acid. The brush method is usually used for longhaired fur-skins. Other acids may also be used for pickling, i.e. formic, acetic, lactic and glycollic. Kaplan, writing in 1971, commented that due to the risks involved with sulphuric acid it was replaced with safer organic acids (Kaplan, 1971). Kaplan also comments that fur-skins dressed using an acid pickle are light and soft but will dry hard if they later become wet and are subsequently dried. Furthermore he comments that acid pickling is useful for opening up the collagen fibres but that a water stable dressing must be used in the subsequent treatment.

An alternative pickling method is to use an alum pickle. This gives a more resistant leather and is less stretchy. Aluminium salts do not enter stable combinations with collagen, unlike chrome salts. Solutions of alum or aluminium sulphate with salt are used and the skins are usually treated by soaking in vats.

Formaldehyde produces a dressing which is not reversed by soaking in water but it was not until the 1920s that this was used in fur-skin dressing. It is used in conjunction with other dressings.

### 15.2.6.7  Chrome tanning

Chrome is only used for processing fur-skins in special cases. Chrome-tanned fur-skins have a higher shrinkage temperature than any other type of dressing, they are resistant to bacterial attack, are durable and may be prepared with considerable softness and pliability. They do not, however, have the correct 'handle' for the fur trade. Chrome-tanned skins are thicker and heavier than acid- or alum-dressed skins and have less stretch when damped. Chrome also imparts a blue/green tone to the skin and fur, and renders them unsuitable for bleaching with hydrogen peroxide. It is therefore rarely used except in cases where they will be dyed with acid or other textile dyes in the temperature range of 60–80°C, which is well above the shrinkage temperature of acid or alum dressing (Kaplan, 1971).

Chrome is now generally used for sheepskins with the wool on or for processing deerskin.

### 15.2.6.8 Oiling

The last process of fur dressing, prior to dyeing, is oiling. Its purpose is to lubricate and prevent adhesion between the fibres of the pelt substance during the drying of the wet pickled fur-skin. Mechanical milling or 'kicking' follows the application of oil or fat to the pelt. This is the more modern method replacing 'foot tubbing'. It is essential that all parts of the pelt are lubricated and the penetration and distribution of the oil have a profound effect on the handle and physical properties of the dressing. The final part of the dressing process involves mechanical treatments to remove excess oil from the fur and the pelt and the stretching of the pelt. The skins are ('drummed') treated with dry sawdust in large revolving drums and then ('caged') treated in revolving cages as part of this mechanical process.

### 15.2.6.9 Beating

Beating the furs may follow which involves beating them with rattan canes to remove the last of the dust. This process may be carried out by machine or by hand. Beating may be done as part of fur processing but it is also done by furriers as part of routine maintenance of furs. The reason is not only to remove dust but also to free the fur fibres, breaking up the adherence of one fibre to another.

It is a process to be carried out with sensitivity as the ferocity of beating a strong pelted fur such as seal would undoubtedly do harm to a delicate fur such as chinchilla. The rhythm of hand beating was described to the author by a furrier as being like a *paradiddle* which is a basic drum roll of alternate double beats.

## 15.2.7 Dyeing

The process of dyeing furs is covered in depth by Austin but for the conservator I have included a brief introduction (Austin, 1922). Austin states: 'Natural furs of the more valuable kinds are above comparison with the majority of dyed furs.'

However, natural colours are not always in fashion and also show wide differences in individuals from the same species or may be of uneven colouring within one individual. Usually the purpose of dyeing fur-skins is to improve their appearance and hence the value. Occasionally furs are dyed in high colours – red, yellow, blue and green – in order to accommodate the demand of fashion, for example the fun furs which appeared in 1966 in order to bring fur to the fashion conscious young. These were most often rabbit but ermine, white fox, beaver, hare, lamb, kid, and squirrel have also been dyed a wide range of colours according to the demands of fashion.

About two thirds to three quarters of furs are dyed. Some such as fisher, leopard and ocelot are never dyed and others such as Persian lamb, astrakhan and caracul are always dyed (Kaplan, 1971; Austin, 1922).

Many of the more valuable furs are 'blended' by skilled application of fast dyes by hand to the tips of the hair only, in order to even out the colour differences and visually blend them together.

Many of the cheaper furs have been dyed and otherwise treated (plucked, sheared, etc.) to imitate furs of the rarer and more costly kind. Marmot, red fox, rabbit, hare, muskrat, squirrel, opossum, and racoon have all been used to imitate others such as mink, sable, marten, seal, chinchilla.

Dyeing furs is a complex process as the hair and pelt have the capacity to take up dye differently with the pelt absorbing dye much more readily than the hair. Different hair types also respond to dye differently with the top hair being more resistant to taking up dye than the underfur. The lustre of the hair can easily be adversely affected by harsh chemicals and dyestuffs and the temperature of the dyebath must not be greater than 35°C in most cases as the pelt will shrink and dry hard. The shrinkage temperature of collagen must be considered for all processes where heat is involved and the dye method used must not extract any of the chemical or other materials used in the initial skin dressing. The dyes also need to have good light fastness and resistance to rubbing.

Traditionally furs are dyed by the brush process whereby only the tips of the upper part of the hair are coloured, or by dipping where the entire fur including the leather is dyed. In many cases the use of both methods is necessary as dipping serves to give the overall required colour and then tipping refines this by treating only the upper part of the hairs.

There are three stages to the dyeing process:

1. Killing.
2. Mordanting.
3. Dyeing or colouring.

### 15.2.7.1 *Killing*

The purpose of killing is to render the fur fibre more receptive to subsequent mordants and dyes. Affinity for dye particles and salt solutions varies from fibre to fibre. The underfur is more receptive than the guard hair and the hair tips of the underfur are more resistant than the basal part. The horny nature of the keratin fibre has much to do with this. The surface of the hair is also covered with a fine coat of fatty material which renders the hair impervious to dye solutions. Alkaline solutions may be used to degrease the hair and to reduce these differences of affinity by hydrolytic breakdown of the cystine S—S linkages. Modern killing treatments also include oxydizing agents or reducing agents as an alternative.

### 15.2.7.2 *Mordanting*

Pretreatment of fur-skins with metallic salts is one of the oldest dyeing techniques. It is accepted as a general rule with certain exceptions that colours produced on mordanted fur-skins are faster than on unmordanted. In addition, dyeings on mordanted fur-skins are more intensive and require less dyestuff.

### 15.2.7.3 *Dyeing or colouring*

Four classes of dyes or colouring materials are used in fur-skin dyeing:

1. Vegetable or 'wood' dyes; materials of plant origin.
2. Mineral or inorganic dyes, which depend upon the precipitation of pigmented metallic compounds.
3. Oxidation or true fur dyes; organic synthetic intermediates (Ursol D to dye black was first marketed about 1894).
4. High-temperature dyes; those borrowed from the textile field and applied at temperatures above those of oxidation dyes and belonging to a range selected from acid, basic, vat, pre-metallized and disperse dyes.

### 15.2.8 Finishing

After the furs have gone through the dyeing process they are dried and finished. Finishing may involve the application of a salt solution to the backs of the skins in order to replace some of the salt lost during the dyeing. They may also be re-oiled prior to drumming with sawdust or sand. Drumming is an important operation in the finishing as the hair receives a polish and the full lustre and brilliancy of the colour is brought out. The final action of finishing is to stretch the furs and render them soft and flexible.

### 15.2.9 Pointing

Pointing is a process which may have been carried out to any of the silky longhaired furs which have been dyed black, but is usually associated with fox. In this process the furs are artificially supplied with the silvery hairs of the badger to simulate the appearance of a natural silver fox. Bachrach (1930) states the process involves individually inserting hairs using a strong waterproof glue. One, two or three hairs are inserted at a time with single-hair work being the most time consuming and therefore reserved for the best skins of highest value. The basal end of the badger hair is dipped into the glue, and blowing aside the fur fibre the end is set as low as possible into the peltry near the skin. The author has found no record of the glue recipe but most likely hide glue was used possibly with the addition of formaldehyde to crosslink the glue and make it insoluble in water (*Figure 15.3*).

Conservators should be sure to identify furs which have been pointed prior to carrying out any cleaning.

### 15.2.10 Making up into garments or accessories

The manufacture of a fur-skin garment involves selection and matching of skins for quality and colour, cutting the individual skins to appropriate size and shape, sewing the skins together, damping the pelts to enable the sewn skins to be stretched to flatten the seams and ease them exactly into the pattern shape required, nailing or tacking on wooden boards to retain that shape, drying them, drumming and caging them, then 'closing', which means assembling, interlining, finishing and lining the completed garment.

A cutter, a machinist, a finisher and a liner are all involved with the making of a fur garment. All are highly skilled but it is the skill of the cutter which is all-important. Cutting is complex work and the skins of each individual species of animal may be cut and seamed in a way that differs from all the others, although there are similarities between them all.

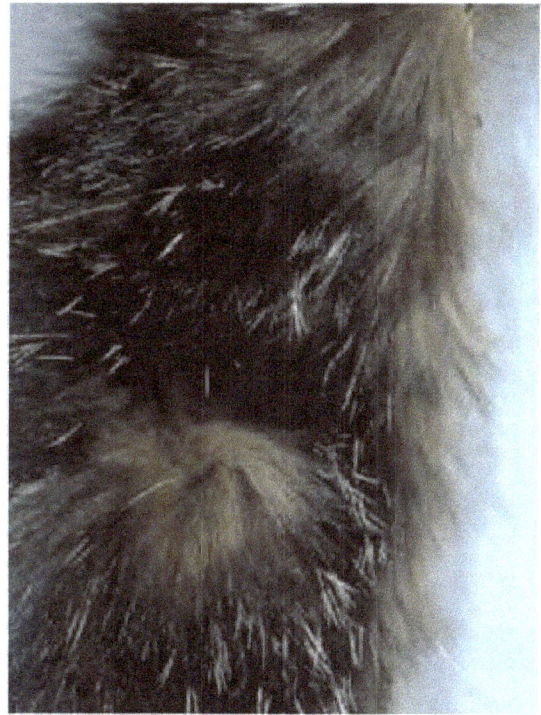

**Figure 15.3** Detail showing pointing.

Rosenberg describes these variations for over 60 of the most popular species but I include here a summary of the basic idea behind the art of cutting (Rosenberg, c.1920).

The process of making up starts with a pattern, in the same manner as making any other garment.

The cutter opens the skins by cutting them through the flank or belly. They are then damped and stretched out flat by hand, then trimmed by having the poor parts, such as heads, paws and sides, cut away. The fur varies from flatness and thinness at the head to fullness and density at the rump. It grows shorter the further it gets from the centre of the back until the belly or side is hardly covered. The more that is cut out the better will be the general appearance and evenness of the finished garment. It will also wear better.

There are two main systems of working which the skins may have undergone in order to manipulate them and join them together in the making of a garment. The first is the 'skin on skin' system and the second is 'dropping' or 'stranding'. The purpose of both is to obtain the necessary length of material required for the garment with the minimum of horizontal seams. An excess of these seams compared to the usual number required for a particular fur is regarded as a fault. Some skins such as leopard or seal are often long enough to be made up without a join.

This author has not found any references to a likely time when 'stranding' was introduced and the technique developed as a method for joining pelts together. Logically, the likelihood is that it was developed after the mid-nineteenth century when fur first began to be used as the outside material for garments, rather than for linings and trimmings only; and from a practical point of view it is likely to date from about 1900 when fur sewing machines were developed. The enormous amount of sewing involved in stranding pelts to make a coat would have been impractical in labour costs and time to undertake totally by hand sewing.

The only mid-nineteenth century mink coat examined by the author to date used the 'skin on skin' method for joining the pelts.

#### 15.2.10.1  *Skin on skin*

This method was used where the skins were too small to be stranded and such an expensive and time-consuming process was not justified. The joining of two skins one above the other means that rump is joined to head. As the hair on the rump is longer than that on the head it follows that the join will show.

The skill of the cutter is to contrive a seam that will show as little as possible (*Figure 15.4*).

#### 15.2.10.2  *Dropping or stranding*

The second method of joining skins manages to avoid cross-seams altogether and it is a method used for the more expensive skins, the value of which will justify the extra work and expense involved. It enables the ideal of long, slender, strands to be achieved from short, wide skins.

The skin is cut vertically down the centre and the cutting proceeds on one of its two halves. Working on one of these halves the first cut is made diagonally from the side of the skin to the centre, that is to say from the left edge of the skin running down to the right edge if it is the left half being dealt with, and from the right edge to the left edge if it is the other half. The cut starts not at the extreme edge of the skin but sufficiently within it to prevent the two parts, severed by this first cut, falling apart and it ends the same way just before reaching the edge.

A second cut is made parallel to the first cut but 6 mm (a quarter of an inch) below it: then another and another until the bottom of the skin is reached.

156  *Conservation of leather and related materials*

**Figure 15.4** Detail showing skin on skin method of joining furs.

**Figure 15.5** Detail showing stranding.

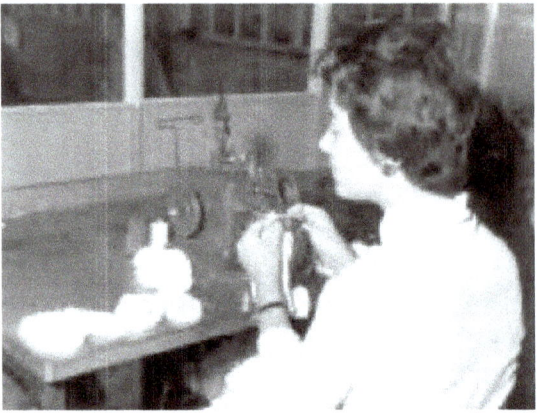

**Figure 15.6** Detail showing fur sewing machine. (Courtesy Stephen Kirsch)

### 15.2.10.3 *Sewing*

A fur sewing machine is not at all like a household sewing machine and is specially designed for the purpose of sewing fur. It has an electric motor and two pedals, one to switch the motor on and off and the other leading to one of the two round cups approximately 6 cm in diameter which are mounted horizontally in front of the machine. When the pedal is depressed the cup is brought slightly forward and when released the cup springs back in contact with the other cup which is fixed to the machine. The machinist completes the cutting of the first strip at both edges and the skin falls into three parts (the upper part, the strip and the lower part). The lower part is put aside and the upper part and first strip are retained. These two parts of the skin are pressed together again, hair side inside, ready to sew along the cut but the first strip is slid 8 mm (three-eighths of an inch) downwards towards the lower edge of the cut. The two pieces are inserted between the cups which having milled edges hold the fur securely. The edges of the fur protrude above the cups and as the machine is set in motion the fur pieces are driven through the machine and the edges are oversewn. At the same time the fur itself is prevented from springing out from between the cut edges by means of a pin used to press it down (*Figure 15.6*).

When the strip of fur has been sewn back to its original position but 8 mm further down this constitutes the first 'drop'. The effect has been to make the skin a fraction longer without moving any hair more than 8 mm from its original position. Once all the 'drops' have been sewn the end result will be to have lengthened the skin considerably but it will be

Depending on the length of the finished strand required, a piece from another skin can be grafted in or if a shorter strand is required then a section of the cut skin may be removed before machining so exactly the right length can be achieved. Next the cut skin is passed to the fur machinist (*Figure 15.5*).

much narrower. Once the second half has been treated the same way and these two halves finally joined together again then a long elegant strand will have been completed.

Some strands will need to be made curved and others made shorter in order to fit into the garment pattern but it is the responsibility of the cutter to place his cuts in such a position that they produce the necessary shape and size strands that when they are all fitted together they will make the finished garment according to the pattern set out.

Once all the strands have been joined together to make the various pattern pieces the raised seams must be flattened and the creases and wrinkles smoothed out. This process involves laying out the pieces pelt side uppermost on a board, wetting out the skin with water and then stretching the pelts and nailing them into position. Wetting will shrink back the pelts and the stretching and nailing process enables all wrinkles to be pulled out and the seams flattened as the pelts are pulled back to size and shape and pinned or nailed in place before being left to dry.

After drying the hair looks flattened so the next process is to drum the pieces for an hour in sawdust to restore the appearance of the fur. The type of sawdust is important and neither oak nor pine is suitable. Oak because of its high tannin content which in the presence of traces of iron could lead to staining of the fur, and pine because its resinous nature could cause stickiness. Beech, birch or poplar is used. The moisture content is particularly important as dry fur fibres, if subjected to mechanical friction, will develop static electrical charges and will cling together. In the presence of oily materials matting may take place and in extreme conditions felting could occur. For these reasons the moisture content of fur should not be allowed to drop below its natural regain of 10–12%.

Finally the pieces are interlined, usually with domette, assembled and joined together, and then the garment is lined and finished (*Figures 15.7* and *15.8*).

### 15.2.11 Plates and crosses

Offcuts and fragments of fur-skins left over after the best parts have been utilized are not discarded. These fragments may be the trimmed-away parts left over, or from parts of the animal of less use, i.e. ears, leg pieces, paws, underparts of the neck, etc. These are sometimes sewn together to make crosses or plates of fur. These large pieces made from the tiny fragments are used in the manufacture of lower

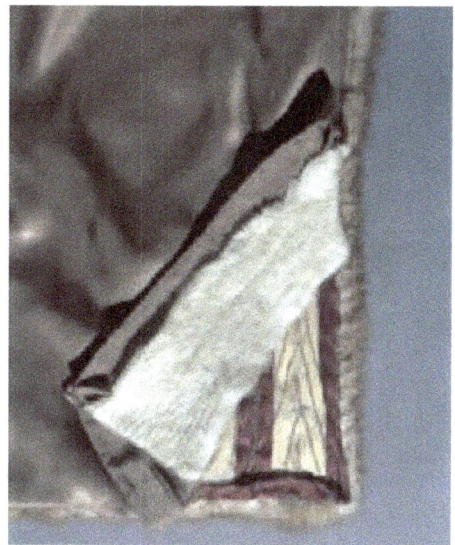

**Figure 15.7** Detail showing lining and interlining layers.

**Figure 15.8** Ermine evening coat *c.*1935–37 French (Paris) (T.272-1976). (Courtesy V and A Images)

priced garments and for linings. In China it is customary to sew these into the shape of a Greek cross. The reason being that it can quickly be manufactured into a garment by doubling it over in half and sewing along the sides and under the arms of the cross which will form the sleeves (Bachrach, 1930).

The lining of a Chinese kesi robe discussed in Section 15.3 was most likely made in this way (Kite, 1991).

## 15.3 Conservation and care

### 15.3.1 Introduction

Few pre-nineteenth century furs from fashionable dress survive due to the ravages of moths, but from whatever date the object originates it is important to identify the species of animals used. It is possible that this information is not included in the registered description of the object.

Important items of dress made from expensive fabrics are frequently altered during their useful lives and an out-of-fashion fur coat may have been altered and refashioned by a skilled furrier into a more up-to-date style. Due to the high cost of the better quality furs this was not and still is not an uncommon practice. Detecting evidence of such alterations in furs is much more difficult for the conservator or historian than it is to detect in a textile garment, and it is likely to be almost impossible without having access to the back of the skins to see the process of sewing and joining the pelts. It is likely that it is only the conservator, during the course of repair treatments, who will have the opportunity to see the backs of the skins. Conservators should therefore be aware of this possibility and attentive to notice any signs of alterations. Additional information regarding the quality of the garment, or origin of the skins, may also be found once the backs of the skins are examined and the cutting and arrangement of pieces can be seen. Skins in the raw state are often marked with distinctive signs to indicate their place of origin. Sometimes owner's initials or trademarks have been used. The Hudson's Bay Company used a range of initials to indicate the place of origin or province from where the animals were trapped which served to indicate to buyers the character of the pelts (Rosenberg, c.1920). Fur dressers and dyers also marked the skins. If discovered by the conservator in the process of their work, these marks should be recorded and retained.

It is also important to note that while sewing machines were pioneered in 1844 and in use from the 1850s, the fur sewing machine did not come into being until 1900, so although the linings of fur garments were machine sewn before this date, the fur-skins were not. Examination of the methods of skin preparation and sewing, style and cut, as well as knowledge of species availability, is therefore essential in order to date a fur garment accurately.

A furrier will clean and repair damaged fur garments in very different ways from those used by conservators. Methods employed by the furrier include cleaning by tumbling in drums with sawdust impregnated with white spirit or hydrocarbon solvents (historically, the aromatic hydrocarbon benzene, $C_6H_6$, was used), removal of damaged areas of skin or whole skins and replacement with new, possibly redyeing or 'blending' and then refinishing with the use of warm irons or gums to bring back the lustre of the fur.

These are not the options available to the conservator when dealing with museum objects and potentially brittle, degraded fur-skins. Museum codes of ethics and practice which govern conservators' professional conduct make such interventions unacceptable, also it is likely that replacement skins to match would not be easily available and totally unavailable under any circumstances if the item concerned had been made from animals from a now endangered and protected species.

A fur garment is an object which may have an important provenance or historic significance in its own right. Damage and evidence of use and wear are as important in furs which have been taken into museum collections as any other fashion item.

### 15.3.2 Species identification

Species identification is difficult but if it is not obvious by the appearance of the fur and a furrier cannot be consulted or is not familiar with a species used in an object which might be over 150 years old, then identification may be done with reference to the morphology of the hair. The scale pattern and cross-section of hair fibres differ from species to species, so microscopic examination of a cross-section and scale pattern may help. However, the morphology of hair will differ within the same species and within an individual animal so good comparative reference materials are essential and a conclusive result may not be obtained easily. Reference to the following publications may be of assistance to the conservator when attempting species identification of furs (Blazej et al., 1989; Novak, 1987; Rogers et al., 1989; Appleyard, 1978; Brunner, 1974).

Recently a method has been developed using mass spectrometry to identify the amino acid combination of each animal species (Liska and Shevchenco, 2003). The method relies on good comparative material and is currently being used by customs officials to identify illegally taken examples of threatened and endangered species. When further developed and the operational time has been reduced it will prove invaluable and a quick method to use, and may also be of use to museum curators and conservators for a positive species identification of materials from animal sources.

### 15.3.3 Damage

The damage a conservator is likely to encounter in furs in museum collections includes:

(a) General soiling due to use, which may include soiling and staining with foodstuffs, make-up, sweat or skin grease before the object was acquired; linings may be particularly soiled around the neckline.
(b) Surface soiling due to an accumulation of dust, and atmospheric pollution, acquired in the domestic environment during a long period of inappropriate storage before the object came into a collection, or due to years of display in unsealed cases in the museum environment.
(c) Wear due to excessive use during the pre-accession life of the object resulting in bald areas on the pelts.
(d) Tears due to accidental damage during wear or due to poor handling after the fur had reached a fragile and unsound state.
(e) Chemical damage or physical damage occurring during skin processing, the effects of which only become apparent with the passage of time; this includes rare but known occasions where the skin has been pared away too deeply during fleshing leading to hair loss. It also includes furs made from pelts taken early in the season when the animal had not completed the growing of the winter coat and the bulbs of the guard hair roots come through the skin. This will result in loss of some of these guard hairs during the dressing process and the fur will have a scanty appearance.
(f) Chemical damage due to the use of inappropriate insecticides. Camphor, naphthalene and PDB (para-dichlorobenzene) are harmful to some colours if used with dyed skins. Naphthalene and PDB are known to have been used in museums as insect inhibitors until the 1970s–1980s, the use of which may be the cause of damage now or in the future.
(g) Light damage leading to the fading of colours or to the yellowing of natural white peltries such as ermine and fox.
(h) Environmental damage, poor or inappropriate storage methods (too hot, too dry, too damp) leading to shrinkage, hair loss, embrittlement, mould and other damage to the pelts.
(i) Insect damage caused by moths (*Tinea* and *Tineola* sp.) or carpet beetles (*Anthrenus* and *Attagenus* sp.).
(j) Damage to linings, either through use and wear or due to the degradation of the lining fabric. Silk and synthetic materials have been used for the linings of fur garments.

### 15.3.4 Conservation methods

Conservation methods used to treat furs must take into account the properties of both hair and skin materials and what has already been done to them during the course of the fur-skin processing and dyeing. When considering items of dress, the linings and any additional trimmings or components of the object must also be taken into account and these may need treating separately, either in situ or they may need to be removed for treatment.

Ethically, stitching should not be unpicked unless absolutely necessary, following usual conservation practice; but it must be kept in mind that in many cases where repair to skins in items of dress is required then access to the back of the skin will be necessary in order to carry out the repair. This will mean that unpicking stitching and resewing parts of the lining and interlining will be unavoidable.

Before any treatment is undertaken, and the background information concerning date and possible provenance has been noted, the fur and skin should be examined closely from the hair side. If there is a smell of naphthalene (mothballs) this should be recorded.

By looking through the hair to the skin it may be possible to tell if the pelt has been dyed. It may not, however, be detectable other than by an expert furrier if the hair has been surface dyed or blended.

It is useful for the conservator having identified the species to know what are the natural colours of the animals concerned. If a fur is an unusual colour for a species or has been prepared to look like a different species, e.g. muskrat to look like mink, then it is reasonable to assume that the pelt has been dyed.

#### 15.3.4.1 Cleaning

Clean fur should be lustrous, so a dull appearance and an unpleasantly dry or a greasy feel to the fur will indicate that it is dirty or dusty. Providing there is not an overall shedding of the hair, a light vacuuming of the fur using low suction and a protective layer of muslin or net over the nozzle of the vacuum will remove surface dust and particulate soiling. Light brushing may also help dislodge particulate soils and insect frass which tends to get trapped at the roots of the hairs and is difficult to remove. If the fur has a greasy feel or there are localized areas of soiling then solvent cleaning may be carried out providing a fume chamber or other appropriate extraction facility is available. Working always in the direction of the hair, swabbing with white spirit BS 245 or Stoddard solvent or petroleum spirit is a useful method of removing surface soiling. The fur should be colour tested before such cleaning to ensure that if it has been dyed the dye is stable in the solvent. The fur should not be soaked with solvent but gently stroked with a solvent-soaked swab lifting off the soiling as work progresses. It may be possible to place absorbent paper below each section of hair to be cleaned and to swab out directly onto this. The fur should be well blotted to remove excess solvent and to speed drying. A large item may need to be treated in sections. An arctic fox fur lining of a large Chinese kesi robe, *c.*1800, was cleaned in this manner (Kite, 1991). Where fur garments have textile linings which require solvent cleaning, the conservator may prefer to clean the lining and the fur as two separate procedures, completing the cleaning of the lining and allowing the solvent to evaporate before cleaning the fur.

Another method of cleaning has been described by Del Re using Stoddard solvent and Vulpex, a spirit soluble soap (Del Re, 1988). This was used for cleaning particularly soiled ermine.

Keratin is stable in polar solvents but if a mixture of deionized water and detergent, or water and IMS (industrial methylated spirit, ethanol and 4% methanol) or IMS alone is used to clean the fur then care must be taken to avoid wetting the skin. Tests must be carried out prior to treatment to ensure that dye is not soluble in whatever solvent is chosen. A particularly soiled bearskin from the uniform of a colonel of the Grenadier Guards was cleaned by sectioning the hair and locally swabbing each tuft onto absorbent paper. First, deionized water was used, followed by a 1% solution of non-ionic detergent, and then to rinse, deionized water was swabbed through followed by a 50:50 mixture of water and IMS. This last mixture was used to remove any remaining soiling. Finally each tuft was swabbed with IMS alone to help remove excess water. Each tuft was then blotted thoroughly, combed and allowed to dry naturally (Kite, 1990). Fur, which has been cleaned using aqueous treatments, may be dried using a cool air blower. On no account should heat or hot air be used.

#### 15.3.4.2 Repair methods

A full-piled skin usually possesses a thin pelt while a thin-haired skin possesses a thick pelt. This guide is useful to the conservator, and if the skin is aged, dry and damaged may provide an indication to the overall strength and fragility of the skin.

A sewn repair to a torn skin is unlikely to be an option for the conservator working with fur garments of fashionable dress so the choice of adhesive and support material must be made with the nature of the skin in mind. The adhesive must not penetrate and impregnate the skin but it must be strong enough to hold firmly to the back of the skin and not peel away when the skin is flexed. The shrinkage temperature of the collagen must be kept in mind when a heat-set adhesive is chosen, and an aqueous adhesive must be able to be used in a form that does not allow the skin to become wet as there is a risk of shrinkage and hardening on drying. The adhesive should also be stable, non-yellowing and able to be removed should the support need to be taken off and the repair redone at a later date. The support material should be strong, light and flexible and ideally have a multi-directional non-woven structure, compatible with the multi-directional arrangement of the collagen fibres of the skin. Leather, gold beater skin, Japanese tissue (*kozo* mulberry fibre and *mitsumata* Edgeworthia papyrifera papers), Holytex, Reemay, Tetex, Pellon, Cerex and nylon gossamer have all been used as support materials for adhesive methods of skin repair.

The adhesives chosen for skin repairs have included Beva 371 (ethylene vinyl acetate), wheat starch paste, a wheat starch paste with the addition of sodium alginate, isinglass (sturgeon glue), Lascaux 360 HV and Lascaux 498 HV (butyl-methacrylate dispersion thickened with acrylic butyl-ester), Jade 454 and Jade 403 (polyvinyl acetate emulsion), Apretan MB Extra (formerly Mowilith DMC2 polyvinyl acetate emulsion), and PVA-AYF (polyvinyl acetate solution).

Several papers have been published describing adhesive methods of repair of skins including Lougheed *et al.* (1983), Dignard and Gordon (1999),

Dickens (1987), Kite (1990, 1991, 1999), Rae and Wills (2002) and Richardson (2002).

Only a few, however, relate to fur items of fashionable dress made up as described in Section 15.2 (Kite, 1990, 1991; Dignard and Gordon, 1999).

In order to carry out an adhesive repair on fur garments of fashionable dress, usually access has to be gained to the back of the skins. Although possible to work from the front, inserting a patch, the author believes that a more satisfactory result will usually be achieved if access can be gained to the back of the pelt. Knowing the construction of linings and interlinings will help the conservator gauge the amount of work needed and time to be allocated in order to unpick stitching of these and then resew them after the repairs have been made to the skins. Linings may also need cleaning and repair and in the worst cases of soiling and damage may need to be removed, cleaned and repaired separately before resewing into the garment.

### 15.3.5 Two case histories illustrating methods

Two case histories reduced from their original published form are included here to illustrate particular conservation problems involving fur and the methods chosen to solve them. For a full account of these case histories and object treatments see the original published works (Kite, 1990, 1991).

#### 15.3.5.1 *The conservation of an ermine lining to a blue felt Paquin cape. French, c.1936 (T123-1980) (Figure 15.9)*

In 1990, conservation was required to the ermine (*Mustela erminea*) lining of the hood from a blue felt cape by the designer Paquin (Kite, 1990). The fur was grubby and in some places the skins had started to split, and in two areas of particular damage the skins had fragmented. At some time these areas had been stuck down to the silk lining of the fur in an attempt to consolidate the skins. The adhesive had become hard but fortunately had not soaked through the skin or impregnated the fur.

The fur lining and attached ermine tie were removed from the cape for treatment and it was necessary to unpick the stitching and partially release the silk lining to the fur to gain access to the back of the skins. The adhesive was soluble in acetone and was easily removed by localized spot treatment using small cotton wool swabs soaked in acetone and a scalpel blade to scrape away the softened adhesive

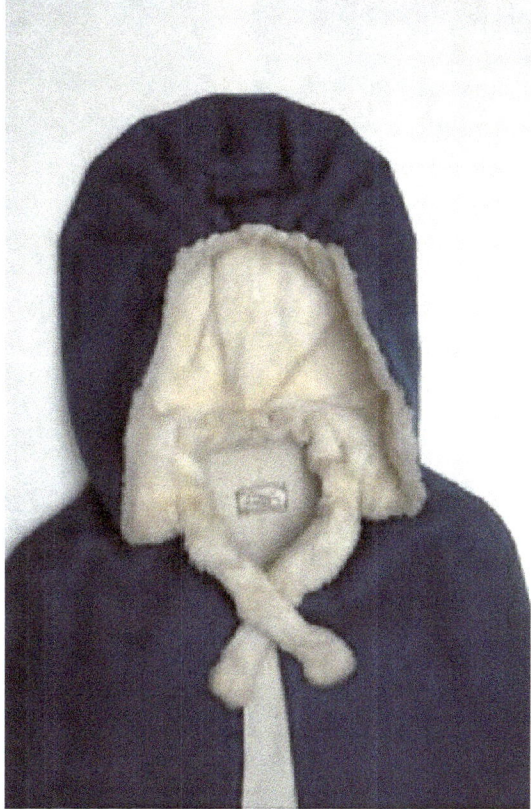

**Figure 15.9** Detail of ermine lining of hood from blue felt cape by Paquin, 1936 (T.123-1980).

from the back of the skin. It was considered inadvisable to flood the skin with acetone due to the risk of removing oils from the skin so its use was kept to a minimum. The general condition of the skins appeared to be good and they were not dehydrated or brittle. However, there was marked yellow staining in the damaged areas. When the cape was worn with the hood up these areas corresponded to the neck and ear regions of the wearer so it was possible that the owner had sprayed herself with perfume and these stains and damage could be directly attributable to the alcohol or other ingredient in the perfume causing the rapid oxidation of fat in the skin and accelerated degradation. The fur was grubby and had some lipstick marks so was cleaned by swabbing with white spirit BS245 onto absorbent paper.

Beva 371 was chosen as the most suitable adhesive for repair so a 15 cm$^3$ of Beva 371 film was prepared by ironing it onto a square of nylon gossamer

162  *Conservation of leather and related materials*

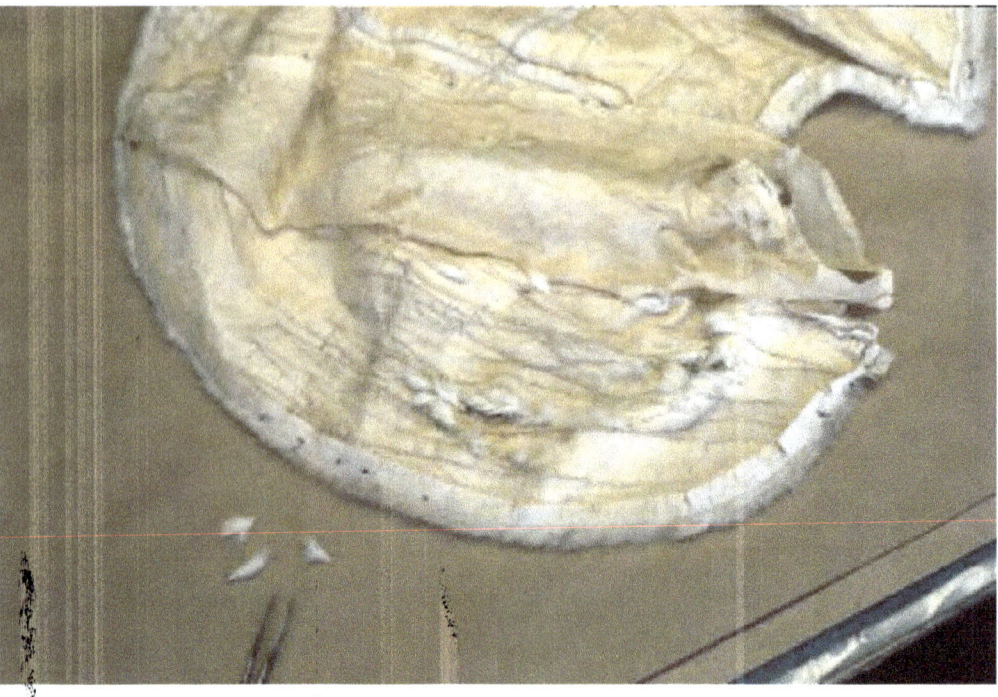

**Figure 15.10**  Detail of ermine lining of hood from blue felt cape by Paquin, 1936 (T.123-1980) showing damage to skins.

**Figure 15.11**  Detail of ermine lining of hood from blue felt cape by Paquin, 1936 (T.123-1980) showing same area after repair. Note yellowing of skin in area of damage.

**Figure 15.12** A Chinese Woman's Dragon Robe Lined with Arctic Fox Fur c.1800–50 (T766-1950).

between two sheets of silicone release paper. A spatula iron set at 70–75°C was used. The damaged fur hood was laid out on a soft board and the tiny fragments of skin were held in alignment using entomological pins stabbed through into the board. By ironing small straps of the Beva film onto each split, gradually the damaged area was repieced and made whole (*Figures 15.10–15.11*).

From the hair side, because there were a few tufts of hair missing, it could still be detected that there had been damage but the skin was again sound. The silk interlining to the fur was resewn in place and the most severely damaged part was covered with a new piece of silk habotai dyed to colour match. The conserved and reassembled fur hood lining was then restitched into the cape together with the ermine ties.

An alternative adhesive treatment can be illustrated with reference to the following brief case history.

15.3.5.2 *The conservation of a Chinese woman's dragon robe lined with arctic fox fur, c.1800–1850 (T766-1950) (Figure 15.12)*

A Chinese woman's dragon robe of yellow plain weave silk with tapestry woven roundels and trimmings, lined with arctic fox (*Alopex lagopus*), required extensive conservation prior to exhibition in 1991. The robe had long sleeves with horseshoe cuffs edged with brown fur; designed to be worn turned back. This fur was identified as mink (*Mustela lutreola*).

The robe and fox fur lining were grubby and in places had a grey appearance with the worst soiling being concentrated at the hem. The fur lining showed extensive insect damage and was splitting and torn. The fur trimmings to the cuffs were worn to the skin in many places and torn.

It was necessary to treat the silk robe, fur lining and cuff fur separately so stitching holding the fur elements was unpicked and the furs were separated from the robe.

Once removed the fur lining could be examined and its condition precisely assessed. An interlining was linked to the skin side of the fur which consisted of layers of paper and a layer of fine yellow silk. This was removed and retained to be sewn back onto the fur after the fur had been conserved. The fur lining was made from hundreds of scraps stitched together to make three distinct bands for each side, which were then stitched together to make up the lining. Although the fox fur was thick the skins were thin and papery and hair follicles could be seen protruding through the underside of the skins. At some time repairs had been made to the skins using what seemed to be a starch paste and an oriental paper similar to that used for the paper interlining. It was not known when nor where this was done but it was significant that the paper used was similar to that used in the construction of the object, suggesting that the two papers were contemporary. These paper repairs were so damaged with insect holes they were no longer fulfilling their purpose so they were removed.

Working in a fume chamber the fur lining was solvent cleaned by swabbing onto absorbent paper using white spirit BS245. The lining was so large it was necessary to clean it in sections waiting for the solvent to evaporate from one area before moving on to the next (*Figures 15.13–15.15*).

The adhesive and method used for the previous paper repairs, although damaged and no longer fulfilling their purpose, had not failed due to the nature of the technique. Moreover, the method was compatible with the oriental nature of the object and had caused no obvious damage or shrinkage of the skin where the patches had been applied. Experiments using wheat starch and various mixtures of wheat starch and sodium alginate were carried out and a number of *mitsumata* and *kozo* papers

164  *Conservation of leather and related materials*

**Figure 15.13** Detail, cleaning the arctic fox fur lining.

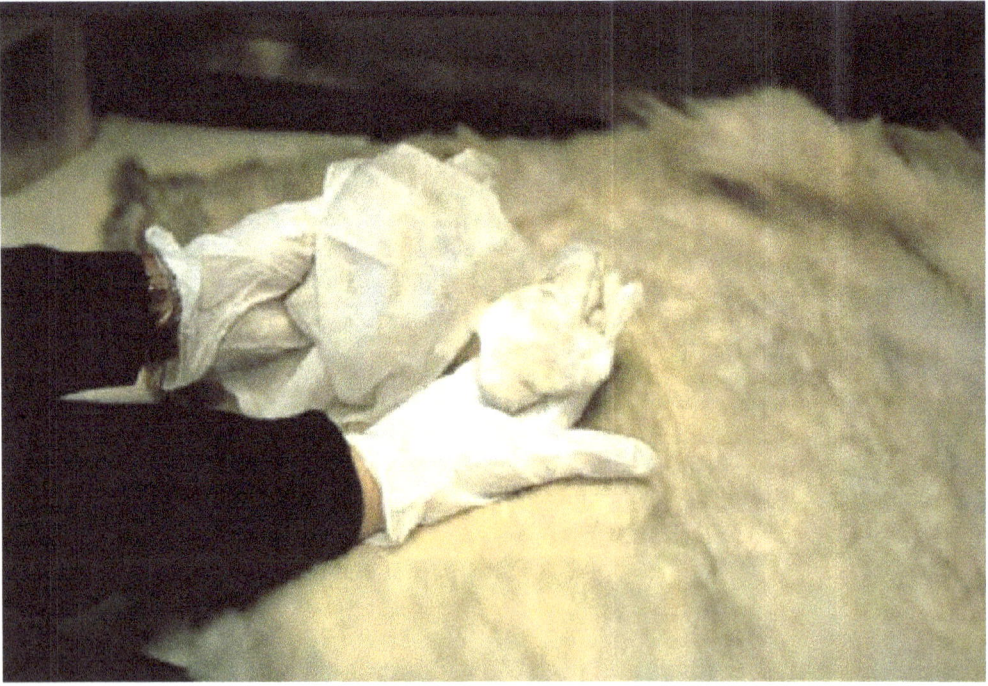

**Figure 15.14** Detail, cleaning the arctic fox fur lining.

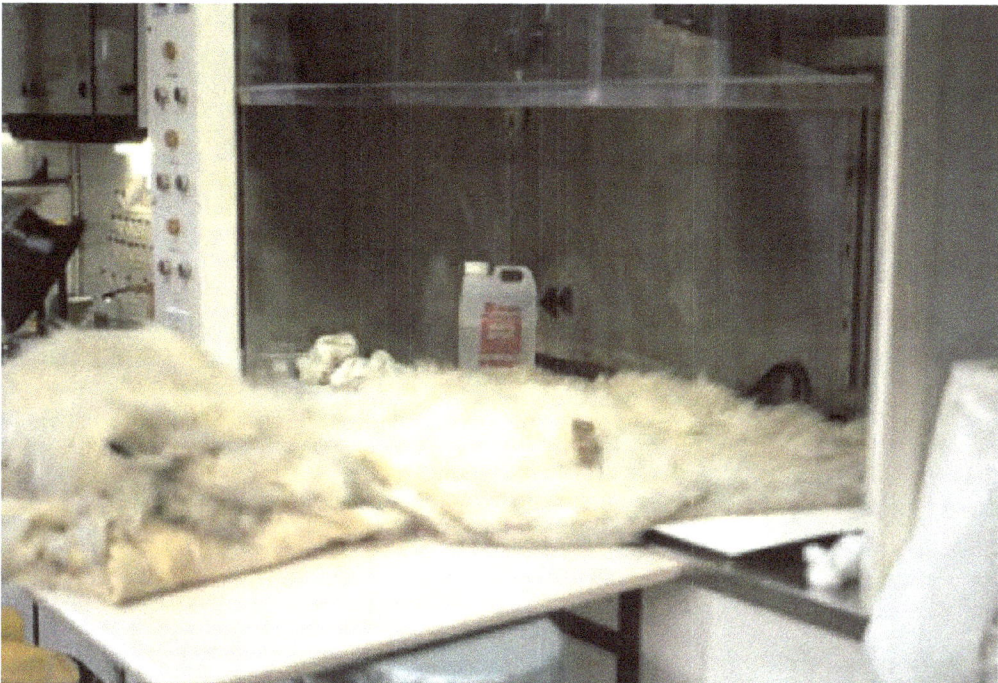

**Figure 15.15** Detail, cleaning the arctic fox fur lining.

were examined for suitability. A *mitsumata* paper was chosen. A rabbit skin was used as the test fur for these experiments as a sample of arctic fox was not available. The most satisfactory mixture and working consistency of the adhesive paste was identified and a suitable method of patch application was perfected on the test fur to ensure that the skin did not become too wet nor the patch, nor skin, shrink on drying. Once the method was perfected the arctic fox lining was repaired by the localized application of patches and a mixture of wheat starch paste and sodium alginate used as dry as possible (*Figures 15.16–15.17*).

The silk robe was cleaned and repaired using traditional stitching techniques used in textile conservation and the garment was reassembled. The fur bands from the cuffs were repaired using Beva film carried on Remay.

### 15.3.6 Freezing tests of adhesives

It is important to know that adhesives selected for use when carrying out conservation treatments to furs will remain stable if the object has to be deep frozen at a later date as part of an insect pest management strategy. In 1990 some tests were carried out. Beva 371, Silicone Adhesive SF2, wheat starch paste and mixtures of wheat starch and sodium alginate in various concentrations had been used for experimental repairs on a dressed white rabbit skin. In order to test the stability of these repairs the skin was deep frozen to −30°C and the results published (Kite, 1992). All the sample repairs had remained stable. Since this time the skin has been frozen for a second time. It is interesting to note that a recent examination of these sample repairs, after 13 years and many occasions of handling for teaching and demonstration purposes, showed all the repairs still to be stable although there was some darkening of the SF2 adhesive. Some of the paper and paste patches have been pulled away and manually tested to destruction over the years but where they have not been so mishandled they are stable and well adhered. The author has not carried out freezing tests on all the adhesives which may be used for the repair of fur-skins, and suggests that this is an area where further work needs to be done.

**Figure 15.16** Detail of damaged skin before repair.

**Figure 15.17** Detail of damaged skin after repair using *mitsumata* paper and a wheat starch paste and sodium alginate mixture.

### 15.3.7 Care of furs

#### 15.3.7.1 *General storage and display*

Fur garments are likely to be heavy so should be stored on substantial well-padded hangers. Enough space should be allowed so that furs are not packed tightly together and crushed. Cotton lawn or Tyvec bags should be provided to cover and enclose each item and a label with the object number and brief description should be fixed to the outside of the bag for easy identification (Heiberger, 2002).

A small photograph of the item would also be useful, included with the label, and fixed to the bag to aid identification and help limit handling. Smaller items and accessories are best stored separately in boxes with the boxes clearly labelled with a small photograph fixed to the outside of the box.

For display a garment should have adequate support so that it is not crumpled or under stress. A mannequin tailored to fit is appropriate for a coat and an accessory should have a purpose-made support or mount.

Although keratin is chemically stable all types of fur are subject to discolouration by light; dark colours may fade and light colours become yellow. Ideally furs are best stored in the dark in a cool dust-free stable environment. For long-term display low light levels of around 50 lux have been suggested in the past. However, in keeping with current practice relating to museum lighting policy for works of art on paper and similarly light-sensitive materials, which includes textiles and furs, together with current theories of risk assessment, an increase in light levels is permitted calculated against less exposure time overall (Derbyshire and Ashley-Smith, 1999; Ashley-Smith, 1999). The RH should be stable at around 45–55% and although water does not harm the fur hairs, humidity levels above 65% can lead to the risk of mould growth on the pelt. Extreme dryness should also be avoided as desiccation can lead to embrittlement and possibly tearing of the skins. If the skins are weakened overall then the sewn lines are particular areas of risk when handling as they have already been physically weakened by the perforations of the stitching. If desiccation occurs in the pelt there is a danger of the collagen fibres shrinking, the follicles opening and the hairs falling out.

### 15.3.7.2 Pests and freezing – cold storage

Regular inspection of storage and display areas and a pest-monitoring programme should be in place where furs are part of the collection. Regular monitoring for insect activity is essential as moths and carpet beetles will eat fur, moths devouring it at the roots then proceeding to perforate the leather. Whenever an addition is made to a collection it is good practice to treat that object for insect infestation as a precaution before putting it into store with the rest of the collection. Before treatment it should be kept isolated until this has been carried out.

There are many publications available in conservation literature on the topic of pest management including Florian (1987), Child and Pinniger (1993), Hillyer and Blyth (1992), Pinniger (1994, 1995), and Norton (1996) The opinion expressed by writers on furs and furriery is that moths are the most damaging risk to furs and that cold storage balanced with a controlled atmosphere is the best remedy (Sachs, 1922; Rosenberg, c.1920; Bacharach, 1930). If the atmosphere is maintained at a high humidity in very low temperatures, moisture will condense into a frost on the garment; but if a dry atmosphere is maintained the cold air will absorb the little remaining moisture causing the skins to dry out and become brittle. It is therefore essential to balance the environmental conditions when furs are kept in reduced temperature storage and specialists in this field should be consulted if this type of storage is to be provided. Maintenance of furs in reduced temperature storage of no more than 8°C (50°F) was recommended by Bachrach as far back as 1930 but few museums have the facility for this. Clothes moths (*Tineola biselliella* and *Tinea pellionella*) will not breed or feed below about 10°C. However, this temperature may not kill larvae for a long time as they probably starve to death. The larvae of carpet beetles (woolly bears) may be more tolerant but the adults will not breed at 8°C.

Until the 1980s treatment for pest infestations was carried out using methyl bromide and ethylene dioxide but since then deep freezing of infested material has become one of the accepted methods and is a straightforward procedure to carry out. Deep freezing to temperatures of −30°C for a period of 72 hours and then repeating the procedure is currently the accepted method of dealing with insect infestations in museums (Florian, 1987). In 2003 it became accepted that it is unnecessary to carry out the second freezing procedure. Anoxic treatments using nitrogen-saturated atmospheres have also been regularly and successfully in use for a number of years (Newton et al., 1996). An alternative for whole collections and large quantities of fur items is to treat them using the thermo-lignum process[3] (Thomson, 1995).

Beating the fur from time to time with sticks is advocated by furriers and is a treatment that goes back to medieval times. William Jurden, Skyner to Queen Elizabeth I, had the essential task of regularly airing and beating the furs and fur-trimmed garments to keep them free of dust, moths and fleas (Arnold, 1988). Although beating is still carried out by furriers today as part of maintenance it is not a suitable treatment for fragile aged skins or museum objects.

## Endnotes

1. An exception was an exhibition held at the Museum of London, November 2000 entitled *Stolen Skins? Fur in Fashion*.
2. Rodents – e.g. beaver, muskrat, nutria, rabbit, marmot, squirrel, chinchilla
   Felines – e.g. leopards, ocelots, cheetah, lynx
   Canines – e.g. wolf, dog, fox, tanuki
   Weasels – e.g. otters, mink, martens, sables, fisher, ermine, wolverine, badger, skunk, genet
   Bear Racoon Group – e.g. racoon, bears, bassarisk (ringtail)
   Marsupials – e.g. opossums, kangaroo, koala
   Ungulates – e.g. sheep (lambs), goats (kids), camel family
   Sundry – others.
3. There are two processes which have been trialled. The first involves raising the temperature gradually from an ambient temperature of 20°C up to 50°C over a period of 11 hours, holding this temperature stable for two hours then allowing cooling back to ambient temperature of 20°C over 10 hours. During this time the relative humidity is strictly controlled and raised from 50 to 60% and back again.

   The second process involves flooding the chamber with argon to give an atmosphere of less than 1% oxygen, then heating to 38°C over a period of 8 hours, holding at that temperature for 3 days and then cooling over a further 8 hours. During this treatment the relative humidity is increased from 50 to 55% and then dropped back to 50%. Both methods are useful for fur-skins but it is the second method that could be useful for degraded fur-skins with very low shrinkage temperatures.

# References

Appleyard, H.M. (1978) *Guide to the Identification of Animal Fibres.* Leeds: British Textile Technology Group.
Arnold, J. (1988) *Queen Elizabeth's Wardrobe Unlock'd.* Leeds: W.S. Maney.
Ashley-Smith, J. (1999) *Risk Assessment for Object Conservation.* Oxford: Butterworth-Heinemann.
Austin, W.E. (1922) *Principles and Practice of Fur Dressing and Fur Dyeing.* London: The Library Press Ltd.
Bachrach, M. (1930) *Fur, a Practical Treatise.* London: Prentice-Hall Inc.
Baldwin, F. (1926) *Sumptuary Legislation and Personal Regulation in England.* Baltimore: Johns Hopkins Press.
Blazej, A. et al. (1989) *Atlas of Microscopic Structures of Fur Skins.* Amsterdam: Elsevier.
Brunner, H. (1974) *The Identification of Mammalian Hair.* Melbourne: Inkata Press.
Child, R.E. and Pinniger, D.B. (1993) Insect Trapping in Museums and Historic Houses. *International Conference on Insect Pests in the Urban Environment,* Cambridge, UK.
Delany, Mrs (1861) *The Autobiography and Correspondence of Mary Granville, Mrs Delany.* Rt Hon Lady Llanover, ed. London: Bentley.
Del Re, C. (1988) Technology and Conservation of Five Northwest Coast Head-Dresses. *Symposium 86, The Care and Preservation of Ethnological Materials,* pp. 53–61. C.C.I.
Derbyshire, A. and Ashley-Smith, J. (1999) A Proposed Practical Lighting Policy for Works of Art on Paper at the V&A. *Preprints of the ICOM-CC 12th Triennial Meeting Lyon, 3 September 1999,* pp. 38–41.
Dickens, J. (1987) The Development of a Treatment to Re-secure Mats of Insect-damaged Fur. *Ethnographic Conservation Newsletter,* **4**, 18–19.
Dignard, C. and Gordon, G. (1999) Metal Ion Catalised Oxydation of Skin: Treatment of the Fur Trim and Collar on a Velvet Cape. *The Journal of the Canadian Association for Conservation,* **24**, 11–22.
Ewing, E. (1981) *Fur in Dress.* London: B.T. Batsford Ltd.
Florian, M.L.E. (1987) Methodology Used in Insect Pest Surveys in Museum Buildings – A Case History. *Preprints of the 8th Triennial Meeting,* pp. 1148–1174. ICOM-CC Sydney, Australia.
Glob, P.V. (1971) *The Bog People.* London: Paladin.
Hahn, H.G. and Weigelt, G. (1967) *Chemische Technologie der Rauchwaren-zurichtung and-veredlung.* Leipzig: [n. pub.].
Heiberger, B. (2002) Caring for Fur at the Museum of London. In *The Conservation of Fur, Feather and Skin* (Margot Wright, ed.). Seminar organized by Conservators of Ethnographic Artefacts held at Museum of London, 11 December 2000, pp. 88–92. London: Archetype.
Hillyer, L. and Blyth, V. (1992) Carpet Beetle – A Pilot Study in Detection and Control. *The Conservator,* **16**, 65–77.
Kaplan, H. (1971) *Furskin Processing,* pp. 13, 113, 137. Pergamon Press.
Kite, M. (1990) A 1740s Horse-Hair Hat, a Busby c1827 and a Fur-Lined Cape c.1936. *Preprints of the ICOM-CC Conference held in Dresden, August 1990,* pp. 645–650.
Kite, M. (1991) The Conservation of a Wet Weather Hat from the Uniform of a Marshal General of the Portuguese Army, and a Chinese Fur-lined Kesi Robe c. 1800. *Preprints SSCR Conference, Paper and Textiles: the Common Ground,* held in Glasgow, September 1991, pp. 105–115.
Kite, M. (1992) Freezing Test of Leather Repair Adhesives. *Leather Conservation News,* **7**(2), 18, 19.
Kite, M. (1998). Fur. *'Core' Conservation and Restoration of Objects of Cultural Patrimony,* pp. 17–20. France. Spring 1998.
Kite, M. (1999) The Conservation of a 19th Century Salmon Skin Coat (626–1905) *Preprints of the ICOM-CC 12th Triennial Meeting Lyon, 29 August–3 September 1999.* pp. 691–696.
Links, J.G. (1956) *The Book of Fur.* James Barrie.
Liska, A.J. and Shevchenko, A. (2003) *Proteomics,* **3**, 19–28.
Lougheed, S., Mason, J. and Vuori, J. (1983) Repair of Tears in Fur Skin Garments. *Journal IIC-CG.* **8 & 9**, 13–22.
Matheson, S.A. (1978) *Leathercraft in the Lands of Ancient Persia.* Vic: Colomar Munmany.
Norton, R.E. (1996) A Case History of Managing Outbreaks of Webbing Clothes Moth (*Tineola Bissielela*). *Preprints of the ICOM-CC 11th Triennial Meeting,* Edinburgh 1996, pp. 61–67.
Novak, M. (1987) *Wild Furbearer Management and Conservation in North America.* Ontario: Ministry of Natural Resources.
Pense, W. (1955) Rauchwaren. In *Handbuch der Gerbereichemie und Leqerfabrikation* (W. Grassman, ed.). Band 111, Wein: Springer.
Pinniger, D.B. (1994) *Insect Pests in Museums,* 3rd ed. London: Archetype.
Pinniger, D.B. (1995) Friends of the Dodo – Recent Advances in the Detection and Control of Museum Insect Pests. *Proceedings of the 11th British Pest Control Association Conference, London 1995.*
Rae, A. and Wills, B. (2002) Love a Duck: The Conservation of Feathered Skins. In *The Conservation of Fur, Feather and Skin* (Margot Wright ed.). Seminar organized by Conservators of Ethnographic Artefacts held at Museum of London, 11 December 2000, pp. 43–61. London: Archetype 2002.
Ribeiro, A. (1979) Furs in Fashion. *The Connoisseur,* December, pp. 226–231.
Richardson, H. (2002) The Conservation of Plains Indian Shirts at the National Museum of the American

Indian, Smithsonian Institution. *The Conservation of Fur, Feather and Skin* (Margot Wright, ed.). Seminar organized by Conservators of Ethnographic Artefacts held at Museum of London, 11 December 2000, pp. 7–24. London: Archetype.

Rogers, G.E. (ed.) (1989) *The Biology of Wool and Hair*. London: Chapman and Hall.

Rosenberg, C.J. (c1910) *Furs and Furriery*. London: Sir Isaac Pitman and Sons Ltd.

Sachs, Captain J.C. (1922) *Furs and the Fur Trade*. London: Sir Isaac Pitman and Sons Ltd.

Samet, A. (1950) *Pictorial Encyclopedia of Furs*. New York: Arthur Samet.

Spindler, K. (1994) *The Man in the Ice*, pp. 135–138. London: Weidenfeld and Nicolson (translated from the German by Ewald Osers).

Spriggs, J.A. (1998) The British Beaver – Fur Fact and Fantasy. In *Leather and Fur, Aspects of Early Medieval Trade and Technology* (Esther Cameron, ed.) pp. 91–101. London: Archetype.

Thomson, R.S. (1995) The Effect of the Thermo-Lignum Pest Eradication Treatment on Leather and Other Skin Products. Report 1995. *Leather Conservation News*, **11**, 11–12.

Thomson, R. (1998) Leather Working Processes. In *Leather and Fur, Aspects of Early Medieval Trade and Technology* (Esther Cameron, ed.). London: Archetype.

Veale, E.M. (1966) *The English Fur Trade in the Later Middle Ages*, pp. 16, 44, 193. Oxford: Oxford University Press.

Vincent, J. (1972) *Structural Biomaterials*. Princeton N.J.: Princeton University Press.

# 16

# The tanning, dressing and conservation of exotic, aquatic and feathered skins

*Rudi Graemer and Marion Kite*

## 16.1 Exotic skins

### 16.1.1 Introduction

Collecting today is increasingly popular and it is therefore not surprising to find beautifully crafted leather articles in museums and other collections, be they public or privately owned. A fair proportion of these will be made from exotic leathers, distinguishable from mammalian leather by their interesting surface texture, such as scales or quill pips and by their fascinating natural markings.

### 16.1.2 Origins and history of exotic leathers

The use of exotic skins dates no doubt back to prehistory, well before the discovery by Europeans of tropical areas of the world, with hunter–gatherer tribes utilizing the skins from reptiles, birds, fish and amphibians for adornments to their clothing or for their head-dresses as well as for covering such articles as drums. These skins were mostly dried and made supple by scraping the flesh side and applying various animal fats, methods that preserve the skin without imparting a tannage.

The supply of exotic skins in bulk, especially to tanners in the United Kingdom, started towards the end of the nineteenth century from countries within the Empire situated within tropical regions. Rain forests, savannas, lakes, rivers and the sea contained wild species with skins large enough to be commercially usable in the manufacture of leather goods. The import of East Indian tanned buffalo hides and goatskins provided the framework for the trade in exotic skins to commence, exported either air-dried, in a raw, wet-salted condition or pre-tanned in a local tannery.

France initially obtained its supplies of exotic skins from its colonial empire within Africa where the rivers and lakes yielded an abundant supply of crocodile skins.

The expansion of the trade soon necessitated searching other areas apart from Africa and the Far East for supplies. Exotic skins were imported from South and North America, and from other tropical or subtropical areas.

### 16.1.3 Uses of exotic leathers

Tanners of exotic skins and manufacturers of finished articles made from these normally understand exotic leathers as those made from either reptiles, birds, fish or amphibians. For any skin to be commercially usable it has to satisfy the following requirements:

(a) The available area of skin must permit a manufacturer of finished leather articles to cut the components with a minimum of waste. It follows that the larger the leather article, such as a handbag, the larger the exotic skin has to be, except where a design is used permitting smaller pieces to be stitched together. This tends to increase labour costs but cheapens the article.

(b) The physical strength and properties of the skin must match the wear and tear and flexing requirements of the finished article. Reptile skins with their tightly woven fibre structure are immensely strong, particularly with a skin which is relatively thick, such as crocodile. The skins of birds, such as ostrich or emu, have a looser fibre structure as have fish which results in a lower tensile strength. To improve flexing of the leather, the tannery will fatliquor the skins. Exotic skins have good resistance to scuffing with lizard skins among the best compared to leather made from domestic animals.

(c) The leather made from exotic skins has to be 'in fashion' – that is to say, it is promoted in the world of fashion as the 'in' material to be used. As such it must be available in fashionable colours and in the desired surface finish, such as bright, shiny, metallic, matt, or dull, such as suede. The tanner has limitations imposed on him by the nature of the skin as well as fashion demand, with crocodile leather for handbags, invariably finished by glazing, giving a brilliant finish. Most reptile leather is finished in this way, but fashion dictates will force the tanner to apply matt finishes or even buff off the surface layer, to give a 'Nubuk' finish. Bird and fish skins were nearly always finished matt after tanning, their fibre structure being too loose to be glazed, but modern finishing systems allow more shiny effects if not the brilliant results obtained from glazing.

### 16.1.4 Preparing the raw skins

The flaying of raw, exotic skins differs from type to type with belly-cut being the norm for many species, leaving the back of the animal in the centre of the leather. Skins which can be back-cut, such as pythons (*Python* sp.) or Teju lizards (*Dracaena* sp., *Tupinambis* sp.) from South America, have their belly scales in the middle of the leather. These are widely used for American cowboy boots with the wide belly scales running down the centre of the boot and vamp. Rattlesnakes (genus *Crotalidae*), in the raw, dried state, are also used, mostly belly-cut in the manufacture of these boots.

Crocodiles and alligators are very heavily ossified on their backs which makes this part of the skin impossible to tan and soften. For this reason only the sides and belly are used in the tanning process (*Figure 16.1*). Lighter ossification is also present in the belly of some species which, if not too severe, can be softened by the tanner with acids. Matt finishing of heavily ossified caiman skins resulted in the 'wild' crocodile popular for making handbags in the 1970s.

### 16.1.5 Tanning and dressing

Tanning exotic skins follows the same basic processes as the tanning of mammalian skins. After thoroughly soaking air-dried skins, they are immersed in a solution of sodium sulphide to remove the keratinous surface layer of the scales, which is the equivalent to the hair and epidermis of mammalian skin. The equivalent with birds are the feathers and with fish the scales.

After bating and pickling with acid follows the important bleaching process which removes,

**Figure 16.1** Crocodile handbag and shoe with snakeskin trim.

if required, the natural pigmentation, a process demanded either by fashion or if the design is not particularly attractive.

With the demand mostly for glazed reptile leathers, the traditional chrome tannage cannot be used or only as a pre-tannage on crocodile leathers. A vegetable or synthetic tannage is applied instead. These plump up the leather enabling it to withstand the high pressures imparted by the cylindrical glazing jack made from glass, agate or steel, which is dragged across the skin by the glazing machine at high speed.

As a result of using these vegetable and synthetic tanning materials certain disadvantages affecting the preservation of the final finished leather manifest themselves. The different finishing methods also affect the fastness to different agents. Of these the following have to be noted.

### 16.1.5.1  *Light fastness*

Collagen, the leathermaking protein, is in itself not fast to light. It will yellow on exposure. This is aggravated by the presence of natural fats. Leather tanned with synthetic tanning materials, even of reasonably good light fastness, will therefore turn yellowish. Reptiles are often tanned with their natural markings emphasized by the use of synthetic tanning materials and will therefore yellow with age. This also affects pastel shades. Most vegetable tanning materials, used mainly for darker colours, will redden on exposure. This is not a problem on colours such as black, dark brown or navy. Light browns, greens or greys will, however, darken. To give these leathers the brilliance in colour, the tanner uses cationic dyes in the finishing coats, notorious for their poor light fastness. Unfortunately pigments that are light fast produce a muddy, dull looking finish when glazed.

No such constraints exist for matt-finished leathers, such as ostrich (*Struthio camelus*), various species of sharks, caiman (*Dracaena* sp.) and occasionally other reptiles. These can be chrome tanned and moderately pigmented to improve light fastness.

### 16.1.5.2  *Water resistance*

Glazed leathers will dull when exposed to dampness. Traditionally, this finishing method, which as its name suggests gives a very high gloss, involves the application of a surface coating based on casein, originally hardened with formaldehyde. This can no longer be used for health and safety reasons. Alternatives do not harden the film sufficiently to prevent water spotting, although a thin spray of nitrocellulose-based lacquer over the finish will improve matters. Matt finishes can be made quite fast to water with nitrocellulose- or urethane-based emulsion sprays, but it should be borne in mind that leather has pores and all finished articles have seams. A detailed account of the production of leather from lizard skins is given in *The Reptile Skin* by Fuchs and Fuchs (2003). This work also provides essential reference material for collectors, curators and conservators seeking to identify species.

### 16.1.6  Conservation

Described above are the practical constraints to the conservation of finished leather articles made from exotic leathers during the twentieth century. Treatment strategies will of course depend on many factors and repair and consolidation methods will be selected as appropriate to the object concerned, the way the skin was processed and the date the object was made. Most likely methods chosen will be drawn from treatments used for other leathers.

Of interest is the conservation treatment undertaken to a Romano-Egyptian cuirass and helmet made from crocodile skin, discussed by Wills (2000). Also, the conservation treatment undertaken to a python skin chair by Sturge (2000).

### 16.1.7  Conclusion

Today the world is quite rightly talking of the vital necessity of the conservation of living creatures. The commercial pressure on some of the species, the spread of humanity as well as the destruction of habitat through the exploitation of timber in the tropical rain forests resulted in the Washington Conference and the setting up of CITES. This organization has been controlling the trade over the past 25 years with licences and prohibiting the trade where species are endangered. Unfortunately some countries have found ways of circumventing these regulations and exotic leather articles are still on sale in shops in 'touristy' areas of the world. Farming of some species has, however, led to particular successes with crocodiles and ostrich, with their leathers and the finished articles, although expensive, freely available.

Exotic leathers with or without their natural markings, in their many colours, textures, metallic effects and finishes have no doubt tempted many customers in the past century to purchase leather goods made from these (*Figures 16.2* and *16.3*). I hope customers of today will find equal pleasure from just watching these beautifully marked and

**Figure 16.2** Nile crocodile skin (*Crocodylus niloticus*).

**Figure 16.3** Waistcoat by Jean Muir *c.*1960s using skin of *Python reticulatus*.

coloured, living animals in a diversity that is perhaps unequalled.

## 16.2 Aquatic skins

The use of fish skin and the skins of aquatic mammals and other marine vertebrates as a leather product dates back some 2000 years when Native Americans were using it to make pouches (Neergaard, 1991; Ingram and Dixon, 1994). In England in the fifteenth century gauntlets and bags were being made from fish skin leather and in the seventeenth century eel and whale were tanned (Ingram and Dixon, 1994). It is recorded that in 1627 a Canadian tanner bought a book covered in cod skin (Ingram and Dixon, 1994).

In the nineteenth century the use of fish skin leather may not have been usual but neither was it uncommon. At a maritime exhibition held at the Westminster Aquarium in London in 1886 a Norwegian exhibitor showed a variety of tanned fish skins among which were flat fish (order *Pleuronectiformes*) prepared for gloves, skins of soles (family *Soleidae*) prepared and dressed for purses, skins of eels prepared and dried for braces and upper leather for shoes made from 'white fish'[1] (Unattributed, 1880). The skins of beluga or white whale (*Delphinapterus leucas*) were being tanned in Bermondsey, London, in 1887 producing a supple, beautiful and high quality leather for use as uppers and in the manufacture of 'bespoke' boots. Laces were cut by hand from the whole length of the skin and are described as unsurpassed in wear (Unattributed, 1888). These laces were the customary choice and quality for army officers' boots. A commercial fish skin tannery was still in operation in Kingston, Surrey, UK, until 1950.

A patent was taken out concerning the making of shoes at Gloucester, Massachusetts, using the skins of cusk or torsk (*Brosme brosme*) and in Colbourn, Canada, an industry was said to be carried on using the skins of 'siluroids' (i.e. catfishes) for glovemaking (Unattributed, 1880).

In the nineteenth century the skins of sharks and whales were tanned and some were used for shoes, but mostly in Europe they were used for covering high quality boxes and cases and for industrial uses.

In the late 1940s approximately 800 000 sharks were caught annually by the Japanese and their skins tanned for the shoemaking industry producing 2 million pairs of shoes or one third of the Japanese annual requirement.

The use of untanned fish skin in the form of shagreen (shark and ray skin) is well known in many cultures, and it has been used for binding ceremonial sword hilts in Europe and by the Japanese for the handles of samurai swords.

In the late nineteenth century and early part of the twentieth century 'shagreen' was used in furniture, most notably by the French cabinetmaker Ruhlman, and on decorative boxes and instrument cases and even in jewellery. The skin of the blue shark (*Prionace glauca*) was used for this purpose (Unattributed, 1880).[2] A fish called chat (*Scyliorhinus canicula*, contemporary name *Squalus catulus*) was used to make a product known as 'peau de roussette' used for cases and other articles but also known as shagreen (roussette is the French name for dogfish) (Unattributed, 1880). Dogfish skin has been widely used as shagreen and has been called 'peau de chien de mer' and 'chien marin'. Another species of dogfish, spotted dogfish or bull huss (*Scyliorhinus stellaris*), was also used under the name of 'roussette tigrée' and in France went under the name of 'Galuchat' when prepared as Monsieur Galuchat had done with the dermal denticles with which they were studded, ground down and smoothed. These were usually dyed green and used to cover sheaths, boxes, compacts and other small decorative fashion items, bags or accessories. It was also coloured black, white or red (Unattributed, 1880). As the skin was rigidly bonded to the article, it did not have to flex and stability was assured as long as the article was kept dry.

The dermal denticles are cartilaginous and are not removed by the normal liming process during the preparation of the skin which is why they were ground down to give a smoother polished surface. The cartilaginous material tends to take dye to a different shade than the collagen giving an interesting two tone effect at the surface.

In the mid-twentieth century, during World War II when occupying forces overran Norway and Denmark and confiscated leather made from animal hides, cod (*Gadus morhua*), eel (*Anguilla anguilla*), plaice (*Pleuronectes platessa*) and catfish (*Anarhichas lupus*)[3] were tanned so local inhabitants could make handbags, belts, harness and footwear (Ingram and Dixon, 1994). 'Neptune' suede, made from the skin of plaice, was dyed bright colours and made into wedge evening sandals. Ten to 12 skins were required for each pair (Ledger, 1982).

### 16.2.1 Fish skin preparation

After World War II an investigation was undertaken into the surviving state of the German tanning industry. The plant owned by Johann Knecht and sons at Elmshorn was undamaged and still commercially producing cod and haddock (*Melanogrammus aeglefinus*) skins as fancy leather for the novelty goods trade. It is recorded that:

*they obtain their skins in a salted condition. They do not keep well unless refrigerated even when salted. They wash the skins and give them a good liming, delime and bate (scales not removed except as they fall off in the process), pickle and one bath chrome tan in a paddle, wash, neutralize, fat liquor and tack as natural chrome, buff flesh and grain, colour flesh side, pigment and emboss. (Schultz and Schubert, 1945)*

This plant was not alone in processing fish skins which were mostly plaice, cod and haddock often obtained from Norwegian curers (Caunce *et al.*, 1946). A detailed method used for processing wet-salted cod skins at Vibo-Schwanheim Lederwerke of Frankfurt-am-Main is given in Caunce *et al.* (1946).

The finishing method describes the skins being embossed with lizard grain on a callender embossing machine, then given three or four applications of colourless cellulose lacquer. This operation was performed by hand using a cloth covered sponge to ensure that only the raised portions of the grain were coated. Finally the skins were dried and hot plated at 50°C very lightly and rapidly on a hydraulic press (Caunce *et al.*, 1946).

### 16.2.2 Structure and identification

Unless the skin has a particular and easily recognizable grain pattern, such as ray skin, or the identifiable scale pattern remains and has not been removed or destroyed during tanning, then close examination of the grain surface using magnification is necessary for a positive identification. Alternatively, identification may be made by sampling, providing this is possible, preparing a cross-section and examination of this. Under magnification the surface of fish skin will appear smooth and devoid of any follicle pattern. A cross-section prepared from a piece of raw salmon skin shows that fish skin has no epidermis and has a

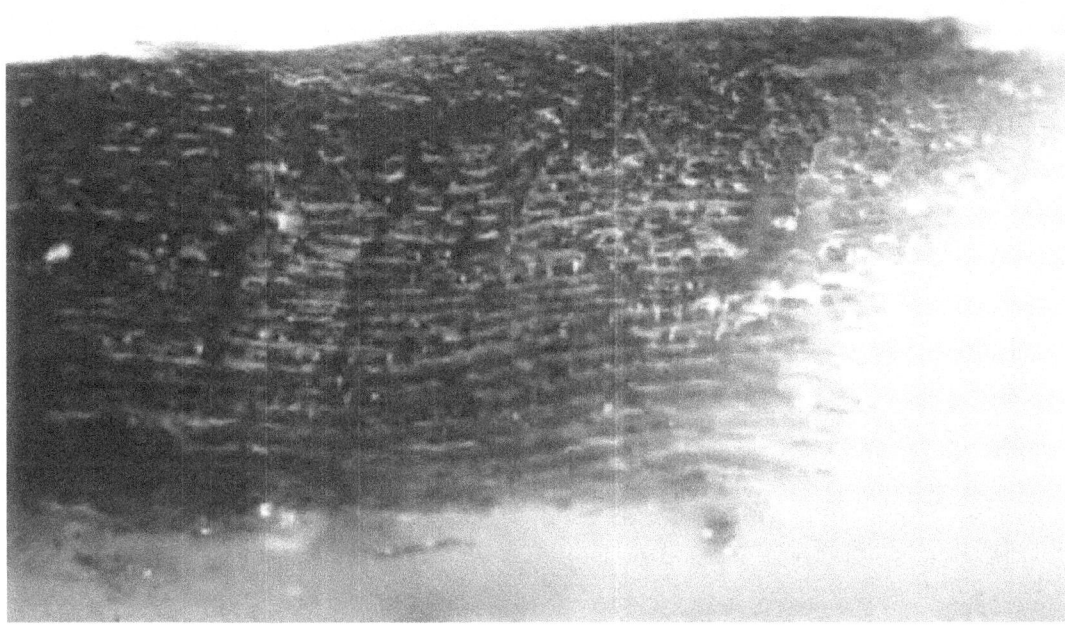

**Figure 16.4** Cross-section of salmon skin.

layered weave structure of horizontally running collagen fibre bundles with a small number running at a slightly different angle. The outer surface layer much like an epidermis is pigmented (*Figure 16.4*). Similar patterns are exhibited by the skins of wolf fish and dog fish (*Figures 16.5* and *16.6*).

Examples of cross-sections for comparison may be found in Whitear (1986).

The sizes and shapes of the leather pieces may also in some circumstances be significant in making an identification of aquatic leather even if the species may not be determined.

The tanning process for fish skin is complex and species specific. Factors such as age and size of the fish also affect the end result. The structure of fish skin is sensitive to enzymatic attack and raw skins will deteriorate rapidly. Fish skin collagen is easily soluble by acids and alkalis, and can be destroyed by anionic detergents which decrease the shrinkage temperature (O'Flaherty *et al.*, 1958). Care has to be taken in handling and storage prior to tanning. Skins can be frozen but during thawing the internal fibres of the skins tend to break so often skins are salted to slow down the deterioration (Ingram and Dixon, 1994).

The shrinkage temperature of fish collagen is considerably lower than mammalian collagen. Native collagen of cod skin shrinks at about 40°C whereas bovine skin requires a temperature of about 65°C. However, the shrinkage temperature varies with the species, with cold water species generally having a lower shrinkage temperature, 38–45°C for cold water fish to 50–56°C for warm water species (Gustavson, 1956). It has been established that the shrinkage temperature has a direct relationship with the hydroxyproline content in the amino acid structure of the collagen. Rockfish (family *Scorpaenidae*) containing the lowest per cent of hydroxyproline (7%) has the lowest shrinkage temperature (33–34°C) while carp (*Cyprinus carpio*) has 11.6% hydroxyproline and a shrinkage temperature of 57°C (O'Flaherty *et al.*, 1958).

Oil tannage does not raise the shrinkage temperature but vegetable tans will raise the shrinkage temperature by about 20°C and chrome by about 30°C.

Synthetic dyes enable the skins to be coloured to any desired shade.

### 16.2.3 Fish skin in ethnographic objects

Fish skin used in clothing is much more likely to be found with the species identified in ethnographic collections but objects are rare (*Figure 16.7*).

176 *Conservation of leather and related materials*

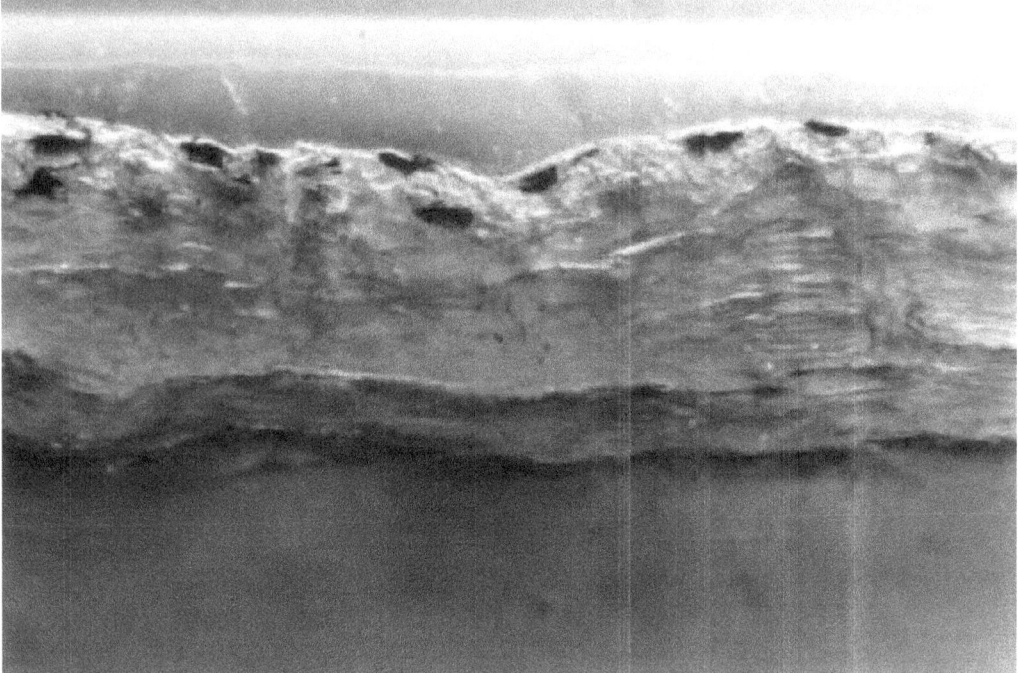

**Figure 16.5** Cross-section of wolf fish skin.

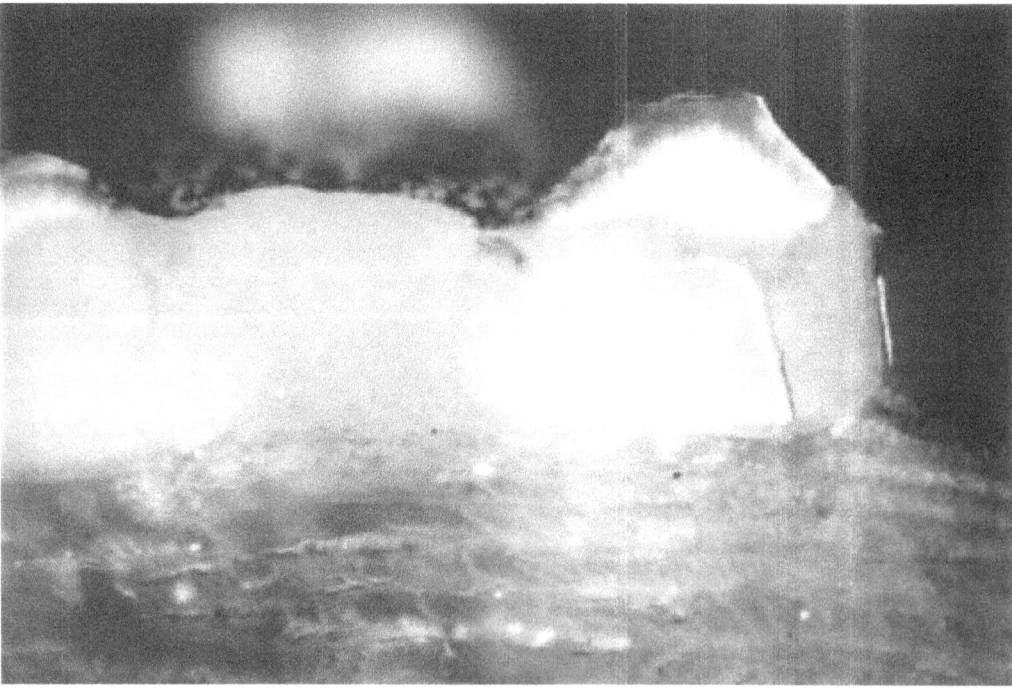

**Figure 16.6** Cross-section of dogfish skin, showing dermal denticle.

**Figure 16.7** Gilyak embroidered salmon skin coat, nineteenth century, showing back view.

Inuit and Aleut peoples from the northwest Pacific rim and other northern Pacific peoples have historically used fish skins for clothing and to make cultural objects. Yupik people of Alaska used fish skins for raincoats, mittens and boots and as bags for water containers.

The methods of skin preparation differ considerably from the commercial methods described above. A method of preparation described by a Yupik woman from southwest Alaska and published by Hickman is as follows. The fish were skinned, soaked in water and scraped, some were scaled and some were not. After scaling they were soaked in urine for between half a day and overnight. The urine from a small boy before weaning was described as best for lighter skins and for tougher skins the urine from an older boy, around the time his voice breaks (Hickman, 1987). The longer the skin is soaked the softer it became. A new bath was then prepared with naphtha soap and aspen shavings then another of clear water. Finally the skin was blotted and laid out on a small board, the outside outermost to dry.

The purpose of soaking in urine was to open up the skin structure. The urea content influences this. Urea is a hydrogen bond breaker and can partially destabilize the collagen. The urea breaks down to give ammonia which acts as a mild alkaline treatment of the skin loosening epidermis and opening up the structure to allow tans to penetrate.

Another description taken from a catalogue published by the Musée de l'Homme in Paris describes a method used by the Ostiak people of Western Siberia to tan the skins of sturgeons and sterletts (*Acipenseridae*), burbot (*Lota lota*), pike (*Esox lucius*), Pacific salmon (*Oncorhyncus* spp) and Kumza (*Salmo trutta*). Translated it reads:

> The skins are prepared in spring and summer by women. The gutted fish is hung for a day or two to dry then split down the stomach to take off the skin which is then made dry by the sun stretched between stakes. Sometimes they are dried by the hearth and here they become covered with soot resulting in a skin which is more water resistant. When dry the skin is rolled in a ball and made supple by beating with a mallet. The softening of certain types is finished by anointing the skin with brains and with fish eggs. (Hickman, 1987)

The brains and fish eggs supply highly emulsified oil which will penetrate into the skins' fibre structure, filming the fibres and preventing the fibres from sticking, maintaining a flexible dry skin. The fat will also make the final leather more water resistant. The shrinkage temperature would not be raised by this tannage.

### 16.2.4 Conservation

For many reasons, frequently due to less than ideal storage, costume accessories come to the conservator in a crushed and misshapen condition requiring simple humidification and reshaping treatment before they are fit to display. They may also need cleaning.

When objects are made from skins it is important to the conservator to know the genus of creature and method of preparation originally used for the leather, before any treatment is carried out. If it is aquatic leather then it is important, if possible, to identify the species and whether it is from warm or cold waters. The significance of the shrinkage temperature and moisture sensitivity is important both for the possible use of humidification treatments for reshaping and for the use of heat set adhesives and aqueous solutions for repair.

The presence and identification of finishes is also important particularly if hydrocarbon solvents are being considered for use in cleaning.

In 2003 there were few published articles concerning the conservation of fish skin and those that are readily available deal with ethnographic materials. The following may, however, be useful to the conservator: Heikkanen (1978), Kite (1999), Murray (1994) and Pancaldo (1996).

## 16.3 Feathered skins in fashionable dress

The general fascination for wearing bird plumage from beautiful, rare and exotic species is documented from Roman times (Doughty, 1975) and the wearing of feathers and bird skins by a fashionable and status conscious society has continued throughout history. Wealthy Romans and Greeks wore the feathers of ostrich (*Struthio camelus*), which is still one of the principal feathers used in the millinery trade. References have been found in the wardrobe accounts of Elizabeth I, to the wearing of swan skins (*Cygnus* sp.) which were prepared by the Queen's Skynner William Jurden (Arnold, 1988).

Many different bird species from all parts of the world have been used since, including herons, rheas, owls, birds of paradise, kingfishers, jays, magpies and many others (Doughty, 1975). By the late nineteenth century the demand by the fashion industries of Europe and America was unprecedented. In a sale held in New York in 1888 30 000 humming birds (*Trochilidae* sp.) were sold in one afternoon (Cunnington, 1953).

During the second half of the nineteenth century many species were becoming at risk of extinction and prominent naturalists and many others in the United Kingdom and America had begun a movement to protect endangered species. This directly led to the Sea Birds' Preservation Act being passed in the UK in 1869. This prohibited the taking of sea birds, which were mainly being used for millinery purposes. Subsequent legislation in the early twentieth century in both the UK and USA followed, radically controlling the taking of many other species of birds. Today, only species of domestic fowls (*Gallus* sp.), peafowls (*Pavo* sp.), pheasants (*Phasianidae* sp.) and a few other farmed species including ostrich (*Struthio camelus*) are commercially available to the feather and bird skin trades.

Accessories made using complete or part bird skins may be found in museums large and small where dress is included in the collections. Items that commonly survive include muffs and possibly trimmings from the eighteenth century, and fans, muffs, collars, trimmings and hats from the nineteenth and early twentieth centuries. Ceremonial and military uniforms also may have feathers and bird skins included. However, it is on hats worn as part of fashionable dress that part and whole birds are most frequently found (*Figures 16.8* and *16.9*).

In the Victoria and Albert Museum dress collection there are collars of grebe skins (*Podiceps cristatus*), a Victorine of male eider duck heads (*Somateria mollissima*) lined with down, muffs of sundry other waterfowl, fans with humming birds (*Trochilidae* sp.) applied to them and many other small bird skin items dating from the nineteenth century. There is an array of whole birds, part birds and bird plumage attached to hats including birds of paradise (*Paradisaea* sp.), humming birds (*Trochilidae* sp.), peafowls (*Pavo* sp.) and many others. Swansdown (*Cygnus* sp.) (skin on) is also to be found (*Figures 16.10–16.13*).

### 16.3.1 Processing

Methods used by Inuit and Native American peoples for processing bird skins have been published (Oakes, 1991; Oakes and Riewe, 1996). Simple methods involving scraping the fat and blood from the skin

*The tanning, dressing and conservation of exotic, aquatic and feathered skins* 179

**Figure 16.8** Greater bird of paradise (female) (*Paradisaea* sp.) on hat c.1940s.

**Figure 16.9** Greater bird of paradise (male) (*Paradisaea* sp.) on hat c.1940s.

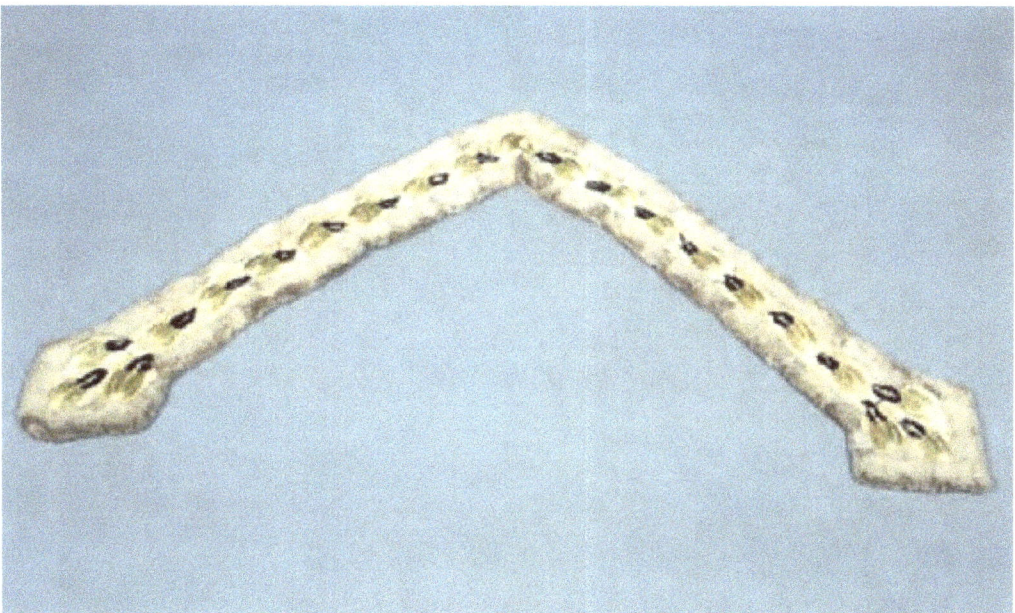

**Figure 16.10** Victorine of male eider duck heads (*Somateria mollissima*) nineteenth century.

180 *Conservation of leather and related materials*

**Figure 16.11** Fan c.1887 with whole hummingbird.

**Figure 16.12** Peafowl muff, hat and tippet, late nineteenth century.

**Figure 16.13** Peafowl muff, showing detail of head, late nineteenth century.

then sucking out the remaining fat were used with alternative treatments including the use of urine, fish fats and roe. Methods of bird skin preparation used by taxidermists have also been published (Gerhardt, 1989). (For recipes see Chapter 14.) However, the author has found no direct references or published works dealing with the processing of bird skins with feathers in situ by fur dressers or tanners from the commercial skin trade. References are made concerning the processing of turkey and ostrich skins in the twentieth century but these had the feathers removed and were prepared as leather and tanned, not tawed or dressed.

It is logical to assume that the methods used for preparing bird skins with feathers in situ intended for use in fashionable dress would be similar to those for furs, using alum, salt and oils and maybe formaldehyde or other pickle. Arsenic and arsenical soaps, including mixtures of arsenic, camphor and soap, have been used in the processing of taxidermy specimens and there are references to arsenic being used to dress feathers for millinery purposes (Kliot and Kliot, 2000). In some cases however, whole or part bird skins may have been simply cleaned and dried.

There is evidence in portraiture, and records in the wardrobe accounts of Elizabeth I, establishing that feathers have been dyed since the sixteenth century, and most likely before this time, but bird skins when used in fashionable dress have usually been chosen for the beauty and natural colours of the feathers and therefore are not dyed. The exception to this is swansdown (skin on) when it has been used to make powder puffs and trimmings which may be found dyed a variety of pale colours. Bird skins have also occasionally been dyed for millinery purposes when the diversity of nature has failed to provide a suitable tone or hue to complement the seasons' fashionable colour palette.

In 1888 Erdmann, a German chemist, took out a patent for dyeing furs and feathers using oxydation dyes. Feathers were first treated in an aqueous or alcohol solution of para-phenylene-diamine then exposed to the slow oxidation of the air. Alternatively they were treated in a second solution with an oxidizing agent. Particularly suitable were ferric chloride, permangenates, chlorates and hydrogen peroxide (Austin, 1922). It is possible these may have been used on bird skins.

Occasionally feathers have been decorated with painted designs, often using metallic pigments to add an extra sparkling decorative element. These are most usually only found on hats.

### 16.3.2 Conservation problems with bird skins

1. In many cases damage may be directly attributable to the original processing and drying, particularly when residual fats were not adequately cleaned from the skin. This can lead to a condition known to taxidermists as 'fat burn' which is rapid deterioration due to the oxidation of the fats in the skin.
2. Bird skins are thin and the feathers protrude into the dermis. Desiccation of the skins is one of the major problems. This can lead to the follicles in which the feathers sit opening as the skin shrinks and the feathers falling out. When this occurs the only effective option is overall consolidation of the skin and securing the feathers using an adhesive.
3. Fading of colours due to display in inappropriate light levels.
4. Physical damage due to wear or poor storage.

A few articles have been published concerning the conservation of feathers including Rae and Wills (2002), Kite (1992, 1993) and Del Re (1988).

Some of these concern objects where the feathers have been removed from the bird skins and the feathers alone have been used to create the object concerned so the treatment of the skin is not an issue. However, Rae and Wills and Del Re discuss treatments for bird skins with the feathers in situ. For more information on the cleaning and conservation of whole birds and part bird skins please refer to Chapter 14.

The use of light brushing and vacuum suction, or organic solvent and aqueous methods of cleaning, are likely to be the usual options considered but since the late twentieth century experiments have been undertaken to test the efficiency of laser cleaning methods for feathers (Lang, 1998/99; Solajic et al., 2002). When further work has been undertaken this could provide a useful alternative to currently available choices.

For general information concerning the care and conservation of bird skins the reader should consult Chapter 14. Additionally, in certain circumstances, much that applies to the conservation of fur-skins may also be applicable to the repair of bird skins. However, it is important to note that degraded and desiccated bird skins may be extremely fragile and the feathers may be very easily disrupted and dislodged from the skin by handling. The skins may also be dry and brittle and tear easily. Expert advice should be sought by the inexperienced conservator before any cleaning and conservation of feathers in bird skins is undertaken, and before a bird skin trimming on an item of dress is dismantled or any stitching is unpicked in order to gain access to the backs of the skins to effect an adhesive repair.

## Endnotes

1. 'Whitefish' is most likely to be referring to the beluga or white whale; however, today it refers to commercially exploited members of the order *gadiformes* (cod and relatives).
2. *Prionaca glauca*. Contemporary references use various common and scientific names, e.g. blue dogfish or 'whaler' or squalus glaucus.
3. Catfish. The name used for the large order, *Siluriformes*, of fishes bearing barbels or 'whiskers' on the head, nearly all of which are freshwater fishes, but also used for the distinctive marine fish *Anarhichas lupus*, also called the wolf-fish which yields an attractively spotted skin.

## References

Arnold, J. (1988) *Queen Elizabeth's Wardrobe Unlock'd*. Leeds: W.S. Maney.

Austin, W.E. (1992) *Principles and Practice of Fur Dressing and Fur Dyeing*. London: The Library Press.

Caunce, A.E. *et al.* (1946) The German Light Leather Industry. B.I.O.S. Final Report No. 1425 item no. 22, pp. 29–30, 184. British Intelligence Sub-Committee. London: HMSO.

Cunnington, C.W. (1953) *Feminine Attitudes in the 19th Century*. London: Heinemann.

Culbertson, C.M. (1922) *The Art of Ostrich Plume Making*. Melvin and Meurgotten, Inc.

Del Re, C. (1988) Technology and Conservation of Five Northwest Coast Head-dresses. Published in *Symposium 86, The Care and Preservation of Ethnological Materials*. C.C.I. pp. 53–61.

Fuchs, K. and Fuchs, M. (2003) *The Reptile Skin*. Frankfurt-am-Main.

Gustavson, K.H. (1956) *The Chemistry and Reactivity of Collagen*. New York: Academic Press Inc.

Hickman, P. (1987) *Innerskins and Outerskins: Gut and Fishskin*. San Francisco Craft and Folk Art Museum.

Hiekkanen, K. (1980) En kvinnodraekt av fiskskinn fran Amuromradet: konservering och undersokning. *Konservering og restaurering af laeder, skind og pergamen*, pp. 321–327. Konpendium fra Nordisk Videreuddannelsekursus 3–14 April 1978. Lund, Sverige: Kulturen.

Ingram, P. and Dixon, G. (1994) Fishskin Leather: An Innovative Product. *Journal of the Society of Leather Technologists and Chemists*, **79**, 103–106.

Kite, M. (1992) Skin Related Materials Incorporated into Textile Objects. *Post-prints of the ICOM-CC Conservation of Leathercraft and Related Objects Interim Symposium. 24–25 June 1992*, pp. 33–35.

Kite, M. (1993) Feathers and Baleen, 2 Keratin Materials Incorporated into Textiles. *Pre-prints of the ICOM-CC Triennial Conference, Washington 1993*, pp. 645–650.

Kite, M. (1999) The Conservation of a 19th Century Salmon Skin Coat (626–1905). *Pre-prints of the ICOM-CC Triennial Meeting, Lyon 1999*, pp. 691–696.

Kliot, J. and Kliot, K. (eds) (2000) *Millinery, Feathers, Fruits and Flowers*. Lacis.

Lang, U. (1998/99) Laser in der Restaurierung. Seminararbeit im WS 1998/1999 an der FHTW. Berlin: Studiengang Restaurierung/Grabungstechnik.

Ledger, F. (1982) *Put Your Foot Down*. Melksham: Colin Venton, p. 158.

Murray, W. (1994) Conservation of a Pair of Fish-Skin Boots. *SSCR Journal*, **5**(3), 13–14.

Neergaard, L. (1991) Fishskin Leather, an Old Idea that's New. *San Francisco Chronicle*, 24 May, p. 2.

Oakes, J. and Riewe, R. (1996) *Our Boots, an Inuit Women's Art*. London: Thames and Hudson.

O'Flaherty, F., Roddy, W.T. and Lollar, R.M., eds (1958) *The Chemistry and Technology of Leather. Vol. II. Types of Tannage*. New York: Reinhold Publishing Corporation.

Pancaldo, S. (1996) Examination and Treatment Report: A Western Eskimo Fish Skin Bag. (Unpublished Report.)

Rae, A. and Wills, B. (2002) Love a Duck: The Conservation of Feathered Skins. In *The Conservation Of Fur Feather and Skin* (Margot Wright, ed.). pp. 43–61. Seminar organized by Conservators of Ethnographic Artefacts held at Museum of London, 11 December 2000. London: Archetype.

Schultz, G.W. and Schubert, A. (1945) An Investigation of the German Leather Industry. Miscellaneous Report No. 2: report prepared by H.Q. Theatre Service Forces, European Theatre Office of the Chief Quatermaster. British Intelligence Objectives Sub-Committee. London: HMSO.

Solajic, M.R., Cooper, M., Seddon, T., *et al.* (2000) Colourful Feathers. In *The Conservation of Fur Feather and Skin* (Margot Wright, ed.). pp. 69–78. Seminar organized by Conservators of Ethnographic Artefacts held at Museum of London, 11 December 2000. London: Archetype.

Sturge, T. (2000) *The Conservation of Leather Artefacts, Case Studies from the Leather Conservation Centre*. Northampton: The Leather Conservation Centre.

Unattributed (1880) *The Boot and Shoe Trades Journal*, 17 January 1880.
Unattributed (1887) *The Boot and Shoe Trades Journal*, 25 June 1887.
Unattributed (1993) Aquatic Leathers. *Leather*, March, 20–21.
Vincent, J. (1978) *Structural Biomaterials*. Princeton, NJ: Princeton University Press.
Whitear, M. (1986) *Biology of the Integument, 2 Vertebrates*. (Bereiter-Hahn J., Matoltsy K. and Sylvia Richards K., series ed). pp. 8–38.
Wills, B. (2000) A Review of the Conservation Treatment of a Romano-Egyptian Cuirass and Helmet Made from Crocodile Skin. *The Conservator*, **24**, 80–88.

# 17

# Ethnographic leather and skin products

*Sherry Doyal and Marion Kite*

## 17.1 Introduction

A book in its entirety could be devoted to the conservation of 'ethnographic leather'; that is skin and leather products of first nation, indigenous or aboriginal peoples. However, this chapter, within the wider context of a volume on a broad range of leather and skin materials, is intended to serve as an introduction, to signpost readers to other published sources. A factor when dealing with ethnographic skin materials is to understand them within the context of the function or purpose of the object concerned and the culture that made it, and the uses to which this object was then put. The sensitive and possibly sacred nature of some first nation and ethnographic objects must be taken into account before any conservation treatment is undertaken.

Material and species identification is of great importance. Hair fibre identification can be of great assistance as can carefully measuring the size of the skins and having some global idea of where the item originated when trying to identify the material (Appleyard, 1978; Chapman, 1982). Parts from now endangered or protected species of the animal kingdom may be incorporated into ethnographic material. Human remains may be part of the collections or may be incorporated into objects.

The aim of this chapter is to highlight factors which may influence decisions made for the conservation treatment of ethnographic leather and skin materials.

## 17.2 Ethics

Certain ethnographic objects have what can be described as a 'silent heritage' relating to the use of the object before it was collected. Within specialist ethnographic collections the spiritual and intangible aspects of these artefacts must be recognized and understood to be an important element of collections management and care. Originating cultures' concerns should be treated with respect and every effort must be made not to compromise them during their handling, storage and care, when they come within a museum context (Patterson and Greenfield, 1998). Certain conservation processes may be thought profane such as handling by one gender, fumigation or the use of non-kosher or taboo repair materials. For example, kosher parchment should be used for Torah repair (Thompson, 1998).

Animist societies believe that the natural world is animated by an immaterial force, a vital breath, a spirit. Objects made from or decorated with skins and fur can possess the stolen talents of the creature whose first skin it was. Objects associated with Shamanism and Totem objects come within this category. Items may have sacred and ceremonial functions which should be understood before conservation treatment is applied. Consultation with the descendants of originating cultures is desirable and can be legally required.

Human remains may fall to the care of a conservator, though this is less common now that indigenous peoples have successfully applied for repatriation of their ancestors. Examples in the past have been South Amerindian shrunken and Maori tatooed heads.

## 17.3 Uses

Ethnographic skin materials may be found in an enormous range of objects; shelters including tipis and

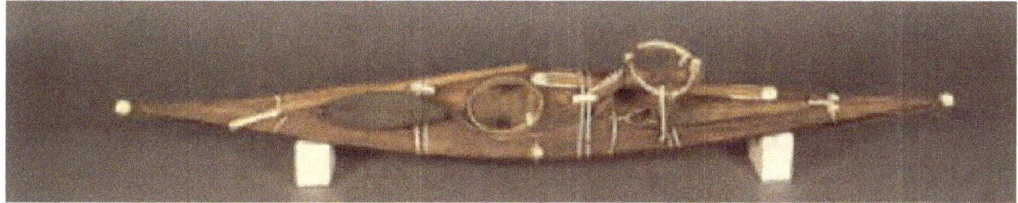

**Figure 17.1** Model kayak, after conservation. Early twentieth century. (Collection of the late Colonel Andrew Croft, DSO, OBE.)

other tents, clothing including gut garments, transport items such as animal harnesses and saddles, or boats such as the kayaks of the Arctic or coracles of Europe. Whole skins are typically used as vessels for water or wine. Membranes such as stomachs are used for bags. Skin is employed in the fabrication of weapons, such as war clubs of stone and raw hide, bolos and knife sheaths. Tools may also be manufactured from skin, for example ray skins are used as rasps by Pacific peoples. Musicians play strung gut or stretched drum skins. Homes are furnished with skins and pelts as floor and bed coverings and in furniture, such as African chairs of state or raw hide-thonged seating of Asia.

## 17.4 Tanning methods

The various methods used for the tanning and preparation of skins are dealt with elsewhere in this volume. Techniques of brain or oil tannage, vegetable tannage and the production of raw hides are all of interest to those working with ethnographic collections. Some skins may have little preparation, for example bird skins may simply have fats scraped or sucked from the skin before they are dried.

The local availability of materials to the peoples of a region give a specific colour and quality to the hides and skins they prepare. This regional distinctiveness can help the specialist provenance and give an attribution to an object. For example, Amerindians use plant materials including barks, roots and seeds for tan liquors, plant additions to brain slurries for oil tanning, and seed cones, leaf and wood for smoking. The plants differ from one tribe to another (Moerman, 1998).

The effect of smoking renders skins more waterproof and can add colour. Different types of fuel were selected to produce differing shades including yellows (juniper leaf), browns (black spruce cones) and reds. Both Amerindian and Sami peoples have used alder smoke to add red colour to leather. These smoked colours can be light sensitive. Sometimes rotten wood was used for smoking. Fungal action on the wood had the effect of removing the cellulose leaving the lignin behind. When burned this released phenols which colour the skin and complex aldehydes such as acrolein which have a tanning action (Mason, 1889).

Not all ethnographic skins are tanned. Often raw hide is used in the manufacture of an object. The fact that green raw hide shrinks and tightens as it dries can be used to good technical effect. The widespread use of raw hide strips to bind the stone and metal heads of axes and other tools to wooden antler and bone handles are examples of this. Raw hide items are hygroscopic and very susceptible to changes in relative humidity. An object can become distorted or even split due to the tensions exerted by raw hide subjected to inappropriate humidity, particularly when combined with other, less environmentally responsive materials.

## 17.5 Construction techniques

Skins can be sewn with sinew or tendon. Post-contact cotton, linen and synthetic threads are used. Seams may be constructed in such a way as to make them waterproof, as those routinely used by Inuit when making footwear (Oakes and Riewe, 1995). Seal gut parkas may have a dried grass incorporated into seams so that they swell on wetting, improving waterproofing.

## 17.6 Decoration

Surface decoration on ethnographic skin objects may be complex and incorporate beads, paint, pigment, quill, applied feathers, applied hair, shells, metal

186  *Conservation of leather and related materials*

**Figure 17.2** Large sealskin bag, hair on, after conservation. Early twentieth century. (Collection of the late Colonel Andrew Croft, DSO, OBE.)

tinklers and sundry other materials. Some large objects with skin or leather coverings may be studded, close nailed, have scratched or impressed decoration, inlays and some will be inherently unstable. Fringes can be used as a decorative effect or for functional purposes; acting as fly swats on clothing, animal trappings or tents or for wicking moisture out of wet garments to assist performance in wet conditions. Fringes and thongs are also worked into cordage from which items such as animal halters are made (Grant, 1972).

## 17.7 Conservation

### 17.7.1 Pre-treatment examination

Expect the unexpected; so-called exotic skins of Europe may be commonplace in the country of origin. Reptile, bird and fish skins and small mammals are all used.

When ethnographic objects are presented for conservation skins and pelts should be examined for signs of wear which can provide evidence of use. Pigment residues on the interior of skin garments may be from body colouring, powder may be evidence of pre-collection cleaning using clays and both may preclude vacuum cleaning as part of the conservation treatment. One author has found remains of lice, of interest to biologists, on horse hair used to decorate a skin garment.

Human remains, may have been consolidated in the field.

### 17.7.2 Poisons – health and safety issues

Ethnographic skins and leathers, particularly those originally acquired by natural history collectors, may have been subject to post-collection treatments to prevent insect and other pest damage and may harbour residual biocides. These can include arsenic compounds, mercuric chloride, DDT (dichlorodiphenyltrichloroethane), naphthalene, paradichlorobenzene, methyl bromide, thymol (3-hydroxy-p-cymene) and ethylene oxide (Johnson *et al.*, 1999). These chemicals can continue to have a deleterious effect on life, including human health. It is suggested that disposable gloves, dust smocks and dust masks should be worn during handling, particularly cleaning. Poisons may also react with objects (Howatt-Krahn, 1987), for example methyl bromide is said to

**Figure 17.3** Small decorated sealskin bag, after conservation. Early twentieth century. (Collection of the late Colonel Andrew Croft, DSO, OBE.)

cause leather products to give off unpleasant odour after treatment (Florian, 1985).

### 17.7.3 Condition

It is important to remember that condition may vary throughout a single skin, either because of the health of the animal or variables in skin preparation. Wear, dryness, cracking, damaged or brittle hair, hair slippage, matting, insect damage, curling or straightening of hair, acidity, shrinkage, damaged or failed seams and splits, rips and tears may all be found in ethnographic skin objects.

It is likely that skins will be acidic due to the degradation or oxidation of the fats and oils used in the initial preparation (Storch, 1987). Metal beads and tinklers which may form part of the object may have reacted with these oils forming acidic corrosion products and causing staining. The acidic nature of the object may have caused sinew or other sewing thread to deteriorate leading to bead loss or the cohesion of the structure to break down. Applied copper decorations or alloys containing copper can cause particular problems. At Exeter City Museums a Native American dagger sheath ornamented with brass dome-headed nails had such an excess of waxy green deposits of cupric ester formation below the domes that the studs were being pushed back out of the leather.

Excess alkalinity of certain glass beads can cause damage to the sewing threads and skin as they deteriorate. This problem is known as 'glass disease' and occurs in beads when the alkali constituents in the glass mixture leach out to the surface of the glass in high relative humidities (Carrol and McHugh, 2001).

### 17.7.4 Cleaning

Initial surface cleaning may be a simple vacuuming, and for furs, a gentle tamping and air-blower grooming. A hair dryer set to cold or compressed air can be used. Fur should be blown in the direction of hair growth. Objects can be cleaned under extraction, in a fume hood or using a vacuum cleaner fitted with a hose and funnel to catch dust raised. Low powered vacuum suction cleaning can be employed assisted with a soft brush such as a squirrel hair mop or baby's hair brush. Before surface cleaning by vacuum suction it is important to inspect furs carefully. The technique of checking or cracking skins, used by Arctic peoples to create ventilated skins (Rahme and Hartman, 2001), by cracking the epidermis so that tiny flaps of skin with fur attached cover a garment, creates a surface which is very vulnerable to damage by vacuum suction. Slippage of fur may be an inherent weakness in a skin resulting from unsuitable methods of preparation. Additionally insect pests may cut fur at the epidermis and this hair may not become detached until disrupted by cleaning. (See chapter on furs for further cleaning information.)

When working with deep furs such as a sheep skin, it can be useful to build a frame with an open weave mesh stretched over it. The skin is placed hair side down over a light-coloured bench or onto light-coloured paper. The flesh side can be tamped by patting gently with the hand or a fly swat. This releases particulate matter from the root of the hair where it may be causing abrasion, without disrupting the lay of the hair. The matter sifts through the screen onto the light-coloured surface were it can be checked for insect debris or an unexpected amount of loose hair, vegetable matter, etc. before it is vacuumed away. Then the skin can be turned flesh side down onto the table and the hair is vacuumed through the screen. If the item is relatively robust this procedure may be repeated until very little soiling is being released.

After initial surface cleaning further cleaning may be required. It is important to trial clean several small areas to help decide on an appropriate level of cleaning.

Aqueous cleaning is rarely suitable for ethnographic leather but mechanical surface cleaning using soft erasers, smoke sponges and chemical sponges is often very effective. Materials need to be cleared from the object by further low powered vacuuming.

### 17.7.5 Solvent cleaning

An author still shudders at the memory of a comment made at the first conservation conference she attended over 25 years ago. A speaker recounted, with some delight, the wonderful scents of berries and smoke released when submitting skin garments to a solvent dry cleaning process. What materials were being extracted? In the intervening years conservators have learnt to do less to the items in our care. We have learnt this from conservators before us who were prepared to admit their mistakes or whose mistakes have only been revealed by the passage of time.

A suitable local solvent cleaning process for furs with oil-adhered soiling would be the grooming of the fur with a white flannel cloth pad moistened with white spirit BS245 or Stoddard solvent wiped lightly over the surface of the fur. The pad should be changed as it becomes soiled. When the solvent has evaporated off the fur can again be vacuumed to remove released soiling. Stains and surface soiling can be spot cleaned using swabs. The technique used is to ring the stain with solvent and work inwards, blotting to remove soiling. (For further information see chapter on furs.)

### 17.7.6 Reshaping

Providing the skin is strong and not brittle and fragmented, areas stiffened by wetting may be softened by manipulation with a blunt tool (such as a bone folder) to loosen and flex stiff collagen fibre bundles. This may be assisted by the penetration of the skin with a 70/30 (v/v) ethanol and water mix to prevent cracking and the skin is gently manipulated until dry (Lougheed *et al.*, 1983).

Historically, reshaping of ethnographic skins has been undertaken using humectants, oils, oil emulsions, organic solvent conditioning and humidity to soften the skins prior to the reshaping treatment (Sully, 1992). In the past overoiling has resulted in residual spews and overall stickiness causing problems years after the treatments were undertaken. At the time of writing the authors' preferred reshaping treatment is to humidify the skin and gradually manipulate the object back to shape as it softens, as it reaches equilibrium in an environment of gradually increasing humidity. When considering humidification one should remember that the thinner the skin the more likely there is to be an adverse reaction to wetting which may easily occur where direct application of humidity is used (with a spray or ultrasonic humidifier, for example). Semi-tanned leathers are hydrophilic (Schaffer, 1974). Humidification should not be rushed and should be undertaken as a gradual, controlled process. Local treatments using humidity membranes such as Sympatex or Gortex can be applied to small areas such as a distorted strap. Unbuffered acid-free tissue paper puffs, conditioned to a higher relative humidity than the object, can be used to pad out areas such as collapsed skin backs to masks. The puff will gradually release the humidity into the object. More puffs are added as the shape is eased out. Whole object treatments can be applied in a humidity chamber. This is, for example, particularly effective for skin garments being brought back to shape for mounting on mannequin figures. A chamber can be as simple as a polyethylene tent using salts or glycerol to set the humidity within the chamber or as complex as fully automated electric systems.

Tears and curling edges may need to be reshaped and flattened, subject to tests, with ethanol before repairs are carried out (Kaminitz and Levinson, 1988). When tears are realigned and objects are reshaped then a method of securing and holding the object in the desired position will need to be found. Glass weights or small sand bags may be used where suitable for flat items. Sand bags have advantages over glass weights as the humidified object can dry through the weight. Glass weights can trap evaporating moisture causing tidemarks. For three-dimensional items, blocking on mounts, unbuffered tissue, crushed nylon net, or inflated polyethelene bags may all be used to hold the object in shape. The object should be allowed to return slowly to ambient conditions while blocked.

### 17.7.7 Mounts/internal supports

Damage has frequently been done to three-dimensional skin objects by either poor storage or display (McNeil *et al.*, 1986; Niinimaa, 1987; Rose and Amparo, 1992). Mounts are ideally made to provide support both in storage and on display; however, storage volume may be much increased by this. A good mount can reveal much about the construction of an object and may make invasive repairs unnecessary. Mounts should be produced to fit the object because skin items may stretch or take up the shape of a poorly shaped support (such as a mannequin with a 1950s exaggerated and pointed bra shape). The mount should provide support to eliminate sagging and stress on the object and to avoid creases.

Successful conservation is often a collaboration between object conservator and skilled mount maker. For hard mounts metal and acrylic sheet working skills are required and for soft mounts sewing and tailoring skills. Just as important is good information about how skin items were used or worn. Consultation with indigena, period drawings, photographs and anthropologists' notes are all valuable assets when devising a mount. The visual material can also be displayed alongside mounted items to further explain an exhibit.

Mannequins may be custom made from Ethafoam plank (Larouche, 1995) or commercial mannequins made of papier mâché can be customized with a saw and then covered with a foil such as Moistop to seal emissions. Either type may then be padded with polyester felt or wadding and covered with cotton calico or cotton stockingette. The latter has the advantage of requiring fewer tailoring skills. The cotton fabrics can be dyed with a conservation quality dye system to give a range of skin or display colours. Heads, arms and legs can be tailored in wadding and cloth over an Ethafoam tube core. Support underpinnings or undergarments can be made by taking a pattern from the clothing and cutting this shape in polyester felt, this has bulk and will stand away from the mannequin to offer support, for a cloak or the skirt of a coat, for example. The felt can be covered with knitted, pile or napped fabrics used to 'key' into the nap surface of the skin garment to provide a friction mount with overall support. Alternatively the mount can be covered with a slippery fabric such as nylon or silk if static or abrasion is a concern. A brass pole system can be used to secure the mount within the case. Brass can be cushioned with Ethafoam or coated by heating and dipping into powdered polyethylene.

For further information on mount making see McNeil et al. (1986), Kaminitz and Levinson (1988), Kite (1994), Kite and Hill (1999), Stephens and Doyal (2002).

### 17.7.8 Mending

As a general principle, mends should be weaker than the object being repaired. When working with material which is subject to dimensional change it is important that repairs rather than the object should fail. Some conservators prefer to use like with like materials, that is they choose materials similar to those of the object so that the expansion and contraction rates of the mend are similar to the object and therefore avoid or minimize tensions between the mend and the object. The temperature and humidity of storage, conservation and display areas and the season in which a mend is made all have impacts on success.

### 17.7.9 Repair supports

Repair supports may be used to give either a suture, patch or full support to an object. When choosing skin products for repairs, criteria to be considered include trying to match the grain, weight, colour and spine direction, and establishing which is the hair or flesh side (Munn, 1985). Thin semi-transparent supports include gold beaters' skin made from cattle intestine. This can be laminated with gelatine (Kaminitz and Levinson, 1988). Membrane from the swim bladders of fish including sturgeon, cod, hake and whiting has been used (Munn, 1985) as has sausage casing. These should be degreased with acetone or ethanol before use and can be pumiced for better adhesion and to cut gloss.

Synthetic web fabrics, non-woven heat-bonded polyester and nylon also may be used. These may be dyed to tone with the object using synthetic dye (Munn, 1985).

### 17.7.10 Sewing

Tacketing or thonging (Munn, 1985), unwaxed nylon dental floss (Kaminitz and Levinson, 1988) and sinew (Kite, 1999) are a few of the materials which have been used for sewn repairs to ethnographic objects.

### 17.7.11 Adhesives

Much has been written about adhesives in other chapters. However, the basic criteria for selecting an adhesive for use on ethnographic objects include viscosity, so that the adhesive does not carry through the skin, flexibility, and that strength should be a little lower than the skin to be repaired. The shrinkage temperature of the skin to be repaired must also be considered. Beva 371, Acraloid B72, polyvinyl acetate, wheat starch paste, pharmaceutical clear gelatine or parchment size are some of the most frequent choices made (Lowenthal et al., 2003). Many more adhesives are of course suitable in certain circumstances (Fenn, 1984; White and Sully, 1992).

Treatment decisions will also be influenced by how the object will be used, mounted, displayed, travelled, or stored.

### 17.7.12 Cosmetic repairs and infills

Infills of similar weight support material can be butt jointed (to create little tension) with a supporting patch below. A semi-transparent overlay or a combination of support and overlay can be employed if the area around the loss is weak or the object so thin that a butt joint has little strength. Synthetic furs can be used in pelts (Nieuwenhuizen, 1998) or, for display purposes, well-matched new fur may be temporarily laid over insect-damaged furs.

For gap fills where leathers have shrunk and split Beva 371, glass micro ballons and dry pigment have been used (Nieuwenhuizen, 1998; Kronthal et al., 2003). Also papier mâché toned with pigments. Bulked fills can be textured using cast impression moulds.

### 17.7.13 Storage

Storage should be cool, dark and well ventilated. Items should be protected from dust by being cased, boxed, placed in enclosed storage or stored under cloth dust covers (Walker, 1988). Storage furniture should be made of materials with no acidic emissions such as stove-enamelled metal. Because of the vulnerability of semi-tanned leathers to water damage a store location away from overhead pipes and water storage is much preferred. Supports are important, especially for heavier skins and furs. To avoid stretching, mounts or rigid supports may be required. Stacking items without dividers should be avoided as an impression of one object may be left on another.

Items folded when supple may stiffen and set in the storage position. This should be considered when packing items. The use of unbuffered acid-free tissue for storage of proteinaceous material should be considered as the alkaline buffering material may leach into the object. Sticky items may need silicone release packaging materials. It should be appreciated that fatty spew may pool in depressions in objects during storage and that dust is attracted to spew.

Storage areas should be checked at least twice yearly for evidence of insect infestation and insect monitoring traps are recommended in storage and display areas. Before entering the store, items may be cycle frozen as an infestation preventive measure (Florian, 1986). Reduced temperature storage has been used for the preservation of oil-tanned items (Fenn, 1985) and as a pest deterrent. However, a breakdown of the cooling system can cause a hazard by exposing items to extremes of relative humidity (Lougheed et al., 1983). It should be remembered that cold stored items should be well supported and brought to ambient temperatures before handling as they may be stiff and more brittle when cold (Carrlee, 2003).

### 17.7.14 Display

In common with most organic materials relative humidities of 45–55% and temperatures of no more than 21°C are suggested for the display of ethnographic materials. Light may fade smoke-dyed leathers and dark furs or cause light furs and repair adhesives to yellow. A light level of 50 lux with an ultraviolet light content of less than 75 W/lm is suggested.

## References

Appleyard, H.M. (1978) *Guide to the Identification of Animal Fibres*. Leeds: British Textile Technology Group.

Carrlee, E. (2003) Does Low-temperature Pest Management Cause Damage? Literature Review and Observational Study of Ethnographic Artifacts. *Journal of the American Institute for Conservation*, **42**(2), 139–141.

Carrol, S. and McHugh, K. (2001) Material Characterisation of Glass Disease on Beaded Ethnographic Artefacts from the Collection of the National Museum of the American Indian. In *Ethnographic Beadwork. Aspects of Manufacture, Use and Conservation* (Margot Wright, ed.). London: Archetype.

Chapman, V. (1982) Scale Impressions of Animal Fibres: A Simple Method. *Conservation News*, **18**, 14.

Florian, M-L. (1985) A Holistic Interpretation of the Deterioration of Vegetable Tanned Leather. *Leather Conservation News*, **2**(1), 1–5.

Florian, M-L. (1986) The Freezing Process: Effects on Insects and Artifact Materials. *Leather Conservation News*, **4**, 6–15.

Grant, B. (1972) *Encyclopaedia of Rawhide and Leather Braiding*. Centerville, MD: Cornell Maritime Press.

Howatt-Krahn, A. (1987) Conservation: Skin and Native-tanned Leather. *American Indian Art Magazine*, Spring, 44–51.

Johnson, J. et al. (1999) Masked Hazard. In *Common Ground, Archaeology and Ethnography in the Public Interest*, Fall, 26–31. Washington: NPC Archaeology and Ethnography Program.

Kaminitz, M. and Levinson, J. (1988) The Conservation of Ethnographic Skin Objects at the American Museum of Natural History. *Leather Conservation News*, **5**, 1–7.

Kite, M. (1994) Support Stands for the Storage and Display of Hats. *The Poster Summary Booklet for the IIC Congress, Ottawa*.

Kite, M. and Hill, A. (1999) Polyester Wadding, Perspex, Vilene, Gatorfoam and Ethafoam used in the Preparation of Supports for 3-dimensional Textiles. *Post-prints of the*

*ICOM-CC Textiles Working Group Symposium Amsterdam, 1994.* London: Archetype.

Kite, M. (1999) The Conservation of a 19th Century Salmon Skin Coat (626–1905). *Preprints of the ICOM-Committee for Conservation 12th Triennial Conference Lyon 1999*, **II**, pp. 691–696. ICOM.

Kronthal, L., Levinson, J., Dignard, C. et al. (2003) Beva 371 and its Uses as an Adhesive for Skin and Leather Repairs: Background and Review of Treatments. *Journal of the American Institute for Conservation*, **42**, 341–362.

Larouche, D. (1995) Intersecting Silhouette Mannequins. *Textile Conservation Newsletter*, Spring supplement.

Lougheed, S., Mason, J. and Vuori, J. (1983) Repair of Tears in Skin Garments. *Journal International Institute for Conservation, Canadian Group*, **8 & 9**, 13–21.

McNeil, K.C., Johnson, J.G., Joyce, D.J. et al. (1986) Mounting Ethnographic Garments. *Curator*, **2**(4), 279–293.

Mason, O.T. (1889) *Aboriginal Skin Dressing: A Study Based on Material in the US National Museum*. Annual report of the Smithsonian Institute, pp. 553–589.

Moerman, D.E. (1998) *Native American Ethnobotany*. Portland, Or.: Timber Press.

Munn, J. (1985) Examination and Treatment of Vellum. In *Recent Advances in Leather Conservation*. Washington: AIC.

Nieuwenhuizen, L. (1998) Synthetic Fill Materials for Skin, Leather and Furs. *Journal of the American Institute for Conservation*, **37**, 135–145.

Niinimaa, G.S. (1987) Mounting Systems for Ethnographic Textiles and Objects. *Journal of the American Institiute for Conservation*, **26**, 75–84.

Oakes, J. and Riewe, R. (1995) *Our Boots – An Inuit Woman's Art*. Toronto: The Bata Shoe Museum.

Patterson, C. and Greenfield, J. (1998) Storage Considerations for Native Arts: A Joint Project between the Denver Art Museum and the American Indian Communities. *Preprints of the Scottish Society for Conservation and Restoration Conference Dundee* (M. Wright and Y. Player-Danhnsjö, eds). SSCR.

Rahme, L. and Hartman, D. (2001) *Leather – Preparation and Tanning by Traditional Methods*. Portland, Or.: The Caber Press.

Rose, C.L. and Amparo, R. de T., eds (1992) *Storage of Natural History Collections: Ideas and Practical Solutions*. Society for the Preservation of Natural History Collections.

Schaffer, E. (1974) Properties and Preservation of Ethnographical Semi-tanned Leathers. *Studies in Conservation*, **19**, 66–75.

Stephens, M. and Doyal, S. (2002) Round the World on a Mannequin. In *Home and Away – Approaches to Textile Conservation Around the World*. UKIC Textile Section.

Storch, P.S. (1987) Curatorial Care and Handling of Skin Materials: Part **II** Semi-tanned Objects. *Conservation Notes*, **18**. Austin: University of Texas.

Sully, D.M. (1992) Humidification: The Reshaping of Leather, Skin and Gut Objects for Display. In *ICOM Committee for Conservation, Leathercraft Group Symposium London* pp. 50–54. ICOM.

Thompson, J.T. (1998) On Restoring Sacred Objects. *Leather Conservation News*, **14**(2), 1–6.

Walker, S. (1988) Using Tyvek in Protective Covers for Artefacts. *Museum Quarterly*, **16**(4), 23–25.

White, S.J. and Sully, D.M. (1992) The Conservation of a Siberian Parka: A Joint Approach. In *ICOM Committee for Conservation, Leathercraft Group Symposium*. London pp. 54–57. ICOM.

## Bibliography

Other works on the preparation, use and conservation of ethnographic skin products include the following.

Anon. (2001) *Arctic Clothing from Igoolik*. London: British Museum.

Burnham, D.K. (1992) *To Please the Caribou*. Ontario: Royal Ontario Museum.

*Conservation Notes*. Canadian Conservation Institute Various Authors
N6/4 Care of Objects Decorated with Glass Beads
N6/5 Care of Quillwork
N8/3 Care of Mounted Specimens and Pelts
N8/4 Care of Raw Hide and Semi-tanned Leather

Fitzhugh, W.W. and Crowell, A. (1988) *Crossroads of Continents – Cultures of Siberia and Alaska*. Washington: Smithsonian Institute Press.

Hickman, P. (1987) *Innerskins and Outerskins: Gut and Fishskin*. San Francisco: San Francisco Craft and Folk Art Museum.

Hobson, P. (1977) *Tan Your Hide! Home Tanning Leather and Furs*. Vermont: Storey Books.

Horse Capture, G.P., Vitart, A., Waldberg, M., et al. (1993) *Robes of Splendor: Native North American Painted Buffalo Hides*. New York: The New Press.

Horse Capture, J.D. (2000) *Beauty, Honor and Tradition: The Legacy of Plains Indian Shirt*. Washington: Smithsonian Institution.

Morphy, H. (1998) *Aboriginal Art*, pp. 160–161, 337, 357, 104. London: Phaidon.

Wright, M. ed. (1998) *Conservation of Leather Skin and Hair*. Conservators of Ethnographic Artefacts. London: Archetype.

Wright, M. ed. (2001) *Ethnographic Beadwork, Aspects of Manufacture, Use and Conservation*. Conservators of Ethnographic Artefacts. London: Archetype.

Wright, M. ed. (2002) *The Conservation of Fur Feather and Skin*. Conservators of Ethnographic Artefacts. London: Archetype.

# 18

# Collagen products: glues, gelatine, gut membrane and sausage casings

*Marion Kite*

This chapter includes some of the less obvious and unusual skin-related materials including gut membranes, glue, gelatine, isinglass, and others which may be the concern of conservators.

These materials may have been used in the production of an object or may be useful materials employed by conservation specialists during the process of conservation of an object.

## 18.1 Animal glues and fish glues

Animal glue is, as the name suggests, derived from animals and has been used since antiquity. There are no exact records telling when and where animal and fish glues were first used but it is known that at least 3500 years ago these glues were used in Egypt and were made by being melted over fire and applied with a brush (Darrow, 1930). From the first century Roman scholar Plinius we learn that two kinds of glue were used, *taurokalla* in Greek, *gluten taurinum* in Latin made of the skins of bulls and *ichtyokalla* made from parts of fishes (Petukhova, 2000).

Connective tissue in skin, bone and other tissue is made up of proteins, principally collagen with small amounts of other components including plasma proteins and elastin.

These proteins are partly hydrolysed on heating in water, acids or alkalis to produce a soluble product generically known as glue. The degraded collagen which goes into solution is concentrated by evaporation until protein solids reach upwards of 45%. They are then cooled and set to gel. The remaining water is then removed and the dried product is ground into particles ready for use. Details of the various processes involved are given by Young (1971).

Animal glues have a number of special characteristics, but are unique in having three phases:

1. liquid when in hot water
2. gel when water nears room temperature
3. solid when water has been evaporated out.

Glues, as used by craftsmen and conservators alike, include a range of products derived from a range of animals and are of varying purity. The quality of an animal glue is based on its gel strength. This is determined by:

1. the source of the collagen
2. the aggressiveness of the treatment applied to extract the collagen
3. the stage at which it is extracted
4. the amount of impurities and lesser proteins included.

## 18.2 Skin glues and hide glues

Skin glues are derived from cattle and sheep skins and are known as hide glues or parchment glues depending on the material from which they are made. They may contain contaminants from skin preservatives or tanning agents. These skin or hide glues have the highest strength of the animal glues. Acid hydrolysis or alkaline hydrolysis can be the method employed to break down or denature the collagen molecules and allow the protein to be brought into a colloidal suspension.

Glue is soluble in few solvents at room temperature but will swell in cold water to form a gel. The gel from animal glues will melt on heating to 30–50°C (Horie, 1987). The pH of hide glue is 6.0–7.5.

Hide glues are used in the manufacture of wood joinery, bookbinding, abrasive papers, gummed tapes, and matches.

## 18.3 Parchment glue and parchment size

Parchment glue was used by medieval craftsmen to strengthen weak parchment and for patching and joining parchments (Reed, 1972).

Over 100 years ago a recipe for preparing parchment glue was given by Watt (1897) and involved boiling 21 grams of parchment scraps soaked in 2½ litres water for 45 minutes, or longer if a stronger size was required.

A general method and recipe used today is to soak fine shavings or pieces of parchment in cold water followed by prolonged boiling. The parchment collagen is degraded to varying degrees from swollen, hydrated collagen fibrils to gelatine (Haines, 1999).

Parchment glue has been used as a consolidant and as an adhesive by conservators from many disciplines and has been used for resizing archival papers that have been washed.

Parchment glue and rabbit skin glues have been used in the production of gessoes and in gilding.

## 18.4 Rabbit skin glue

True rabbit skin glue is made from rendered down rabbit skins and like parchment glue goes readily into solution. Rabbit skin glue has many applications in fine art and especially in gilding. It is flexible but its structural strength is weak so is not used for gluing any material under stress.

## 18.5 Bone glue

Bone glues are derived from the bones, sinew and cartilage of most farm animals. Bone glues are considered inferior to skin and parchment glues. Typically the pH = 5.8–6.3 (Hubbard, 1977).

Bone glues are commercially used as carton and box adhesives.

## 18.6 Gelatine

Gelatine is a purified form of hide glue. It can be rendered insoluble by reacting it with trivalent metal ions, tannins or aldehydes. The production of gelatin is normally brought about by giving skin and other collagen containing materials treatment with either lime or acetic acid, followed by heat.

Gelatine prepared with alkali has slightly different properties to that prepared with acid. For the production of high quality gelatin of uniform composition, the skin of immature pigs is used, treated with acetic acid followed by heating (Haines, 1999).

Since the nineteenth century gelatine solutions have been used widely for adhering and consolidating objects from many disciplines. Mercuric chloride or formaldehyde may have been used as an additive to inhibit mould growth (Horie, 1987). If so, then the mercuric chloride would be detectable using EDXRF analysis and the formaldehyde would cause crosslinking and the gelatine to be insoluble.

From the late 1880s until well into the 1930s, costume and accessories were decorated with sequins. (*Figure 18.1*). Frequently these were made from

**Figure 18.1** Evening dress by Callot Soeurs *c*.1922 embroidered with pearlized gelatin sequins and glass beads. (Courtesy V and A Images)

194  *Conservation of leather and related materials*

**Figure 18.2a** Detail of gelatin sequin before steaming.

gelatine which was dyed or treated with any one of a number of coatings to simulate other materials such as jet, mother of pearl, iridescent or coloured glass and a range of different coloured metals. Cellulose nitrate was one of the coatings used for pearlized sequins (Paulocik, 1988). Gelatine is very reactive to changes in humidity. A dramatic rise will often cause the surface of gelatine sequins to 'bloom' and go dull and an increase of as much as 25% moisture content of gelatine at room temperature can result in the gelatine swelling and becoming rubbery (Horie, 1987). Direct application of water will cause gelatine to swell alarmingly (*Figures 18.2a* and *18.2b*). Aqueous cleaning methods for textiles where gelatine sequins have been used are therefore unsuitable. The use of hydrocarbon solvents is also not usually possible as organic solvents will dissolve cellulose nitrate and may harm the other finishes or coatings (Kite, 1992).

## 18.7 Fish glue

There are two types of fish glue. The highest quality product is known as isinglass and is made almost entirely from the swim bladders of fish. The glue extracted from the swim bladders of sturgeon (*Acipenseridae* sp.) is known as sturgeon glue. The swim bladders are opened, washed, stretched and dried to produce sheets of this protein. Fine sheets may be produced by pressure rolling after the material has first been softened in cold water. These may be shredded or flaked.

The second type of fish glue is a much less expensive product and is extracted from heads, skins and skeletal waste from cod (*Gadus* sp.), haddock (*Melanogrammus aeglefinus* sp.) and mackerel (*Scomber scombrus* sp.) and other fish as a by-product of the fresh and frozen food markets.

In medieval Europe fish glues were used as a size for parchment and as a medium or binder for paint. The property of fish glue to adhere well to the porous surface of parchment made it a useful material for the illumination of manuscripts with gold leaf and painting. In China it was used as a binding medium by calligraphers.

It has been used by conservators as an adhesive and consolidant.

## 18.8 Gut membrane

Gut membrane is prepared from the outer or peritoneal coat of the caecum of bovines and is also

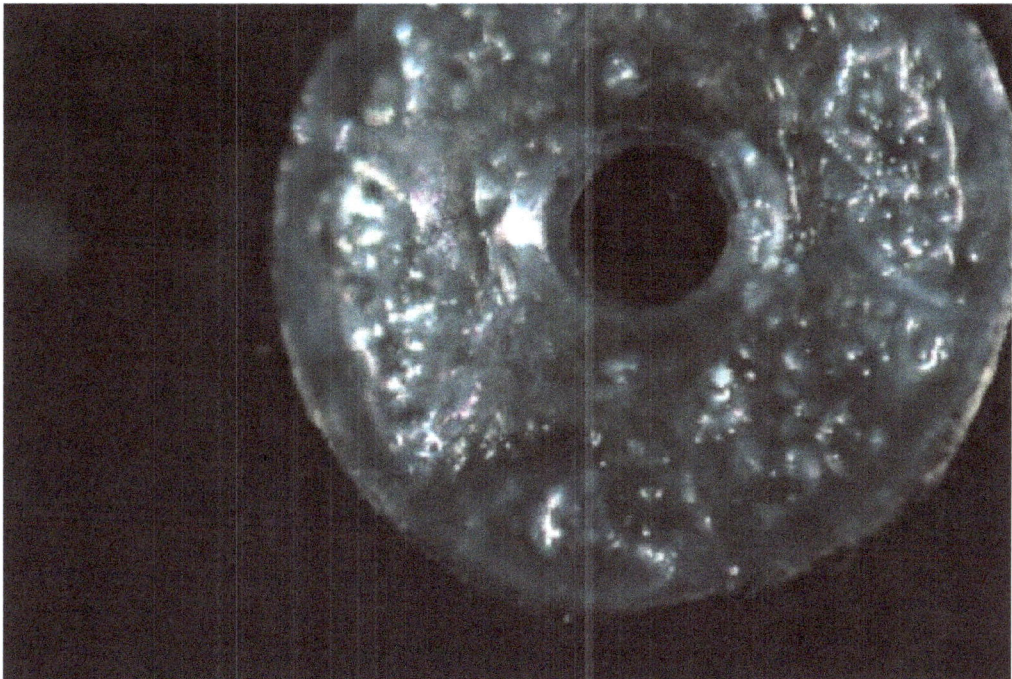

**Figure 18.2b** Detail of gelatin sequin after light steaming, showing swelling.

known as goldbeaters' skin. It was originally used as a separation layer during the beating out of gold when making gold leaf.

Goldbeaters' skin is thin, light, strong and a fully transparent sheet material composed of collagen.

The method of processing is described by Watt (1897) and includes treatment by soaking in a potassium hydroxide solution, scraping, washing, stretching and treatment with alum.

The swim bladders of certain fish such as sturgeon or cod have also been used to prepare an alternative material to gold beaters' skin. These, however, are more yellow in colour and opaque.

In the eighteenth century gut membrane was used to make the first balloons for manned flight. It was also used in the construction of the first aeroplanes in the early days of aviation history.

In conservation it is used as a support material where a thin, strong, lightweight support is required and is often used by bookbinders for the repair of parchment. It has also been used for the repair of other skin materials (Kite, 1999).

In textiles it is found as a substrate for some types of metal threads. Beaten-out gold leaf was applied to sheets of gut which were cut into strips forming the filaments which were then spun round a textile core. These membrane threads were combined into the structure of silk fabrics. They are often to be found in Italian silks of the fourteenth and fifteenth centuries (Kite, 1992) (*Figures 18.3 and 18.4*).

This type of gold thread had the advantage of being light and flexible and much of it could be incorporated into the structure of a comparatively lightweight silk. Gold leaf applied to membrane was a cheaper metal thread as much less of the precious metal was needed in its manufacture. Darrah (1987) has shown there is evidence of a gilded silver layer on some membrane indicating that gold and silver were beaten out together from a metal block before being laid onto the membrane; also concentrations of copper have been found, possibly indicating a triple layer of metal on the membrane in some cases.

Textile conservators must be aware of the usage of this type of metal thread and know at which period or date to expect to find them and where on a textile to look for them.

## 18.9  Sausage casings

Historically sausage casings were made from natural intestines of bovines. Gut dressing is complex and

196  *Conservation of leather and related materials*

**Figure 18.3** Fragment of fourteenth century Italian ecclesiastical fabric from a cope woven with membrane metal threads.

**Figure 18.4** Detail of *Figure 18.3*, showing membrane metal thread spun around a flax thread core.

involves 11 distinct operations in the processing, including scouring, turning inside out, putrid fermentation, scraping, washing, inflating, drying, measuring, and sulphuration. The process is fully described by Watt (1897).

Prior to the identification and outbreak of bovine spongiform encephalopathy (BSE) in the late twentieth century, sausage casings were also made from the flesh splits of mature cattle skins. The flesh splits were processed mechanically and chemically to break them down to form a solution. Acetic acid was used to partially degrade them and the gel obtained was used to form the casings by means of extrusion which were then precipitated in a saline solution and then washed and dried.

Apart from their obvious culinary uses sausage casings are used for backing and infilling parchment as an alternative to goldbeaters' skin.

## References

Darrah, J.A. (1987) Metal Threads and Filaments. *University of London Institute of Archaeology Jubilee Conference Preprints*, pp. 211–221.

Darrow, F.L. (1930) *The Story of an Ancient Art, from the Earliest Adhesives to Vegetable Glue*. Lansdale, PA and South Bend: Perkins Glue Company.

Haines, B.M. (1999) *Parchment, the Physical and Chemical Characteristics of Parchment and the Materials used in its Conservation*. Northampton: The Leather Conservation Centre.

Horie, C.V. (1987) *Materials for Conservation*. Oxford: Butterworth and Co. (Publishing) Ltd.

Hubbard, J. (1977) Animal Glues. In *Handbook of Adhesives* (I. Skeist, ed.), New York: Van Nostrand Reinhold.

Kite, M. (1992a) Gut Membrane, Parchment and Gelatine Incorporated into Textile Objects. *The Paper Conservator*, **16**, 98–105.

Kite, M. (1992b) An Overview of Skin-related Materials Incorporated into Textiles. *Postprints of the ICOM Leather and Related Objects Group Interim Symposium, June 24 and 25, 1992*, pp. 33–35.

Kite, M. (1999) The Conservation of a 19th Century Salmon Skin Coat (626–1905). *ICOM-CC Triennial Conference Lyon, 29 August–3 September 1999*, pp. 691–696.

Paulocik, C. (1989) Of Buttons, Beads and Sequins, Beware of Water Sensitive Embellishments. Abstract of paper presented at Harper's Ferry Conference in Washington, November 1988. *Leather Conservation News*, **5**(2), 6–7.

Petukhova, T.A. (2000) History of Fish Glue as an Artist's Material: Applications in Paper and Parchment Artifacts. *The Book and Paper Group Annual. Vol. 19*. American Institute for Conservation.

Reed, R. (1972) *Ancient Skins, Parchment and Leather*. London: Seminar Press.

Watt, A. (1897) *Leather Manufacture, a Practical Handbook of Tanning, Currying, and Chrome Leather Dressing*. London: Crosby Lockwood.

Young, H.H. (1971) *Glue, Animal and Fish in Adhesion and Bonding* (N.M. Bikales, ed.). New York: J. Wiley and Sons.

# 19

# The manufacture of parchment

*B.M. Haines*

The process of transforming animal skin into clean, white sheet material suitable for writing has changed little over the centuries. The essential stages of the process are as follows.

## 19.1 Temporary preservation

Immediately the skin is removed from the animal it is vulnerable to attack by bacteria. Inevitably there will be a delay before the start of parchment processing. To hold the skin in good condition during such a delay requires some form of temporary preservation. In warm dry climates the skin can be air dried, but in more temperate regions salt is liberally applied to the raw skin. The salt maintains the skin in a sound condition while it is held for sale and transported to the user.

## 19.2 Soaking

The skin is immersed in a large volume of cold, clean, sometimes running water for 48 hours, to wash away blood and dirt, extract the salt and rehydrate the skin. The soak may be carried out in either stone or wooden vats and at intervals the skins are moved in the water either by stirring by hand or by the use of revolving wooden paddles.

## 19.3 Liming

The skins are immersed in a suspension of slaked lime for eight days (up to 16 days in cold weather) during which time the skins are gently moved about. The alkalinity of the lime attacks the keratin of the hair and the epidermis, loosening them and allowing them to be scraped from the skin.

With sheepskin, the process is somewhat different. In order to prevent the wool being damaged by the alkali it is loosened by a paste of slaked lime that is applied to the flesh surface of the skin. The alkali penetrates through the skin loosening the wool at the roots, which is then pulled from the skin, sorted into different grades, washed, dried and sold to the wool merchant. The skin is immersed in a lime liquor for a further period, after which it is split into two layers. The grain layer is tanned usually by the vegetable tanning process to produce skivers and the flesh split converted into parchment.

## 19.4 Unhairing and fleshing

The skins are unhaired by placing them grain uppermost, over an upright, convex curved, wooden board, termed the 'beam'. The loosened hair and epidermis are scraped off using a curved, blunt-edged knife. The grain surface is further scraped to remove residual hair and surface grease. The skin is then reversed on the beam and fat, muscle and loose flesh layers are cut from the flesh surface. The skins are usually returned to the lime for several days, after which they are washed in water for one or two days. Today, machines are employed to remove hair and flesh.

## 19.5 Drying

The quality of the parchment depends on the careful control of the drying. The skin is suspended within a rectangular wooden frame. It cannot be nailed to the frame, as the shrinkage that occurs during this

While the skin is still wet it is vigorously scraped on both surfaces, using a two-handled crescent shaped knife (*Figure 19.1*). During this operation the skin is kept wet by repeated applications of water. Throughout the process the skin is continually retensioned by means of the pegs. Finally the fully taut skin is air dried. The rate of drying has to be controlled: exposure to direct sunlight or fast movement of air can cause the parchment to become damaged by too high a temperature and/or too rapid drying.

When the skin is considered to be sufficiently dry, both surfaces are scraped and shaved, again using the crescent-shaped knife. The thick regions of skin are pared from the flesh surface in order to obtain a sheet of uniform thickness. Unless the parchment is intended for book covers, the grain surface is also shaved to remove the gloss that is characteristic of an intact grain. High gloss is considered undesirable in a writing parchment.

After shaving, the dry parchment is cut from the frame and the surfaces rubbed smooth with powdered pumice.

**Figure 19.1** Paring the surface of the skin.

stage of the process would cause the skin to tear away at the nailed points. Metal clamps are also inadvisable, as contact with the metal can stain the parchment. The method adopted allows for the frequent retensioning of the skin throughout the drying period. Every few centimetres around the edge of the skin a small smooth pebble or a ball of rolled-up newspaper is wrapped in the soft skin tissue and the knob so formed is secured by cord to adjustable tensioning pegs in the wooden frame. These have been likened to the tuning pegs of a violin. An alternative to the rectangular frame is a circular frame made of a flexible wood such as willow. The limed skin is suspended within this frame using tensioning cords in the same manner and the frame itself adjusts to the tensions that develop as the skin dries.

## Bibliography

Haines, B.M. (1999) *Parchment: The Physical and Chemical Characteristics of Parchment and the Materials Used in its Conservation*. Northampton: The Leather Conservation Centre.

Hamel, C. de (1992) *Scribes and Illuminators*. London: British Museum Press. (Medieval Craftsmen)

Kennedy, C.J. and Wess, T.J. (2003) The Structure of Collagen within Parchment – A Review. *Restaurator*, **24**(61).

Reed, R. (1972) *Ancient Skins, Parchment and Leather*. London: Seminar Press.

Rück, P. (1991) *Pergament*. Sigmaringen: Jan Thorbecke Verlag.

Ryan, K. (1987) Parchment as Faunal Record. *Museum Applied Science Center for Archaeology Journal*, 4, 124–138.

Tomlinson, C. (1864) *The Useful Arts and Manufactures of Great Britain*, 2nd ed. London: SPCK.

# 20

# The conservation of parchment

*Christopher S. Woods*

## 20.1 Introduction

This chapter describes the nature of parchment and its characteristics, in particular those that have a direct influence on its preservation and conservation. The different forms in which parchment has been used are summarized, and features of these that influence preservation are included. Also included is a review of conservation treatments and how they have developed. The intention is not to go into significant depth in any of the areas covered, nor to recommend any specific practice (although a cautious and sensitive approach is encouraged). Instead, the purpose is to introduce the reader to the subject and offer some direction for more detailed study or guidance.

Parchment has been used in a number of ways since it was first produced. A precise definition of parchment is probably: animal skins that have been dried under tension to produce a stiff, sheet material with a flat, even surface. Skins of this type have been found dating from around 2800 years ago. Parchment produced using a lime bath in its preparation, as described in Chapter 19, appears in recipes written as early as the eighth century (for example, the Lucca manuscript, Codex 490 ff. 21–25) (Reed, 1972: 33) and Byzantine parchment of the sixth century has been defined as limed skin (Wächter, 1962). Parchment-type skins have been used in a variety of ways, but the great majority of parchment made over the centuries has been used as a manuscript substrate (see *Figure 20.1*). As a consequence of this predominance, this chapter devotes most attention to the conservation of parchment in manuscript formats, but it also includes a summary of parchment used in other formats. The intention is to make conservators and others aware of the likely places where this type of skin material can be found and to alert them to its characteristics.

There is a large body of published material on the topic of parchment and its variations, in several languages, covering many aspects of science and conservation and published over a period of more than 150 years (not including the many recipes for production described over the last 1300 years). In preparing this chapter the author has reviewed over 100 texts from the nineteenth century to the present, most of them English language publications and translations. The author regrets that many more could not be included, but hopes that the references and citations will be useful as a select bibliography.

## 20.2 Parchment production and use

The English word 'parchment' is understood to derive from the French *parchemin* and the Latin *pergamentum*, which in turn derives from the name of the Hellenistic city Pergamon of Mysia (Bergama in modern Anatolia, western Turkey), once believed to be the birthplace of parchment making under the second century BC King Eumenes II. According to Pliny (himself quoting earlier sources), *charta pergamene*, or 'paper of Pergamon', was supposedly invented in response to an Egyptian embargo on the supply of papyrus to the city (Diringer, 1982). However, the oldest examples of parchment appear to be fragments of camel skin found in the Hebron region of modern-day Jordan, dating from around the eighth century BC and much earlier than that which we associate with Pergamon (Reed, 1972: 118, 277). Skins are recorded as being used for writing purposes as early as the Egyptian Fourth Dynasty (c.2700 BC) (Ryder, 1991).

**Figure 20.1** There are many millions of sheets of parchment stored in archives across western Europe in the form of deeds and charters, dating from about the tenth century AD to the twentieth. The above English document, from the reign of Queen Anne, is very typical, being a composite object made of printed and manuscript inks on limed sheepskin, with red shellac applied seals, a blue paper duty stamp with metal strip, and other endorsements. (Courtesy Dorset Archives Services)

Material from these periods is understood to have been produced by stretching and drying skins, sometimes with the addition of vegetable or fruit infusions that had a light enzymatic or tanning effect (Reed, 1972: 72–120; Haran, 1991). The Dead Sea scrolls from the Qumran caves are perhaps the most well known of such material and are dated at around 2000 years old. Having been tanned, these might be described more accurately as leathers, but for the fact that they have probably been dried under tension to produce a writing surface. This type of skin, referred to in Aramaic as *gᵉwil*, was not usually 'split' in the way that we may associate with most parchment, but remained the full thickness of the hide.[1] The hairs were removed and the hair side was used for writing since it was smoother than the largely untreated flesh side (Vorst, 1986).

From the second half of the first millennium AD, liming appears to have become a common feature of parchment making, with the possible exception of Hebrew Talmudic parchment, which seems to have begun to be made using a liming process from about 1100 AD (Vorst, 1986). The untanned skins of sheep, goats, calves and deer have all been used to make limed parchment and there remains some inconsistency in common terminology used for any or all of these. In this chapter the term parchment is used for all of these different hide types and the term vellum is used in its precise form as parchment made from calfskins. The term membrane is also used from time to time to mean a single sheet of parchment of any form.

It is not clear where or how skins began to be limed in preparation for parchment production, whether it began in Islamic or European culture. It is argued that Arabs may have adopted the practice from European sources and added it to their existing procedures (Haran, 1991). It has been pointed out that the use of an alkaline process was easier to control in colder climates, where the process could continue for longer periods without the rapid deterioration of the fibres that would be experienced in temperatures above 85°F, a temperature regularly exceeded in the Arab countries of Asia Minor, and where a plentiful supply of cold water for cooling the skins may not have been readily available (Vorst, 1986).

The use of lime in the preparation of parchment skins had the advantage of opening up the skin fibres in such a way that they could be pulled more effectively into a flat weave structure. It also produced a pale grey or strong white colour, as compared to the light brown or yellow colour produced by tanning processes. An even, flat surface and white colouring both lend themselves well to manuscript production. Paper began to be made in the Islamic world several centuries before the Europeans (around the seventh century AD). Papyrus had been more commonly used for Islamic writing material, parchment being used for specialist purposes. A rapid expansion in the demand for writing materials from the eighth century AD onwards required a more readily available material. Paper could meet this demand, and by the middle of the tenth century AD it was in widespread use throughout the Islamic world (Loveday, 2001). In western European countries, parchment appears to have been the only writing material used until paper began to be common in the fourteenth and fifteenth centuries, and it continued to be used for many purposes for another 400 years after that time.

The method of producing limed parchment outlined in Chapter 18 is a fairly standard approach, used in similar ways for several centuries and until modern times. The period when parchment use was predominant in Europe, and when its production qualities were probably at their highest, was arguably between the tenth and sixteenth centuries AD. During this period it was used for ecclesiastical texts, for legal and financial records and those of the royal courts, and for the covers of books. Its production in the United Kingdom during the last three centuries has been most commonly intended for use as legal documents, Acts of Parliament, certificates and awards, for drumskins and parts of other musical instruments, and for specialist works of art. The use of parchment is believed to have spread to India in the tenth century AD, probably via Muslim invasions. However, it never really took hold as a writing material and is more commonly found as a substrate for paintings and ink drawings and in musical instruments, dating from about 1600 AD onwards (Gairola, 1958; Lokandum and Chaudray, 1991). Parchment-like material was also used to make shadow puppets, pots and vessels and more recently in linings for shoes, as a support in clothing and in some sports equipment (Lokandum and Chaudray, 1991).

The three basic layers of a skin are the epidermis or hair side, the middle corium layer, and the inner flesh side. These are different for each animal type. For fine writing purposes, it is the corium layer, in particular from sheep, goats and very young calves, which provides the best surface characteristics. There remain differences between each side of each layer, however. The hair side of the corium has a slightly harder and shinier surface than the flesh side, which has a softer, smoother surface, and is often more fibrous in appearance. When preparing a skin for parchment, its use informs the extent to which a side is scraped, or which layer is split off for use, to produce either a 'hair-split' or a 'flesh-split' skin. The 'hair-split' has more of the epidermal (hair) layer present, the flesh side having been scraped away to a greater extent. A 'hair-split' parchment is often thicker, has a harder surface on the hair side and can be strong enough to be used for book coverings, for example. A 'flesh-split' skin has had more of the epidermal side removed and is a softer, often thinner material suitable for use as a writing material.

In developing its use as a writing material, different surface characteristics were used to meet specific needs. As mentioned above, very early parchments were not split; the only surface smooth enough to receive writing was the hair side. After several centuries, the surface qualities began to be refined. Both sides were scraped, with more of the hair side removed, and the thin, smooth material produced by later processes could receive ink on either side and on both at the same time, which became an essential feature for book production. However, western European documents such as indentures, charters and deeds, etc. were usually written only on one side, almost invariably the flesh side.

The standard use of the flesh side for document writing appears to have had an effect on the different qualities of the sides of this kind of parchment material over succeeding centuries. Whereas both surfaces of earlier, refined parchments were good enough to write upon, the hair side of later ones used for documents is so shiny and hard that it would have been more difficult to use it for writing. This difference might be because the preparation of much early European writing parchment was intended for use in ecclesiastical settings (where texts were often bound and both sides required). The gradual increase in demand for parchment in legal and court settings resulted in a separate, secular industry producing and supplying material more suited to these different purposes. Over the centuries, the production methods and quality of parchment produced for standard documents altered, particularly between the seventeenth century and the late nineteenth century. Later skins tend to be more rigid, the hair side can have a pronounced waxy feel and they have slightly different behavioural characteristics (Calabro et al., 1986).

Finely split, limed vellum was used in western Europe as a writing material from early times. It appears, however, that sheep and goat skins gradually became the more common writing parchment materials, particularly those used for secular purposes from about the fourteenth century. Vellum from this period onwards is perhaps more commonly associated with book coverings. For this purpose it was not split to the same extent as sheep or goat writing parchments. The hair-split provided the maximum strength and resistance needed to make a robust protection for text blocks. Its use as a book covering was commonly in a format without boards known as 'limp binding' and from the eighth century to the seventeenth a wealth of styles were used in this fashion throughout Europe, in leather as well as vellum (Scholla, 2003). These limp bindings were used for manuscript texts, first on parchment and subsequently on paper.

In the UK, stationery bindings (blank books used for accounts, for example) were commonly produced in limp format from the fifteenth to the eighteenth century, and frequently vellum covers were added to gatherings of paper records once a suitable quantity was reached. From around 1700 vellum was also used as an adhered covering on boards for ledgers and other 'hardback' stationery bindings. In early nineteenth century England, for example, church registers of baptisms, marriages and burials became standardized, the first such consisting of printed pro-forma pages sewn into boards covered with adhered vellum, including green vellum (see below).

The flesh-split was also used for these and other coverings, for both manuscript and printed texts. A flesh-split covering can often be distinguished by a paler colour and somewhat grainy appearance, compared with the common yellow–brown and shinier surface of the hair-split covers. Flesh-split covers have often become weak and torn, especially along the joint and spine exposed to sunlight. Frequently, older vellum and sometimes writing parchment (often musical manuscript or legal documents) were reused for the covering of manuscript texts. The finer parchments of this nature are frequently found to be damaged, being unsuited for this purpose.

Surface characteristics of parchment sheets were altered further by the preparation used by the scribe or illuminator, such as 'pouncing' with powdered pumice, the application of light vegetable oils, or polishing. Different methods of preparation were used in different parts of Europe (Bykova, 1993). Surfaces that were intended to receive layers of pigment or gold leaf were frequently prepared further, sometimes coated with a mineral 'ground' such as gesso or bole. These different surface characteristics all affect the ways in which parchment responds to its environment.

Parchments were rarely coloured for use, although manuscript membranes were sometimes whitened further with chalk. Skins sometimes had a patterning or colour caused by the presence of blood in the vascular system. If the animal had died of natural causes, or if the carcass was deliberately beaten before being flayed, iron and other residual blood products remained in the skin and membranes causing strong patches of brown 'staining'. Byzantine and Arabic parchments intended for particular purposes were sometimes dyed yellow (with saffron) and in rare cases a blue–black, using dyes from the murex shellfish (Bloom, 1989). In eighteenth century Europe, vellum was sometimes dyed with copper to produce 'green vellum'. This was most commonly used for stationery binding covers, for wallets and for small notebooks that often incorporated a pouch or pocket.

Other uses for both limed and other untanned, split-skins include its incorporation in western clothing (Kite, 1992) (see *Figure 20.2*), in furniture (especially the work of Carlo Bugatti) (Giovannini, 1999; Munn, 1989), as a decorative surface covering for eighteenth century telescopes (often green vellum), and as the support material for seventeenth and eighteenth century fans (Hermans, 1992). American Indian and Inuit cultures used split sealskin for kayaks and seal gut for clothing and bags and in African cultures a similar material was used for musical instruments (Dignard, 1992; Kaminitz and Levinson, 1988).

Ethnographic artefacts from all over the world often incorporate untanned skin material, such as shrunken heads, mummified bodies (those which do not involve tanning-type preparations) and costumes decorated with pelts or even whole animals. Many of these various items may not constitute what we commonly think of as parchment, but their qualities as untanned collagen membranes give them similar characteristics to parchment, and their conservation frequently involves similar considerations to those used for manuscripts (see Chapter 17).

## 20.3 Chemical, physical and deterioration characteristics

Skins are not homogeneous structures. They are described as displaying morphogenesis, which in

204  Conservation of leather and related materials

**Figure 20.2**  Two examples of garments employing parchment in their composition. On the left, an English bodice from c.1660 decorated with parchment lace. On the right, a Guipure work chasuble from Piedmont dating from 1730–40. The embroidery on the chasuble uses metal thread worked over parchment shapes and cord.

simple terms means that they are formed by living organisms from standard building blocks (skin collagen molecules) into varied, complex shapes, influenced by the many changing circumstances of an animal's existence and growth (Harrison, 1981). This complex variety continues to be present in skins that have been made into parchment and the manner of its production causes this material to display a pronounced capacity to deform and change in different conditions. These are key features of the physical characteristics of parchment, and ones that need to be borne in mind by all who must care for heritage objects that contain or consist of parchment.

There are significant differences between skins that have been limed and those that have been tanned. The strong alkali used in a lime bath causes collagen to undergo a significant degree of molecular change that is different from the changes brought about in tanned leather. The tanning process causes the collagen molecular matrix to become crosslinked, so that sites that would otherwise be available for molecular bonding are linked to their neighbours, producing a more robust structure with a higher shrinkage temperature than fresh skin. By contrast, the hydrolysis and racemization of molecules in parchment skins causes the matrix to be expanded and weakened, leaving it with a lower shrinkage temperature than fresh skin (Haines, 1987). Polar sites available for hydrogen bonding are exposed and the whole matrix becomes highly hygroscopic.

The salting, liming and scraping or shaving process used in the production of a high quality writing parchment results in the removal of plasma proteins, mucopolysaccharides, keratin, elastin and fats, leaving an almost pure collagen fibre network with about 13% water content under normal conditions and residual lime of about 1.6% (Haines, 1994). Skin, and especially limed skin, is labile in heat and will shrink in high temperatures, particularly if this is associated with water or high relative humidity. With a shrinkage temperature of between 55 and 60°C in new material, and as low as room temperature in old, decayed examples, parchment is very sensitive to some of the uses and treatments that present a lower risk to other organic materials such as paper and some textiles (Haines, 1999). Because parchment consists almost entirely of collagen, and collagen is highly sensitive to bonding with polar molecules such as water, it is water and aqueous solutions that place the material at greatest risk of damaging changes. Non-polar or low polarity solvents will not react with polar collagen. These solvents do have the capacity to dissolve fats and oils, and for most western parchment, this is not a problematic feature, since such fats and oils have been removed completely during the salting, liming and washing processes. Early parchment was sometimes prepared with light vegetable oils, however, and these might be adversely affected by such solvents.

Limed skins will absorb and adsorb water in humid environments, and lose most of this water when conditions are drier, which means they are very sensitive to fluctuating environments. This sensitivity manifests itself as a change in rigidity from a relatively rigid state in dry conditions to a markedly limp state in damp conditions. These changes are accompanied by significant expansion and contraction across all dimensions. In humid conditions, new parchment has been shown to be able to take up 10% of its weight in additional water in one hour and, although absorbed water is lost quite rapidly in drier conditions, additional adsorbed water (water bonded to molecules) takes longer to be lost and may remain indefinitely (Woods, 1995; Berardi, 1992). During these changes, unrestrained parchment deforms considerably, demonstrating a typical curling and, in large sheets, some reversion to the natural undulations of the hide. It has been shown that temperature and humidity fluctuation can cause dimensional changes in parchment of nearly 4.5% (Berardi, 1992).

Parchment exposed to high temperatures and fluctuating conditions is vulnerable to a breakdown in the fibres in the form of a gelatinization. Examples of parchment items exposed to high temperatures in dry conditions are shrunken and hard, brittle and often stuck together where they are folded, for example. Hot, wet conditions invariably cause such items to blacken and fuse until any adjacent faces are inseparable and have the appearance of having been melted together (see *Figure 20.3*). The same effect can happen over long periods of time at lower temperatures and in dry conditions, as has been noted of fragments such as those found in caves in Qumran and Hebron, which would have experienced fluctuations and repeated drying and condensation effects over many centuries (Kahle, 1986).

Wetting and subsequent uncontrolled drying of a parchment membrane will cause it to shrink and become transparent. This shrinkage can be uniform, in the case of a small item, for example, or patchy depending on the degree of wetting in any area and also in its position on the skin. Areas around the shoulder, flanks and butt, for example, tend to have a more coarse grain structure and can often be more translucent than other areas even in a good condition parchment. Transparency tends to

**Figure 20.3** Fifteenth and sixteenth century account and inventory rolls, previously stored against a brick basement wall. Repeated episodes of moisture and probably some heat from hot water pipes, over a long period of time, have been the most likely cause of the ends of these rolls to fuse together, harden and blacken. The parchment in the rest of each roll is good, but this damage prevents them from being opened without very serious intervention. (Courtesy Dorset Archives Service)

be caused by the three-dimensional shrinkage in the skin bringing wetted fibres and fibrils into contact with one another. The increased sites for bonding prevalent in limed skins cause the fibres and fibrils to stick to each other if they are not kept apart under tension during drying. Transmitted light at the surface will not be scattered but will penetrate the skin, giving it a translucency or even transparency. This characteristic has been exploited in the past to make 'window parchment' and is evident in raw hide skin articles such as shadow puppets (Reed, 1972: 143; Chaudray et al., 1991).

Parchment membranes that have been affected by water or heat in localized areas, especially when folded, can present significant problems for opening and preservation. Local patches of shrunken skin cause a wide region of distortion radiating from them across the healthier skin (Hassel, 1999). A less dramatic form of shrinkage and hardening is even more common. Membranes that have been folded, rolled or crumpled for a long time, or tightly bound in a book, can have become sufficiently rigid as to make opening difficult or risky. Items in this condition are often damaged further by overenthusiastic attempts to open them out, leading to tears and delamination. Objects that contain parchment components and which are exposed to liquid water can be distorted subsequently by the shrinkage of the parchment elements. Examples include clothing and also vellum-covered books, especially those with the vellum pasted down onto boards and adhered to the spine of the text block. This can happen even in humid environments (see *Figure 20.4*).

It has been noted that 'limp vellum' binding styles (with no boards or adhesion) cause less damaging distortion to their text blocks, especially when exposed to water. For this reason and several others, the format has been recommended as a conservation binding style, whether using vellum or another material, even for those items that may previously have been bound in a different form (Reed, 1991: 217–220; Fitzsimmons, 1986). Notwithstanding the proven positive features of such bindings, heavy limp vellum covers can also be found to have become shrunken and distorted over time, frequently exposing the text block and resulting in wear to the weaker paper, especially at the fore edge (see *Figure 20.5*) (Craft, 2003; Clarkson, 1982). The choice of quality of the covering material, and which sewing support to use with it, are important features of these structures.

As well as being used as a total covering, whether limp or with boards, vellum and parchment has been used commonly for structural and decorative features in bindings and coverings, including 'tacketing' (decorative sewing), protective vellum corners for cloth- and leather-covered bindings, and for thongs and ties. The shrinkage of vellum thongs in a sewing structure can be the cause of distortion of the spine of a text block, similar to that described above.

Parchment membranes are as susceptible as any other material to becoming dusty, dirty and stained. Stains that arise because of a spillage of an aqueous solution are problematic because the most effective way of reducing them (using another aqueous solution) would probably cause further damage: inks and pigments would be at risk and the parchment would be affected by the wetting and particularly by its subsequent drying. Like paper and some textiles, surface dust and dirt can also become ingrained if a dirty membrane becomes damp or wet, and even low polarity solvents such as alcohol and acetone can lead some surface dirt into the fibre structure.

A common deterioration problem is microbiological infestation. Parchment membranes contain varying amounts of partially or completely gelatinized collagen fibre, especially at the surfaces. This material, and the collagen fibre matrix itself, is an attractive medium for fungi and bacteria, which will flourish in prolonged damp conditions (Kowalik, 1980). The effects vary depending on the length of the episodes of growth. Minimal growth can be observed in almost all old parchment, manifesting itself as a slight complexion of tiny, faint black, brown or purple speckles under the skin surface. Badly affected membranes, with much surface growth, will have become weakened and fibrous; their surfaces flaked and delaminated, losing inks or pigments in the process. Infestations can cause livid staining of purple, bright yellow and pink. At its very worst, a parchment membrane can be completely destroyed by mould, rendered to dust.

Inks and pigments are often the first casualty of the steady degradation of a surface by mould or by expansion, contraction and distortion. Heavily applied pigments and gold leaf crack and flake off through the action of this movement and are undermined by microbiological activity. Manuscript inks often do not penetrate deeply into the fibres and are easily rubbed off. Nineteenth century parchment in particular seems to have a harder surface and inks do not always adhere strongly, particularly if exposed to water. The recipe of an ink also affects its capacity to adhere or penetrate. Carbon inks are very vulnerable and this will have become apparent in early times, leading to the more common use of iron gall inks. These in turn can be unstable and burn through their substrate (Wächter, 1987a). Parchment appears to be more resilient to this acidity than paper, perhaps because of its high calcium carbonate content deposited by the liming process, which helps to neutralize acids. This same resilience may protect parchment from other sources of acidity such as modern airborne pollutants. Inks containing verdigris (copper acetate) can penetrate strongly and pigments of this nature can produce what is referred to as 'copper green' decay, where the fibres of parchment are weakened considerably (Fleming, 1983).

The fading of iron inks is a problem common to paper and parchment manuscripts alike. Since at least the early seventeenth century, gallic acid (e.g. derived from gallnuts soaked in white wine) was used to intensify faded inks (Petrus, 1619). Unfortunately, after a few years the treated area discoloured to

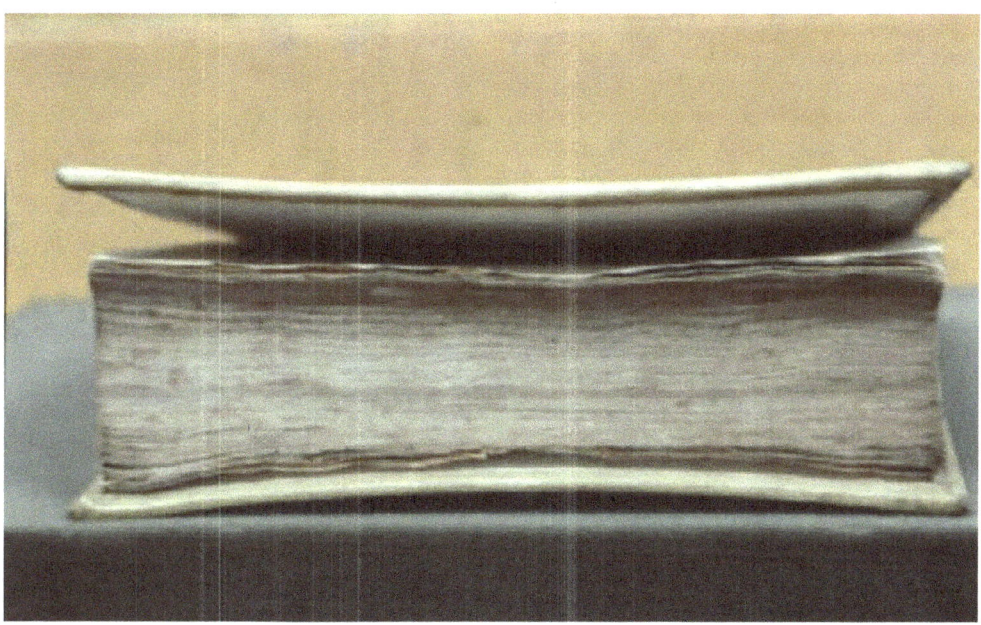

**Figure 20.4** A good example of warped boards on a vellum-covered volume, the distortion being caused by the shrinkage of the vellum following damp conditions. This warping exposes the text block to dust and makes handling difficult. (Courtesy Bodleian Library)

**Figure 20.5** Typical examples of English manuscript stationery bindings of the seventeenth and eighteenth centuries. These are registers and accounts bound up coarsely into volumes using heavy hair-split vellum for their covers. The vellum has in most cases shrunk over the years, exposing the paper text blocks, which in turn have become damaged. (Courtesy Dorset Archives Service)

brown and caused the writing to be illegible. In the nineteenth century and the first half of the twentieth century a variety of intensifying solutions were recommended for both specialist and domestic application, such as tea or tannic acid, ammonia, ammonium sulphide neutralized with lime water, and potassium ferrocyanide.[2] Most of these solutions had damaging properties. One 1930s observer reported the application of an aqueous solution of ammonium hydrosulphide to intensify the ink, followed by a mixture of tannic and acetic acids to make the reaction 'permanent' and counter the effects of the first solution. It was noted that the manuscripts subjected to this treatment 'were not very reassuring in appearance, having great, discoloured blotches' (Smith, 1938a).

Perhaps the most extraordinary of the various mixtures employed were ferrocyanate solutions which discoloured to a bright blue (of the kind used in 'blueprint' plans) and frequently obliterated the text (Fuchs and Mrusek, 2002). More destructive still, the strongly acidic solutions have caused the gradual hydrolysation and destruction of the fibres of the substrates to which they were applied, making them brittle and crumbling. While paper seems to have suffered from this decay more quickly and dramatically than parchment, the calcium residue in the latter has not been sufficient to resist the action indefinitely.

Although parchment is very attractive for certain forms of mould and bacteria, it appears to be a little less immediately attractive to insects and other, larger pests. Parchment membranes can be grazed upon by silverfish and nibbled by rodents, but perhaps they do not provide sufficiently digestible nutrition, for this damage seems to be more common and extensive in paper. Mice and rats will use parchment, like other available materials, for making their nests, so examples can be found that are damaged by the urine and faeces of these creatures. Insect damage seems to be most commonly associated with marks caused when the skin was part of the living animal, such as warble-fly holes in vellum. These holes were regularly repaired as part of the preparation of sheets for writing and often they include aspects of historical production that are valued by the scholar and conservator, rather than requiring a remedy (see *Figure 20.6* and Clarkson, 1992).

The format in which parchment is used is often the cause of its damage. The sensitivity of parchment to humidity results in distortion to a membrane where one area is restrained or covered and another is not. This happens in a normal way by the use of parchment in certain forms, for example where the pages of an early manuscript have been turned causing a deposit of oils and dirt from the fingers of the readers.

**Figure 20.6** Many kinds of hole can be found in parchment items. These pages from an English register of baptisms, marriages and funerals of the early seventeenth century, include: old stitching holes, a hole that existed in the skin before it was prepared as parchment and repaired with a piece of older parchment before use, and holes caused by more recent insect damage. (Courtesy Dorset Archives Service)

The fore edge and especially bottom corners of manuscript volumes frequently display this soiling and the action has sometimes worn the skin and made it weak and fibrous. The more oily areas in particular will not respond as readily to atmospheric changes as the surrounding skin because the oils block the passage of moisture to the fibres, leading to distortion around these areas. Distortion can also be found around repair patches or close to the gutter of a tightly bound book and it is common where a membrane has been adhered to a surface under tension and part of it has subsequently come free. This frequently leads to splits and running tears; it can even pull a whole membrane apart (Soest, 1989).

Many archival parchment formats are sealed documents, such as royal charters and grants, indentures and deeds. The structure and format of a parchment document can often be determined by the manner in which it has been sealed and so, given the enormous quantity of such items in existence, we cannot describe the characteristics of parchment without making some reference to this very common format. Seals rarely affect the chemical stability of parchment, but especially for the more common, pendant examples, the behaviour of sealed items will be influenced by the presence of the weight of the seal or many seals hanging from one edge. Seals can be affected by the hygroscopic nature of parchment and also by its calcium content, depending upon their composition (there are many different materials used in different formats). Seals should not be removed from documents, however, since to do so is considered a fundamental breach of their integrity and provenance. Sealed documents should be treated holistically, as composite items, rather than as two separate historical and material features.

In summary, it may be said that there are two varieties of format in which parchment presents characteristics of which the conservator must be aware. One format is loose or unrestrained sheets or leaves of parchment, the other is parchment that is adhered or partly adhered to another, less flexible material. Both formats are profoundly affected by the capacity of parchment to absorb and release atmospheric moisture, and the changes in size and shape, or tensions, that this fluctuation causes. Further variations or influences are common to both of these two basic formats, such as animal type, method of production, fibre structure, the site on the hide from which a membrane has been cut, its surface and laminar characteristics, the effects of inks and other applied coatings, the extent of surface damage and fibre strength, etc. Add to these many variations the fact that each animal skin is unique, and it can be understood that every piece of parchment must be approached with consideration of its own, complex, individual characteristics.

## 20.4 Display and storage

The display of parchment membranes has been a common subject for publication. The adverse affects of excessive tension and adhesion to a rigid mount have been observed above. To avoid the risk of tearing caused by the natural tendency of parchment to shrink in dry conditions, various successful methods have been devised over the last 25 years which involve the application of strings or even springs around the edges of a membrane, rather like a trampoline, which are then attached to a mount in favour of direct attachment of the skin. These approaches include the use of linen threads, which stretch and shrink in opposition to the skin, allowing the membrane to maintain an appearance of even tension during fluctuating conditions (Clarkson, 1987). Japanese hand-made paper strips have also been used for a similar purpose, based upon the tendency of these long-fibre papers to accommodate the expansion and contraction of the skin, and also an ingenious method of using strips of polyester sheet (Mylar or Melinex) cut into the form of 'springs' (Pickwoad, 1992; Norman, 1993).

Light levels for parchment on display are the same as for most organic materials, in particular those with coloured pigments. Ferrogallic inks and mineral pigments fade in strong light, so levels of 50 lux, with no ultraviolet, are important (Myers, 1980; BS5454:2000).

Safe storage conditions for parchment are widely recognized as being dry (a stable point between 45% and 60% relative humidity), to avoid the growth of mould (BS5454:2000). Drier conditions can be acceptable for long-term preservation, as may be evinced by the great longevity of some extant membranes. However, periods of use in more humid conditions will cause the expansion and humidification of this exceptionally hygroscopic material and rapid changes of environment should be avoided. Temperature control is of great significance in maintaining a stable humidity and it is widely recognized that lower temperatures reduce damaging hydrolytic activity. The use of individual protective packaging made from acid-free and buffered cards and papers is recommended, which also has the advantages of maintaining dark conditions and improving environmental stability. Reduced handling is advisable, for reasons given above.

## 20.5 Conservation treatments

Common features in the deterioration of parchment can be summarized as: shrinkage and brittleness, distortion, tears and holes, mould damage and the weakness or loss of inks and pigments. Over the last century many different approaches, techniques and materials have been used to remedy these problems. The one thing they have in common is that they are all different; no complete sequence of treatment or recipe of solution is repeated in quite the same way by the different authors reviewed for this chapter, but there are common approaches or technical 'themes' that can be discerned. In reviewing these approaches as trends over time, one can observe a gradual movement towards more minimal intervention in recent years. The rate of change in this trend varies geographically and it has its exceptions, but it is discernible and will form a part of the conclusion of this work.

The trends can be divided into two strands: the different types of material used to remedy problems and the different approaches or attitudes towards remedial treatment. During the mid-twentieth century there was a general trend away from using age-old methods that involved compatible or sympathetic materials, towards the exploration and use of modern synthetic materials. However, techniques for the use of these materials continued to be guided by what might be called the old fashioned craft production approach – the intention to remedy all problems and return the imperfect item to its former condition; serviceable and 'as good as new'. In western Europe from the 1930s to the 1980s there appears to have been an expectation that modern chemical science would provide the answers to problems and help conservation 'operatives' to remedy every imperfection (BS4971:1973). This scientific approach becomes less evident when we review western conservation texts in the 1980s and 1990s,[3] but from the same period we find many published tests and recommendations for the use of a wide variety of chemical treatments emanating from the old Soviet Union countries.

Since the 1990s there appears to have been a gradual return to the use of more compatible materials, common experience having shown modern synthetics to have undesirable characteristics. At the same time, approaches and techniques have also changed, becoming less invasive, the craft production approach being no longer considered necessary or appropriate. There have been several useful reviews of practice over the years and reference is made to these below. The most exhaustive review of approaches used up until its time of publication, Dr Reed's *Ancient Skins, Parchments and Leathers* of 1972, referred to in the end notes, provides essential historical information on the nature and production of parchment, and also a detailed insight into the many and varied methods and recipes sought by 'restorers' to remedy every conceivable imperfection in a membrane. Another fascinating source is L. Herman Smith's contemporary description of the practices undertaken in European archives of the 1930s, published in *The American Archivist*, also referred to in the endnotes. As well as revealing the attitudes of their times, these reviews are invaluable as a record of the many types of treatment and material used in the past, evidence of which a modern conservator might find when examining a previously restored membrane. The different types of treatment commonly required or encountered are grouped here into key activities.

### 20.5.1 Mould and fumigation

Mould growth on parchment has been treated in a number of ways. The fluctuation of environmental conditions in temperate climates has the tendency to allow for episodes of mould growth to be countered by periods of dry conditions that kill off the mould, while leaving residues that are reactivated when conditions become damp again. The increase in availability of chemical fungicides during the nineteenth and twentieth centuries has resulted in a number of attempts to kill mould on parchment and paper items by the use of various chemicals. These have included: sodium pentachlorophenate (Santobrite), isopropyl-metacresol (Thymol), sodium ortho-phenyl-phenate (Topane), formaldehyde, DDT, hydrogen cyanide, methyl bromide, ethylene dioxide, p-chloro-m-cresol (or 4-chloro-3-cresol) and quarternary ammonium salts (such as Catamin-AB).[4] The use of most of these has not been satisfactory because of serious risks to health and to the environment, in some cases the residues of these may continue to pose a risk to users of the items. Some have also caused heavy discolouration (pink or blue) of the parchment (Kowalik, 1980). In addition to these fungicides, methods of disinfecting membranes using alcohol solutions have also been explored (Voronina *et al.*, 1980).

The growth of mould is a risk for all waterlogged organic materials and parchment is no exception. Ways of avoiding this risk have been explored, including freeze-drying and gamma irradiation. Freeze-drying for parchment has been shown to be feasible but is not without its own risks and adverse effects (Parker, 1993). Gamma irradiation has been shown to be deleterious for paper and parchment, although there have also been positive views expressed for its use in large-scale emergencies (Adamo *et al.*, 2001; Sinco, 2000). The drying and surface cleaning of parchment items, followed by maintenance in dry conditions for storage and use, continues to be the most non-invasive and effective approach to controlling growth (Craig, 1986).

### 20.5.2 Cleaning methods

Surface cleaning methods for parchment tend to follow similar practices as those used for paper. Popular methods in the first half of the last century included the use of pellets of sticky bread, rolled or pressed on the surface; wiping with cotton wool or chamois leather moistened with water; or swabbing with benzene (Reed, 1972; Haslam, 1910). Most of these approaches were gradually replaced by the use of rubber or vinyl erasers, or just soft brushes for loose dirt, partly because of concerns about surface abrasion and the uncertainties over the deposit of material attractive to fungi, and especially arising from an increasing understanding of the risks associated with the application of liquid water. Wet methods have been replaced with alcohol solutions applied locally with cotton buds, which introduce more controlled cleaning with less risk to the surface and to the highly hygroscopic collagen fibre.

There are other risks associated with surface cleaning parchment. When applying dry techniques it is easy to ingrain soft dirt into the fibres, sometimes in streaks. Sadly, in attempting to avoid this, it is common to apply more vigour at the outset, which in turn risks damaging surface fibres. It is also essential to try to avoid damaging inks or pigments, so unfortunately the application of erasers around writing, for example, can lead to an unattractive penumbra of dirt around the letters. Various forms of solid eraser (sliced or grated) and sponge rubber are commonly in use, especially those incorporating a narrow point for accuracy (Szcepanowska, 1992). Restricting cleaning to gentle brushing and removal of spots of loose surface dirt appears to be the safest intervention.

The use of laser technology for surface cleaning has been explored in all areas of conservation, including for paper and parchment. For some otherwise good condition material, the removal of ingrained dirt by this means may be less damaging than traditional methods. Reservations about the risks to organic materials such as parchment are not unfounded, and it has been recognized that care in application is critical to the success of laser cleaning of parchment without causing damage (Kautek *et al.*, 1997; Cooper *et al.*, 2000). These experiments have been revealing and contribute to knowledge about potential methods and limitations of conservation. It is reasonable to question whether it is ever necessary to remove dirt so thoroughly, particularly in cases where little or no adverse affect may be caused by its presence.

In similar vein, it should be noted that although there have been examples of the use of bleaching agents to remove or reduce discolouration in parchment, these are not common. Methods have included hydrogen peroxide, lemon juice and even onion juice, and complex mixtures involving soaps, ammonia and chloramine B.[5] In the paper and manuscript field there has been a gradual trend away from bleaching processes as conservators have learned about the risks in their application, and perhaps because it has been recognized as unnecessary or inappropriate to 'brighten' items simply in order to access text that is already legible. Equally, the use of solutions to intensify faded inks, as described above, appears to have been abandoned. Although such chemical methods were still being explored as late as 1981 (Flieder, 1982), the use of ultraviolet light to reveal texts had begun to replace these at least as early as the 1930s. Smith, in his tour of European archives at that time, observed the new use of UV equipment in some institutions (including a malodorous cupboard at the Bodleian Library where users would operate a lamp (Smith, 1938b)), as well as reluctance in some to use UV. In referring to the latter he comments: '… they do not take into account that the use of ultra-violet light entails no risk whatever to the manuscript (something which certainly cannot be said of chemical reagents, no matter how carefully they may be applied)' (Smith, 1938a). Steady advances in light technology and digital image enhancement in the late twentieth and early twenty-first century have made it possible to discern writing against or underneath stains or added text, without intervention (Mairinger, 1981–82; Fuchs, 2002).

### 20.5.3 Humidification and softening

As described above, parchment is commonly found to have become less flexible or even hardened, preventing a sheet from being viewed. In the case of large documents, for example, the rolling and creases can become set, causing the skin to become spring-like and difficult to hold open without risk. An unassisted researcher opening a long-unused, multi-membrane account roll can appear like a person wrestling with an octopus. In the past, it was commonly held that a roll like this should be flattened permanently, often involving pasting it down onto a solid acrylic sheet lined with a woven polyester fabric, or a board covered with waxed paper (Cockerell, 1958). After drying, the membrane would be peeled off the lining material, or vice versa (sometimes leaving offset ink behind), revealing something resembling an embossed skin table mat. These days there is perhaps more willingness to support users in accessing difficult items with care, in favour of reformatting the items to meet all comers, and certainly a greater appreciation of the need to preserve original formats (Clarkson, 1992). When an entire collection consists of difficult but original formats, the most sensitive (and cost-effective) approach may be to humidify gently those that are intractable, copy the collection, return items to their original formats, package them and place them in a suitable storage environment (Lindsay, 2003). This is equally applicable to single items as to collections (Woods, 2002).

Methods of humidification vary and tend to be determined by whether it is necessary for the general relaxation of a whole membrane or just localized treatment to a crease or a cockled region. People have long recognized the capacity for parchment to become limp in humid environments and this has been exploited, either by placing a membrane into a humidity chamber or into a 'damp pack'. The precise amount of moisture in an enclosed space necessary to relax different weights of membrane has been considered (Woods, 1995). Earlier methods of humidification involving damp packs tended to place the parchment at greater risk of getting wet on the surface, and the introduction of different kinds of humidity chamber must in part have been intended to avoid this risk.[6,7] In recent times, the use of 'Goretex' (a specialist fabric incorporating a barrier membrane that allows water to pass through only as a vapour) has become common as a means of humidifying a chamber, or used in a pack in such a way that liquid cannot come into contact with the skin (Buchanan, 1993; Singer, 1992). Also common is the use of ultrasonic humidifiers or nebulizers to introduce a fine mist into a chamber (Quandt, 1996).

When humidifying parchment, especially whole sheets, the principal concern is to avoid any application of liquid water if possible. The energy of water is significant, arising from its strong polar charge (Kremen and Southwood, 1960). The reduced shrinkage temperature inherent in parchment makes it vulnerable to damage by the application of polar liquids, as well as when a membrane is dried without restraint against shrinkage (Haines, 1999). The high energy of water can exceed the energy holding decayed collagen molecules in an expanded structure, causing molecules to collapse and the fibres to shrink, even at room temperature. The structural or molecular condition of collagen is not easily determined by visual examination and the gradual deterioration of collagen fibres has been shown to lead to a 'pre-gelatine' stage in which wetting or even high humidity can cause apparently fibrous material to dissolve into a gelatinous state (Larsen, 2002).

Even if this 'pre-gelatine' stage has not been reached, the surface tension of water as it evaporates exerts a pull between fibrils and fibres, bringing them together and consequently allowing them to bond to each other. Tensioning the skin or applying a water-soluble, low-polarity solvent such as alcohol can reduce this effect, allowing the water to be removed without bringing the fibres into contact. Tensioning literally holds the fibres apart as the water evaporates. Alcohol mixes with water producing an azeotropic solution with a much lower surface tension than the water alone. The mixture evaporates without exerting a pull on the fibres, leaving the skin dry and largely unaffected.

Rather than risk wetting an old but otherwise sound membrane, and having subsequently to remove the water under controlled conditions to minimize any further damage, it may be safer to use an environment with a high relative humidity and to allow the membrane to condition at its own rate. Even after gentle humidification, however, the natural movement of the skin as it loses moisture may need to be restrained to avoid distortion. Various methods have been used to restrain or tension skins during drying, including the use of successive changes of dry blotting paper under boards and weights, and commonly by 'pegging out' a skin to place it under tension, in a similar fashion to its original production (see, for example, *Figure 20.7*). The risks associated with the use of presses for drying have long been recognized (Cockerell, 1958). Vulnerable inks and pigments can be offset, fibres can be pinched or squashed together to cause hardening and translucency, and the prolonged presence of moisture can permit mould to grow. As can be imagined, there are also risks of damage in the use of excessive tensioning when pegging out skins, particularly those with holes or tears; it should be remembered that a damp skin will undergo significant shrinkage on drying, so allowance for this movement must be made.

Drying a moist skin between absorbent sheets and under a board with no additional weight is a minimal form of restraint. It will not result in a very flat, even finish; instead some natural undulation and channels along old folds will be maintained. In many cases this may be preferable since it avoids unnecessarily changing the appearance or structure of an item, especially if it is to be returned to its original folded or rolled format. Where a more even surface is important, the 'pegging-out' approach is most effective and there have been many variations in this method described over the years (see *Figure 20.7*).[8]

Localized humidification has been undertaken safely by using features of the above. Gore-tex can be used in the form of a small poultice, for example, and ultrasonic humidifiers can be applied in a limited

**Figure 20.7** A common approach to tensioning a parchment membrane. Several methods based on this simple principle have been developed over the years to control the rate of tension produced by the shrinkage, for example the inclusion of elastic between the peg and the clip, or the use of Velcro strips instead of pegs. (Courtesy Bodleian Library)

area (Lee, 1992; Singer, 1992; Quandt, 1996). These humidifiers produce a light steam, which is airborne droplets of water, so this approach applied to the surface of a skin can pose some risks both to skin and inks if it is not undertaken with care. As well as the risks associated with the application of liquid water, thicker skins in particular can be vulnerable to laminar scission and fibre damage when stretching or tension is applied. This can be caused by incomplete humidification of the skin, when only the surface fibres are damp and flexible but the internal fibres remain dry and rigid (Haines, 1994).

The characteristics of alcohol described above have been used in local applications of an alcohol and water solution in which the alcohol predominates (70 or 80%). Alcohol alone can do little to remove creases; it requires the input of water to make fibres plastic and deform under manipulation. The particular penetrative and lubricant characteristics of isopropanol (propan-2-ol or isopropyl-alcohol) in contrast to other alcohols have been known for many years and have been described in the context of parchment conservation (Reed, 1972: 127; Ellement, 1987; Viñas, 1987). Local application of very small quantities of 80/20 ispropanol and water, followed by light tension applied with the fingers, can reshape a membrane with the minimum of intervention. Moisture is carried deep within the skin so that humidification is more effective and uniform. However, while most iron gall ink is found to be stable in such a solution, pigments may not be, and shellac seals are softened and discoloured by contact with alcohols. For these reasons, an alcohol solution may not be suitable for relaxing an entire skin, although such a method has its uses when trying to soften a shrunken membrane (Viñas, 1987).

Hardened and heavily shrunken skins are usually the result of fire or flood damage, and sometimes by the ill-advised application of hot air to dry out wet skins. The differences between wet and dry shrinkage have been described in a conservation context quite recently. The search for remedies has a longer history. In 1924 C.G. Matthews, a public analyst in England, published attempts to open and flatten folded documents that had been baked inside metal safes which had been in a fire. The most successful approach to opening them was also the most minimal: a gradual humidification with steam for half an hour, removing condensation to avoid wetting, and a few hours spent conditioning in this moist environment. With admirable restraint, Matthews was content to observe that although 'there was not a very close resemblance to the original parchment ... it was distinctly legible, including the signature' (Matthews, 1924).

Matthews had observed that humidified skin in this state shrinks back and hardens again when it dries, and so to keep it softened for longer he applied glycerol. This was only a temporary expedient, however, because glycerol eventually dries out and does not leave behind a residue that can hold the fibres in their expanded state (Wächter, 1962; Calabro et al., 1986). Collagen fibres shrink when their shrinkage temperature is reached, but at temperatures above 85°C the molecules undergo a fundamental change. A coating of glycerol is insufficient to alter the characteristics of a skin so severely affected.

By 1969 attempts had been made to remedy the problem of shrunken parchment, seeking perhaps to overcome the effects described above. These attempts resulted in the use of ever more complex recipes, usually involving urea (carbamide), to try to force apart the collapsed molecules and microfibrillar structure, and to introduce a longer lasting replacement for the previous hydrogen bonding that kept the molecules apart in an expanded state (Belaya, 1969; Yusopova, 1986). However, in a remarkably similar set of circumstances as those described above, in 1970 restorer S. Cockerell reported using these new solutions to soften documents from a fire-damaged safe. After employing a mixture of urea, alcohol and water for up to 96 hours for each item, the skins were softened with varying success. Many remained distorted and all were vulnerable to tearing during the process. Cockerell observed: 'The object of the exercise was achieved, though it is not known yet if the documents will remain stable' (Cockerell, 1970).

Cockerell and others at that time may not have been aware that, while they can look similar to the naked eye, dry heat damaged parchment has undergone a very different change to that which has become hardened and shrunken by the effects of water and uncontrolled drying. In the case of fire damage, for example, it appears that some effects cannot be reversed and it may be better do the minimum necessary to retrieve information and accept that skin affected in this way may never be the same again. The treatment of the blackened and charred areas of burnt parchment might also be considered in the same vein, although it has not stopped some people from trying mixtures of dammar wax and alcoholic methylpolyamide (PFE 2/10) and even melted amber oil with spermaceti wax in benzol, all of which have been observed to have undesirable characteristics (Yusopova, 1980).

The shrinkage and hardening of water-damaged parchment is a problem that can be ameliorated to some degree. The differences in the behaviour of collagen fibres that have become hardened by excessive heat, compared with those affected by water, have been described (Hassel, 2003). The fibres and molecules of a wet and dried skin will have become weakened but the hardening may often be the result primarily of surface tension shrinkage. Humidification of such material can allow an area to be stretched and retensioned to make writing legible, for example (Hassel, 1999). Even if the skin has become very hard there still may be the scope to introduce a 'lubricant' in between the still intact fibres and fibrils and with it a new molecule which, if it is large enough and will not itself break down or evaporate, can remain after initial drying to keep the structure in an expanded state, preferably with the capacity to equilibrate with its surrounding environment.

The use of urea and glycerol for this purpose has been demonstrated to be unsatisfactory (Calabro et al., 1986). Greater success has been achieved with the use of low molecular weight (200) polyethylene glycol (Viñas, 1987; Calabro et al., 1986). The process takes a very long time (months) to complete but appears to cause the adsorption of this relatively large molecule into the structure, even to the extent of reintroducing opacity as well as (a rather limp and peculiar) flexibility. PEG 200 does not dry out in the way other solvents will and it effectively replaces lost hydrogen bonding and bound water in the skin with its own, artificial form of 'hygroscopy'. The very slow rate of change that the skin undergoes means that it may be many years before items treated with PEG display any evidence of change that differs from a 'natural' membrane and it may not be possible to attribute such a change to the PEG rather than the long-term effects of the shrinkage that the skin had suffered before its treatment. As has been noted, however, the use of isopropanol in preparation for immersion in PEG so often results in sufficiently beneficial softening effects as to render the PEG treatment unnecessary (Viñas, 1987). Although the useful characteristics of isopropanol and PEG have been reported widely, some interest in urea and glycerol has continued (Dobrusina and Visotkite, 1994; Stancievicz, 1996).

As described already, it is not always easy to identify how far advanced is the decay of the fibres in a parchment membrane. The physical characteristics of a 'healthy' parchment are self-evident – the skin is strong, dense, opaque and flexible, even if it has tears or holes. Decay associated with a hardened and shrunken skin or a fibrous structure can be misleading and difficult to diagnose with accuracy. It is not obvious how low its shrinkage temperature has reached and humidification or semi-aqueous treatments can result in a greater level of shrinkage and brittleness than might have been anticipated.

Before the late 1990s, it had been possible to test the shrinkage temperature of new material in a laboratory setting, but this method was not applicable for historic manuscripts or artefacts (Haines, 1987). Subsequent advances in testing have provided conservators with the means to identify the shrinkage temperature of a few fibres removed from a skin, by microscopic observation of the contraction of the fibres, and at the time of writing, a major research project is under way in Europe, which seeks to provide a method of evaluating the extent of decay of collagen in a skin (Larsen, 2002).

### 20.5.4 Consolidation of weak parchment

Consolidation techniques have been applied both to the treatment of flaking media and to decayed parchment substrate. It is commonly found that the weakening of one is closely associated to the condition of the other, particularly when mould has been an active cause of decay. Equally it may be observed that, even for very specific, localized pigment reapplication, the search for a suitable consolidant necessarily involves an appreciation of its effect upon both. This is especially the case for flaking manuscript ink because the weakness frequently occurs uniformly throughout the text and it is often necessary to consider the application of a coating to most or all of a membrane. For an item with a healthy parchment substrate, this presents the conservator with a very difficult dilemma.

For parchment which itself has become weak and friable, the most common approach has been the application of a very light gelatin solution, frequently made from parchment itself. Gelatinous solutions have been used in a variety of recipes and for different purposes for centuries (Reed, 1972: 220). The advantage of using such a solution is deemed to be its compatibility with the parchment itself. A 'size' made from ground parchment shavings, or bits of fish swim bladder (isinglass), contains pieces of fibre and connective tissue (elastin) that make it an effective consolidant and adhesive. The membrane being treated will have contained similarly gelatinized fibres when new. Weakness and loss of cohesion across a skin may even be attributable to a loss of this substance through the activity of mould or bacteria.

Various methods of applying gelatinous consolidants have been described, from direct brushing onto the surface or under lifting inks, to the application of a fine, alcoholic mist (Quandt, 1991; BS4971:2002). Vacuum suction tables are often used during this process, to control the application of the consolidant, to hold the membrane flat during treatment, and to control its drying. Some of these treatments have been developed for use on membranes with significant losses, to infill holes as well as to consolidate a fibrous skin (Bëothy-Kosocza et al., 1990).

The greatest risk in using a parchment size as a consolidant for friable skin is of shrinkage and brittleness being caused by the application of too much size or too much water, usually because the precise concentration of such a solution is not easy to control. There is also a risk to pigments: parchment size may maintain a degree of alkalinity from the lime content of the skin, which can adversely affect the nature of a coloured medium. For this reason some people prefer to use a consolidant made from isinglass or a purified gelatin (Quandt, 1996). There is a range of purified gelatins produced from hides of different animals, with alpha- or beta-form amino acids (depending on whether the hides were reduced in an acid or alkali process). These gelatins are produced with different Bloom numbers, which means that the strength of the resultant gels in solution is different, with the higher Bloom number producing stronger gels. This variation provides a measure of control and precision in mixing tailored solutions for different circumstances. However, the essential requirement for a partially aqueous solution remains, and it may be this that prevents the use of gelatin for consolidation in some cases.

With the notable exception of the Vatican Library, many European archives and libraries in the 1930s moved away from the tried and tested use of gelatinous solutions for consolidating weakened membranes (Smith, 1938c). From this time until the mid-1970s a common approach was the use of a plasticized synthetic solution in non-aqueous solvent. This extended also to the lamination of membranes between supporting layers of material (dealt with in more detail below). Various synthetic solutions were used and most were found to be unsatisfactory or even disastrous. In the 1930s, several institutions were reported as having routinely used solutions of cellulose nitrate or cellulose acetate ('Zapon' and Cellon) or copal-resin varnish (Kopallack) (Smith, 1938). The instability and destructive capacity of these solutions had already been observed by 1937 (Grant, 1937). It is perhaps not surprising therefore to find no reference to their use beyond the 1950s. A non-aqueous solution of amide polymer (Nylon) was a popular consolidant from the 1950s to the 1970s and its use was reported in a variety of contexts (see also below).

The decay properties of many synthetic solutions caused them to be rejected for use in the conservation of organic materials generally, including parchment. However, synthetic consolidants continued to be used in later years with the availability of commercial acrylic powders or solutions, alone or added to fibre pulps (such as Plexigum) (Sievers, 1985). The application of a uniform coating of a synthetic plastic polymer to the entire surface of a skin changes its physical and hydroscopic characteristics, even causing it to be hydrophobic (Wouters et al., 1995). There might be circumstances where such an excessive intervention may be the only way of saving a crumbling item with inks or pigments that would be unstable in a gelatin solution, for example, and it is worth noting that parchment that has become so severely decayed has lost much of its hydroscopic nature anyway. Conservators would be understandably reticent, however, to 'seal' a membrane inside a plastic medium, for example; experience has shown us that such an approach can subsequently be found to be regrettable (Smith, 1938; Wouters et al., 1992).

Other non-aqueous consolidants for fibrous parchment have been tried, including hydroxypropyl cellulose ethers (the Klucel range) (Boyd-Alkalay and Libman, 1997). The long-term behaviour of these is uncertain (see also section 20.5.5). Early attempts at a minimal intervention by placing weakened fragments between layers of glass or plastic sheet proved problematic (Boyd-Alkalay and Libman, 1997). The most sustainable approach to the preservation of weakened parchment would appear to involve non-invasive means (e.g. surrogacy, packaging and stable storage), and if necessary remedial treatments with known and compatible characteristics.

For the most precious items, it may be appropriate to place parchment inside a case or enclosure without oxygen. In 1974 Dr Reed set out a design for a glass enclosure containing nitrogen gas, but it is not known whether any of these were ever used (Reed, 1972: 208). More recently, the use of anoxic environments for organic materials has been explored using a variety of large- and small-scale systems (Stiber, 1988; Carrió and Stevenson, 2003).

### 20.5.5 Consolidation of inks and pigments

Several different solutions have been tried and evaluated to consolidate flaking inks and pigments. These have included mixtures of natural and synthetic gums, waxes, polyvinylacetate (PVA) or cellulose derivatives (Marconi, 1962; Yusopova, 1980; Guiffrida, 1983). Most reviewers over the years have expressed a preference for a gelatinous consolidant.[9] However, there have been several attempts to find suitable consolidants that avoid risks associated with gelatinous and other aqueous mixtures, in particular for the consolidation of heavily pigmented illuminations where the use of aqueous solutions may be damaging, and for supporting areas damaged by copper green decay. Unsaturated fluorocarbon polymer derivatives such as fluoron F-24L and fluorolon H6 mixed with vinylacetate solution (CVEB) have been tried but caused discolouration (Bykova, 1993; Yusopova, 1980) (fluorocarbon polymers include Teflon, a well-known non-stick coating). Soluble nylon (commercial products such as Maranyl) dissolved in alcohol was recommended and its use commonly reported.[10] Unfortunately it was soon found to discolour and become irreversible.[11] Also tried were alcohol solutions of copolymers of vinylacetate with ethylene, mixtures of cellulose ethers in methylene chloride and methanol, and acrylic polymers such as Paraloid B-72 (Bykova, 1993; Tanasi et al., 1985).

The use of cellulose ethers continues to be tried, such as aqueous solutions of methylcellulose (Tylose) and hydroxypropylcellulose (Klucel G) in a non-aqueous solvent. The stabilization of some flaking pigments using Klucel G in ethanol has been described as beneficial as a pre-treatment before the use of gelatine, especially where the pigment will be adversely affected by even slight presence of an aqueous consolidant (Quandt, 1996). Tests of both of these have had mixed results, some showing undesirable characteristics (Chahine et al., 1991; Botti et al., 1996).

Localized coating of parchment fibres with such plastics may cause an imbalance of hydroscopic properties across a skin, leading to deformation as described earlier. However, it is also frequently the case that a heavily pigmented area, often with a layer of gesso or other ground below the pigment, is already behaving differently to surrounding areas and this difference may have contributed to the flaking of the pigment. Equally the decay of collagen can cause a significant reduction in its hydroscopic characteristics. Coating an area like this may make no significant change to the characteristics across the skin, but such a change may not be apparent until after the application.

Dimensional characteristics will be different for each individual item and no general rule can easily be applied for local consolidation treatments.

## 20.5.6 Repairs and supports

As with so many other conservation treatments for parchment, methods applied to support weak parchment, fill holes and repair tears have been many and varied. The repair of tears and holes was an important aspect of preparing skins for use, so many examples of how this was done can be found when examining manuscripts of all ages. The variations of approach have employed both adhesive and non-adhesive techniques, both in production and in conservation. The adhesives used have been varied and their characteristics are a major feature of any consideration of treatment.

Methods of repair used in centuries past were principally twofold. Edges were joined or rejoined by stitching them together (using raw hide strips or linen or hemp twine) or gelatine-based glues were used, alone or mixed with other constituents. The rejoining of torn edges has also been undertaken in the past using only acetic acid (Reed, 1972: 224). The acid melts the fibres at the edge and provides tacky gelatinous surfaces that will stick together. However, the continuous action of the acidity, the damage caused in application and long-term discolouration of the parchment make it inappropriate for use.

Stitching as a remedial conservation treatment continued until quite recently. It might be described as being derived from early craft production methodology, having been used in the scriptorium to prepare membranes. Methods are described in several published sources.[12] Its use as a conservation treatment appears to have become uncommon from the mid-1980s, probably because the puncture of original material ceased to accord with contemporary ethics. It seems to have been restricted to situations in which an adhered repair was not deemed to be sustainable (Munn, 1989).

Clearly the opportunity to reuse existing holes in a membrane may make stitching acceptable, and of course the rebinding of an old codex is a context in which the use of stitching is appropriate. Wherever original stitching exists it is usually recorded in detail and may be preserved in situ if it poses no threat to the item.

Old patches and adhered joins are also preserved and recorded. They are either found as an essential part of membrane preparation (and so text may run over them) or can be an important feature of the historic structure or use of the item. For these reasons, the separation of joins or removal of patches is rarely undertaken except when they place the item at risk of continuing damage. The most common such exceptions involve the reversal of poor quality repairs made in fairly recent times (probably during the last 50 to 100 years). The perceived risk often arises from the decay properties of an adhesive and/or its application, or when the patch itself is causing damage or covering text.

The qualities of repair adhesives have been the subject of much consideration over the years. Early adhesives were usually gelatine based, often a thick version of parchment size. They were frequently mixed with honey and vinegar to adjust their properties for use and many further variations on the theme have been reported, which include the addition of egg white, gum Arabic and starches (Reed, 1972: 220–224). Modern refinements have been made to the additives, using synthesized or purified versions, such as sorbitol and acetic acid (Cains, 1983).

During the twentieth century it became common practice in UK institutions to use starch pastes instead of gelatine glues and this had its own effect on how the membranes were treated. Starch is not a high tack adhesive and the difficulty in applying a patch, for example, necessitated the excessive wetting and flattening of both old and new parchment in order to ensure that the patch maintained an even contact with the membrane surface while the paste dried out (Woods, 1995). It became the practice to adjust the surface of the new parchment (and sometimes the old) to make this easier. This involved abrading ('buffing') the surface with sandpaper so that it became fibrous, the fibres ensuring good absorption and adhesion. This process was even mechanized in one UK institution, where a sanding machine was made to increase production of prepared skins for lining and patching (Smith, 1938). Membranes were wetted excessively and so dried out to a stiff, sometimes shrunken and brittle state. Excess starch was commonly sanded off to improve the appearance of the finish. Because the original membrane had to be stuck down with paste onto a sheet to hold it flat during the process, this sanding was an essential feature of the procedure to remove starch from the reverse. Items treated in this way have a rough, hardened feel, their lateral dimensions are usually smaller than before treatment, they are often more translucent than previously, and sometimes they have 'blistered' areas where the skin has dried inconsistently.

Once practitioners had learned this method it was often applied routinely, even to skins that had only small tears or holes. Despite the presence of published alternatives from the 1930s onwards, it was only in the late twentieth century that this practice in the UK began to be replaced by more compatible and minimal methods, and when it was more widely recognized that excessive intervention was unnecessary for the majority of parchment manuscripts.

Collagenous glues such as gelatine continue to provide compatible, sympathetic and minimal options for the adhesion of tears and patches and their method of application has been described in detail. The high tack properties of gelatine make it possible for only the smallest amount of adhesive to be applied and because it does not display syneritic properties (it does not release water as a liquid as it sets), the parchment does not become wet in the process of adhesion. Care is needed to avoid using too much gelatine, and to use it at the correct application temperature to avoid penetration and consequent translucency. Only very small amounts are necessary and are applied in such a way that the glue has dried and become tacky at the time the surfaces are brought into contact. This combination of features results in an adhesive that sets almost immediately and without the excessive contraction during drying that occurs with so many other colloidal solutions such as starches and the cellulose ethers, and which in turn avoids the risks of distortion around a patch or repair. Various substances have been added to gelatine adhesives, in order to make the adhesive liquid at room temperature. These include natural products such as honey and flour, and more recently the cellulose ethers. There does not appear to be any published comparative information describing the long-term behaviour of these mixtures.

Other adhesives have been tried as an alternative to gelatine, such as commercial heat-activated adhesives, proprietary starch pastes, methyl polyamide mixtures, and PVA emulsions.[13] Given the widespread recognition of the risks associated with heat in contact with parchment, it seems in retrospect surprising that heat-activated, pressure-applied adhesives should, in the past, have been used in the treatment of this material, and perhaps not surprising that they are no longer advocated (Powell, 1974). PVA emulsions have been used with a variety of patching materials such as goldbeaters' skin and Japanese paper.[14] The PVA emulsions do not display the hydroscopic properties usually required for treatments to parchment and are commonly understood to have undesirable decay characteristics, including irreversibility. In spite of the fact that newer PVA polymers have been developed that are intended to be reversible (e.g. ethylvinylacetate or EVA), and in the absence of any long-term ageing data in this context, it is reasonable to question their longer-term properties. It has been suggested that tears might be readhered using minute strings of a Nylon material activated by ultrasound. While the material might be reason for concern, the use of ultrasound instead of heat to melt a dry-state adhesive momentarily is worthy of further investigation (Mayer, 2002).

The reversibility properties of an adhesive are an important consideration. It may be possible to effect a successful repair to parchment using a very dry starch paste, for example. However, the reversal of an aged starch adhesive invariably involves the application of water. A patch may be successfully removed by the application of humidity alone, but removal of the adhesive residue, where it has been determined that this continues adversely to affect the behaviour of the membrane, can require the application of liquid water, so placing the membrane at further risk. The advantages of collagen as the basis for an adhesive, in addition to its ageing properties and compatibility, include its inherent molecular capacity to be broken down and subsequently to undergo a partial reformation. This means a collagenous source can readily be melted and dissolved in hot water but parts of its peptide structure can reassemble as it sets (Veis and Cohen, 1960). Once set, it can subsequently be softened in conditions of high humidity (such as in a humidity chamber or pack).

Various materials have been used for patches and other means of filling lacunae. Similar to the consolidation referred to above, the infilling of holes and weakened areas has frequently been undertaken using a thick suspension of a pulped material. Parchment glue (thicker parchment size) has been used in centuries past for filling small holes, sometimes mixed other organic constituents such as casein and egg white, and sometimes substituted by a solution of reduced and acid-dissolved collagen (Reed, 1972: 222–226). In more recent times, various methods have been described using suspensions of parchment fibres, alone or with the addition of paper fibres, and in aqueous or water/alcohol solutions. These methods are frequently employed on a suction table or even in a leaf-casting machine, usually masking off all but the area to be filled (Wouters et al., 1995).

The most common and ancient method of infilling larger holes has been the use of new parchment applied with a gelatine-based glue. There have been variations during the last 75 years, such as the common use of starch pastes as the adhesive in UK archives as described above, and the use of Japanese papers for patches in many European archives. Compatibility and a very long precedence of stability are sound reasons for using parchment, adhered with gelatine. However, new parchment is much more rigid compared with an old and possibly weakened membrane, and softer, more flexible infill materials are justifiable. Japanese paper, while being a very different material, has the advantage of being long fibred, stable, soft and, if used carefully, has the capacity to absorb atmospheric moisture without causing distortion to the membrane. Parchment can be pared to make it thin, softer and translucent, so for sound membranes it is still a suitable material. Its excessive whiteness when new can be overcome by the use of skins toned during production (usually with coffee) or subsequently (Cains, 1983).

In the past, weak areas have most often been supported with a finely split parchment or the prepared lining of bovine intestine, known as 'goldbeaters' skin'. Like parchment, goldbeaters' skin has been used for many hundreds of years and if applied with a small quantity of gelatine glue can remain stable, sympathetic and strong for very long time indeed – even over several centuries (Wächter, 1962). It is almost completely transparent and extremely thin. A similar material made from fish swim bladder has also been used as an alternative (Cains, 1983; Munn, 1989). Both of these materials continue to be used and will be found referred to in many of the references noted herewith, which include some very detailed descriptions of ways in which they have been applied.

Although repairs using goldbeaters' skin have withstood the test of centuries, for much of the twentieth century there was a trend in trying to find new ways of supporting torn or weak skins, including lamination with materials such as papers and plastics such as polyvinylchloride (Reed, 1972: 232–233). Japanese tissues or papers were used in France from at least the 1920s and in Dutch archives its use can be traced back to 1858 (Smith, 1938). The thin paper was adhered to both sides of a weak membrane using a wheat starch containing alum, and then the fibres sandpapered off to reveal the surface and writing. In France and Germany, a transparent paper, referred to as 'parchment paper' and probably similar to a 'glassine' paper, was used from about 1927 (Smith, 1938).

From at least the 1930s until the 1980s, silk chiffon was widely used in UK institutions to cover weak paper documents. The practice, called simply 'silking', was sometimes extended to parchment, although its use for this material was not so common as for paper. It involved the pasting of a layer of silk chiffon to the written faces of a manuscript, often in combination with a backing layer and patches (and sometimes edge 'framing') using new parchment.[15] For larger membranes linen was the preferred alternative to new parchment, being a cheaper material. The fine silk gauze was used to ensure that text was visible.

The use of silk as a support was slow to disappear (it was still being used in some archives until at least the early 1980s), even though it had been observed to discolour and decay (Reed, 1972: 232). It was substituted with a synthetic alternative referred to as 'bonded fibre fabric', an alkali-modified cellulose tissue (Viscose) commonly used for producing tea bags (Cockerell, 1958). In spite of the problems of bringing heat into contact with parchment, a few practitioners in the 1970s chose to use paper tissues which they coated with commercial emulsions that could be activated by heat and pressure (Powell, 1974).

Lamination with the use of plastic films was common for paper manuscripts from the 1930s to the 1960s and its use was extended to parchment in some institutions in Europe. In Germany, cellophane was used for a range of documents in the 1930s, although it was observed to cause the manuscripts to have an unattractive wrinkled surface. At the same time in the United States a cellulose acetate film was in use, but it does not appear to have been applied to parchment. In Europe in the 1950s a self-adhesive polyvinylchloride (called Mipofolie) was sometimes used for parchment. These PVC films proved to discolour or cause deterioration and later had to be removed (Wächter, 1987b).

Another common alternative to goldbeaters' skin, used from the 1970s if not before, was a form of reconstituted collagen. It was commonly purchased from the food industry in tubular form and produced for sausage casings (Reed, 1972: 226–229). Lengths would be slit to produce a sheet, washed, dried and pounced, patches cut to size and then pasted with starch or parchment size over the affected area of parchment (Guiffrida, 1983). It was thought appropriate because of its translucency and the fact that it was derived from skin materials (it actually consisted of about 50% collagen, 20% glycerol, some methylcellulose, 1.5% inorganic matter, 0.5% fat) (Woods, 1995). However, its appearance and yellowing colour, and the comparative qualities of goldbeaters' skin, have caused it to fall into disfavour.

The undesirable characteristics of these various methods and materials had been described already by the 1970s. Referring to the use of starch, silk, paper tissues and sausage casings, Dr Reed pointed out in 1972 that: 'Although it is possible to restore volumes in serious need of repair to *forms* which look "as good as new", it must be remembered that in twenty years' time they will require further attention.' He goes on to describe how difficult they are to reverse, which could have been as much the result of inappropriate adhesives as the laminate support materials themselves (Reed, 1972: 231).

## 20.6  Conclusion

In times past, parchment has been a precious commodity, in particular as a vehicle for dissemination of thought to the privileged few who could read and write. Over the passing centuries it became more and more commonplace, although still prized as the most appropriate material for records or agreements and contracts to be kept for generations. Paper has replaced it for all but the most exceptional writing and printing purposes, as have other materials where it was used in clothing, book covers and so on. The rapid decline in production of parchment during the twentieth century means that it is now largely a material of the past and this may make it precious to us once more.

Parchment could be described as a contradictory material. It has characteristics that make it sufficiently robust to have withstood many centuries, often in fairly good condition where use has been low. It can resist acid hydrolysis and can be strong enough to be used as a hardwearing book covering. However, it is damaged by water, heat, mould and regular and insensitive handling, common features in the existence of any item over centuries. The information and decoration that it usually carries is easily lost. Most notably (in this context) it can be damaged by ill-advised conservation treatments. It has often been seen as just another kind of paper, or even leather, and suffered from inappropriate categorization and lack of awareness of its nature and idiosyncrasies.

In some respects, parchment could be used to characterize the remarkable variety and excesses of the development of conservation in the twentieth century. In common with so many other areas of conservation, as conservators have been able to learn in ever-greater detail about its nature and peculiarities, so their understanding of the importance of minimal intervention and sensitivity has grown and this is to be welcomed.

## Acknowledgements

The author wishes to thank the various kind people on three continents who responded positively when seeking sources; Marion Kite, Trevor Miles, Robert Minte and Sabina Pugh for helping to find images and for comments to the text, and most especially Ange Tack of the Bodleian Library for patiently and cheerfully hunting down so many of the texts listed below.

## Endnotes

1. The term 'split' is used for a skin reduced in thickness either by literally splitting or delaminating layers with a knife or machine, or by shaving and scraping the sides. It is not easy to tell the difference between a skin that has been literally split apart from one that has been scraped down, especially as the surfaces on both kinds were subsequently shaved further in final preparation, and sometimes also by the scribe or other user. For further information on the different ways that skins have been reduced in thickness see: Reed, R., *Ancient Skins, Parchments and Leathers*, Seminar Press, London 1972, pp. 54 and 121; and Haines, B., *Parchment*, Leather Conservation Centre, 1999, pp. 3–7.
2. For examples of different methods used, see: Haslam, J., *The Book of Trade Secrets, Recipes and Instructions for Renovating, Repairing, Improving and Preserving Old Books and Prints*, London 1910, p. 43; Langwell, W., *The Conservation of Books and Documents*, London 1957, p. 48; and Smith, Manuscript Repair in European Archives, I Great Britain, *The American Archivist*, Vol. I, Number 1, 1938, p. 16.
3. Compare *ibid.* with the Foreword of BS4971:1988 *Repair and allied processes for the conservation of documents*, BSI, London 1988.
4. See, for example: BS4971:1973, *Recommendations for repair and allied processes for the conservation of document, Part 1. Treatment of sheets, membranes and seals*, BSI, London 1973, p. 5 and Bond, M., *The Record Office House of Lords, Report for 1952*, London: Record Office, 1953.
5. For respective examples see the following: Reed, R., *Ancient Skins, Parchments and Leathers*, Seminar Press, London 1972, p. 215; Wächter, O., The Restoration of the Vienna Dioscorides, *Studies in Conservation*, **7**, 1962, p. 25; and Yusopova MV Conservation & Restoration of Manuscripts and Bindings on Parchment, *Restaurator*, vol. 4, Number 1, 1980, p. 62.
6. Examples of old damp pack methods include: Reed, R., *Ancient Skins, Parchments and Leathers*, Seminar Press, London 1972, p. 211; Powell, R., Case History of Repair and Rebinding of an Eighth Century

Vellum Manuscript, in Smith, P. (ed.) *New Directions in Bookbinding*, **22**, 1974, p. 181; and, Guiffrida, B., Book Conservation Manual, Part Four: The Repair of Parchment and Vellum in Manuscript Form, *The New Bookbinder*, **3**, 1983, p. 31.
7. Detailed descriptions of humidity chambers for parchment include: Cains, A., Repair Treatments for Vellum Manuscripts, *The Paper Conservator*, **7**, IPC 1983, pp. 20–21 and Clarkson, C., A Conditioning Chamber for Parchment and Other Materials, in *The Paper Conservator*, **16**, IPC 1992, pp. 27–30.
8. For examples ancient and modern, see the following: Cockerell, D., Condition, Repair and Binding, in Milne, H., Skeat, J., *Scribes and Correctors of the Codex Sainaticus*, British Museum, London 1938, pp. 84–85; and Mr Cockerell's 'stretching frame' was designed to carry out both humidification and stretching at the same time. Its use at the Public Record Office, London, and the revealing staff perception of its limitations are described in Smith, L., Manuscript Repair in European Archives, I Great Britain, *The American Archivist*, Vol. I, Number 1, 1938, p. 7; Cockerell, S., *The Repairing of Books*, London 1958, pp. 73–74; and Burns, T. Bignell, M., The Conservation of the Royal Charter and Great Seal of Queen's University, *The Paper Conservator*, **17**, IPC 1993, p. 9.
9. For example, see: Wächter, O., The Restoration of the Vienna Dioscorides, *Studies in Conservation*, **7**, 1962, p. 25; Guiffrida, B., Book Conservation Manual Part Four: The Repair of Parchment and Vellum in Manuscript Form, *The New Bookbinder*, **3**, 1983, pp. 36–38; and Quandt, A., Recent Developments in the Conservation of Parchment Manuscripts, *The Book and Paper Group Annual*, **15**, 1996, pp. 1–2.
10. For examples, see: Gowers, H., Treatment of a Manuscript, *Museums Journal*, Vol. 58, March 1959, p. 280; Report of the Annual Repairers' Meeting in *Society of Archivists Journal*, Vol. III, Number 3, 1966, pp. 153–154; BS4971:1973, *Recommendations for repair and allied processes for the conservation of document, Part 1. Treatment of sheets, membranes and seals*, BSI 1973, p. 7 De-acidification of Paper 7.1; and Powell, R., Case History of Repair and Rebinding of an Eighth-century Vellum Manuscript, in Smith, P. (ed.) *New Directions in Bookbinding*, **22**, London 1974, p. 181.
11. For a review of the use of soluble nylon see: *Paper Conservation Catalog*, AIC Book and Paper Group 1994, pp. 23–24.
12. For example, see: Cockerell, S., *The Repairing of Books*, London 1958, p. 71; Guiffrida, B., Book Conservation Manual Part Four: The Repair of Parchment and Vellum in Manuscript Form, *The New Bookbinder*, **3**, 1983, pp. 27–28; and Cains, A., Repair Treatments for Vellum Manuscripts, *The Paper Conservator*, Vol. 7, IPC 1983, pp. 18–19.
13. For an example of the use of commercial emulsions and PVA, see: Powell, R., Case History of Repair and Rebinding of an Eighth-century Vellum Manuscript, in Smith, P. (ed.) *New Directions in Bookbinding*, **22**, London 1974, p. 181.
14. For example, see: Bienvenida, V., Procedencia: patrimonio cultural manuscrito miniado, en pergamino, s xvi (Processes used on an illuminated 16th-century parchment manuscript), *Centromidia* 1984.
15. For a description of these methods see: Smith (1938), Manuscript Repair in European Archives, I Great Britain, *The American Archivist*, **I**(1), 5–8.

# References

Adamo, M., Brizzi, M., Magaudda, G. et al. (2001) Gamma Radiation Treatment of Paper in Different Environmental Conditions: Chemical, Physical and Microbiological Analysis. *Restaurator*, **22**(2), 107–131.
BS5454:2000 (2000) Recommendations for the Storage and Exhibition of Archival Documents. London: BSI.
BS4971:1973 (1973) Recommendations for the Repair and Allied Processes for the Conservation of Documents. Part 1: Treatment of Sheets, Membranes and Seals. London: BSI.
BS4971:2002 (2002) Repair and Allied Processes for the Conservation of Documents – Recommendations, 13.5 Consolidation. London: BSI.
Belaya, I. (1969) Softening and Restoration of Parchment in Manuscripts and Bookbinding and Instructions for the Softening of Parchment Manuscripts and Bookbindings. *Restaurator*, **1**(1), 20–48, 49–51.
Bëothy-Kozocsa, L., Sipos-Richter, T. and Szlabey, G. (1990) Parchment Codex Restoration Using Parchment and Cellulose Fibre Pulp. *Restaurator*, **11**(2), 95–109.
Berardi, M. (1992) Why Does Leather Deform? Some Observations and Considerations. *Leather Conservation News*, **8**(1), 12–17.
Bloom, J.M. (1989) The Blue Koran. In *Les Manuscrits du Moyen-Orient: essais de codicologie et de paléographie* (Deroche, F., ed.). pp. 95–99. Istanbul and Paris.
Botti, L., Impagliazzo, G., Montovani, O. et al. (1996) Investigation of Some Polymers for the Protection of Paint Films. In *Proceedings of the International Conference on Conservation and Restoration of Archive and Library Materials*. pp. 563–581 Rome: Instituto Centrale per la Patalogia del Libro.
Boyd-Alkalay, E. and Libman, E. (1997) *Conservation of the Dead Sea Scrolls at the Laboratory of the Israel Antiquities Authority*. pp. 211–213. London: IPC Conference Papers.
Buchanan, A. (1993) IPC Humidification Meeting. *Paper Conservation News*, **66**, 11–12.
Bykova, G.Z. (1993) Medieval Painting on Parchment: Technique, Preservation and Restoration. *Restaurator*, **14**(3), 188–190.
Cains, A. (1983) Repair Treatment for Vellum Manuscripts. *The Paper Conservator*, **7**, 17–22.

Calabro, G., Tanasi, M.T. and Impagliazzo, G. (1986) An Evaluation Method of Softening Agents for Parchment. *Restaurator*, **7**(4), 169–180.

Carrió, V. and Stevenson, S. (2003) Assessment of Materials used for Anoxic Microenvironments. In *Conservation and Science 2002*. Proceedings of the conference held Edinburgh 2002, pp. 32–38. London: Archetype.

Chahine, C., Rottier, C. and Ruoy, D. (1991) Effects of Adhesives on the Mechanical Properties of Parchment. *Sauvegarde et Conservation des Photographies, Dessins, Imprimés et Manuscrits*, pp. 139–146. Paris.

Chaudray, M., et al. (1991) In *Pergament: Geschiste, Struktur, Restaurung, Herstellung* (Rück, P., ed.). pp. 315–322. Sigmaringen: Jan Thorbecke Verlag.

Clarkson, C. (1982) *Limp Vellum Binding – and its Potential as a Conservation Type Structure for the Rebinding of Early Printed Books*. Hitchin: The Red Gull Press.

Clarkson, C. (1992) Rediscovering Parchment: The Nature of the Beast. *The Paper Conservator*, **16**, 5–26.

Clarkson, C. (1987) Preservation and Display of Single Parchment Leaves and Fragments. In *Conservation of Library and Archive Materials and the Graphic Arts* (G. Petheridge, ed.). pp. 201–209. London: IPC.

Cockerell, S. (1958) *The Repairing of Books*. London: Sheppard Press, pp. 74, 75, 230.

Cockerell, S. (1970) A Note on the Conservation of Some Parchment Documents Damaged by Fire. *Business Archives*, **33**, 18–19.

Cooper, M., Spoertun, S., Stewart, A. et al. (2000) Laser Cleaning of an 18th Century Parchment Document. *The Conservator*, **24**, 71–79.

Craft, A. (2003) The Conservation of Limp Vellum Books at the National Archives. *ARC*, **172**, 11–13.

Craig, R. (1986) Alternative approaches to the Treatment of Mould Biodeterioration – An International Problem. *The Paper Conservator*, **10**, 27–30.

Dignard, C. (1992) Tear Repairs of Skins with Minimal Access to their Backs: The Treatment of a Kayak. *Leather Conservation News*, **7**(2), 1–8.

Diringer, D. (1982) *The Book Before Printing*. New York: Dover, pp. 170–171.

Dobrusina and Visotkite (1994) Chemical Treatment Effects on Parchment: Properties in the Course of Ageing. *Restaurator*, **15**(4), 208–219.

Ellement, P. (1987) A Note on the Structure of Vellum and the Effects of Various Solvents. In *Conservation of Library and Archive Materials and the Graphic Arts* (G. Plenderleith, ed.). pp. 199–200. London: IPC.

Fitzsimmons, E. (1986) Limp Vellum Bindings: Their Value as a Conservation Binding. *Restaurator*, **7**(3), 125–142.

Fleming, L.E. (1983) Conservation of Oriental Miniatures. *Eastern Pictorial Art Conservation News*, **2**, 12.

Flieder, F. (1982) L'analyse et la revelation chimiques des encres metallo-galliques. *Restaurator*, **5**(1–2), 57–63.

Fuchs, R. (2002) The History of Chemical Reinforcement of Texts in Maunscripts – What Do We Do Now? *Care and Conservation of Manuscripts*, **7**, 159–170.

Fuchs, R. and Mrusek, R. (2002) New Methods of Reflectography with Special Filters and Image Processing Techniques. *Care and Conservation of Manuscripts*, **7**, 212–213.

Gairola, T.R. (1958) *Journal of Indian Museums*, **14–16**, 43–45.

Giovannini, A. (1999) *Die Restaurierung von Zwei mit Pergament Überzogenen Möbeln von Carlo Bugatti*, pp. 61–65. IADA Preprints.

Grant, J. (1937) *Books and Documents, Dating, Permanence and Preservation*, p. 194. London: Grafton.

Guiffrida, B. (1983) Book Conservation Manual Part 4: The Repair of Parchment and Vellum in Manuscript Form. *The New Bookbinder*, **3**, 30–37.

Haines, B. (1987) Shrinkage Temperature in Collagen Fibres. *Leather Conservation News*, **3**(2), 1–5.

Haines, B. (1994) The Physical and Chemical Characteristics of Parchment, Casings, Goldbeater's Skin and Gelatin. In *Conservation for the Future* (C. Woods, ed.). p. 26. Dorset: AIM Proceedings. Society of Archivists.

Haines, B. (1999) *Parchment*. Northampton: Leather Conservation Centre.

Haran, M. (1991) Technological Heritage in the Preparation of Skins for Biblical Text in Medieval Oriental Jewry. In *Pergament: Geschiste, Struktur, Restaurung, Herstellung* (P. Rück, ed.). pp. 35–43. Sigmaringen: Jan Thorbecke Verlag.

Harrison, L. (1981) Physical Chemistry of Biological Morphogenesis. *Chemical Society Review*, **10**(4), 491.

Haslam, J. (1910) *The Book of Trade Secrets. Recipes and Instructions for Renovating, Repairing, Improving and Preserving Old Books and Prints*. London: Haslam and Co., pp. 10, 11.

Hassel, B. (1999) Conservation Treatment of Medieval Parchment Documents Damaged by Heat and Water, pp. 253–256. *Preprints of the 9th IADA Congress*.

Hassel, B. (2003) Heat Damaged Parchment. *Papier Restaurierung*, **3**(4), 31–38.

Hermans, J. (1992) *Fans on Skin: Their Conservation and Storage*. Postprints of the ICOM Committee for Conservation Leathercraft and Related Objects Group interim meeting, pp. 42–44.

Kahle, T.B. (1986) State of Preservation of the Dead Sea Scrolls. *Nature*, **321**, May 8 and citing Plenderleith, H.J. (1955) *Discoveries in the Judean Desert 1. Qumran Cave 1*.

Kaminitz, M. and Levinson, J. (1988) The Conservation of Ethnographic Skin Objects at the American Museum of Natural History. *Leather Conservation News*, **5**(1), 1–7.

Kautek, W., Pentzien, S. and Krüger, J. et al. (1997) Laser Cleaning of Ancient Parchments. *Lasers in the Conservation of Artworks*, pp. 69–78. Vienna.

Kite, M. (1992) Gut Membrane, Parchment and Gelatine Incorporated into Textile Objects. *The Paper Conservator*, **16**, 98–105.

Kowalik, R. (1980) Decomposition of Parchment by Micro-organisms. *Restaurator*, **4**(3–4), 200–208.

Kremen, S. and Southwood, R. (1960) The Influence of Hydrogen Bonding on the Solvent Dehydration of

Hides and Skins. *Journal of the American Leather Chemists Association*, **55**, 24–40.

Larsen, R. (2002) Improved Damage Assessment of Parchment. *Care and Conservation of Manuscripts*, **7**, 29 and plates III and IV.

Lee, L. (1992) The Conservation of Pleated Illuminated Vellum Leaves in the Ashmole Bestiary. *The Paper Conservator*, **16**, 12.

Lindsay, H. (2003) Preservation Microfilming and Digitisation at the London Metropolitan Archives. *The Paper Conservator*, **27**, 47–57.

Lokandum, B. and Chaudray, M. (1991) Parchment in India – Some Observations. In *Pergament: Geschiste, Struktur, Restaurung, Herstellung* (P. Rück, ed.). pp. 315–322. Sigmaringen: Jan Thorbecke Verlag.

Loveday, H. (2001) *Islamic Paper – A Study of the Ancient Craft*. Don Baker Memorial Fund, pp. 12–23.

Mairinger, F. (1981–82). Physikalische Methoden zur Sichtbarmachung verblasster oder getilgter Tinten (Physical Methods of Making Faded or Obliterated Inks Visible). *Restaurator*, **5**(2), 45–56.

Marconi, B. (1962) Use of Wax for Fixing Flaking Paint. *Studies in Conservation*, **7**(1), 17–21.

Matthews, C.G. (1924) *Notes from the Reports of the Public Analysts*, pp. 516–517.

Mayer, M. (2002) Ultrasonic Bonding. *Papier Restaurierung*, **3**(4), 19–22.

Munn, J. (1989) Treatment Techniques for the Vellum Covered Furniture of Carlo Bugatti. *The Book and Paper Group Annual*, **8**, 27–38.

Myers, M. (1980) Storage of Leather Objects. *In Conservation and Restoration of Leather, Skin and Parchment*. Copenhagen.

Norman, D. (1993) Mounting Single Leaf Parchment and Vellum Objects. *V & A Conservation Journal*, **9**, 10–13.

Parker, A.E. (1993) Freeze Drying Vellum Archival Materials. *Journal of the Society of Archivists*, **14**(2), 175–188.

Petrus Maria Caneparuis (1619) *De atramentis cuiuscunque generic*, p. 179. Venice. (Ad intraurandas literas ex antiquitate fere abolitas – How to Freshen up Letters from Antiquity that are Almost Erased.)

Pickwoad, N. (1992) Alternative Methods for Mounting Parchment for Framing and Exhibition. *The Paper Conservator*, **16**, 78–85.

Powell, R. (1974) Case History of Repair and Rebinding of an 8th Century Vellum Manuscript. In *New Directions in Bookbinding* (P. Smith, ed.), **22**, 181.

Quandt, A. (1991) The Documentation and Treatment of a Late 13th Century Copy of Isidore of Seville's Etymologies. *The Book and Paper Group Annual*, **10**, 164–195.

Quandt, A. (1996) Recent Developments in the Conservation of Parchment Manuscripts. *The Book and Paper Group Annual*, **15**, 1–12.

Reed, R. (1972) *Ancient Skins, Parchments and Leathers*. London: Seminar Press.

Reed, R. (1991) Some Thoughts on Parchment for Bookbinding. In *Pergament: Geschiste, Struktur, Restaurung, Herstellung* (P. Rück, ed.). pp. 217–220. Sigmaringen: Jan Thorbecke Verlag.

Ryder, M. (1991) The Biology and History of Parchment. In *Pergament: Geschiste, Struktur, Restaurung, Herstellung* (P. Rück, ed.). pp. 25–33. Sigmaringen: Jan Thorbecke Verlag.

Scholla, A. (2003) Early Western Limp Bindings: Report on a Study. *Care and Conservation of Manuscripts*, **7**. Museum Tusculanum Press, pp. 132–158.

Sievers, J. (1985) Fehlstellenergänzung bei Pergament durch Anfasern. *Maltechnik-Restauro*, **19**, 63–66.

Sinco, P. (2000) The Use of Gamma Rays in Book Conservation. *Nuclear News*, 38–40.

Singer, H. (1992) The Conservation of Parchment Objects using Gore-Tex Laminates. *The Paper Conservator*, **16**, 40–45.

Smith, L.H. (1938a) Manuscript Repair in European Archives II: The Continent; France, Belgium and the Netherlands. *The American Archivist*, **1**(2), 51–57, 64–67.

Smith, L.H. (1938b) Manuscript in European Archives I: Great Britain. *The American Archivist*, **1**(1), 6–19.

Smith, L.H. (1938c) Manuscript Repair in European Archives II: The Continent: Italy. *The American Archivist*, **1**(2), 70–75.

Soest, H. van (1989) *Researching the Conservation and Restoration of Parchment*. ICOM Arbeitsgruppe Leathercraft and Related Objects, International Leather and Parchment Symposium, pp. 94–103.

Stancievicz, J. (1996) The Conservation of a Hebrew Manuscript Number 802238, an Evaluation of the Method. *Restaurator*, **17**(2), 66–67.

Stiber, L. (1997) Thomas Jefferson's Rough Draft of the Declaration of Independence: The Evolving Condition of an American Icon and Implications for Future Preservation. In *The Proceedings of the Fourth International Conference of the Institute of Paper Conservation* (J. Eagan, ed.). pp. 53–63. London: IPC.

Szcepanowska, H. (1992) The Conservation of 14th Century Parchment Documents with Pendant Seals. *The Paper Conservator*, **16**, 86–92.

Tanasi, M.T., Vallone, L., Impaglazzo, G. *et al.* (1985) A propositio di un intervento strordinario di restauro: recupero di un documento in pergamena dell'archivo do stato di Torino. (An Exceptional Restoration: Recovery of a Parchment Manuscript from the Torino State Archives.) *La Conservazione delle Carte Antiche*, **5**(9–10), 37–45.

Veis, A. and Cohen, J. (1960) Reversible Transformation of Gelatin in the Collagen Structure. *Nature*, **186**, 720–721.

Viñas, V. (1987) The Use of Polyethylene Glycol in the Restoration of Parchment. In *Conservation of Library and Archive Materials and the Graphic Arts* (G. Plenderleith, ed.), pp. 195–197. London: IPC.

Voronina, L., Nazarova, O. and Petushkova, Y. (1980) Disinfection and Straightening of Parchment Damaged by Micro-organisms. *Restaurator*, **4**(2), 91–98.

Vorst, B. (1986) Parchment Making – Ancient and Modern. *Fine Print*, **12**(4), 209–210.

Wächter, O. (1962) The Restoration of the Vienna Dioscorides. *Studies in Conservation*, **7**, 23–25.

Wächter, O. (1987a) The Malicious Metallic Inks. *Maltechnik-Restauro*, **23**(3), 19–23.

Wächter, O. (1987b) The Delamination of the Carolingian Evangelary from the Treasure of the Cathedral of Essen. *Maltechnik-Restauro*, **23**(2), 34–38.

Woods, C. (1995) Conservation Treatments for Parchment Documents. *Journal of the Society of the Archivists*, **16**(2), 223, 228–229.

Woods, C. (2002) The Role of Modern Reprographics as a Tool for the Conservator. *Care and Conservation of Manuscripts*, **7**, 84–91.

Wouters, J., Gancedo, G., Peckstadt, A. *et al.* (1992) The Conservation of the Codex Eykensis: The Evolution of the Project and the Assessment of Materials and Adhesives for the Repair of Parchment. *The Paper Conservator*, **16**, 68.

Wouters, J., Peckstadt, A. and Watteeuw, L. (1995) Leafcasting with Dermal Tissue Preparations: A New Method for Repairing Fragile Parchment and its Application to the Codex Eyckensis. *The Paper Conservator*, **19**, 7.

Yusopova, M.V. (1980) Conservation and Restoration of Manuscripts and Bindings on Parchment. *Restaurator*, **4**(1), 58–61.

# 21

# Conservation of leather bookbindings: a mosaic of contemporary techniques

## 21.1 Introduction

*Randy Silverman*

Successful conservation of leather bookbindings, unlike the treatment of many other types of cultural artefacts intended primarily for exhibition and storage, requires books be returned to a functional state. Traditional leather rebacking – the repair of damaged hinges and headcaps with new leather onto which original leather is reattached – produces aesthetically appealing results due to leather's ability to blend new and old materials sympathetically. While still a stock option, the disadvantages of using leather as a repair material, including chemical instability of vegetable-tanned skins and loss of physical strength resulting from overparing, have forced conservators to re-evaluate this choice over the past 20 years. What has emerged is a variety of conservation alternatives that achieve functionality, durability, and aesthetic compatibility more quickly than traditional methods. Most significantly, contemporary conservation options for treating leather bookbindings tend to minimize the intervention, ensuring repairs preserve the integrity of historic books.

The following short essays reflect crosscurrents in conservators' innovative approaches to reattaching boards, repairing joints and headcaps, and consolidating desiccated book leather. Collectively, they reflect a common interest in substituting cellulose for collagen when choosing a repair material to increase the chemical stability and physical durability of the resultant work. These developments include a range of bookbinding techniques, including: Cains' joint tacketing; Ruzicka, Zyats, Reidell, and Primanis' work with consolidants, Puglia and Anderson's solvent-set book repair tissue, Etherington's Japanese paper hinge repair, Minter and Brock's cloth hinge repair, and Zimmern and Minter's development of board slotting machines. They also address the problem of treating inherent vice in historical bookbinding leathers, especially in skins produced during the past 200 years, through the use of chemical consolidants such as Klucel G, SC6000/Klucel and Lascaux 498 HV. The synergism underlying these trends reveals the creativity afoot in book conservation internationally.

## 21.2 Binding solutions to old problems

*Anthony Cains*

### 21.2.1 Introduction

In his tribute to the historian Maurice Craig, the late William O'Sullivan recalled searches made together in the Long Room at Trinity College, Dublin, looking for rare bindings (O'Sullivan, 1992; Cains, 2000). During that time, they sheltered their discoveries in a seventeenth century armoire in the office of the librarian H.W. Parke. He wrote: their objective was to give some measure of protection to a few of the better-preserved fine bindings, which, like those on all Long Room books, had suffered severely from the polluted atmosphere of the city since Victorian times. O'Sullivan also made observations on the relative condition of different skin covering materials, comparing the robust, early seventeenth century vegetable-tanned calf to the mid-eighteenth century calf which 'crumbled'; the 'morocco' which resisted pollution better than the calf, and vellum which weathered best of all.

It will be observed, however, that vellum, long exposed to light and heat, was often found to be brittle and cracked, while tawed skin survived largely intact although its exposed surface would erode with the loss of the grain layer.

Referring to the manuscript collection in his charge, O'Sullivan lamented the large-scale rebinding events of the eighteenth and nineteenth centuries, the decay of those bindings despite their relatively sheltered existence in the Manuscript Room, and the low standards of twentieth-century commercial binding that led the librarian to suspend all binding and repair in 1950. The work of conserving early Irish manuscripts at Trinity College Dublin (TCD) begun in 1952 by Roger Powell, and the establishment of the Conservation Department (1972–74), were due in no small measure to Dr O'Sullivan's efforts.

The Long Room project to survey, clean and conserve the collection was started in 1979–80, and the oak sash windows were then covered with ultraviolet (UV) filtering acrylic sheets in addition to the holland blinds (Cains and Swift, 1988). In preparation for this task, a number of consolidants and leather dressings were tried on typical examples of historic vegetable-tanned leathers, particularly calf and sheepskin. The project needed a treatment that contained neither water that would darken the degraded leather nor oil that would stain the text paper, and one that could be handled by the intelligent but unskilled hands of college students employed during summer vacation. Of the several cellulose ethers that Margaret Hey introduced in the TCD Library Conservation Department workshop for paper sizing/consolidation trials, Klucel G (hydroxypropylcellulose) in ethanol proved the most effective, and provided the essential process and the foundation for the project.

### 21.2.2 Klucel G

As the product guide notes (Hercules Incorporated, 1979), Klucel has excellent binding properties. Following the publication of a brief report on the use of Klucel G as a consolidant for degraded leather (Cains, 1981), prompt feedback was elicited from colleagues: Dr Anthony Werner (a TCD graduate) stated the process proved effective in the treatment of Polynesian grass skirts using isopropanol as the solvent, while bookbinders working in an archive commented it worked well on reverse calf and sheep. Dr Nicholas Pickwood noted some problems with skins darkening, possibly due to the Klucel G solution being overexposed to the atmosphere before being applied to leather by his (volunteer) staff. In their product guide (Hercules Incorporated, 1979: 7), the Hercules Company note the following:

*Moisture Absorption: Klucel absorbs moisture from the atmosphere, as do other water-soluble materials. The amount absorbed depends on the relative humidity and temperature of the environment. As packed, moisture content of all grades does not exceed 5% by weight, and is generally between 2% and 3%. It is suggested that Klucel be stored in tightly closed containers and in a dry atmosphere to prevent any increase in moisture content.*

*Equilibrium moisture content at 50% RH and 73 F. (23 C.) – 4%*

*Equilibrium moisture content at 84% RH and 73 F. (23 C.) – 12%*

Instructions for dissolving Klucel G are also given in the product guide. The solvent is agitated, preferably in a laboratory mixer, and the product added slowly to the vortex. During the mixing, the powder may coagulate but these gel lumps will eventually dissolve to form an even, viscous solution.

As a further precaution, one may elect to use a dry alcohol and to desiccate the Klucel G powder. For the Long Room project, however, the solution was prepared using industrial-grade denatured ethanol (IMS 100 @ 20°C toilet grade) containing only a trace of water, a solution that has proved reliable over the years.

The writer believes the source of the water causing the darkening of leather following an application of Klucel G relates to the environmental conditions in which the work is undertaken. As an example, two equally degraded, mid-nineteenth century volumes bound in half calf were removed from a very dry storage environment at TCD. In preparation for a summer class to be taught in Italy, one of the volumes was treated with Klucel G in the TCD workshop which satisfactorily consolidated the leather without its becoming darkened. Several days later, in the warm, humid environment north of Rome, the same Klucel G solution was used to treat the second volume, but this leather darkened perceptibly. Because of this and similar experiences, it is recommended that subject material be preconditioned prior to treatment in a drying cabinet or air-conditioned room at a relative humidity less than, say, 40%. The moisture content (EMC) of the subject leather can be checked with a suitable moisture meter (Aquaboy type) in a range of less than 6%.

The project was conducted in a closed-off cabin area in the Long Room, a relatively dry and stable environment compared with the conservation workshop. In monitoring the project, blackening of the leather was not observed except when too much flour paste had previously been applied to secure torn flaps or labels. Blackening is frequently seen on old rebackings when a paste-wash has been applied in an attempt to consolidate the powdery leather.

### 21.2.3 Application of Klucel G

Sufficient Klucel G 5–10–15 g/l is shaken and decanted into a beaker for the day's work in hand. While it can be applied with a well-used, 55 mm wide, Japanese brush of the type called Shiro-Ebake, these brushes are rather expensive and may be substituted with a regular one-inch paint brush. With the fore edge down on the bench, brush the solution onto the spine and caps, which will be immediately absorbed by the areas lacking grain. On the sides where the grain is intact, surplus Klucel G is removed with a cotton wool pad worked in a circular or in a figure-of-eight motion (rather like applying French polish to wood). This serves the purpose of cleaning the cover and leaving it without smears. The board edges and turn-ins are then treated, using the opportunity also to remove soil and size the brittle paper areas of the pastedowns and flyleaves. This is best done using another brush and container of Klucel G for paper only (5 g/l). If you are not familiar with the leather of the subject, spot-test an area before applying the consolidant.

The gains made by the treatment, other than cosmetic improvements, occur in the reading room where the treated material is cleaner to handle and reduces the soiling of reading cradles and padded supports. In the conservation workshop, consolidation of desiccated leather provides the foundation for other leather treatments, such as facing, and does not interfere with them.

### 21.2.4 Facing degraded leather

A review of the author's paper on facing, given at the Institute of Paper Conservation Manchester 1992 conference (Cains, 1992), is now timely. Regarding the casting of the tissue: by reducing the water by half when diluting the two acrylic dispersions, a fatter resin layer results which sticks and detaches more readily than the standard heat/solvent set recipe given.

Standard tissue: Plextol M 630 one part; Plextol B 500 two parts; and water six parts.

Facing tissue: Plextol M 630 one part; Plextol B 500 two parts; and water three parts.

The latter should not be used on endpapers because too much resin is present to be discharged by simply swabbing with acetone. It may only be used if the subject paper can be immersed in acetone to dissolve the residual resin. This new proportion was determined after a trial on Victorian leather by John Gillis of the TCD workshop.

After applying the Klucel G to the leather, it should be allowed to dry thoroughly before adhering facing tissue to the areas of the binding to be lifted and underlaid with new conservation leather (Cains, 1994, Figures 5 and 6). The main purpose of the facing technique is to remove the spine leather intact whatever its condition (Cains, 1981), to gain access to the sewing structure for repair (see joint tackets below) or disbinding, and to prevent loss and fragmentation of the cover during repair.

### 21.2.5 Technique

For raised bands, cut pieces of tissue to fit the panels and separate strips to overlay the bands. If the joint is to be underlaid, the band strips should stop short of the joint by 6 to 8 mm; this area will be overlaid by the new leather but underlaid at the panels.

The spine is dampened, not flooded, with ethanol and each piece of facing is activated with ethanol. As it softens, it is applied to the surface of the leather and pressed home with a pad of cotton wool and then firmly pressed with the fingers. When the facing has dried superficially, but the leather is still moist, the process of lifting can commence. The idea is to perform the 'surgery' while the system is soft and flexible.

The preferred tool for lifting the leather is a bookbinder's paring knife of the German pattern made from a high-speed steel, machine hacksaw blade. The profile and rigidity of the blade is important for controlling the process. Flexible, disposable blades are not recommended.

Following the removal of the spine in one piece – or if this proves difficult, lifting of each panel and one-band segment, which should be identified by numbering on the tissue – the flesh side is levelled by sanding and paring. What often happens with ancient calfskin is the grain layers will separate from the flesh layer, in which case one may elect to remove it entirely or reattach it, particularly if this separation is localized. The flesh may be consolidated with Klucel G. If one chooses to remove a degraded and damaged fleshing, it may be peeled away leaving the (tooled) grain layer secured to the facing. In preparation for replacing the spine onto the new leather, every scrap of tissue not supporting leather is cut away with a scalpel. The spine can be glued directly to the dry rebacking leather. A better approach would be to mount the spine onto a suitably coloured Japanese paper, allow time to dry, pare, and then mount onto the new rebacking leather. This is a valuable approach if the surviving grain is rather lace-like, and provides a support and release layer for the future. Before facing the entire spine, the labels should be faced and removed. In the case of only partial detachment of the grain layer (Cains, 1994, Figures 5 and 6), the flesh should be glued down and allowed to dry before adhering the grain layer to its support. Only when all work has been completed and thoroughly dried is the tissue removed with acetone or xylene.

### 21.2.6 Treatment of the boards

When applying tissue to areas of the board to be repaired and underlaid with new leather, care must be taken to avoid tide marks from localized application of ethanol. It is recommended the tissue be activated with ethanol and applied without pre-wetting the leather. It may be pressed into place with a pad dampened with the solvent and then fingered to secure.

### 21.2.7 Adhesives

The traditional adhesive for new leather is flour (wheat starch) paste, a thick mixture (12% weight/volume) thinly applied. Moisture from the new leather must not be allowed to come into contact with the exposed flesh side of the binding's original leather. Strips of thin polyester must be inserted at every point of possible contact. After the work has dried overnight and the necessary paring/sanding of the edges completed, the flaps are glued or pasted down. For the sides, one may use paste; for the spine, a paste and ethyl vinyl acetate/polyvinyl acetate (EVA/PVA) mixture, say, 10% EVA for a durable and flexible bond is recommended. The object is to use an adhesive with the right working characteristics applied in a controlled way.

### 21.2.8 Offsetting

To control the amount of adhesive used, apply it evenly to a slip of polyester which is then inserted under the flap. The flap is pressed gently onto the adhesive, the polyester removed, and the flap pressed firmly onto the board. The spine and detached components can be pressed onto a film of adhesive applied to the surface of a (lithographic) paring stone. Of course, a brush may be used but the offsetting technique does reduce the need to lift and flex the leather flap. An artist's fan brush is also a popular tool for this purpose. Again, the adhesive must be allowed to dry completely before removing the facing. Fragments of grain remaining on the facing tissue are accurately cut from the tissue; the lacunae glued, and the fragment replaced. Some workers have used animal glue/gelatine (12%) for reattaching the grain layer to its substrate and as a local consolidant. A traditional bookbinding adhesive is thin animal glue to which a small quantity of paste is added to slow the drying time. After final cosmetic retouching with dye or watercolour, the repaired cover is given a sparing application of leather dressing. The late Dr Ron Reed gives a number of suitable recipes in his famous book (Reed, 1972). Because the wax/oil, British Museum-type dressing interferes with the facing process, application was discontinued in the Long Room project, although a recipe was given in the booklet documenting that work (Cains and Swift, 1988).

### 21.2.9 Board attachment

The joint tacket method of board attachment was first published in 1988 (Cains and Swift, 1988), the year of an important conference held at Corpus Christi College, Cambridge (Cains, 1994), and is one of several repair methods designed in the TCD workshop or in collaboration with colleagues.

Bernard Middleton was probably the first to illustrate a method for adding multiple threads to the existing cord or thong sewing supports to create a new slip (Middleton, 1972). This approach represents a family of methods using new thread or strands of cord lightly glued to the original supports and then secured by oversewing or resewing all along at several points down the spine. The new fibre should not be attached by oversewing directly to the original stations without the assurance that plenty of residual strength still remains in the original system.

### 21.2.10 Helical oversewing

Text blocks that are intact and unbroken but with cords or vegetable-tanned leather thongs totally degraded can be resewn in situ onto new supports. After lining the cleaned-off spine with pasted Japanese paper and thin fabric, centres of the sections are marked with slips of paper and new endbands sewn on as best practice recommends to provide slips as board attachment points. The original endband cores can be drilled and oversewn onto the text block. New supports are lightly glued to the original stations and oversewn onto the spine through random but regularly spaced intervals through the selected centre, or sewn all along for several gatherings at the joints and at intervals between. Oversewing is achieved by piercing the centre fold from inside and following the awl point with needle and thread back into the fold and exiting to the spine at the other side of the support. A centre, four or more gatherings down, is opened and the process repeated, creating a neat spiral of thread that embraces the original sewing threads, the linings, and the new supports.

The more usual approach is related to Middleton's method and has been illustrated in the postprint of the Corpus Christi Conference (Cains, 1994: 129, Figure 8). For large volumes, a substantial slip needs to be made of multiple threads. These are sewn around the original supports and through the centres of several sections at spaced intervals to distribute the load, starting at the lowest section. As one moves toward the joint, the lower threads are embraced to form a slip. In the illustrated example noted above, the slip has been passed through the first original lacing tunnel and then returned to the joint, divided and then tacketed over itself with the ends fanned out. To reinforce the first and last fold, a loose guard of slunk parchment or tough handmade paper may be inserted before sewing.

### 21.2.11 The joint tacket

The joint tacket system is also illustrated in the same publication (Cains, 1994), but is not described in as much detail as the variation of this technique published by Robert Espinosa and Pamela Barrios (Espinosa and Barrios, 1991). The joint tacket was the product of a moment's improvisation to deal with a book urgently needed for an exhibition. The original method was to drill a tunnel at the base of each saw kerf of the recessed sewing; the drill angled to exit at a point about 10 mm below and through the leather of the spine – or, more usually, having removed the spine leather, through the new lining.

A threading hook is formed by grinding a slot in one side of the eye of a needle, which is then inserting into a pin vice (to be discussed later). The hook is inserted from the spine to the joint, and about a foot of thread is passed through, forming a loop at the spine. (The reverse of this threading direction is considered better 'engineering' by my colleagues.) The two ends of the thread are passed through the loop and the hitch tightened. The geometry of the board attachment system was devised with conscious reference to eighth century binding practice.

When drilling the bifurcated lacing tunnels in the board, it is preferable to drill from the pastedown side. Mark the centre of the lacing station at the joint edge of the pastedown, then mark the centres of the two angled tunnels in relation to it, and drill from that side to the board joint edge. Drilling this way ensures a more symmetrical triangle will result. The threads are hooked through the tunnels and the ends wound together two or three times and tightened to form a tacket. Espinosa suggests a square, or reef knot.

The threads are cut off leaving a short length on each side of the tacket (or knot) which are frayed out, pasted down, and boned or lightly hammered after drying. The tacket form of twisted thread gives a flatter profile than a knot, but if a totally level result is required, a 'V' channel may be cut between the two tunnels following Carolingian practice.

### 21.2.12 Drills

Modern twist drills are often too short in relation to the gauge needed for our work, and can be brittle or tend to clog. The chuck of the flexible-shaft drill should be of the collet type; even the smallest of the Jacobs key chucks is too bulky for many jobs requiring acutely angled holes. The Foredom Electric Company (16 Stony Hill Road, Bethel, CT 06801, USA, tel. 01-203-792-8622; http://www.foredom.com/) offers a splendid range of tools, foot-controlled motors and handpieces.

### 21.2.13 Making the needle drill bit

Traditional flat drill bits are easily made from bookbinder's needles or other sources of hardened carbon steel wire (such as balance staff and pivot wire for clock making). We use wire in 0.9, 1.2 and 1.6 mm diameters in 7.6 cm (3 inch) lengths which is cut down to the length required.

The point of the needle is cut off with wire nippers and again at the base of the eye; the cutting edge is formed at this, the widest point of the needle below the eye. Secure the needle in a pin vice. The famous L.S. Starrett Company (1993) no. 240 series range is recommended with collet chuck capacity of 0–1.4 mm (A) and 0.64–1.9 mm (B). A bench-mounted 'hobby' grinder with 15.2 cm (6 inch) diameter wheel can be used, but great care must be taken to avoid blueing (softening) the carbon steel of the needle. When grinding with this

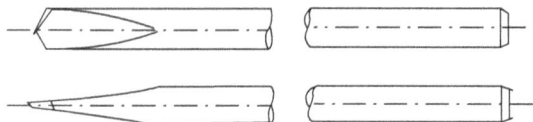

**Figure 21.1** The flat drill.

machine, hold the needle to the wheel for a second and then remove it for a few seconds to air cool without changing the position of the hand (a sort of pivoting action), and repeat until a hollow ground bevel has been made on one side. Turn the needle 180 degrees and repeat, being careful to ensure the symmetry of the ground edges. The drill point is ground with a few light touches of the grindstone or formed on a fine (Norton India) oil stone. The drill point should be central and the two bevelled cutting-edge angles equal. Inspect with a linen glass (lupe) to check the symmetry and angles compared to *Figure 21.1* given here.

## 21.3 Leather Conservation – bookbinding leather consolidants

*Glen Ruzicka, Paula Zyats, Sarah Reidell and Olivia Primanis*

### 21.3.1 Introduction

The repair of books and leather bookbindings is unlike that of stationary leather objects. Books are functioning structures, and, as such, conservation treatment in most libraries and archives must respond to physical demands for access to the text. Repairs made to an original binding are typically required to be flexible and strong. The joints of the book covers must hinge properly, and move easily, while still effectively attaching the text to the covers. Spines must be mobile. The materials used in book conservation must be long lasting and flexible. The use of stiff fill materials, such as coloured waxes, is impractical if the volume is to be used.

The challenge of conserving both the form and function of leather-bound books is daunting. As leather deteriorates, it becomes powdery and loses flexibility. This is especially typical of nineteenth and twentieth century vegetable-tanned leather used to cover books because of the types of tannins and acidic chemical processes employed to stabilize the animal skins. Often there is ample justification to simply house a deteriorated book in a polyester jacket and a box, and, if severely deteriorated, to rebind it preserving the original binding. But, if it is repaired with the intention that the book is to be used, conservators must address the powdery leather and detached book covers and spines.

### 21.3.2 ENVIRONMENT Leather Project

The ENVIRONMENT Leather Project (Larsen *et al.*, 1996) developed a helpful procedure for determining whether a leather binding should be treated with a consolidant. This procedure offers a practical method for characterizing the deterioration of leather by comparing a sample with images of four increasingly desiccated leather samples. *Figure 21.2(a)* recreates the Project's fibre samples portraying the fibres in a new leather sample as thick and long. *Figures 21.2(b)–21.2(d)* depict leather fibres in increasing intervals of deterioration, reduced in length and width, until eventually degrading to a powder. The ease or difficulty with which the few discrete fibres can be teased from the leather cover for examination also aids in assessing their condition.

### 21.3.3 Consolidants

As part of the repair process, book conservators can treat deteriorated leather bindings with the application of a consolidant. Solvents in dressings affect surface finishes by wetting, swelling, and deforming the leather, as well as dissolving or dislocating the original tannage components (McCrady and Raphael, 1983). The effect of consolidants is most apparent in changes that sometimes occur to the optical characteristics and flexibility of bookbinding leathers. Nevertheless, the application of a consolidant has become a practical response to the immediate problem presented by powdery leather.

Traditionally, leathers were treated using oil dressings to maintain their surfaces and keep them flexible. However, it has been shown that excess oil added to aged leather replaces the chemically bound moisture content of leather, actually causing further desiccation in the fibres (Hallebeek and Soest, 1981). Not only does oil dressing impede further conservation treatment, but it has been shown to weaken leather (McCrady and Raphael, 1983). Because of this, many book conservators sought alternatives to traditional leather dressings.

In the last 25 years, the most widely used consolidant for deteriorated leather ('red-rot') has been Klucel G (hydroxypropylcellulose). Klucel G is prepared as a 1–2% solution in ethanol or isopropyl alcohol.

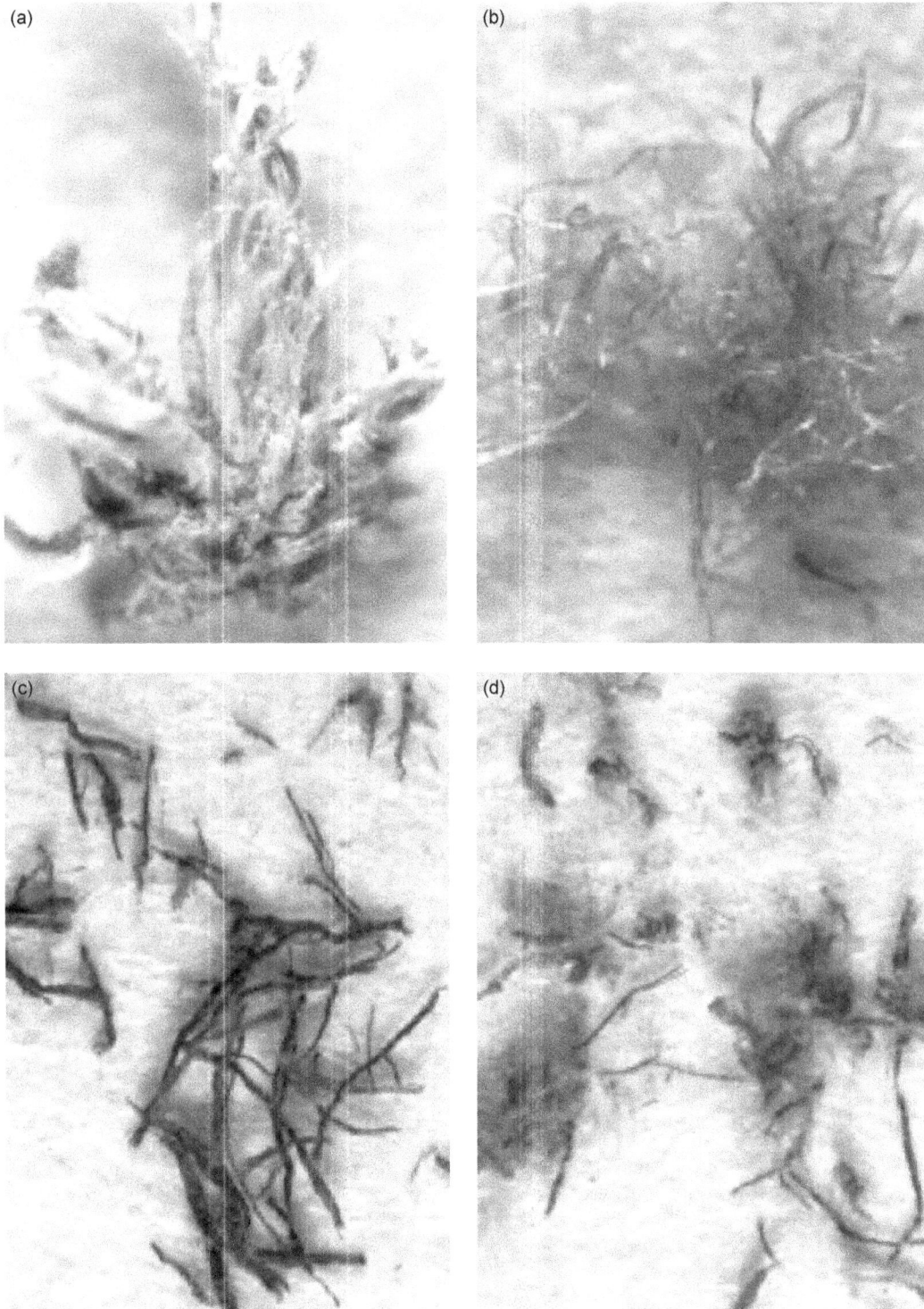

**Figure 21.2(a–d)** Fibre samples at the various levels of deterioration (4 × magnification).

Applied directly to deteriorated leather, Klucel G is usually immediately effective in consolidating the powdery leather dust with minor change in the appearance of the leather. In 1984, following the Foundation of the American Institute for Conservation (FAIC) Leather Conservation Intensive held in Harpers Ferry, West Virginia, book conservators became aware of the limitations of Klucel G (Haines, 1985). According to Pieter Hallebeek (Central Research Laboratory for Objects of Art and Science, Amsterdam), the film formed by Klucel G is known to break down in the presence of sulphuric acid, a by-product of leather deterioration. A 1991 *Journal of the International Institute for Conservation* entry by the Canadian Conservation Institute (CCI) referenced in Hein (1997) states that both Paraloid B72 and Klucel G-F have high stress values and should only be used on semi-rigid materials because stress fractures easily form. In light of this and other research (Haines, 2002), concern about the flexibility of Klucel G over time prompted conservators at the Conservation Center for Art and Historic Artifacts (CCAHA) in Philadelphia to search for alternative coatings.

Limited research in leather surface treatments to date has failed to provide clear guidance in selecting an appropriate treatment. In the 1980s, a new coating introduced by the Leather Conservation Centre (University College Campus, Boughton Green Road, Northampton, NN2 7AN, England; tel. 01604 719766), SC6000, seemed promising. SC6000, an acrylic polymer and wax emulsion dressing, imparted some strength and protection to the leather's surface. In practice, however, it was found that SC6000 did not sufficiently penetrate the grain layer of the skin. The comparatively strong wax and acrylic film that formed on the grain surface of deteriorated leather caused the grain layer to split from the corium. The finish of the original SC6000 was not only too strong for weakened skins, but also imparted an unwanted shine to the leather.

Conservators at the CCAHA experimented with a mixture that would improve the penetration of SC6000. Equal parts (1:1:1) of SC6000, Klucel G in a 2% ethanol solution, and ethanol blended together proved to be a practical solution. The SC6000/Klucel mixture proved effective even on very deteriorated leathers and satisfactorily penetrated the grain of the leather. An advantage of this mixture is that both the strength of the coating is diluted and the shiny finish toned down. Not only is the new strength more appropriate for aged bindings, but the gentle sheen looks more presentable. The SC6000/Klucel mixture was affectionately dubbed the 'red-rot cocktail'.

Because powdery, deteriorated leather is the endpoint of acid hydrolysis, it is irreversible. Treatment can only try to halt its further progress and reinforce or support the damaged structure (Jackman, 1982). Protecting the leather from air pollutants, which increase the hydrolysis reaction, can slow the process. The wax component of the SC6000/Klucel mixture does provide a level of protection for the leather. In less severe cases of red rot a mild humidification of the affected leather through a light application of methyl cellulose in water, or even thin paste, adds moisture and some flexibility to the skin. A coating of the lightweight SC6000/Klucel mixture applied to the humidified leather helps retain the added moisture, as well as provides a fairly durable surface. The coating does not penetrate into the skin appreciably, unlike oil dressings, which can cause further desiccation of leather (Zyats, 1996). It has been demonstrated that the SC6000/Klucel G mixture holds up well in recent research on coatings for leather (Haines, 2002).

The light sheen of the SC6000/Klucel mixture, combined with the protection afforded by the wax and acrylic in the emulsion, also works very well to coat acrylic-toned paper repairs used on damaged leather bindings. Paper repairs, if left untreated, generally have fragile surfaces and can become nappy with use. The SC6000/Klucel mixture effectively prevents this sort of abrasion, while the sheen of the finish visually integrates the paper repair with the original surface.

In general, the SC6000/Klucel mixture has been a very useful consolidant in the conservation treatment of fragile and aged leathers. The mixture can also be used successfully as a surface treatment for new leather and paper repairs, including those applied to parchment bindings and documents. Using the SC6000/Klucel mixture on added materials has become an integral part of CCAHA repair procedures.

## 21.4 Solvent-set book repair tissue

*Alan Puglia and Priscilla Anderson*

Solvent-set book repair tissue is an experimental treatment under development at Harvard University Library. The technique uses prepared Japanese paper that is colour toned and coated with acrylic adhesive. The dry acrylic adhesive is reactivated with non-aqueous solvents of relatively low toxicity, such as ethanol and isopropanol. The technique was developed for batch treatments of leather-bound, special collection materials and is a variation of

Etherington's Japanese paper hinge repair (see 21.5 below). The potential advantages of solvent-set book repair tissue include: quick application; strength and stability of the repair materials; ease in matching repair tissue to binding substrate; and reversibility of the adhesive. Also, a solvent-based adhesive is less likely to darken or more seriously damage aged leather as is common with water-based adhesives.

Acrylic adhesives as a general class retain some reversibility with either organic solvents or heat. Lascaux 498 HV (Lascaux Colours & Restauro, Alois K. Diethelm AG Farbenfabrik, Zürichstrasse 42, CH-8306 Brüttisellen, Switzerland; tel. +41-1-807 41 41; http://www.lascaux.ch/english/restauro/index.html) is currently used in the painting conservation field for lining original canvases to new support linings. Accelerated ageing and testing of Lascaux 498 HV indicates that although the adhesive does yellow somewhat, it does retain solubility (Duffy, 1989). Nonetheless, the use of this adhesive in this application should be considered experimental since its interaction with leather has not been tested.

### 21.4.1 Preparation of the repair tissue

The tissue selected for this technique should be a strong, long-fibred, 100% kozo paper. It should be reasonably transparent so the colour of the volume's leather will show through the finished repair. A paper that has proven successful is KTLG from the University of Iowa's Center for the Book Paper Facility (100 Oakdale Campus #M109 OH, Iowa City, IA 52242-5000, USA; tel. +1-319-335-4410; http://www.uiowa.edu/~ctrbook/), a smooth-finished kozo tissue about 0.05 mm in thickness.

The tissue is toned with thinned liquid acrylic paints to lightly tint the tissue resulting in a highly translucent material. It is only necessary to match the general colour family of the leathers, not to exactly replicate the hue and saturation.

Spread a thin, even layer of the Lascaux 498 HV adhesive onto polyester film by squeegeeing through a screen. If the resulting layer is too thick, reduce it by offsetting the adhesive onto another sheet of polyester film, or use a finer screen. A thin coat on a repair tissue is more flexible, conforms more closely to the leather surface, and is less likely to cause weak leather to delaminate than a thicker coating.

Lightly mist both the tissue and the fresh adhesive layer with water. Slowly drop the tissue onto the adhesive, avoiding wrinkles and bubbles. Surface tension between the two wet surfaces encourages full contact; pressing is not necessary. Allow the tissue and adhesive to air dry completely. The dried tissue may be left on the polyester film for convenient cutting and handling. If bubbles do occur, use a warm (not hot) tacking iron on the dried tissue to join it to the underlying adhesive layer.

### 21.4.2 Leather consolidation

As mentioned above by Cains, consolidation of deteriorated leather may be required to prepare the surface to accept and hold a sound repair. As an alternative to Klucel G, one part Lascaux 498 HV adhesive dissolved in six parts isopropanol yields a clear, slightly viscous solution that may work well as a consolidant. Preliminary experiments indicate that the Lascaux solution requires only one coat to consolidate most leathers, it leaves the leather flexible, and it provides a strong bond with the repair tissue. Though the use of Lascaux as a consolidant is experimental, its favourable working properties encourage continued research into the use of acrylics as consolidants.

### 21.4.3 Repair technique

Cut and shape a repair strip from the dried tissue with a scalpel and peel it off the polyester film support. Reactivate the adhesive film by brushing the adhesive side with either alcohol or the diluted acrylic adhesive, working on a non-absorbent surface. Place the repair tissue over the damaged area. If the repair is over a joint, gently lift the board immediately after applying the tissue to locate the natural crease line of the joint. The adhesive has sufficient slip to allow rapid adjustments to the placement of the strip, and then sets in under two minutes. After drying overnight, any lifting areas are easily touched up with a warm tacking iron. Additional toning with acrylics or watercolours, and/or surface finishing as described by Etherington may be undertaken if necessary.

### 21.4.4 Reversing solvent-set tissue repairs

Removing tissue applied with acrylic adhesives over weak, damaged leather should be approached with care. Lascaux 498 HV can be softened with alcohols and/or heat. Brush small amounts of solvent to one edge of a repair and slowly lift and work back the repair from the leather. Alternatively, a heated microspatula can be used to lift the repair.

### 21.4.5 Conclusion

Solvent-set book repair tissue has been used successfully to rejoin detached boards, to reinforce split joints and breaks in the spine, and to rebuild end caps. It is well suited for quick repairs and batch treatments. It is also useful for more complex treatments and may be combined with other repair methods, such as tacketing, when rejoining detached boards of larger volumes. The technique should be considered experimental until suitable testing of leathers treated with acrylic adhesives can be performed.

## 21.5 Split joints on leather bindings

*Don Etherington*

Faced with the problem of determining the best use of the conservation dollar, preservation administrators are always looking for ways to minimize the labour-intensive procedures that have prevailed over the last 50 years. While working at the Library of Congress, I was instrumental in developing the concept of 'phased preservation', a technique that protected, rehoused, and supported material en masse with, at times, some minimal treatment.

In pursuing this philosophy over the last 20 years, I developed various techniques that have been adopted extensively throughout the United States. The latest technique is the use of Japanese paper for reattaching or supporting weakened joints of leather bindings, particularly of nineteenth and twentieth century vintage, on books no larger than 25.4 cm (10 inches) high and 3.8 cm (1.5 inches) thick (Etherington, 1995).

The procedure is relatively simple and particularly effective on books that have a tight spine that would generally require skilled expertise and extensive time to repair. Anyone who has contemplated rebacking a tight spine, especially on a thin book with raised bands, will appreciate the problems associated with this type of work.

Many research library special collections have large groups of nineteenth and twentieth century bindings bound by French and English trade houses with detached boards or weakened joints, both inside and out. This is caused by the choice of poor-quality leather at the outset and the trade practice of paring leather very thin for ease in working and to satisfy aesthetic tastes. Usually, the spine itself is intact and the sides of the boards are in good condition; it is only at the joint that the damage is apparent.

To alleviate the time-consuming practice of lifting the leather spine and the leather from the sides of the boards, a Japanese paper strip is placed over the joint that extends slightly over the spine and the boards. Another strip of Japanese paper serves to strengthen the inside joint. The paper used for the outside is a very strong, solid-dyed paper with good tear strength; for the inside, natural-coloured Japanese paper is used to sympathetically match the endpapers or textblock. If it is necessary to match the original colour of the leather cover, some dyeing of the coloured paper may also be attempted.

Ideally, the dyed paper as produced by the manufacturer can be used, as there are currently some 35 or so colours available. The strips of Japanese paper for the inside are attached to the text block before attempting the outside repair. This is to make allowance for the ease of opening at the joint. The other portion of the inside hinge will be attached to the board at a later stage.

A strip of Japanese dyed paper is cut to size using a technique that produces a slightly feathered edge. An expeditious approach is to use a sharpened bone folder dipped in the standing water used to prevent PVA brushes from drying out. This mixture gives a well-defined line for tearing the paper strips. The strip is generally no more than 6.3 mm (1/4 inch) to 7.9 mm (5/16 inches) in width, and extends about 12.7 mm (0.5 inches) beyond the length of the boards.

The boards are placed in position on the book and a weight is placed on top. Paste or a mixture of rice starch paste and Jade 403 (or similar PVA) is used to attach the strip across the joint, rubbing down lightly with the palm of the hand so the paper sinks into the undulations of the damaged leather and across the edges of the raised bands. The feathered edge of the strip blends into the leather nicely. Let dry for an hour and then turn in the strip at the head and the tail. In most instances, I turn it down only to the height of the square of the board and then cut off the repair paper near the edge of the endpaper. If a leather binding exhibits red rot, treat the leather with Klucel G (as described by Cains in 21.2). This treatment is necessary, as books with friable red rot will reject the Japanese paper strip.

The Japanese paper is now attached to the inside of the board. This attachment can either be over the original endpapers or slid beneath them. I generally let the structure of the book indicate what is possible. Obviously, attachment under the original endpapers (both the free-fly end and the board paper) is the more sophisticated method. In general, if the need to lift the inside board paper is purely cosmetic, the added cost should be evaluated carefully.

After the book has been repaired, the leather and the repair strip are given a light surface coating consisting of a 1:1 mixture of Klucel G and SC6000 (Klucel G is a leather consolidant produced by the Hercules Chemical Company and is obtainable from the usual conservation supply houses; SC6000 is a plasticized acrylic polymer and is available through the Leather Conservation Centre, England). The paper repair is thus sealed, and a satin finish is obtained. If a higher shine is desired, a second, light application of pure SC6000 is appropriate, followed by a light buffing with a soft cloth. If the book is valuable and heavily gold tooled on the spine, an option for improving the aesthetic finish of the repair is to remove areas of the strip that may be covering tooling, and lifting the edge of original leather labels so the Japanese paper can be slipped underneath.

For books that are broken only at the joints, the rationale for making repairs with a strong Japanese paper instead of a pared strip of leather comes down to one word: 'strength'. The application of two strips of Japanese paper, one outside and one inside, tends to create a strong board attachment to the spine, and is a method that results in minimal intervention to the original binding. The technique normally takes about an hour.

## 21.6 A variation on the Japanese paper hinge – adding a cloth inner hinge

*Bill Minter*

While the reattachment of leather-covered boards with Japanese paper has been shown to be strong, there are times when it may prove inadequate. A variation of this practice is to attach a cloth inner hinge in a technique similar to one presented by David Brock at the AIC Book and Paper Group meeting in Albuquerque, New Mexico, in 1991.

The boards can be prepared in the usual manner, as described by Etherington (in 21.5). However, prior to attaching the board, the pastedown should be lifted to accept a cloth inner hinge that will be sewn to the text block. A strip of unbleached cotton muslin, cut on the bias, is adhered to the shoulder. Instead of positioning the cut edge at the top of the shoulder, in this variation the cut edge is positioned at the bottom of the shoulder with the free cloth extending upward.

This cloth is then mechanically secured by sewing in a unique manner with a so-called linen-thread 'staple'. Holes are stabbed (or drilled, as described by Cains in 21.2) through the cloth, through the shoulder, and through the leather in the following manner. Two holes, about 5 mm apart, should be stabbed in the least obtrusive location, such as along raised bands, avoiding any gold tooling. While stabbing, use a piece of rubber or plastic to support the leather, thereby eliminating the possibility of chipping the spine leather where the needle stabs through. Use as many pairs of holes as are needed for the book, and stab two additional pairs of holes near the head and tail. A piece of linen thread is passed through the pair of holes with the ends on the outside of the spine, and both ends are cut to about 20 mm. Secure the thread inside the shoulder on the cloth with a dab of adhesive. All pairs of holes are 'stapled' in the same manner.

Now, cut the ends to about 4 mm, fray out/fan out, and paste the tails to the leather. Fold the cloth along the shoulder and attach the board with the Japanese paper as described by Etherington. (Note: I prefer to use an undyed 100% kozo fibre Japanese paper because some dyed papers contain unstable groundwood pulp.) When the boards are secured with the Japanese paper, the cloth inner hinge can be inserted under the pastedown and finished in the usual manner. This cloth hinge can be hidden with a strip of Japanese paper. While retouching the Japanese paper on the outside, those frayed-out pieces of thread can also be retouched.

## 21.7 Split-hinge board reattachment

*David Brock*

The following is a technique I have been using at Stanford University for the past several years to reattach the boards of leather-covered, tightbacked books sewn on raised bands (Brock, 2001). It involves lifting the spine leather, board leather, and pastedowns at the head and tail only.

The steps are as follows:

1. At the head and tail spine panels, make a cut through the leather near the base of the band, going from shoulder to shoulder.
2. Lift the spine panels.
3. Pastewash the spine to remove any deteriorated spine linings and adhesive.
4. Paste a light Japanese paper lining onto the spine.
5. When the Japanese paper lining is dry, secure loose or broken endbands with thread.
6. Depending upon the size of the book, weight of the boards, and movement of the spine, adhere one or two cotton or airplane linen linings with PVA. The cloth should be cut on the bias and extend 12.7 to 25.4 mm (0.5 to 1 inch) beyond the shoulders, and go from near the base of the band to the end of the spine. If the spine has endbands, extend the cloth onto the endbands to help anchor them a little more securely to the textblock. It is important to work the PVA well into the cloth, as the strength of this mend

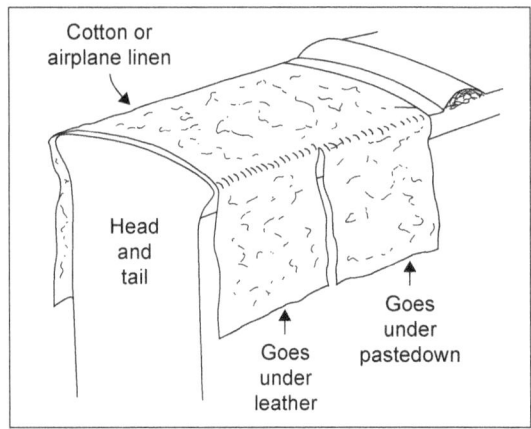
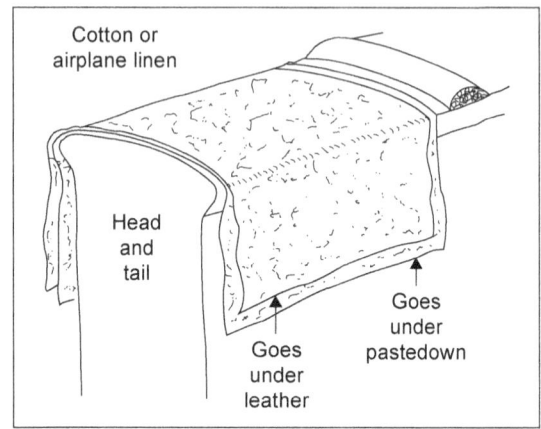

**Figure 21.3** Single spine lining.

**Figure 21.4** Double spine lining.

depends, in part, upon a strong bond between the cloth and the text spine. The cloth can be dampened slightly during bonding to achieve greater adhesion.

7. Determine now if further spine linings are needed to support the spine when the book is opened.
8. Lift the leather and pastedowns of the boards at the head and tail.
9. If only one cloth lining has been used, cut the extensions in half (see *Figure 21.3*).
10. Fray out the edges of the extensions.
11. Place the boards in position on the book. If one spine lining has been used, adhere the outer half of the extensions to the boards with PVA, going underneath the lifted leather. When using two spine linings, adhere the whole extension (see *Figure 21.4*). Allow to dry.
12. Open the boards and glue (PVA) the remaining extensions to the boards, going underneath the lifted pastedowns. When dry, readhere the pastedowns.
13. Before gluing down the lifted leather, adhere a wet-torn strip of Japanese paper over the cloth showing in the joint. I usually use a heavyweight Uda. This strip should begin slightly underneath the lifted board leather and extend a little way onto the spine. This will hide the weave of the cloth and add strength to the joint. Colour the Japanese paper to match the leather with artist's acrylics mixed in a little methyl cellulose. Mixing the acrylics in methyl cellulose slows down their quick drying time and makes their application easier.
14. Re-adhere the lifted board leather and spine panels.
15. Open the boards and paste a strip of suitable-weight Japanese paper in the hinge area, running from head to tail and from the base of the shoulder to the top of the board. This adds to the strength of the attachment and gives a neater appearance to the inside of the covers. The narrow line of paper that shows in the joint when the boards are closed can be coloured with the acrylic paint/methyl cellulose mixture to match the leather.
16. If the joint edges of the spine panels at the head and tail are visually distracting, a narrow strip of Japanese paper can be adhered just over the edges and then coloured to match the leather.

This method has been successful on board reattachments for both large and small books, with a few variations required depending upon the weight and size of the book. While this technique doesn't entirely replace leather rebacking in my conservation work, I use it much more often than traditional rebacking.

## 21.8 Board slotting – a machine-supported book conservation method

*Friederike Zimmern*

### 21.8.1 Introduction

Board slotting is a very useful approach for repairing leather bindings, especially for nineteenth century hollow backs. It also works well for tight back bindings, even those with raised bands, if the original spine can be effectively removed.

Board slotting was conceptualized by Christopher Clarkson in the late 1970s while working at the Library of Congress and later perfected while working at West Dean College (Clarkson, 1992). Consequently, the method was established at the Bodleian Library in Oxford in 1994 under the guidance of Edward Adcock (Simpson, 1994b). After working at the Bodleian Library in 1996, I was able to improve the method by designing and constructing a new board slotting machine in co-operation with an engineering company. This project was conducted in 1998 as a senior thesis at the Book and Paper Conservation Program at the Academy of Art and Design in Stuttgart, Germany (Zimmern, 1999).

### 21.8.2 The method

The board slotting technique is divided into three steps:

1. Treatment of text block
2. Treatment of boards
3. Reattachment of text block and boards.

### 21.8.3 Treatment of the text block

The first step is to detach the original spine, if it has not become separated already, and to remove damaged or loose spine linings. Paste a piece of pre-washed cotton fabric to the spine of the text block that is cut the length of the spine and overlaps either shoulder by approximately 2.5 cm. The cotton lining can be mechanically secured to the spine by sewing through the first and last sections. Any loose sections are reinforced by catching them up with the sewing thread. Next, a hollow tube cut to the height of the boards can be pasted to the spine. Paste a second lining of appropriately dyed cotton fabric to the hollow tube. Turn in this fabric at head and tail of the hollow and allow it to dry. After drying, paste the two fabrics together at their flaps (see *Figure 21.5*). If no hollow is used, turn in the second fabric at head and tail, and paste only the overlapping parts to the flaps of the first fabric. As a result of the pasting, two tongues will emerge that are taped down temporarily to a thin cardstock insert to remain flat following drying. The flaps should be well dried, preferable overnight, so they become stiff enough to insert into the slots in the boards.

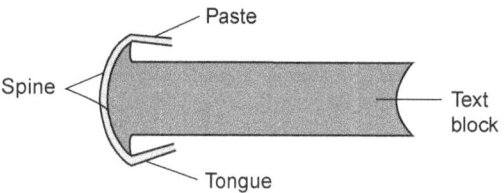

**Figure 21.5** Schematic diagram of the text block. The covers are taken off and a cotton fabric is pasted onto the spine. A second cotton fabric is worked over the first, but pasted together only at their flaps to form a tongue.

The question as to whether or not a hollow is necessary has to be answered individually, book by book, and depends upon how the book opens. If the spine construction is very weak, a spine lining and/or hollow may be appropriate; conversely, if the spine is quite tight, no additional spine linings or hollows should be used. However, a very light hollow can be made from thin Japanese paper that will be helpful in adjusting the new spine lining later on. If there is a piece missing from the original spine, a cardstock fill can be positioned under the second fabric on top of the hollow.

### 21.8.4 Treatment of boards

A milling machine is needed to create the slot in the spine edge of the board in preparation for insertion of the new material. The depth of the slot, which can be between 5 and 8 mm deep, is dependent upon the size and the strength of the boards. The slot, which is cut at an angle, only extends the height of the boards so the head or tail edges of the cover remain undamaged. The cutting begins almost directly underneath the original board leather and extends to the middle of the board thickness. The angle ensures the boards' function will not change and the book will open well after the repair is completed. The precise degree of this angle depends upon the board thickness and is easily calculated from the thickness of the board and the depth of the slot (see *Figure 21.6*).

### 21.8.5 Reattachment of text block and boards

The pasted cotton tongue is used to reconnect the text block and cover. Stiffened by the paste, the tongue is cut to the respective depth of the slot. The slot in the board can be opened a bit with a spatula to make insertion of the tongue easier. With a syringe, starch paste is fed into the opening and smoothed. Old book boards absorb quite a bit of paste

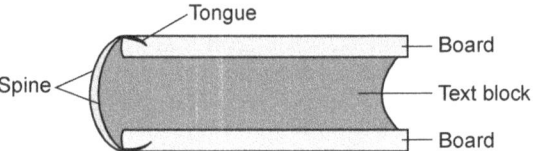

**Figure 21.6** Schematic diagram of a slotted book board. The slot starts close to the leather and continues to the middle of the board's thickness. Its depth can vary between 5 and 8 mm. The angle depends upon the board's thickness.

**Figure 21.7** Schematic diagram of a book treated with the board slotting method. The boards are reattached by pasting the fabric into the slot.

so an adequate amount must be injected. The tongue is then placed into the slot with the help of a spatula. The book is pressed carefully and left to dry. Finally, the original spine is cleaned of old spine linings and attached to the new cotton spine (as described by Cains in Section 21.2) (see *Figure 21.7*).

### 21.8.6 The board slotting machine

Board slotting requires a machine. Christopher Clarkson's first prototype was a commercially available milling machine that required a few adjustments to be adapted to board slotting. This concept worked well, but the setup was laborious, making the whole system relatively complicated and time consuming in practice. Part of my project was to develop a machine that would streamline the method. In collaboration with the engineering firm Becker Preservotec (now ZFB – Zentrum Fur Bucherhaltung GmbH, Mommsenstr. 7, D-04329, Leipzig, Germany; tel. +49-341-25989-0; http://www.zfb.com), a board slotting machine that met this objective was developed. In contrast to the machine in use at Oxford, the present model is simpler to handle and therefore saves preparation time (see *Figures 21.8* and *21.9*).

### 21.8.7 Scientific analyses

The board slotting method, as with any other type of conservation treatment, is only as good as its repair material. Therefore, the newly inserted materials required testing for their strength and stability.

#### 21.8.7.1 *Fold endurance test*

The fabric inserted into the board slot is required to flex at the joint each time the book is opened. For this reason, the fabric should possess very good flexible strength. Consequently, fold endurance tests were carried out with linen and cotton fabrics. The following three materials were tested: Aerolinen 155 g/m², Aerocotton 145 g/m² (Samuel Lamont & Sons Limited, Ballymena, North Ireland), and Aerocotton 100 g/m² (Friebe Luftfahrt-Bedarf GmbH, Mannheim, Germany). Three samples of

**Figure 21.8** The milling machine Emco FB-2, used for the board slotting technique by the Bodleian Library.

each material were tested with the grain direction of the fabric parallel to the folding. Each material was tested until the fabric tore, and the average of the three samples was taken. The test machine was an apparatus manufactured by the firm Louis Schopper of Leipzig, and was designed to bend materials 180 degrees under tension. This approach is a non-standard methodology, but the data are useful for comparing the folding characteristics of the three materials with the other.

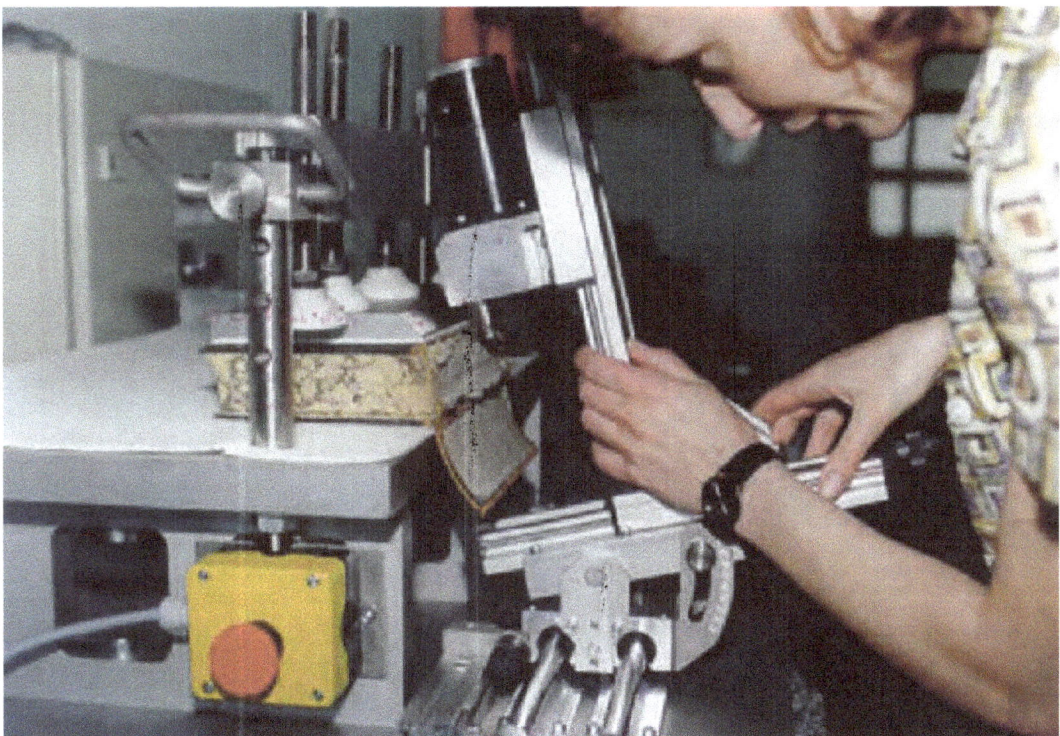

**Figure 21.9** The board slotting machine developed by Becker Preservotec in cooperation with the Academy of Art and Design, Stuttgart.

21.8.7.1.1 *Results*

| | |
|---|---|
| Aerolinen 155 g/m² | 16 000 folds |
| Aerocotton 145 g/m² | 90 000 folds |
| Aerocotton 100 g/m² | 71 500 folds |

The results show cotton has significantly better folding strength than linen, and even the thinner cotton fabric proved much stronger than the linen. Though linen has better tear resistance and tensile strength than cotton, linen fibres are inflexible and break much earlier than cotton fibres. Fold tests against the grain direction reduced the number of folds of each material by at least one third. No tests were carried out with the fibres diagonal to the folding movement.

21.8.7.2 *Tensile strength test*

A second parameter to be tested was the strength of the adhesive within the slot. Three different types of adhesives were tested: polyvinyl acetate, gelatine, and wheat starch paste. Under specially designed conditions, each adhesive was used with fabric samples that were subsequently pulled out of their board slots. This testing was done at the Institut fuer Textilchemie, Denkendorf, and conducted according to German standard 'DIN 53857'. Three samples of each adhesive underwent this test and the average was taken of the resultant data.

21.8.7.2.1 *Results*

| | |
|---|---|
| Polyvinyl acetate | 24.8 kg |
| Gelatine | 22.7 kg |
| Wheat starch paste | 19.5 kg |

These results demonstrated that PVA and gelatine were slightly stronger than paste. Nevertheless, wheat starch paste exhibited sufficient adhesive power to hold the fabric inside the slot, even when used in conjunction with heavy books. Because of its open (e.g. drying) time and well-known permanence, paste emerged as the preferred adhesive.

### 21.8.8 Dyeing with reactive dyes

The first lining of cotton fabric shows inside the book once the repair is completed, so it should

240  *Conservation of leather and related materials*

**Figure 21.10(a & b)** Detail image of two books treated with the board slotting method by the author. (Ioh. Pomtow (ed.) (1885). Poetae Lyrici Graeci Minores, vols 1 and 2.)

blend with the colour of the pastedowns and flyleaves. Cotton fabric normally has a cream colour that provides a sympathetic match for most flyleaves. The second fabric lining is visible in the outer joints and should therefore match the original colour of the book cover. This is always a difficult task to achieve because spine and board coverings tend to fade differently due to their dissimilar exposures to light. Therefore, a dyeing procedure with reactive dyes was adapted from the Bodleian Library (Simpson, 1994a) and modified slightly (Zimmern, 1999). Lightfastness tests were used and the dyeing procedure found to be acceptable. Still, the dyeing procedure proves to be very time consuming, and it would be advantageous if a commercial firm could be identified willing to produce a good, standard cotton fabric (like Aerocotton) and dye it in several tones with light- and water-stable dyes (*Figure 21.10*).

### 21.8.9 Conclusions

In contrast to other methods, board slotting has the following advantages:

1. Gilding and decoration on the cover as well as on the board edges are not affected by the process.
2. Original leather parts on the boards and the pastedowns are not damaged.
3. The method is applicable for degraded leather because the leather itself requires no manipulation.
4. In contrast to manual board slotting, slotting with the machine is much more exact and even. While slotting with the machine, parts of the cardboard cover are removed preventing the slot from swelling due to the newly inserted fabric.

As a result, one can say that the board slotting method is fast, economical, and able to return books to a durable and usable condition without changing the integrity of the object. With experience, it becomes apparent that board slotting does not change the appearance of the book significantly, and can be effectively combined with other methods, such as Japanese paper hinge repair for small books, as described by Etherington, Minter and Brock (21.5, 21.6 and 21.7 above respectively).

The board slotting method enriches the already well-known possibilities for reattaching book boards. It is hoped that it will be used more in the future.

### 21.8.10 Acknowledgements

Without the help of many people, this project would not have been possible. Sincere thanks are due to: Prof. Dr G. Banik, Ing. E. Becker, Dr A. Blüher, Ing. H. Dallmann, Ing. H-P. Gaibler, Prof. Dr G. Hardtmann, B. Hassel, Dr J. Migl, D-E. Petersen, Dr V. Trost, Prof. Dr J. Szirmai and Dr W. Wächter.

## 21.9 A variation on the board slotting machine

*Bill Minter*

There is a very nice hand tool from FEIN Power Tools, Inc., USA (1030 Alcon Street, Pittsburgh, PA

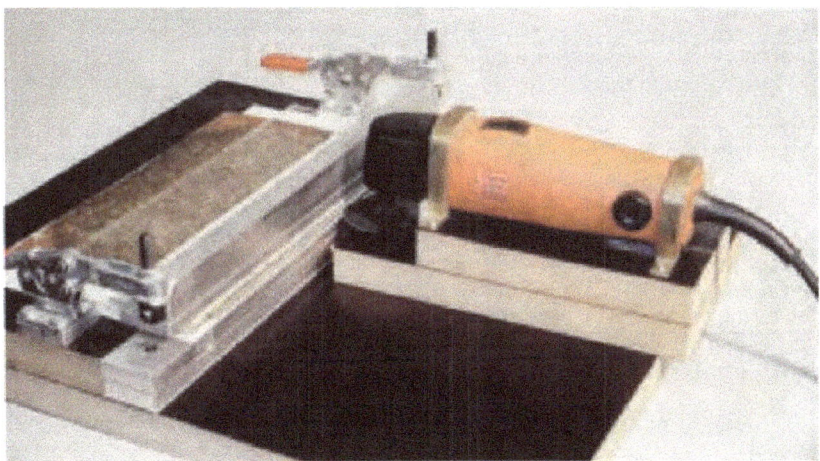

**Figure 21.11** Minter's mounted FEIN MultiMaster with oscillating saw blade for board slotting is advanced by hand to create a slot in the edge of the binderboard which is clamped in the jig.

**Figure 21.12** Mayer's mounted FEIN MultiMaster for board slotting from the University Library of Graz, Austria.

15220, USA; tel. 01-412-922-8886; http://www.feinus.com/) that can be adapted to board slotting. The beauty of this tool is that it is very safe to operate. The FEIN MultiMaster tool (MSXE 636-2) oscillates a saw blade in such a manner that it will cut and slot binderboard, but will not cut human skin. The tool is mounted on a platform and advanced by hand into and along the edge of the board while it is secured in a fixture (*Figures 21.11* and *21.12*).

A variation on this idea also employing the FEIN MultiMaster was subsequently developed by Manfred Mayer from the University Library of Graz, Austria, demonstrating the suitability of this tool for building an affordable board slotting machine.

## References

Brock, D. (2001) Board Reattachment. *Abbey Newsletter*, **24** (April, no. 6), 93.

Cains, A. (1981) Preparation of the Book for Conservation and Repair. *New Bookbinder*, **1**, 11–25.

Cains, A. (1992) A Facing Method for Leather, Paper and Membrane. In *Conference Papers, Manchester 1992* (S. Fairbrass, ed.), pp. 152–157. Institute of Paper Conservation.

Cains, A. (1994) In-situ Treatment of Manuscripts and Printed Books in Trinity College, Dublin. In *Conservation and Preservation in Small Libraries* (N. Hadgraft and K. Swift, eds), pp. 127–131. Parker Library Publications.

Cains, A. (2000) The Long Room Survey of Sixteenth- and Seventeenth-century Books of the First Collections. In *Essays on the History of Trinity College Library Dublin* (A. Walsh and V. Kinane, eds), pp. 53–71. Dublin: Four Courts Press.

Cains, A. and Swift, K. (1988) *Preserving our Printed Heritage*. Trinity College Library.

Clarkson, C. (1992) Board Slotting – A New Technique for Re-attaching Bookboards. In *Conference Papers Manchester 1992* (S. Fairbrass, ed.), pp. 158–164. Institute of Paper Conservation.

Duffy, M.C. (1989) A Study of Acrylic Dispersions Used in the Treatment of Paintings. *Journal of the American Institute for Conservation*, **28**(2), 67–77.

Espinosa, R. and Barrios, P. (1991) Joint Tacketing; A Method of Board Reattachment. *Book and Paper Group Annual*, **10**, 78–83.

Etherington, D. (1995) Japanese Paper Hinge Repair for Loose Boards on Leather Books. *Abbey Newsletter*, **25**(3), 48–49.

Haines, B. (1985) Bookbinding Leather Conservation. In *Recent Advances in Leather Conservation: Proceedings of a Refresher Course Sponsored by FAIC, June 1984*. Foundation of the American Institute for Conservation of Historic and Artistic Works.

Haines, B. (2002) *Surface Coatings for Binding Leathers*. Northampton: Leather Conservation Centre.

Hallebeek, P. and Soest, H.A.B. van (1981) Conservation of Caskets and Furniture Covered with Leather. In *ICOM Committee for Conservation, 6th triennial meeting, Ottawa, 21–25 September 1981: preprints* (vol. 4), p. 2. International Council of Museums.

Hein, J. (1997) Leather Treatment Procedures: An Excerpt from an Annotated Leather Bibliography. *Leather Conservation News*, **13**(2), 2.

Hercules Incorporated (1979) *Hercules KlucelS, Hydroxypropyl Cellulose, Chemical and Physical Properties*. Hercules Incorporated [Hercules Plaza, 1313 North Market Street, Wilmington, DE 19894-0001, USA; tel. (302) 594-5000; http://www.herc.com/].

Jackman, J. (1982) *Leather Conservation: A Current Survey*. Northampton: Leather Conservation Centre.

Larsen, R., Vest, M., Poulsen, D.V. and Kejser, U.B. (1996) Fibre Assessment. In ENVIRONMENT *Leather Project: Deterioration and Conservation of Vegetable Tanned Leather* (R. Larsen, ed.), pp. 113–120. Royal Danish Academy of Fine Arts, School of Conservation.

McCrady, E. and Raphael, T. (1983) Leather Dressing: To Dress or Not to Dress. *Leather Conservation News*, **2** (December), 2.

Middleton, B. (1972) *Restoration of Leather Bindings*. American Library Association.

O'Sullivan, W. (1992) Binding Memories of Trinity Library. In *Decantations: A Tribute to Maurice Craig* (A. Bernelle, ed.), pp. 168–176. Dublin: Lilliput Press.

Reed, R. (1972) *Ancient Skins, Parchments and Leathers*. London: Seminar Press.

Simpson, E. (1994a) Dying Aerolinen and Aerocotton with Reactive Dyes. *Paper Conservation News*, **71**, 7–9.

Simpson, E. (1994b) Setting up a Board Slotting Programme. *Paper Conservator*, **18**, 77–89.

L.S. Starrett Company (1993) *Catalogue no. 28*. Third edition. The L.S. Starrett Company [121 Crescent Street, Athol, MA 01331, USA; tel. 01-978-249-3551; http://www.starrett.com/].

Zimmern, F. (1999) Board Slotting. *Eine Maschinenunterstützte Buchrestaurierungsmethode*. Deutsches Bibliotheksinstitut, Dbi-Materialien, 184.

Zyats, P. (1996) *The Effects of Coatings or Dressings on Degraded Bookbinding Leather*. Graduate student project. Winterthur/University of Delaware Program in Conservation.

# 22

# The conservation of archaeological leather

E. Cameron, J. Spriggs and B. Wills

## 22.1 Introduction

### 22.1.1 The archaeological context

It is a paradox that leather survives under extreme conditions and that desiccated and waterlogged contexts both provide excellent conditions for salvage. Preserved organic objects reveal much about the past, not least because the majority of cultural artefacts were (and still are) made from organic materials. The survival of a representative selection of such ancient material is unusual, and where it occurs it illuminates a culture in a special way. Ätzi, the Bronze-age ice-man discovered in the Tyrolean Alps in 1991, possessed a range of garments and utensils that revealed a sophisticated use of different skins (Spindler, 1994).

The infrequency of such preservation may lead to misinterpretation; skin products from the Pharaonic period have given the ancient Egyptians an undeserved reputation for excellent leather technology (Driel-Murray, 2000). Extant evidence which is common in some cultures but rarely found in others (perhaps more sophisticated) may thus distort our cultural perspective. The balance can be redressed by the identification, preservation and interpretation of organic remains, including leather, even (perhaps especially) if these are fragmentary and incomplete. Detailed investigation and conservation is both possible and rewarding.

Methods used in the retrieval and treatment of archaeological finds are designed to protect the integrity of the remains, however insubstantial, in order to preserve their scientific value. The process, starting with on-site retrieval and recording, progresses to research and conservation, but as conservators have an input at each of these stages this chapter sets out to explain their involvement and show what they aim to achieve.

### 22.1.2 Leather technology and material culture

Methods of skin preparation and ways of working leather, or applying it on composite objects, normally follow cultural traditions which are influenced by availability of materials and by the character of the local economy. Study and analysis of evidence of materials, methods and traditions in archaeological leatherwork not only helps to indicate its probable origin and date but can also contribute significantly towards a better understanding of the culture and economic background from which it comes. The manner in which leather was originally processed is an important line of enquiry.

#### 22.1.2.1 Tannage

Hardwearing vegetable-tanned leathers, fine alum-tawed skins, oil-tanned and untreated skins all have their place in developing societies depending on geographical location and technological progress, rather than date. All of these types might survive dry burial and/or peat bogs, but only vegetable-tanned leathers commonly survive wet anaerobic deposits of a more general nature. For this reason collections of archaeological leather from European sites are strongly biased towards those prepared by vegetable tanning. Recent analyses show that tannage can be identified, or sometimes inferred, when evidence is handled sensitively (Groenman-van Waateringe et al., 1999; Driel-Murray, 2002; Mould et al., 2003). The identification of material evidence for tanning from archaeological remains is hampered by problems, some of which are pointed out by Cameron (2000), Mould et al. (2003) and Wills (2002).

#### 22.1.2.2 *Currying*

Currying, which involves the working of oils into the grain surface of vegetable-tanned leather and the paring down of hides to a standard thickness, rarely leaves any evidence (but see Mould et al., 2003). Consequently we know little of the history of currying or of the relationship between early tanners and curriers in England or elsewhere.

#### 22.1.2.3 *Leatherworking*

Apart from the half-moon knife and diamond-shaped awl few hand tools of medieval or earlier date are exclusive to leather workshops. Many, such as traditional hog's bristle sewing needles, may have been of organic materials that do not normally survive burial. Since the character of leatherworking can vary widely between different cultures all kinds of observations relating to technique, levels of craftsmanship and artistic skill are useful for typological classification. This extends to the embellishment of leather in various ways such as tooling, embroidery, gilding or the attachment of decorative fittings. Obscure methods used for moulding and hardening leather (so-called *cuir bouilli*) for the production of medieval armour as well as other items of a more domestic nature continue to arouse curiosity (Cameron, 2000; Dobson, 2003).

#### 22.1.2.4 *Leather dyes and pigments*

Dyes and pigments do not often survive burial but examples include two sheaths with red pigment from London (Cowgill et al., 1987: MOL cat. Nos 479 and 487), dyed leather from Kerma, ancient Nubia (Wills, 2002) and a red-painted girdle from York (Mould et al., 2003: cat. no 15872).

#### 22.1.2.5 *Environmental and social questions*

When leather is found in large quantities, typically in Europe from waterlogged medieval urban contexts, there are opportunities to analyse the social setting by linking artefact studies with other forms of environmental evidence. Patterns of animal husbandry such as seasonal or age-related butchery and connections with the urban food supply can be revealed through bone evidence (McGregor, 1998). In studying leather it is sometimes possible to identify species, animal age (approximately) and to observe its general condition while variability in the leather assemblage may answer questions relating to standards, quality control and industrial organization. Large collections of shoes can also be used for population studies, including statistical analyses of foot sizes and evidence for foot defects. Even the frequency of cobbling, the term for patching and repairing worn items to extend their useful life, can offer fascinating insights into social habits and practice.

## 22.2 Wet leather

### 22.2.1 Condition

The condition of excavated leather is due to a combination of pre- and post-depositional factors, such as method of manufacture, in-use wear and tear, and extent of decay during burial. Other factors, such as imperfect tanning, and the animal species and body area that the leather derived from, are also important. For these reasons, leather is found in all states of preservation, but if it survives at all, it should be possible to preserve it intact.

Apart from splits, tears and other types of physical damage, a curious type of degradation regularly encountered in waterlogged leather is that of delamination where the leather falls apart into two separate layers. This condition, widely reported to affect archaeological leather of all periods, sites, object types and animal species, has sometimes been blamed on poor tannage, but is more likely to be the result of partial degradation at a point of weakness of the collagen fibre weave in the leather structure (Spriggs, 2003; Ganiaris et al., 1982).

Waterlogged leather is typically almost black in colour having become saturated with humic matter from the burial medium and the products of reactions between ferrous ions and tannins. Leather from marine environments can be found in a variety of paler brown colours. Leather which has been heavily compacted in the ground becomes ingrained with silt, clay and other materials which obscure the surface and can be difficult to remove. Iron salts and encrustations can also disfigure surfaces, and marine concretions from shipwreck sites can become so firmly attached to leather as to obscure the true form and nature of objects (Jenssen, 1987).

The occasional finds of bog bodies, both in the UK and Denmark, demonstrate that raw hide can survive in wet contexts although in these cases the human skins are thought to have been preserved by more than one factor. These include partial tanning in the tannin-rich bog deposits, the bacteriocidal properties of low pH, and the crosslinking of collagen fibres in skin by sphagnan (a tanning agent found in sphagnum moss and its peat derivatives) (Painter, 1991; Turner and Scaife, 1995).

### 22.2.2 Preserving wet leather before treatment

#### 22.2.2.1 Storage

To avoid irreversible shrinkage and embrittlement caused by uncontrolled drying, wet archaeological leather requires wet storage from the moment of its discovery. Leather finds, still dirty from the ground, can be double wrapped in polythene with a little water, and stored in plastic boxes or tubs with airtight lids (Spriggs, 1980; Watkinson and Neale, 1998). These boxes should themselves be stored in cool (1–5°C), dark conditions. Unless the leather is due for conservation soon after discovery, the boxes must be checked every month or so, to ensure that their contents are still wet, and free of fungal growth. Opinions are divided on the use of fungicides. While Edwards and Mould (1995) advocate the disposal of leather left untreated after six months, broad-spectrum fungicides such as Panacide and Adesol are employed to good effect in many institutions, given adequate safety and disposal procedures. Cold, but not frozen, storage may also be recommended to maintain less degraded leather in good condition until it can be treated. The use of waterproof and fade-proof labelling for wet-packed materials is essential – Tyvek (spun-bonded polythene) labels with solvent-based waterproof markers proving the most adaptable (Jones et al., 1980). Marine leather should be desalinated by passing it through changes of fresh water soon after discovery, the rate of salt removal being monitored with conductivity or chloride meters until all trace is removed (Pearson, 1987).

Leather should be handled as little as possible in the wet state owing to its fragility and the weight of attached soil. Composite objects, such as complete shoes, require support on a lightweight base of some waterproof material, such as Correx (fluted plastic) board to prevent the various elements falling apart. Most leather will originally have been made up into the finished object by stitching with flax or hemp thread, which may have degraded or disappeared altogether.

#### 22.2.2.2 Washing and recording

Most waterlogged leather is physically sound enough to be washed manually using a gentle stream of running water and soft brushes. Very fragile objects, which could be damaged even by gentle brushing, are sometimes cleaned ultrasonically (in a laboratory), supported on fine plastic mesh.

Leather is often bagged on site in context groups and after cleaning it is sorted into categories such as objects and object parts; offcuts and leather production debris; and featureless scraps. The advantages of recording leather before treatment are threefold – security against loss of identification, a basic visual record of the collection, and measurements (against which shrinkage that might occur during treatment can be checked). The commonest method is to trace around the wet leather in pencil onto plastic drafting film. By tracing around the object again with red crayon after treatment any shrinkage can be calculated as a function of either dimensional or area change (Ibbs, 1990). Condition and technological detail may be added to the tracings using pencil and coloured crayons, following a series of conventions (Goubitz, 1984; Spriggs, 1987). These drawings can serve as a quick visual way of auditing the collections and their condition. Many small but important and interesting details may be observed during the cleaning and drawing of wet leather and simple drawings are often the best way of recording them (*Figure 22.1*). Radiography can be useful for recording composite leather objects, such as hobnailed Roman shoe and sandal soles, and also for examining leather with mineral concretions from marine contexts (Spriggs, 1987).

#### 22.2.2.3 Assessment

Leather is assessed before treatment and the assessment report acts as a permanent record. The purpose of an assessment is:

(a) To make a list of the collection and provide a basic record of what the collection contains in terms of range of object types and manufacturing waste.
(b) To evaluate the importance of the collection in terms of its information potential and therefore research value.
(c) To record its general condition, the range of preservation and to identify any particular pathologies.
(d) To identify the resources (time, equipment, materials) required to carry out conservation work and future collections' care needs.

#### 22.2.2.4 Selection

Waterlogged leather is not found so often that the application of a selection and disposal procedure

**Figure 22.1** Technical drawings of cleaned waterlogged leather are made before treatment as a permanent record of the leather 'as found'. (Courtesy York Archaeology Trust)

need be applied and so there is normally a presumption that all preserved leather be kept. On the very rare occasions that extremely large assemblages of leather have been found, the decision has occasionally been taken not to treat the waste pieces of hide from shoe-making and shoe-mending, but such decisions are normally made in consultation with a leather specialist and museum curator.

### 22.2.3 Past treatments

Old treatments for wet leather are rarely well documented but are of interest to present-day conservators faced with old collections that appear to be in poor condition, actively degrading and in need of retreatment (Williams et al., 1999). The progressive introduction of conservation techniques is charted here by following the development of leather conservation at the York Archaeological Trust (YAT) (Spriggs, 2003).

Large assemblages of archaeological leather, discovered in the deep anoxic deposits that underlie much of the modern city of York, have been systematically conserved, studied and published over the past 30 years by the York Archaeological Trust. Its methods and materials have tended to mirror techniques being employed at other laboratories, both in the UK (Ganiaris et al., 1982) and in Scandinavia (Peacock, 2001). The emergence of international specialist research groups such as the ICOM Conservation Committee Working Group on Waterlogged Organic Archaeological Materials have assisted in raising standards and creating collaborative research projects.

#### 22.2.3.1 Solvent dehydration techniques

Early treatments for waterlogged leather involved the gradual replacement of water with oil, such as castor oil. The first methods employed at YAT were based on the staged replacement of water in the leather with a solvent of a lower surface tension in which various proprietary leather dressings might also be dissolved. These treatments, developed from contemporary published methods (Reed, 1972; Waterer, 1972), were applied to several assemblages of leather from York during the 1970s and the results were considered successful at the time.

The practice at York was to pass the leather through several baths of acetone until, by measuring the density of the acetone with a floating hydrometer, dehydration was judged complete. The leather was then passed through a tank of proprietary leather dressing, dissolved in an appropriate solvent, such as mineral spirits (also known as white spirit) and 1,1,1-trichloroethane (Genklene), both being miscible with acetone. Of the dressings used, products such as 'Guildhall Leather Dressing', 'British Museum Leather Dressing' (BMLD) and 'Pliantine' were used, the latter becoming, for a while, the favoured material at the YAT laboratories (Spriggs, 2003). These were all based on lanolin with varying quantities of other wax or oil constituents in hexane solvent. The dressings were applied both by immersion and by subsequent brushing or wiping on, in order to plump up the surfaces and improve the rather dry appearance. Although some shrinkage was always experienced with these treatments (typically 5–10% linear reduction), the properties of texture, feel, flexibility and weight were all considered satisfactory. However, there was a feeling of unease at the time that the dressings being applied had been formulated for use on historical and ethnographic leathers, rather than archaeological leather. Only in retrospect, many years later, have these treatments been found to have had a seriously deleterious effect on the objects (see below).

Another promising product was Bavon which had been developed for use in the leather processing industry as a lubricant and waterproofing agent for chrome-tanned leathers. Based on a mix of alkylated succinic acid and mineral oils it was available in two forms: Bavon ASAK/ABP, the solvent-based version, and Bavon ASAK 520S, its emulsified form (Peacock, 2001). Tests of the water-based treatment were unsatisfactory. Severe shrinkage and embrittlement occurred, and a white deposit of emulsified oils remained on the surface after freeze-drying which was very difficult to remove. However, test results on the use of the solvent-based ABP version were more promising (Rector, 1975), and a system using Bavon ASAK/ABP was developed and used extensively in York until the advent of freeze-drying (see below). This involved pre-treatment with dilute hydrochloric acid or EDTA (to remove silt and lime concretions and to improve colour) followed by solvent dehydration and then immersion for several days in dilute baths of Bavon ASAK/ABP and lanolin. After removal from the bath pure Bavon ABP was rubbed into the surface and the leather was then set aside to dry in a fume cupboard (Spriggs, 2003).

The quantities of leather being excavated in York and elsewhere in the UK by the late 1970s, coupled with mounting concerns about health and safety and the cost of solvent-based treatments, led to increased efforts to find alternatives which did not involve the use of organic solvents.

#### 22.2.3.2 *Freeze-drying from glycerol and PEGs*

A method of treating wet leather based on heated baths of polyethylene glycol (PEG) was developed at the British Museum (Muhlethaler, 1973). However, owing to the necessity of providing heat, both to keep the PEG molten and to drive off the water, the leather tended to shrink and warp badly, and tests at York resulted in the process being abandoned there (Spriggs, 2003).

Reports of successful treatments involving vacuum freeze-drying spurred conservators at YAT to experiment for themselves (Organ, 1958; Sturge, 1973; Rosenquist, 1975). Early tests showed that good results could be obtained with both glycerol and the lower molecular weight PEGs, once the optimum quantity of these materials had been introduced into the leather. Based on these successes, a freeze-drying unit was acquired in 1979, making YAT one of the first archaeological laboratories in the UK to use the technique on a routine basis (Spriggs, 1981).

### 22.2.4 Present-day conservation treatments

If archaeological leather is allowed to dry naturally from the wet state, contractile forces will cause it to shrink dramatically and become stiff and hard, a state which is difficult to reverse. The aim of any stabilization treatment for wet leather is to render it dry while preserving its size and shape, preferably by doing as little to the leather as possible (minimal intervention). The conservation process should also aim to render the leather suitable for occasional handling, study and storage in normal museum conditions through improving its physical strength and flexibility and its resistance to damage from environmental fluctuations. Further work might involve reshaping crushed pieces and other measures undertaken to improve the appearance of important items wanted for exhibition.

Freeze-drying is normally preceded by treatment with a water soluble compound such as glycerol or PEG 400. In this two-stage process, just enough glycerol or PEG is used to give dimensional stability during freeze-drying and also to ensure some flexibility and cohesion to the leather structure. It also acts as a cryoprotector as it protects the leather structure from damage from ice-crystal formation and expansion during rapid cooling prior to the freeze-drying process.

In vacuum freeze-drying ice crystals in the frozen product are encouraged to sublime at reduced pressure by applying a vacuum. This drying overcomes the natural forces of surface tension inherent to liquid water, and can be made to happen quite rapidly (Adams, 1994). Freeze-drying is quick and cost effective. It allows batch treatment as opposed to each piece being dealt with separately at every stage of the treatment – a great boon if dealing with large quantities (*Figure 22.2*). Non-vacuum freeze-drying requires simpler equipment but can also give good results although it is much slower (Storch, 1997).

The appearance of freeze-dried leather is dry and fine detail of decoration, manufacture and wear is well preserved and highly visible.

Glycerol and the lower molecular weight PEGs act as cryoprotectors during freeze-drying, and as lubricants and humectants afterwards. Further dressings, such as PEG 1500, can be applied to the surface

**Figure 22.2** Preparing a load for accelerated freeze-drying: a temperature probe is being inserted into a sample waterlogged leather piece. (Courtesy York Archaeology Trust)

for cosmetic purposes or as consolidants. Any leather item that is fragile or likely to shed fibres may require further consolidation in order to strengthen it.

22.2.4.1 *Reshaping and reconstruction*

Leather reconstruction is a delicate and time-consuming business normally undertaken by conservators for specific purposes, but as it requires much direct handling of the leather the potential for causing damage is high. Many leather items, mainly shoes, have been reconstructed at York for various exhibitions and on each occasion time spent in identifying the best techniques and materials for joining, lining, supporting, stitching and padding out these objects produced fine results (Spriggs, 2003) (*Figure 22.3*).

Large shoe components and other finds of sheet leather, often recovered in a crushed, bent or creased state, sometimes need to be reshaped. These can be pressed between sheets of card or Correx board after

**Figure 22.3** Freeze-dried and reconstructed leather footwear from the Coppergate excavations, York. (Courtesy York Archaeology Trust)

the glycerol pre-treatment so that the new shape becomes 'set' during freeze-drying.

Freeze-dried leather that has been pre-treated with glycerol will accept adhesives readily and reshaping is also possible by partial rehumidification. One problem regularly encountered is that of creases in the leather that, despite best endeavours, persist through the treatment process. These can be removed to an extent by applying gentle tension to the leather after relaxing it in a high-humidity environment. The possibility of rehumidifying, and so relaxing, the leather structure is one of the advantages of employing a hygroscopic lubricant such as glycerol in the initial treatment (Peacock, 1983).

### 22.2.4.2 Marine leather

Once it has been desalinated, leather from marine environments may normally be treated very similarly to any other sort of waterlogged leather. However, the leather may be disfigured with marine concretions and iron staining which may need to be removed. Mechanical techniques using scalpels, modified engraving tools and dental descalers (applied with great care) may be effective but chemical cleaning techniques are often necessary too. Chelating agents, mineral and organic acids as well as other compounds have been used in the past (Jenssen, 1987) but many of these have an unknown effect on the leather, and some are very hazardous to health. Other reagents, such sodium di-thionite and ammonium citrate are used for iron-stain removal and research has demonstrated their effectiveness on marine leather prior to freeze-drying (Godfrey et al., 2002).

### 22.2.4.3 Composites and special items

Leather objects incorporating metal fittings, such as shoe and belt buckles, studs and harness attachments, are frequently found on excavations. Conservation treatments and storage environments for leather and metals tend to be incompatible and leather is normally given priority. At York it was felt that the solvent-based treatments for leather would not adversely affect metals, whereas the aqueous PEG treatments were likely to be more cause for concern, especially for iron. In a YAT survey of six leather items with iron fittings, the iron on the three that had received solvent-based treatments had survived well. But the iron fittings on the three treated with glycerol and freeze-dried showed signs of post-conservation corrosion damage. A further ten items with copper alloy, tin and lead/tin alloy attachments had not suffered any obvious post-conservation corrosion damage (Spriggs, 2003).

In archaeological deposits where Roman leatherwork is preserved, the soles of *carbatinae* and other hobnailed shoe and sandal types present a particular conservation problem. The partial corrosion of the hobnails may often impregnate and embrittle the surrounding leather of the multi-layered sole structures, often to the extent that they are liable to separate and fragment and the hobnails lose their positions during the cleaning process. X-radiography of the sole units before cleaning will at least record the original positions of the hobnails. After that, only very careful support during handling, and custom-made packaging after stabilization, will protect the sole from disintegration (Spriggs, 1987).

The use of water soluble iron corrosion inhibitors, such as Hostacor IT, are used at some centres for both the temporary storage and water-based treatment of leather/iron composites (Starling, 1987; Degrigny et al., 2002). Variable levels of protection seem to be obtained and a search to find the best inhibitor and the optimum conditions for its use continues.

Raw hide and other non-vegetable-tanned hides occasionally survive in waterlogged contexts. As an example, a crumpled mass of skin complete with hair was recovered from Viking levels at 16–22 Coppergate, York, in 1978. It was identified as being calfskin and analysis of a fragment using infrared spectrophotometry showed that it was heavily impregnated with Stockholm tar (softwood resins), the bactericidal properties of which may well explain its survival. The hide required repeated immersions in a chelating agent to release the folds, a process that was only partially successful, followed by treatment with PEG 1500 and freeze-drying (Spriggs, 2003). Most of the recent finds of bog bodies have been preserved using a similar system of PEG impregnation followed by freeze-drying (e.g. Omar et al., 1989).

#### 22.2.4.4 *Comparisons of techniques*

It is always instructive to compare the results of dissimilar treatments to the same material. In York, a recent examination of some of the pieces treated with Bavon ASAK/ABP in the 1970s concluded that the results were not inferior to the freeze-dried product. Despite varying widely in condition as-found, many pieces are still soft, pliable, dry in feel and appearance, and a rich brown colour. There has been no undue crumbling or fragmentation, and the original packaging is unstained. Dimensional shrinkage was between 3 and 10%.

Adverse comments have been made about the freeze-drying of leather, both in terms of the unnecessary complexity of the freeze-drying process and equipment (Goubitz, 1997), and, more importantly perhaps, concerning the condition and appearance of the dried leather (Swann, 1997). Most archaeological conservators, although convinced that freeze-drying is the most reliable technique currently available for treating waterlogged leather, would welcome new developments in this field.

#### 22.2.4.5 *New approaches*

Conservators are continually searching for more effective treatments and substitutes for PEG 400 and glycerol. A novel technique developed from medical research and used to preserve histological specimens is that of 'plastination' (Smith, 2003). This method involves the dehydration of waterlogged leather through baths of acetone, which is then replaced with silicone oil (under vacuum). This oil can be polymerized into a solid either by introducing a catalyst vapour (under vacuum) or by applying liquid catalyst to the skin surface. The treatments are reported to give a dimensionally stable, flexible and aesthetically pleasing result, but there are concerns about the long-term stability of the polymer, as well as the fact that it is not reversible (Baltazar et al., 1996).

## 22.3 Dry leather

### 22.3.1 Condition

Dry-preserved leather is often found in, though not exclusive to, hot and arid regions of the world. Temperate regions also provide examples of protection from moisture, such as Roman shoes from a burial in Southfleet, Kent (British Museum registration numbers PRB 1836, 0213.17–19). The shoes were preserved inside a child's stone coffin although an associated textile was 'reduced to tinder' (Rashleigh, 1803; Walker, 1990). In good condition, the purple colour, form and stitching, including gold stitching, was largely intact (*Figures 22.4* and *22.5*). Medieval and later shoes which have been concealed in buildings also show an exceptional degree of preservation. Leather in consistently dry burial conditions has achieved a degree of stability within its environment. Collagen, the principal component of leather, is inflexible in the desiccated state; it is unaffected by fungi, bacteria and hydrolysis, the major

252  *Conservation of leather and related materials*

**Figure 22.4** Roman shoes from Kent, third century AD.

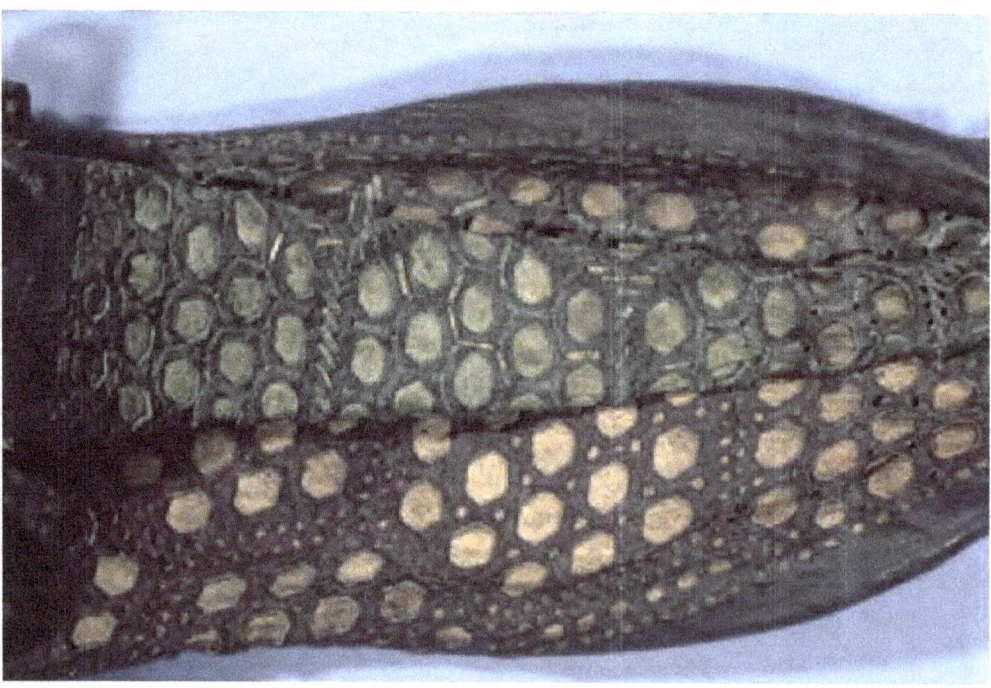

**Figure 22.5** Detail of third century AD Roman shoes from Kent, showing reticulation and gold thread.

**Figure 22.6** Newly excavated post-Meroitic leather, c.350 AD.

elements of decay (Cronyn, 1990). Desiccation allows leather to keep some of its distinctive qualities; colour, form, surface finish, tannins and, occasionally, flexibility. Less ideal contexts result in a loss of elasticity and fragmentation. Ancient desiccated leather may be found in remarkably good condition, or, at the other extreme, so brittle that it cannot be handled without damage (*Figure 22.6*).

#### 22.3.1.1 *Deterioration*

Change, however, does occur over time; much of the water content (typically 15% for new leather) is lost under desiccating conditions, leading to inflexibility and permanent damage. 'Leather' that is lightly cured ('pseudo-tannages' such as skins treated by smoking, dressing with mineral earths, salt, animal fats or oils) may survive better than vegetable tanned. In general, relatively unaltered collagen survives much better than highly modified skin or leather, which are variable in their stability (Horie, 1990).

If skin is better preserved than leather in dry archaeological and historical contexts, this may be because of its physical as well as its chemical structure. The compacted fibres of a dried skin allow less interaction with the environment than in the more open structure of leather. In effect, a smaller surface area admits fewer pollutants.

Consistently dry environments, predictably, preserve organic materials more successfully than intermittently wet contexts such as those subject to rain, ground water, or close contact with a decaying body (Wills, 2001). High temperatures and RH cause hydrolytic degradation resulting in shrinkage, fibre collapse, embrittlement and cracking (Cronyn, 1990); the more frequent the cycle of wetting and drying, the greater the damage. Water dissolves tannins, which when redeposited on the surface cause darkening and hardening of the leather. Deteriorated leathers are especially vulnerable, because the damaged fibres are more susceptible to collapse.

Vegetable-tanned leathers absorb sulphur dioxide and follow a path of acidic decay often called 'red rot', where fibre scission leads to the leather disintegrating ultimately to a reddish fibrous powder. This pattern of decay is more common in late nineteenth century and twentieth century leathers, but may be found in more ancient leathers. To compare the deterioration patterns of ancient leathers with those of more recently naturally aged leathers, Larsen analysed seven small samples of Egyptian, Nubian and Coptic origin. Due to the small amount of samples, only amino acid analysis and measurement of hydrothermal stability were undertaken. In general, the significant oxidative changes of the collagen from the samples was found to follow the same pattern and lie in the same range as the newer historical leathers (Larsen, 1994, 1995). The Nubian leather gelatinized completely on contact with water at 20°C. Examination under the microscope of the Egyptian leather showed that it had lost the fibre structure, and was turned into a hard and amorphous mass, like dry glue. This resembles the behaviour of some samples of the ancient Roman and Egyptian leathers at the British Museum, which show gelatinization on wetting.

### 22.3.2 On-site retrieval

#### 22.3.2.1 *Planning for excavation*

The retrieval of desiccated leather on excavation can be a surprise, however well prepared the expedition. Personal preparation is vital. Along with assembling the required materials and tools and consulting the available literature, useful advice is to be had from those with relevant personal experience. Practical treatments should be considered and tested, if possible, in advance of excavation. Consideration (Schofield, 1996) should be given to:

(a) which materials, chemicals, solvents, packaging or other facilities are obtainable locally;
(b) how the local climate and area affects planned treatments and procedures;
(c) the destination of excavated material;
(d) the degree of conservation/restoration required.

#### 22.3.2.2 *Limitations on transport of solvents and materials*

Solvents and resins in solution may not be transported by air, and alcohols will not be readily available in countries under Sharia law. The limited weight of tools and materials it is possible physically to carry out to excavation (by air or otherwise) encourages careful choice. It may be necessary to use resins that serve more than one conservation purpose.

#### 22.3.2.3 *On site; initial procedures*

On-site lifting procedures, as well as transport, storage and handling, depend on the size and condition of the object, its intimate association with another structure (such as a body) and its condition (Cruickshank, 2001). When lifting fragmentary items, all pieces should be kept in their original relationship as far as is possible. Fragile items should be supported. Newly excavated pieces, though apparently dry, will contain some moisture. These should be kept initially in a sealed container and allowed to dry slowly (or equilibrate to local conditions) over a period of days.

Composite artefacts, e.g. a quiver containing arrows, require particular care. The arrows may first be removed singly so that these do not subsequently damage the quiver itself when lifting. Each object dictates its own approach; requiring an understanding of construction, areas of weakness and the relationship of different materials.

#### 22.3.2.4 *Handling*

Protective gloves and a mask should be worn to avoid contamination of (and also from) the material at all times. This is of particular importance if DNA analysis is planned. Some material is so fragile that handling without further breakage is almost impossible. This may dictate consolidation at an early stage (see below).

#### 22.3.2.5 *Packing and transport*

The mode of transport, destination and condition of the roads will influence decisions on packing. Fragile material may be supported in Plastozote (polyethylene foam) cut-outs, each tailored to the fragment shape, packed in Perspex boxes.

#### 22.3.2.6 *Storage*

Dry leather is best stored in dry, even, cool conditions. It should be boxed to protect from dust and to buffer from temperature and RH fluctuations, and padded/supported as necessary to avoid damage. Good labelling is essential to protect from unnecessary handling.

### 22.3.3 Recording procedures

Standard conventions in recording, assessment and selection apply (see above). Initial cleaning, and possibly stabilization, following the usual conservation procedures, may be necessary before diagnostic processes can proceed. Drawings and technical information records are useful (*Figure 22.7*). Desiccated leather may be examined and tested to identify and record the following features (Wills, 2002):

(a) Tannage identification
(b) Surface features, hair and species ID
(c) Seams, structural and decorative elements, impressions, etc.
(d) Colour
(e) Leatherworking tools
(f) DNA analysis.

#### 22.3.3.1 *Presence of tannage*

Ancient dry leather materials may carry evidence of vegetable tannage or other skin preparation (Daniels, 1997; Thickett and Wills, 1996) even after millennia. Two thirds of leather samples from a Nubian cemetery site (Kerma Ancien and Kerma Classique, c. 2500 BC– 1750 BC), for example, showed the presence of vegetable tannage (Ryder, 1984).

#### 22.3.3.2 *Vegetable tannage test*

A simple test for the presence of a vegetable tan is based on the reaction of iron salts with tannins. A 2% freshly made solution of ferrous sulphate ($FeSO_4$) w/v in distilled water, applied to a small area gives a blue–black stain if positive. The colour change is irreversible, so this test should be carried out on a hidden area. If the material is for study only, it may

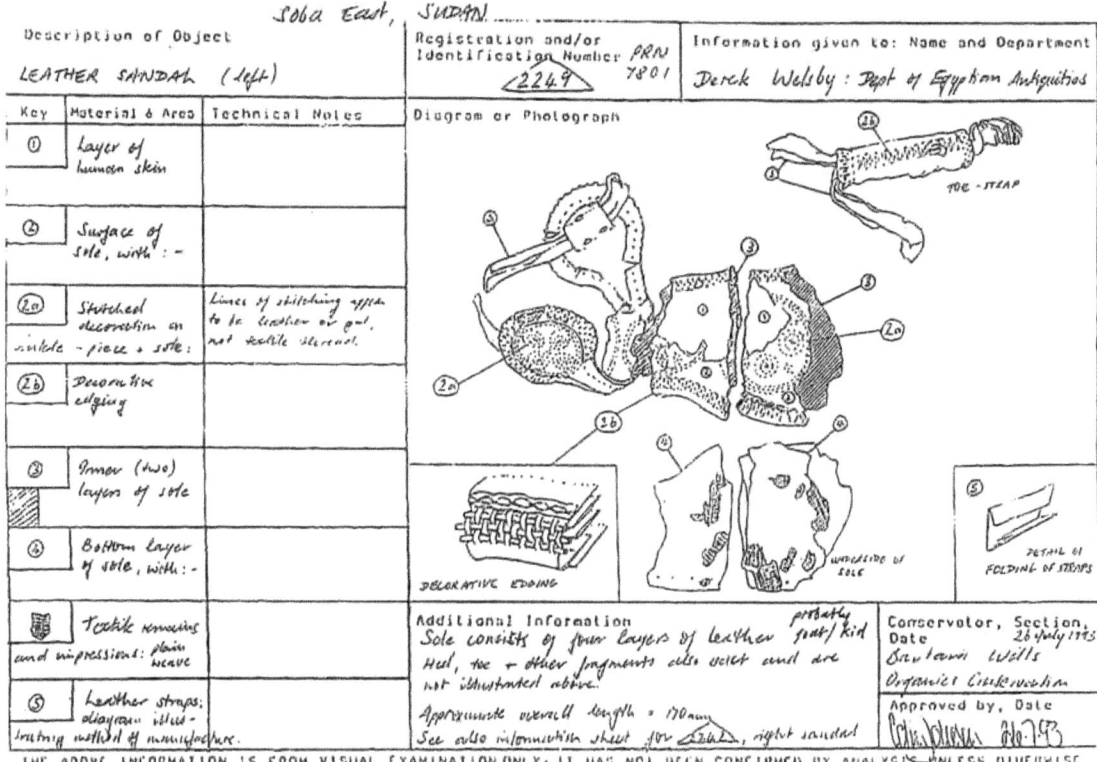

**Figure 22.7** Technical record of an early medieval Nubian sandal.

be done with a fine sable brush in situ on an object. If for display, a few fibres may be removed to be tested under magnification. As a further refinement, the type of tannin may be identified by the application of a drop of vanillin in alcohol, followed by a drop of concentrated hydrochloric acid. A positive result is bright red in colour, showing condensed vegetable tannins. No reaction demonstrates a hydrolysable tannin (Daniels, 1993). It should be noted that the test results will not show on very dark leathers, and that a positive result demonstrates the presence of tannins, not necessarily tannage. Testing a cross-section of leather is of greater value, showing whether the tannin is superficial. If so, it may relate to a vegetable colouring agent such as pomegranate juice or madder. Care is required with very decayed material, as water may rapidly convert aged skin to gelatine, or, with acidic leather, activate residual sulphates to create sulphuric acid.

Tannin tests carried out on some early medieval Nubian (Welsby, 1998) leather samples in the British Museum (Thickett and Wills, 1996) showed varying results. Extracts were tested separately with ammonium ferric sulphate solution for the presence of tannin, and aluminon (the triammonium salt of aurintricarboxylic acid) solution for the alum. Thin caprine leather gave some positive results for vegetable tan, and the thicker thongs (more likely to be raw hide or simply tanned) gave a range of results, including positive for the presence of alum. It should be noted that any possible background contamination with alum had not been taken into account, so these results should be interpreted with discretion.

22.3.3.3 *Surface features and cross-section*

Information is revealed by examination under magnification of the grain pattern, presence of hairs, decoration, tooling, evidence of wear, deposits, staining and abrasion. Species identification is often possible. A cross-section of the skin gives further diagnostic detail. If hair remains, fibre analysis is possible. The presence or absence of a woolly winter coat may indicate the season of death (Ryder, 1987). The thickness of the leather is of interest, giving some indication of the species and age of the animal if intact, and leatherworking techniques if split.

#### 22.3.3.4 Seams, structural and decorative elements, impressions

The range of features (type of seam, stitching, perforations, nails, knots, bindings, etc.) is recorded in a systematic manner (Goubitz, 1984; Spriggs, 1987). Seams stitched when wet are recognizable by distortion created by the pull of the thread.

Materials adjacent to the leather may leave interpretable impressions. In examples of excavated Kerma leather, where no textile remained, the indentations of a plain-weave textile were evident. Weave analysis was possible; four pieces of leather revealed thread counts varying from 10 to 12 threads by 12 to 14 threads per $10\,mm^2$ (Welsby, 2001).

#### 22.3.3.5 Colour

Colour may be natural, applied pigment, dye, or the result of decay. Dye and pigment analysis techniques might be accessed and adapted from textile analysis protocols (Smith, 2000).

#### 22.3.3.6 Leatherworking tools

The methods of cutting, stitching and finishing give clues to the tools used. This may become significant if the tools themselves have not been found.

### 22.3.4 Present-day treatments

#### 22.3.4.1 Considerations prior to conservation treatment

The end purpose (study or display) and the end place (storage/display in the country of origin or elsewhere) will affect the type of treatment. This should be discussed and methods considered early in the project. Samples should of course be taken before treatments begin to safeguard the future potential for analysis. Desiccated leather is, however, occasionally found in such good condition that intervention is neither necessary nor desirable (Volken, 2001).

#### 22.3.4.2 Cleaning

Conservation training is needed to understand and apply treatments successfully, and this includes apparently straightforward cleaning techniques. The standard range of non-aqueous cleaning treatments (dry brushing, vacuuming, Wishab sacrificial rubber, chemical sponges, etc.) should be tested and applied as appropriate.

#### 22.3.4.3 To flex or not to flex

It may be helpful to consider here the arguments for and against 'restoring' suppleness to leather. Most ancient dry leather has lost flexibility. Archaeological leather on display or in storage is static; it need not be supple. Methods of introducing permanent flexibility into old leather have sometimes limited the object's lifespan, and dressings and leather softeners can be damaging (see below). Temporary pliancy is, however, useful when reshaping a three-dimensional object (Wills, 2000), and such methods will be considered more fully below.

#### 22.3.4.4 Dressings

The above paragraph leads to discussion of another misconception; the 'feeding' of leather. It is not hungry, and does not need food. Dressings lubricate working leather structures such as harness, but can be damaging to aged leather. Dry leather pieces treated with glycerine and thymol in the Louvre in the 1980s had, when re-examined, turned from red to black (Montembault, 2001).

Leather dressings damage in the following ways (Raphael, 1993):

(a) darken many lighter colour leathers
(b) encourage mould and biological attack
(c) form fatty spews over the surface
(d) oxidize over time to stiffen the material
(e) wick into surrounding materials
(f) soften original finishes and decoration
(g) cause dust to accumulate
(h) impede future conservation treatments
(i) contaminate the material for future scientific analysis

to which may be added:

(j) provide a route of ingress for contaminants and catalysts for further decay.

#### 22.3.4.5 Consolidation as a first-aid measure on site

Consolidation is sometimes necessary to strengthen a fragile object in situ on site. It allows lifting and subsequent safe transport, handling, recording and reconstruction of an item that might otherwise be lost or damaged. Further treatment may well be necessary. If first-aid consolidation is required, it becomes particularly important to remember to take samples for analysis before treatment.

#### 22.3.4.6 *Consolidation: a range of resins*

The ideal leather consolidant has not yet been found. Conservation standards require such a material to confer resilience, strength, stability, flexibility, reversibility, and retain colour and surface qualities. Resins currently or recently used in dry leather consolidation include those based on acrylic resins such as Paraloid B72, Paraloid F10, Pliantex; those based on cellulose such as Cellugel, Klucel G, and others such as a polyvinyl butyryl (Mowital B30H). A wider range of potential consolidants has been explored in the past, including gelatine, polyurethanes, polyvinyl acetate emulsions, epoxy resins and a polyvinyl acetate/ethylene copolymer (Vinamul 3252).

Though many of the above resins are useful in specific areas, new materials are still sought for conservation treatments in the studio and on excavation. Cyclododecan (Jagers and Hangleitner, 1995; Hilby, 1997), a temporary consolidant of fragile material, is worth investigation for use on site.

Some conservators have found it beneficial to treat dry leather as wet; to rehydrate it and then treat as waterlogged (Volken, 2001). This applies only to robust and recently excavated material.

#### 22.3.4.7 *Consolidation in hot climates*

Dry leather from hot, arid sites, and which remains stored or displayed in the country of origin, is subject to the potentially damaging extremes of that climate. Conservation treatments should take this into account and be modified to fit those challenging conditions (Wills, 2001). The more familiar conservation materials used in temperate climates (such as Paraloid B72 which has a $T_g$ of 40°C) may have to be rejected in favour of those stable at higher temperatures with a higher glass transition temperature ($T_g$).

#### 22.3.4.8 *Reshaping*

Reshaping is sometimes required to help comprehend a three-dimensional object. It is a delicate and subtle process, not to be undertaken without appropriate knowledge and experience. Robust, newly excavated leather items may be most pliable immediately after excavation, and so are best reshaped gently as soon as possible. Others, too fragile to manipulate, require padding to stabilize their present shape. Reshaping may also be done off-site, back in a conservation studio, or later still in the case of long-stored material.

Misshapen organic objects often retain a memory of their shape during their 'working' life. This assists the process of recalling the original form, along with cultural knowledge (Sully, 1992). Reshaping a rigid, deformed leather item involves the (often temporary) admission of a plasticizing agent such as moisture, to the structure of the skin, causing it to swell. The process may be carried out before or after consolidation (Wills, 2000).

#### 22.3.4.9 *Repair, backing and reconstruction*

Where it is necessary to make joins, a range of acrylic and other conservation adhesives is available. To reinforce the joins, suitable backing materials include Japanese kozo (mulberry fibre) papers and non-woven polyester fabrics. Solvent reactivation of adhesive films (or the adhesive-impregnated backing material) often works well. To make fills, glass microballoons and pastes derived from cellulose fibres have been used. A displayable leather object may additionally need mounting and support.

For a more extensive and detailed discussion of the conservation of desiccated archaeological leather, please refer to Wills (2002).

### 22.4 Mineralized leather

#### 22.4.1 Condition

Metals such as iron and copper alloy corrode during burial and the products of their corrosion dissolve in the ground water. Iron readily forms insoluble complexes with tannins, but leather buried in very close proximity to corroding metals and saturated by the water is also affected by the solute precipitating within the skin structure (Turgoose, 1989). Minerals most commonly encountered in leather are oxides and phosphates of iron, and carbonates and chlorides of copper. High mineral content in waterlogged leather is indicated by metal staining and decreased flexibility (Hovmand and Jones, 2001). In the course of mineralization metal ions are precipitated either inside collagen fibre bundles (*replacement*) or outside them (*cementation*). Under certain conditions mineralization is progressive and skin structures can first undergo cementation followed by a second stage during which the fibre structures themselves are replaced. This does not noticeably retard the hydrolysis and dissolution of the protein element, but mineralized skin or leather sometimes survives burial when unaffected parts of the same object, further from the source of metal ions, does not. Completely mineralized leather is rigid, crumbly and ochre coloured with no visible organic content, yet some features of its original structure may still be preserved (*Figures 22.8* and *22.9*). This type of evidence is

258  Conservation of leather and related materials

**Figure 22.8** Detail of a mineralized decorated knife sheath (seventh to eighth centuries AD). Grave 5352, St Mary's Stadium, Southampton, UK. (Courtesy of Institute of Archaeology, Oxford)

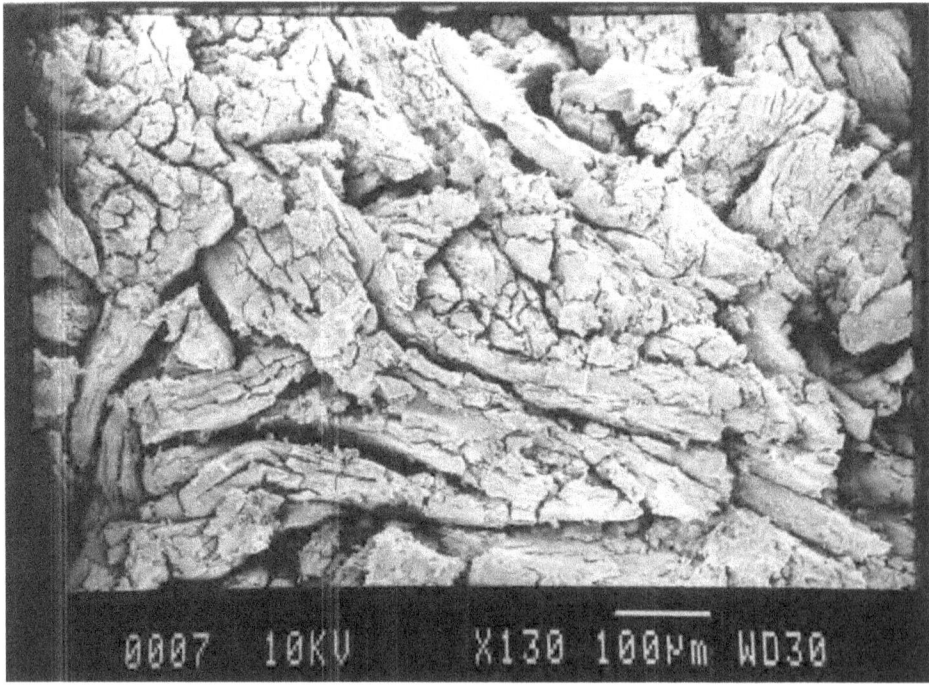

**Figure 22.9** Scanning electron micrograph of mineralized collagen fibres from a knife sheath (sixth to seventh centuries AD). Grave 4269, Buttermarket, Ipswich, UK.

prone to physical damage through poor handling, abrasion and inadequate conditions of storage.

## 22.4.2 On-site retrieval

Completely mineralized leather is most often found in aerated acidic conditions which promote the corrosion of metals. The existence of mineralized leather may, however, be disguised beneath bulky corrosion products so it is best practice to retrieve finds complete with adhering deposits. Small items such as knives and sheaths up to 200 mm in length can usually be lifted straight into a box, but larger or complex finds require support from beneath and at the sides to prevent collapse. One advantage of this approach is that transient evidence at the edges is held in position until it can be examined and recorded in a more controlled setting. Different ways of lifting fragile finds are described by Watkinson and Neal (1998).

Totally mineralized organic remains do not normally undergo change following excavation but they are at risk if the metal with which they are associated undergoes fresh corrosion or if they are not protected from physical damage, including unnecessary handling. Both are avoided by using standard packaging regimes (Watkinson and Neal, 1998). Iron that has not totally oxidized before excavation and is not subject to environmental control afterwards can be expected to corrode afresh. If left unchecked the gradual disintegration of the metallic element is likely to cause remains of mineralized leather to fall away.

For those involved in on-site retrieval it will be essential to the specialist to know the orientation of finds in the ground, such as which way up, and the spatial relationship of finds groups. In some cases the specialist should be invited on site.

## 22.4.3 Recording

The extent of mineralized deposits within soil blocks, or around metal objects, can be recorded by radiography, a standard procedure in the recording and assessment of finds. X-rays can reveal multi-layered constructions, such as the scabbards of swords, and the existence of straps or rivets before any cleaning takes place.

It is likely that a limited amount of soil removal may then be necessary in order accurately to identify the organic materials present and to assess their condition. A picture will gradually emerge of the research value of the remains and the required level of recording. When they are seen at magnifications of ×25 or even ×1000, details of skin structures preserved in mineralized leather vary in quality according to the circumstances of their original formation and subsequent history. The make-up of the mineral compounds and their sequential deposition can reproduce structures in fine and precise detail, or not at all as the case may be. The characterization of mineralized organic materials in general is specialist work. On significant finds and when more than one type of material is present (such as wood, textile and leather) some of the cleaning as well as the recording should be completed by the specialist concerned to ensure that nothing important is overlooked or removed.

Features indicative of skin or leather can be preserved in mineralized remains to such a degree that its identity, even to species, may be obvious. Grain patterns and hair shafts may be recognized as well as evidence of craft work such as stitching and tooling (Filmer-Sankey and Pestell, 2001: pl. XXXVI, XXXVIII, XXXIX). However, if the grain surface has been lost underlying deposits can be searched (using microscopy and scanning electron microscopy) for evidence that the remains represent a skin product of some kind (Cameron and Edwards, 2003).

Mineralized leather tends to be regarded as unpromising material for organic analyses (such as for DNA, tannins or dyes), but the same may not be true of partially mineralized leather.

## 22.4.4 Treatment

Partially mineralized leather most commonly comes from anaerobic waterlogged contexts and feels inflexible and crystalline and attempts to flex it can result in cracking. Following excavation there is a danger that the mineral content may undergo volume change through oxidation and that this could result in distortion of its surviving connective tissue and further chemical breakdown through changes of pH. Methods of reducing mineral content by using sequestering agents, and important considerations in deciding whether to do so, are discussed by Ganiaris *et al.* (1982) and Hovmand and Jones (2001).

Completely mineralized leather, being generally friable and brittle, is normally found in aerobic deposits and is already oxidized. However, it often gives the impression of extreme weakness, and fear of it crumbling away normally initiates a strong desire to consolidate. For research purposes this is discouraged because it prevents further examination or sampling for analysis. Arguments for consolidation, to allow transport, display and handling,

## 22.5 Long-term storage of archaeological leather

### 22.5.1 Storage requirements

#### 22.5.1.1 *Storage of dried and desiccated leather*

Dry leather is best stored in moderately dry, even, cool conditions; it is generally content in the compromise environment that suits both metals and organics (c.16°C and 40%RH). The use of glycerol or PEG 400 (both hygroscopic) in the freeze-drying treatment of wet leathers makes this all the more necessary.

Optimum environmental conditions for the storage and display of partially mineralized leather depends on the proportional balance between organic and mineral and on the relative importance assigned to each. Cool, dry environments to minimize chemical change are likely to be beneficial, especially as so many mineralized leathers contain metallic elements, such as iron hobnails in the soles of Roman shoes, which tend to corrode if levels of relative humidity rise above 20%. Oxygen scavengers such as RP System (MGC) in sealed containers or packages are useful for storing combinations of materials with differing RH requirements.

#### 22.5.1.2 *General points on storage*

(a) box to protect from light, dust and to buffer from temperature and RH fluctuations
(b) pack in suitable containers, padding as necessary to support fragile material
(c) good labelling essential to protect from unnecessary handling and inadvertent disposal
(d) regular monitoring of conditions
(e) set and check pest traps
(f) regular condition surveys (e.g. Sully and Suenson-Taylor, 1998).

### 22.5.2 Condition assessments of treated leather

Condition audits on collections are a core function of good collections care, and it will be found necessary sooner or later to repack old collections of conserved leather to bring them up to modern standards (Edwards and Mould, 1995). A risk assessment should be undertaken under Health and Safety Executive regulations (2000) and appropriate safety precautions taken when handling items showing signs of past mould growth, and to protect against dust inhalation. Collections condition surveys are a means of assessing the effectiveness of packaging and storage conditions, as well as the success of the conservation treatment. Published surveys of leather collections at York (Spriggs, 2003), the University of Trondheim (Peacock, 2001), and the Museum of London (Sully and Suenson-Taylor, 1996) used rather different survey techniques, yet each recognized that the conditions under which the collections were being kept needed to be improved in a number of ways. At York it was noted that much of the leather appeared to be inherently fragile, and required better packaging to prevent fragmentation. The widespread effect of biodeterioration (mould growth) was ascribed to high humidity in storage areas and overcrowded boxes. Finally, the leather that had been treated before 1976, using the early solvent and dressing-based treatments, was moist and sticking to its packaging.

### 22.5.3 Old collections/retreatments

A surprising amount of dry archaeological leather, of varying ages and in different states of preservation, is to be found in museums and stores. This ranges from material that had always been dry, to that which was once wet or waterlogged. Old conservation and excavation records, where these exist, are helpful when re-evaluating these collections. The leather itself may show evidence of past interference such as dressings or repairs. Early conservation reference works give clues to possible old treatments if no records are found. *The Conservation of Antiquities and Works of Art* (Plenderleith and Werner, 1956) and *A Guide to the Conservation and Restoration of Objects Made Wholly or in Part of Leather* (Waterer, 1972) are of value. Any subsequent conservation treatments undertaken must take into account earlier treatments.

## 22.6 Purpose of treatment: a call for clarity

The conservation of archaeological leather would benefit if we defined more clearly the two functions of an object after excavation. Its first function is as a resource for the information it carries; the second is as a displayable object. In its first role, as an object of research, its keepers are hedged about with restraints which limit its chemical contamination and protect its physical integrity. In effect, while its potential is assessed, records made, specialist opinions sought and samples taken the environment is adapted to the requirements of the material. It may not be possible, or convenient, to maintain this situation indefinitely and a point may be reached when the object must be adapted to suit the environments of museum stores and galleries. This change of status, marking the transition of an object from its first to second functions, is in many respects a one-way process in that conservation treatments at this level are designed around permanency and exhibitions. Sometimes this involves treatments that can never be completely reversed. Most archaeological leather in museum collections has already passed through this second stage because it has been freeze-dried, consolidated or treated in some other way. Conservators in the future need to be respectful of the research potential of freshly excavated leather and appreciative of its visual impact on display. The two functions require different types of conservation skills and treatments, performed in the correct order, but in doing so the various needs of the researcher and museum curator can be met equally.

## 22.7 Conclusion

Leather is an extraordinary material because it carries its entire history, including that of the original animal, in its structure and on its surface. It has a technological richness that few fully appreciate, with unexplored zones waiting to be researched, such as the history of the development of vegetable tanning in Europe and the Near East, and an evaluation of archaeological skin and leather as a resource for genetic studies through DNA analysis (Wayne *et al.*, 1999). Among archaeological finds, where so much that endures is of metal or ceramic, the discovery of organic materials is both exciting and demanding, especially when evidence of craft or industry also emerges from the same context. No one can predict where and when leather will be found but organized retrieval followed by sympathetic handling and treatment will affect the quality of the published results and broaden our view of past cultures.

## References

Adams, G. (1994) Freeze-drying – Art or Science? In *A Celebration of Wood* (Spriggs, J., ed.), pp. 49–54 Wetlands Archaeological Research Project **8**.

Baltazar, V., Lyons, W. and Murray, A. (1996) Plastination as a Consolidation Technique for Archaeological Bone, Waterlogged Leather and Waterlogged Wood. In *Twenty Second Annual Association of Graduate Programs in Art Conservation Student Conference Papers* Queen's University, pp. 4–10. Ontario.

Cameron, E. (2000) *Sheaths and Scabbards in England AD400–1100.* Oxford: British Archaeological Reports, British Series 301.

Cameron, E. and Edwards, G. (2004) *Evidence of Leather on Finds from Anglo-Saxon Cemeteries at St Stephen's Lane/Buttermarket, Ipswich, Suffolk 1987–88 and Boss Hall, Ipswich 1990.* Portsmouth: English Heritage Centre for Archaeology Report 23/2004.

Cowgill, J., Neergaard, M. de and Griffiths, N. (1987) *Knives and Scabbards: Medieval Finds from Excavations in London.* Museum of London.

Cronyn, J. (1990) *The Elements of Archaeological Conservation.* London: Routledge.

Cruickshank, P. (2001) Leather Shrouds ... a Rare Find in Jordan. In *Leather Wet and Dry: Current Treatments in the Conservation of Waterlogged and Desiccated Archaeological Leather* (B. Wills, ed.), pp. 63–70. London: Archetype.

Daniels, V. (1993) *Evaluation of a Test for Tannins in Leather.* British Museum Department of Conservation Internal Report no. CA 1993/2.

Daniels, V. (1997) *Analysis of Excavated Leather Samples from Kerma.* British Museum Department of Conservation Internal Report no. CA 1997/48.

Degrigny, C., Baron, G., Christodoulou, P., Tran, K. and Hiron, X. (2002) Conservation of a Collection of Waterlogged Composite Rifles Dating from the 17th Century Recovered from the Brescou II Marine Site. In *Proceedings of the 8th ICOM Group on Wet Organic Archaeological Materials Conference* (P. Hoffman, J. Spriggs, T. Grant, C. Cook, and A., Recht, eds), pp. 399–412. Stockholm (Bremerhaven).

Dobson, C. (2003) As Tough as Old Boots? A Study of Hardened Leather Armour, Part 1: Techniques of Manufacture, pp. 55–77. *Art and Arms Florence, City of the Medici, Conference Proceedings May 30th–June 1st 2003.*

Driel-Murray, van C. (2000) Leather and Skin Products. In *Ancient Egyptian Materials and Technology* (P.T. Nicholson and I. Shaw, eds), pp. 299–319. Cambridge: Cambridge University Press.

Driel-Murray, van C. (2002) Ancient Skin Processing and the Impact of Rome on Tanning Technology. In *Le Travail du Cuir de la Préhistoire à nos Jours* (F. Audoin-Rouzeau, and S. Beyries, eds), pp. 251–265. Antibes: APDCA.

Edwards, G. and Mould, Q. (1995) *Guidelines for the Care of Waterlogged Archaeological Leather.* English Heritage Scientific and Technical Publication, Guideline **4**.

Filmer-Sankey, W. and Pestell, T. (2001) *Snape Anglo-Saxon Cemetery: Excavations and Surveys 1824–1992.* East Anglian Archaeology Report 95.

Ganiaris, H., Keene, S. and Starling, K. (1982) A Comparison of Some Treatments for Excavated Leather. *The Conservator*, **6**, 12–24.

Godfrey, I., Kasi, K. and Richards, V. (2002) Iron Removal from Waterlogged Leather and Rope Recovered from Shipwreck Sites. In *Proceedings of the 8th ICOM Group on Wet Organic Archaeological Materials Conference* (P. Hoffman, J. Spriggs, T. Grant, C. Cook, and A. Recht, eds), pp. 439–467. Stockholm (Bremerhaven).

Goubitz, O. (1984) The Drawing and Registration of Archaeological Footwear. *Studies in Conservation*, **29**, 187–196.

Goubitz, O. (1997) What is Wrong with Freeze-drying? In *ICOM Working Group on the Treatment of and Research into Leather, in Particular of Ethnographic Objects* (P. Hallebeek, ed.), pp. 36–37. ICOM interim meeting, Amsterdam, April 1995. Paris: ICOM.

Groenman-van Waateringe, W., Kilian, M. and Londen, H. van (1999) The Curing of Hides and Skins in European Prehistory. *Antiquity*, **73**, 884–890.

Health and Safety Executive (2000) Management of Health and Safety at Work. In *Management of Health and Safety at Work Regulation 1999. Approved Code of Practice and Guidance L21*. HSE Books.

Hilby, G. (1997) Cyclododecan, a Volatile Binding Medium. *Restauro*, **2**, 96–103.

Horie, C. (1990) Deterioration of Skin in Museum Collections. *Polymer Degradation and Stability*, **29**, 109–133.

Hovmand, I. and Jones, J. (2001) Experimental Work on the Mineral Content of Archaeological Leather. In *Leather Wet and Dry* (B. Wills, ed.), pp. 27–36. London: Archetype.

Ibbs, B. (1990) Shrinkage Calculations. *Conservation News*, **41**, 7.

Jagers, E. and Hangleitner, M. (1995) Flüchtige Bindemittel. *Zeitschrift für Kunsttechnologie und Konservierung*, **2**, 385–392. Worms.

Jenssen, V. (1987) Conservation of Wet Organic Artefacts, Excluding Wood. In *Conservation of Marine Archaeological Objects* (C. Pearson, ed.), pp. 122–163. London: Butterworths.

Jones, A., Jones. J. and Spriggs, J. (1980) Results of a Marker Trial. *Conservation News*, **11**, 6–7.

Larsen, R. (1994) Deterioration and Conservation of Vegetable Tanned Leather. In *STEP Leather Project. Protection and Conservation of European Cultural Heritage Research Report* **1** (R. Larsen, ed.), pp. 165–179. European Commission, Directorate-General for Science, Research and Development, The Royal Danish Academy of Fine Arts, School of Conservation, Denmark.

Larsen, R. (1995) The Mechanisms of Deterioration. In *Fundamental Aspects of the Deterioration of Vegetable Tanned Leathers*, PhD thesis. The Royal Danish Academy of Fine Arts, School of Conservation, Denmark.

McGregor, A. (1998) Hides, Horns and Bones. In *Leather and Fur: Aspects of Early Medieval Trade and Technology* (E. Cameron, ed.), pp. 11–26. London.

Montembault, V. (2001) Treatments of Archaeological Leather in France. In *Leather Wet and Dry: Current Treatments in the Conservation of Waterlogged and Desiccated Archaeological Leather* (Wills, B. ed.), pp. 45–50. London: Archetype.

Mould, Q., Carlisle, I. and Cameron, E. (2003) *Leather and Leather-working in Anglo-Scandinavian and Medieval York*. The Archaeology of York AY17/16. York: The Council for British Archaeology, pp. 3227–3231.

Muhlethaler, B. (1973) *Conservation of Waterlogged Wood and Leather*. Paris, pp. 25–72.

Omar, S., McCord, M. and Daniels, V. (1989) The Conservation of Bog Bodies by Freeze-drying. *Studies in Conservation*, **34**(3), 101–109.

Organ, R. (1958) Carbowax and Other Materials in the Treatment of Water-logged Palaeolithic Wood. *Studies in Conservation*, **4**, 96.

Painter, T. (1991a). Preservation in Peat. *Chemistry and Industry*, 17 June, 421–424.

Painter, T. (1991b). Lindow Man, Tollund Man and Other Peat-bog Bodies: The Preservation and Antimicrobial Action of Spagnan, a Reactive Glycuronoglycan with Tanning and Sequestering Properties. *Carbohydrate Polymers*, **15**, 123–142.

Peacock, E. (1983) The Conservation and Restoration of Some Anglo-Scandinavian Leather Shoes. *The Conservator*, **7**, 18–23.

Peacock, E. (2001) Water-degraded Archaeological Leather: An Overview of Treatments used at Vitenskapsmuseum (Trondheim). In *Leather Wet and Dry: Current Treatments in the Conservation of Waterlogged and Desiccated Archaeological Leather* (B. Wills, ed.), pp. 11–25. London: Archetype.

Pearson, C. (1987) *Conservation of Marine Archaeological Objects*. London: Butterworths, p. 115.

Plenderleith, H.J. and Werner, A.E. (1956) *The Conservation of Antiquities and Works of Art*. London: Oxford University Press.

Raphael, T. (1993) The Care of Leather and Skin Products: A Curatorial Guide. *Leather Conservation News*, **9**, 1–15.

Rashleigh, P. (1803) Account of a Further Discovery of Antiquities at Southfleet in Kent, in a Letter from the Rev. Peter Rashleigh, Rector of that Place, to the Rt. Hon. Sir Joseph Banks, Bart. K.B.P.R.S. and F.S.A. *Archaeologia*, **14**, 221–223, pl. 38. London: Society of Antiquaries.

Rector, B. (1975) The Treatment of Waterlogged Leather from Blackfriars, City of London. *Transactions of the Museums Assistants Group*, **12**, 33–37.

Reed, R. (1972) *Ancient Skins, Parchments and Leathers*, p. 267. London: Seminar Press.

Rosenquist, A. (1975) Experiments on the Conservation of Waterlogged Wood and Leather by Freeze-drying. In *Problems in the Conservation of Waterlogged Wood* (A. Oddy, ed.), pp. 9–23. National Maritime Museum Monographs and Reports, **16**.

Ryder, M. (1984) Skin, Hair and Cloth Remains from the Ancient Kerma Civilisation of Northern Sudan. *Journal of Archaeological Science*, **11**, 477–484.

Ryder, M. (1987) Sheepskin from Ancient Kerma, Nothern Sudan. *Oxford Journal of Archaeology*, **6**(3), 369–380.

Schofield, G. (1996) From Fjord to Wadi. *Scottish Society for Conservation and Restoration Journal*, **7**, 17–18.

Smith, A. (2000) Down Among the Molecules: Chemical Analysis in the Study of Historical and Archaeological Textiles. In *Textiles Revealed; Object Lessons in Historic Textile and Costume Research* (M. Brooks, ed.), pp. 149–150. London: Archetype.

Smith, W. (2003) *Archaeological Conservation Using Polymers*. Texas Agriculture and Mining Anthropology Series 6.

Spindler, K. (1994) *The Man in the Ice*. London: Weidenfeld & Nicolson.

Spriggs, J. (1980) The Recovery and Storage of Waterlogged Material from York. *The Conservator*, **4**, 19–24.

Spriggs, J. (1981) The Conservation of Timber Structures at York – A Progress Report. In *Proceedings of the ICOM Waterlogged Wood Working Group Conference, Ottawa* (D. Grattan, and J. McCawley, eds), p. 149. Canadian Conservation Institute.

Spriggs, J. (1987) Aspects of Leather Conservation at York. In *Recent Research in Archaeological Footwear* (D. Friendship-Taylor, J. Swann, and S. Thomas, eds), pp. 43–46 Association of Archaeological Illustrators and Surveyors Technical Paper **8**.

Spriggs, J. (2003) Conservation of the Leatherwork. In *Leather and Leather-working in Anglo-Scandinavian and Medieval York* (Q. Mould, I. Carlisle, and E. Cameron, eds), pp. 3213–3221. The Archaeology of York AY17/16.

Starling, K. (1987) The Conservation of Wet Metal/Organic Composite Archaeological Artefacts at the Museum of London. In *Conservation of Wet Wood and Metal* (I. MacLeod, ed.), pp. 215–219. Proceedings of the ICOM Waterlogged Wood Working Group Conference, Freemantle.

Storch, P. (1997) Non-vacuum Freeze-dry Treatment of Two Leather Objects. *Leather Conservation News*, **13/2**, 15–17. Minnesota Historical Society/ICOM.

Sturge, J. (1973) *The Conservation of Wet Leather*, unpublished dissertation for the Diploma in Conservation, Institute of Archaeology, London.

Sully, D. (1992) Humidification: The Reshaping of Leather, Skin and Gut Objects for Display. In *The Conservation of Leathercraft and Leather Objects* (P. Hallebeek, M. Kite, and C. Calnan, eds), pp. 50–53. London: ICOM Symposium.

Sully, D. and Suenson-Taylor, K. (1996) A Condition Survey of Glycerol Treated Freeze-dried Leather in Long-term Storage. In *Archaeological Conservation and its Consequences* (A. Roy, and P. Smith, eds), pp. 177–181. Preprints of the IIC Congress, Copenhagen.

Sully, D. and Suenson-Taylor, K. (1998) An Interventive Study of Glycerol Treated Freeze-dried Leather. In *Conference Proceedings of the 7th ICOM Group on Wet Organic Archaeological Materials* (C. Bonnot-Deconne, X. Hiron, and Q. Tran, eds), pp. 224–231. Grenoble.

Swann, J. (1997) Shoe Conservation: Freeze-drying Problems? In *ICOM Working Group on the Treatment of and Research into Leather, in Particular of Ethnographic Objects* (P. Hallebeek, ed.), pp. 34–35. Interim Meeting, Amsterdam, April 1995. Paris: ICOM.

Thickett, D. and Wills, B. (1996) *Report on First Season Gabati Leather*. British Museum Department of Conservation Internal Report 96/15/O/8.

Turgoose, S. (1989) Corrosion and Structure: Modelling the Preservation Mechanisms. In *Evidence Preserved in Corrosion Products: New Fields in Artefact Studies* (R. Janaway and B. Scott, eds), pp. 30–32. United Kingdom Institute for Conservation Occasional Paper **8**.

Turner, R. and Scaife, R. (eds) (1995) *Bog Bodies: New Discoveries and New Perspectives*. London: British Museum.

Volken, M. (2001) Practical Approaches in the Treatment of Archaeological Leather. In *Leather Wet and Dry: Current Treatments in the Conservation of Waterlogged and Desiccated Archaeological Leather* (B. Wills, ed.), pp. 37–44. London: Archetype.

Walker, S. (1990) *Corpus Signorum Imperii Romani, Great Britain* II, Fascicule 2, pp. 57–58. Catalogue of Roman Sarcophagi in the British Museum. British Museum Publications.

Waterer, J. (1972) *A Guide to the Conservation and Restoration of Objects Made Wholly or in Part of Leather*. London: Bell.

Watkinson, D. and Neal, V. (1998) *First Aid for Finds*. UKIC Archaeology Section and Rescue.

Wayne, R., Leonard, J. and Cooper, A. (1999) Full of Sound and Fury: The Recent History of Ancient DNA. *Annual Review of Ecology, Evolution, and Systematics*, **30**, 457–477.

Welsby, D. (1998) Soba II: Renewed Excavations Within the Metropolis of the Kingdom of Alwa. *BIEA Memoir*, **15**, 182–185. London: British Museum Press.

Welsby, D. (2001) Life on the Desert Edge; Seven Thousand Years of Settlement in the Northern Dongola Reach, Sudan. *Sudan Archaeological Research Society*, **7**, London. 450–457.

Williams, E., Harnett, L., Bonnot-Diconne, C. and Barthe, J. (1999) Castor Oil and PEG 400 Followed by Freeze-drying: Re-assessing Treatment Methods for Waterlogged Leather. *Leather Conservation News*, **15**(2), 1–10.

Wills, B. (2000) A Review of the Conservation Treatment of a Romano-Egyptian Cuirass and Helmet Made from Crocodile Skin. *The Conservator*, **24**, 80–88.

Wills, B. (2001) Excavating Desiccated Leather: Conservation Problems on Site and After. In *Leather Wet and Dry: Current Treatments in the Conservation of Waterlogged and Desiccated Archaeological Leather* (B. Wills, ed.), pp. 51–62. London: Archetype.

Wills, B. (2002) Windows into Ancient Nubian Leatherwork. In *Le Travail du Cuir de la Préhistoire à nos Jours* (F. Audoin-Rouzeau, and S. Beyries, eds), pp. 41–64. Antibes: APCDA.

# 23

# Case histories of treatments

*Staff of the Leather Conservation Centre*

This chapter offers a selection of case histories of treatments which were undertaken to a variety of objects at the Leather Conservation Centre, Northampton, UK, during the ten years prior to publication of this book. It is not intended that they be read as an instruction manual of how to treat similar objects, but rather to show what has been done to particular objects in the circumstances which prevailed at the time.

To this end these case histories are published in the form of conservation reports. It is intended that they will provide examples of treatments which will expand on the information given in Chapter 13.

The objects have come from a variety of sources including museums, historic houses and private clients. In some cases, where the objects belong to private clients, they are still in regular use so treatments have been chosen with this in mind. Recommendations for aftercare have also been included in cases where appropriate.

## 23.1 The Gold State Coach. 1762

The Royal Collection
LCC Ref: 359–95. Conservator: **Theodore Sturge**
Date: February 1996

### 23.1.1 Description

This is the coach that is used on the most important royal occasions such as coronations. It was built in 1762 and has a leather roof. Originally the leather was silvered and then lacquered with a crimson glaze which was built up in five layers. The effect would have been for it to glitter with a red gold colour in the sun. Since its original decoration it has been repainted seven times. The existing top coat consists of gold leaf which has been varnished. On top of this there is a laurel leaf decoration in a dark brown glaze which was sealed with a further heavy varnish. In all, the leather has had 45 layers of decorative finish applied to it, to give a total of eight decorative surfaces. As a result, the paint layer is about 1 mm thick.

The paint was sectioned by Jo Darrah of the Victoria and Albert Museum's Conservation Department. She has suggested the following chronology for the decorative surfaces:

1. 1762 George III's coronation – Crimson glaze on silver leaf.
2. 1789 George III's recovery – Crimson glaze on pink ochre.
3. 1820 George IV's coronation – Red glaze on white.
4. 1830 William IV's coronation – Red glaze on dark red.
5. 1837 Victoria's coronation – Red glaze on red.
6. 1862 Albert, the Prince Consort's death. – Black.
7. 1887 Victoria's Golden Jubilee – Red glaze on crimson.
8. 1901 Edward VII's coronation – Gold leaf, heavy varnish.

### 23.1.2 The problems and the options

Following a very hot summer, cracks in the roof became significantly worse. A major split had opened up in the front quadrant where the leather had given way. There was a smaller split in the rear quadrant. In addition, there were numerous small cracks in the surface where the underlying leather was starting to break down as the weakened fibres pulled apart. There was some loss of paint associated with these cracks and splits.

There were two treatment options. One was to dismantle the roof of the coach and to replace the leather with new. This would have involved more than just taking off the original leather. All the decorative gold cherubs, etc. would have had to have been removed to give access to the edges and the fixing points. This would have been very complex and time consuming. Instead, it was decided to repair the existing roof.

### 23.1.3 Treatment

The surface was cleaned with mineral spirits. It was further cleaned later as part of the consolidation procedure (see below).

The leather had been stretched over a timber roof structure but was not attached with adhesive. As a result, it was possible to insert a patch of polyester sail cloth behind the larger of the splits (*Figure 23.1*). This fabric was chosen because it has good dimensional stability, resistance to rot, and would not allow the adhesive to penetrate from one side to the other. This was important because, if at all possible, the leather and its patch should not become attached to the timber underneath. If it had, it would have restrained further movement. This could have led to undesirable stresses should there be further changes to the storage conditions, even if these were fairly modest. The edges of the split had lifted slightly so the leather was humidified by laying Aquatex, a semi-permeable membrane, over the painted surface. Wet blotting paper was then laid over this. There were sufficient cracks in the adjacent paint work to allow the water vapour which passed through the membrane to relax the leather into its correct plane. While the leather was still humid, Vinamul 3254 was inserted between it and the sailcloth and sandbags were laid on top to hold it in place as it dried. Vinamul 3254, an ethylene vinyl acetate copolymer emulsion, was chosen for its toughness and relative fluidity, which allowed it to be easily inserted into the rather tight gap between the sailcloth and leather. The smaller split proved too tight to treat in this way and was simply filled, along with the other gaps, as described below.

The cracks in the paint work were consolidated with a 5% solution of Paraloid B72 in toluene. This was flooded into the cracks and the surplus was cleaned away with swabs moistened with toluene. This had an additional useful effect in that the overall paint surface was further cleaned. The aim was not only to consolidate the cracks and the loose

266  *Conservation of leather and related materials*

**Figure 23.1**   Major split in roof and cracks in the paint surface. The sailcloth has been inserted.

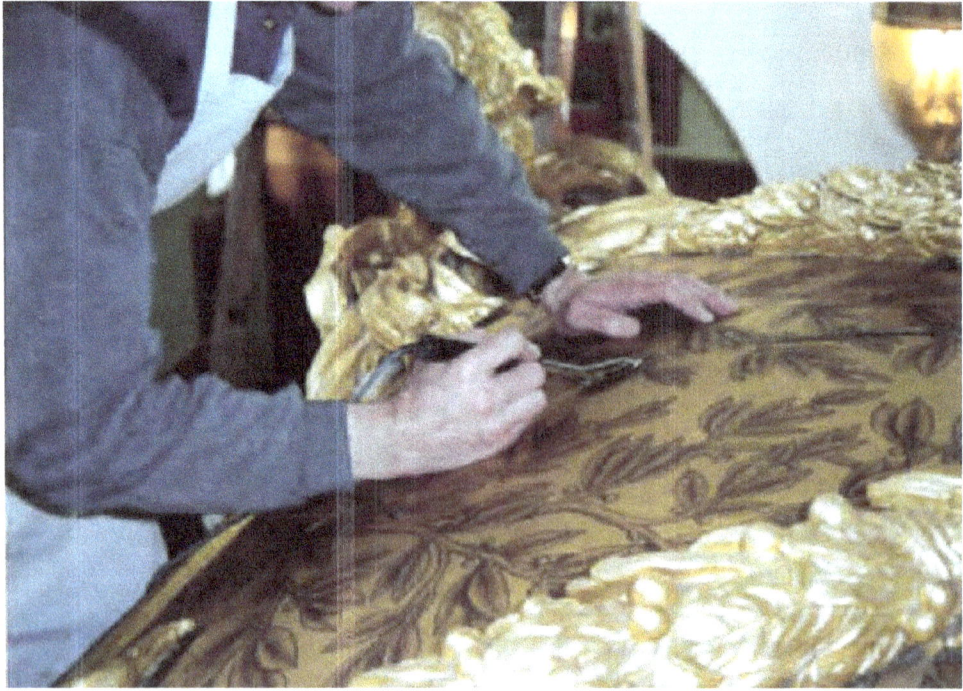

**Figure 23.2**   Filling the split with Beva 371 and a heated spatula.

**Figure 23.3** The finished repair. Inpainting still to be carried out.

paint, but also the underlying leather. In the areas under the cracks, the leather fibres were starting to pull apart and it was hoped that the Paraloid resin would bond them together, reducing the extent of the splitting.

The gaps left where the edges of the two larger splits did not come together completely, the larger cracks, and the areas where the paint had been lost, were filled with solid Beva 371. The Beva was made up using flake metal powders so that the colour was similar to that of the gold finish. It was applied with a heated spatula (*Figures 23.2* and *23.3*).

The inpainting of the fills and the final varnishing of the whole roof were carried out by Royal Mews staff. The varnish filled the cracks further, and probably increased the roof's resistance to water, should it ever be used for a state occasion on a wet day.

## 23.2 Dog Whip – believed to be eighteenth century

Private collection
LCC Ref: 354–95. Conservator: **Theodore Sturge**
Date: September 1995

### 23.2.1 Description

This dog whip comes from a church in Chesterfield where it was used to chase dogs out before services. It has a wooden handle with a leather grip and a raw hide thong. The thong is attached to the handle with a raw hide loop bound to the wood with cord.

The thong, although distorted, was intact, but the loop attaching it to the handle had broken. The wood was broken under the grip and the end was missing. The leather grip had dropped off and there was evidence of past woodworm damage, but this was old and inactive (*Figure 23.4*).

The whip was not actively deteriorating, but there was a risk that loose sections would be lost. In addition it was not in a fit state to display. As a result of these needs the work was more akin to restoration than to conservation.

### 23.2.2 Treatment

The length of the missing wood was estimated and a new section was made in pine (*Figure 23.5*). The break in the original wood was sealed with a coat the Vinamul 3254 (EVA vinyl acetate copolymer dispersion adhesive). This gave a soluble layer which could be broken down in future should removal be necessary. The break was very uneven so it was filled with Ronseal wood filler, a polyester-based filler. The filler also acted as an adhesive between the old and the new wood. The new wood was toned down to match the original with wood stain and water colours.

The broken raw hide loop at the top was replaced with new raw hide. The cord binding was replaced as the original was found to be too weak to use again (*Figure 23.6*). The new cord binding and raw hide were toned down with Winsor and Newton watercolours.

The leather grip was humidified to allow reshaping. Water was introduced by wrapping Aquatex, a semi-permeable membrane on fabric, round the leather and then putting wet paper towel around this. Water vapour, but not liquid water, penetrated the membrane and relaxed the leather. The leather was reattached using a three to one mixture of Lascaux dispersions 498 HV and 360 HV (thermoplastic butyl methacrylate copolymer dispersions thickened with acrylic butyl ester). The adhesive was applied at either end, with the centre section left free. The grip was given a final light finish with British Museum Leather Dressing (lanolin, beeswax and cedar wood oil in hexane).

The raw hide thong was humidified and straightened (*Figure 23.7*).

Case histories of treatments 269

**Figure 23.4** Before conservation.

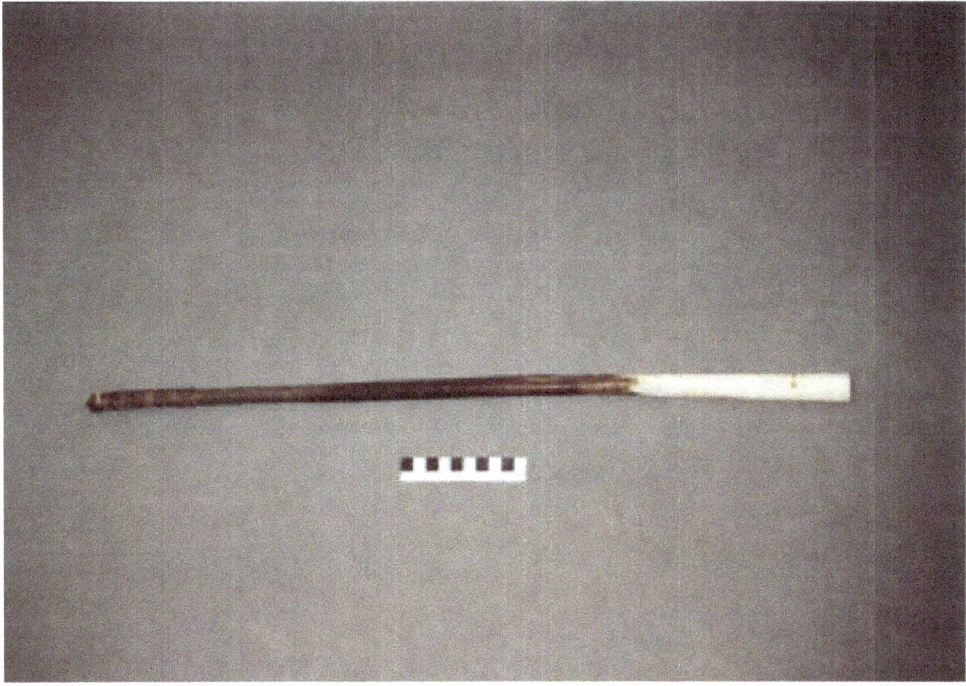

**Figure 23.5** New timber spliced onto handle.

270  *Conservation of leather and related materials*

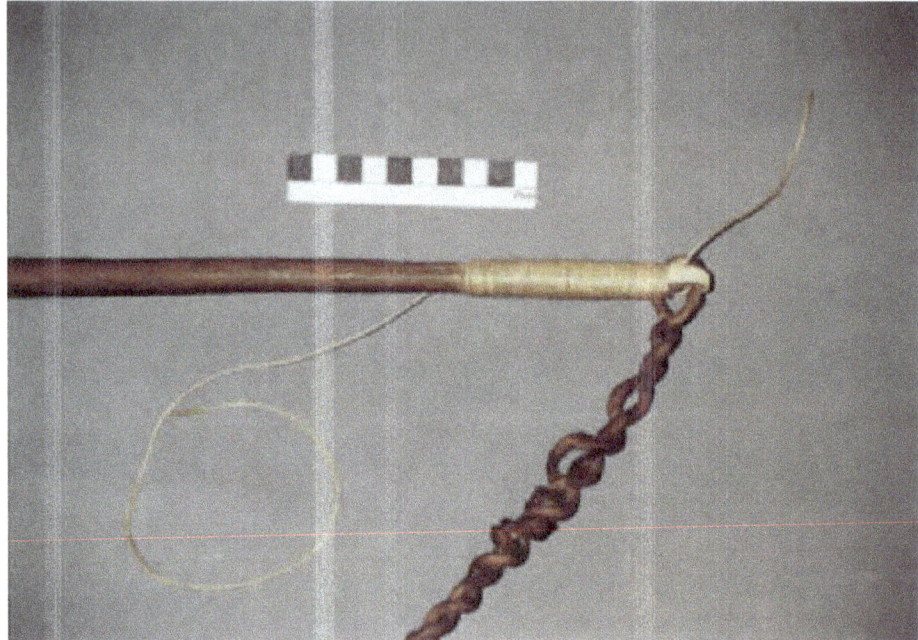

**Figure 23.6** Replacing the broken raw hide loop on the tip.

**Figure 23.7** The finished whip.

## 23.3 Fire Bucket

Private collection
LCC Ref: 549–99. Conservator: **Theodore Sturge**
Date: August 1999

### 23.3.1 Description

This fire bucket was made for the Westminster Insurance Company, and probably dates from the eighteenth century. It bears the date 1717 but this relates to the foundation of the insurance company rather than to its manufacture. It is 29 cm high and 28 cm in diameter across the top.

The main body of the bucket is made from two substantial pieces of vegetable-tanned hide. The sides are made from a single piece with a vertical sewn seam, and the base from a separate piece, also sewn. The rim is reinforced with a half round timber lath over which leather has been moulded and then sewn. Two leather loops, which pass down the outside of the top edge and behind the lath, are sewn to the main body. These contain iron rings which, in turn, hold the handle. The interior of the bucket is coated with pitch and the outside painted black. The painted crest of the company is on the front, as is the date 1717.

The bucket was in poor condition (*Figure 23.8*). Almost half of the lath was missing. Part of the leather covering for the lath was hanging off, and about half was missing. Adjacent to this, there was a hole in the main body of the bucket and about 100 cm$^2$ of leather was missing. The handle itself was made with multiple pieces of leather sewn together. This was intact but some of the stitching had broken down. The loops of leather attaching the iron rings for the handle to the bucket were in a fragmentary state and had lost their strength.

### 23.3.2 Treatment

The whole bucket was surface cleaned with Wishab sponges followed by damp swabs. The crest was cleaned with isopropyl alcohol. As with all painted surfaces, great care needed to be taken when cleaning with solvents and careful testing was vital. Some fire buckets, but not this one, are decorated with transfers rather than paint. If this is the case, it is likely that very little cleaning will be possible.

The leather adjacent to the area of loss was consolidated to give a sound surface to enable the repair to be attached. A 10% solution of Paraloid B67 in

**Figure 23.8** Before conservation.

white spirits was chosen as a high level of flexibility was not needed, and the high glass transition temperature of this acrylic resin ensured that it did not have a tacky surface. The same solution was used to consolidate the paint adjacent to the lost leather and the pitch on the interior.

The missing section of wooden lath on the rim was rebuilt using laminated balsa wood. The layers were shaped using a violin maker's bending iron and bonded together with the Lascaux acrylic dispersion 498 HV. The laminated balsa was then shaped to the correct profile (*Figure 23.9*). Balsa worked well as it was easy to bend and cut to shape and the laminated structure was quite strong compared with balsa wood on its own. Once the lath was in place, the loose leather was reattached with more of the Lascaux adhesive. The missing sections of leather in the main body of the bucket were filled with solid Beva 371 to which pigment had been added. This was built up and bonded to the original using a heated spatula (*Figure 23.10*). The original plan was to use vegetable-tanned hide but it proved impossible to mould it satisfactorily and cut to fit it to the complex shape. The exposed lath was covered with new vegetable-tanned calfskin (*Figure 23.11*). The lath and surrounding leather were covered in cling film to keep them dry while the new leather was thoroughly wetted and moulded around the lath. When dry, the new leather held its shape and was trimmed to size and bonded into place with the

272  *Conservation of leather and related materials*

**Figure 23.9**  The new section of lath for the rim built up in balsa wood.

**Figure 23.10**  Building up the missing section with solid Beva 371 and a heated spatula.

**Figure 23.11** New calfskin moulded to fit the rim.

Lascaux dispersion. The original had a very heavy layer of paint on it and when the new leather was painted to match the grain texture showed through. The thick paint on the original was therefore replicated by working solid Beva into the surface of the new leather with a heated spatula. The Beva fill was then painted to match the original using artists' quality acrylic paints with a final coat of gum arabic toned with Winsor and Newton watercolours to give a suitable glazed finish.

The weak sections of the handle were reinforced by inserting new vegetable-tanned calf between the layers and then replacing the missing stitching using the original holes.

The bucket was to go into an environment where it was likely to be handled by untrained staff so it was very likely that it would be lifted by its handle. The original disintegrating loops used to hold the handle to the main body were so damaged that it was felt a satisfactory, strong repair was impossible. As a result they were replaced using new leather stitched into place through the original holes. Although the work was not for a museum, the owner still asked for a minimum level of intervention consistent with the handling demands of the object. Full inpainting of the missing areas of the crest was therefore not carried out. Instead, only a

**Figure 23.12** The finished bucket.

minimum consistent with making the image readable was done. This was carried out with Winsor and Newton watercolours.

The exterior of the bucket was lightly finished with Renaissance micro-crystalline wax polish and the handle was lightly dressed with Pliantine (*Figure 23.12*).

## 23.4 Fireman's Helmet

Private collection
LCC Ref: 549–99. Conservator: **Theodore Sturge**
Date: August 1999

### 23.4.1 Description

This early helmet, one of four, is probably eighteenth century. It is made from degrained, vegetable-tanned cattle hide, except for the hatband which is vegetable-tanned sheepskin.

The main shell has been moulded in two halves and then sewn together. The joint was thickened to increase its strength, with a rope-like material. The wide brim was sewn into place. Inside there was a label, apparently giving the name of the owner. Each hat had a different name.

The main body of the helmet was in good condition, but the brim was in poor condition (*Figure 23.13*). Its pH was 2.5, and it was suffering from severe red rot. Diluting the test solution ten times gave a pH of 3.5, indicating that damaging strong acids were present. The hatband was also suffering from red rot.

### 23.4.2 Treatment

Initially, the whole hat was cleaned with a soft brush and a vacuum cleaner. Further cleaning was carried out with Wishab sponges. These soft sponges crumble with the dirt sticking to the fragments which are brushed and vacuumed away. However, they need to be used with care, as on a fragile surface they can cause damage and remove flakes.

The red rot on the rim and hatband was treated with a 1% solution of aluminium alkoxide in mineral spirits. The leather was saturated and then the excess was blotted off with paper towels. This latter process is important, as if there is an excess left on, an insoluble white deposit can remain on the surface. This is very difficult to remove once it forms. After treatment the pH had increased to 3.8, well above the figure of 3.0 usually regarded as the pH below which leather is unstable.

The fragile rim was consolidated with a 3% solution of Pliantex, a flexible acrylic resin, in toluene. Although this consolidated the body of the leather, it did not hold the flakes in place. These were reattached by inserting an acrylic dispersion, Lascaux 498 HV and 360 HV in a 3:1 mixture, under each loose flake, and then pressing them into place. Because it is in the form of a dispersion rather than a solution, the adhesive does not soak into the leather. As a result, it holds the flakes well.

The hole in the rim was filled with new vegetable-tanned calfskin coloured with Sellaset dyes. This was butt jointed to the original leather. As this was not a strong join it was reinforced with suitably coloured Stabiltex (polyester) fabric. Beva film was used as an adhesive. This was first applied to the Stabiltex, which was then cut to size and applied to the join. The bond is made with an electric-heated spatula. Stabiltex and Beva film were also applied across the splits in the rim. Stabiltex was chosen for its strength and relative transparency, which made a discreet repair possible.

The areas on the rim where the grain surface had been lost were coloured with Sellaset dyes. This was applied locally with a brush to tone the damage in with the surrounding colour. A final light finish was applied to the entire helmet using Renaissance microcrystalline wax polish (*Figure 23.14*).

Case histories of treatments 275

**Figure 23.13** Before conservation.

**Figure 23.14** After conservation.

## 23.5 Leather Lion

Private collection
LCC Ref: 397–96. Conservator: **Ian Beaumont**
Date: October 1996

### 23.5.1 Description

This lion was made by H.E. Jackman in 1950. It was designed to sit on a mantelpiece or desk. The core was a coarse white material, probably plaster of Paris. This was reinforced with a steel wire armature. The vegetable-tanned leather had then been moulded over the core, presumably while wet. There were some joins in the leather under the chin and around the mane. The leather was a natural brown in colour.

The leather was, in general, in good condition. It had been handled over the years, leaving a smooth surface and some slight soiling. The main damage was to the end of the tail where a section of the leather and core were missing (*Figure 23.15*) and to the front left paw which was also missing (*Figure 23.16*). There was also some damage to the back left paw.

**Figure 23.15** Before conservation, note damage to tail.

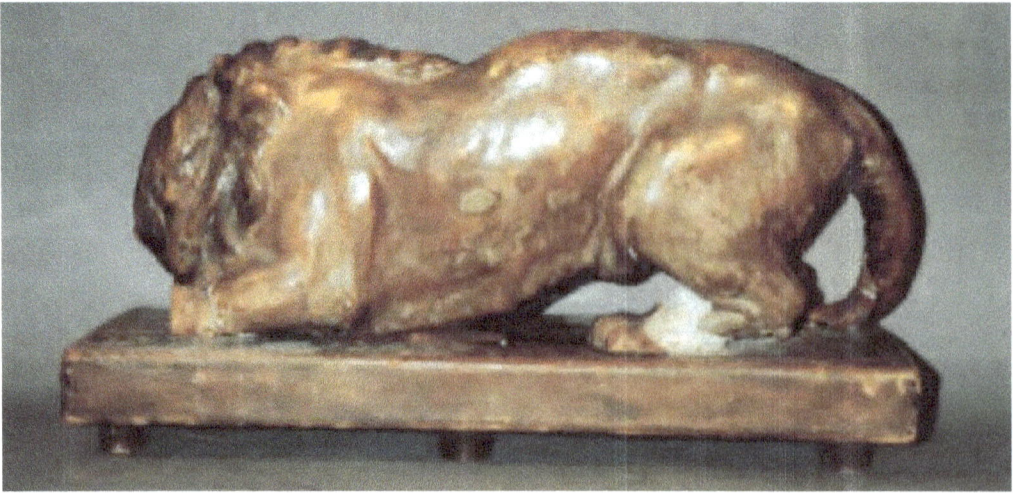

**Figure 23.16** Before conservation, note missing front paw and damage to rear paw.

Case histories of treatments 277

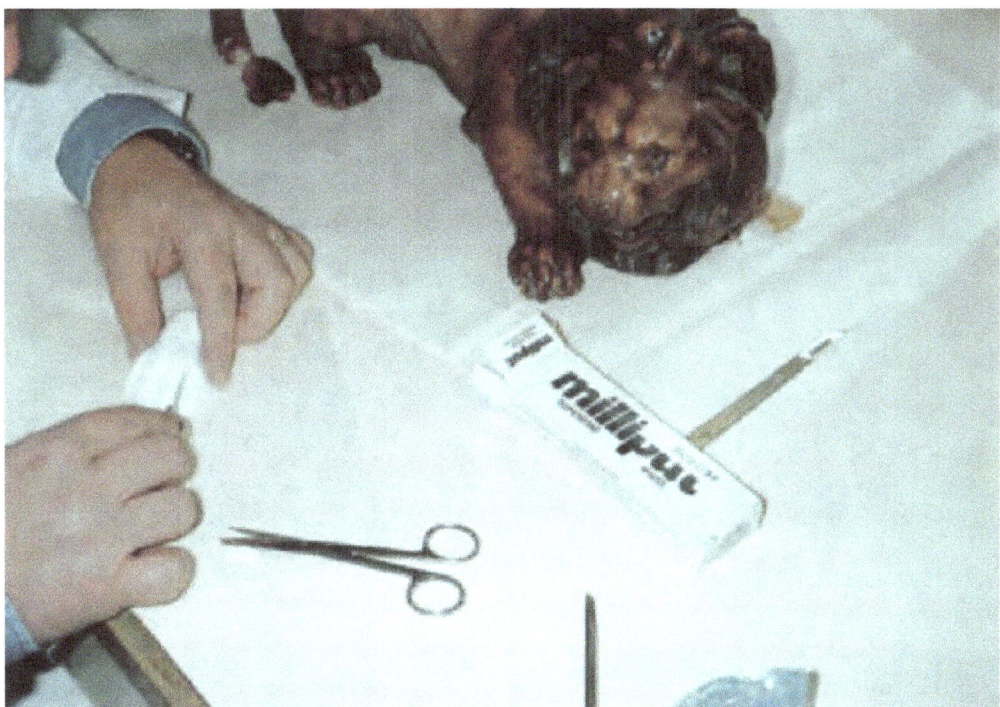

**Figure 23.17** Modelling a new core for the front paw.

**Figure 23.18** After conservation.

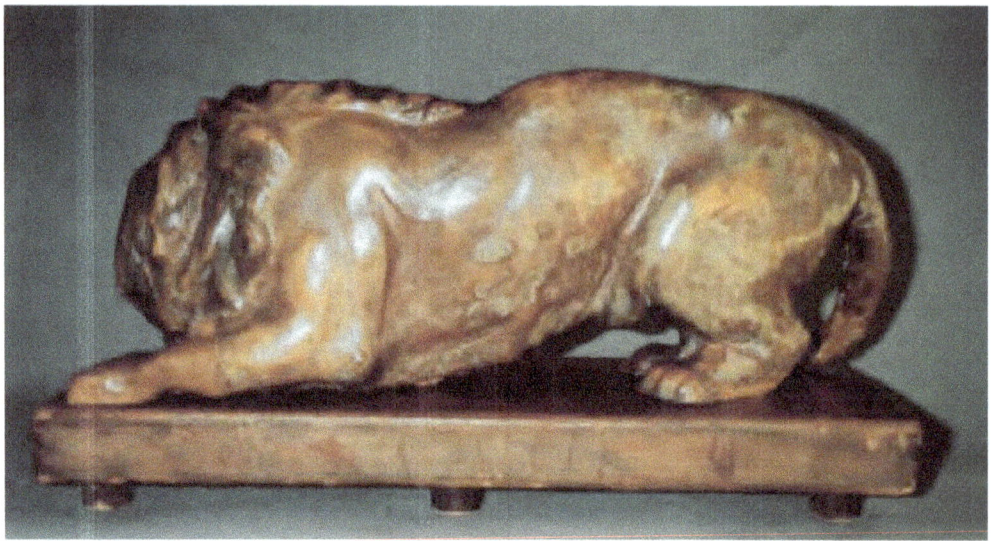

**Figure 23.19** After conservation.

### 23.5.2 Treatment

The missing sections of the core were built up with Milliput, an epoxy putty (*Figure 23.17*). The surface of the original plaster was friable so it was consolidated with 5% Paraloid B72 in toluene to give a sound surface for the Milliput to bond to. The Milliput was chosen as it bonded well to the original core and it is easy to model as it cures. The shape of the paw was outlined on the base and this was used as a guide.

The areas where the leather was missing were re-covered with new vegetable-tanned calfskin. When wet, vegetable-tanned leathers usually mould well. In this case, the leather used was very thin, 0.8 mm, and it took the impression of the underlying core very well. It was even possible to work it into the claws of the lion so that they showed on the front. It is helpful to continue to work the leather as it dries. When part dry it often reaches a stage where it is almost 'cheesy' in consistency and it can be pushed into shape with tools such as bone folders. Once dry, the shape will be held.

The dry moulded leather was attached to the core with a mixture of acrylic emulsion adhesives, Lascaux 498 HV and 360 HV, in a 3:1 ratio.

The whole lion was lightly cleaned and polished with Renaissance micro-crystalline wax polish. The new sections were toned down by adding dry pigments to the wax. This was applied only to the new leather. Pigments in wax were chosen, rather than dye, as it was a surface patination that was required rather than an in-depth colouring of the leather (*Figures 23.18* and *23.19*).

## 23.6 Sedan Chair

Private collection
LCC Ref: 416–97. Conservators: **Theodore Sturge and Nina Frankenhauser**
Date: July 2000

### 23.6.1 Description

This eighteenth century sedan chair was conventionally constructed with a wooden frame, the outside of which was covered in vegetable-tanned, black, patent leather calfskin. The inside was covered with textiles. It was approximately 170 cm high, 81 cm wide and 94 cm deep (*Figure 23.20*).

The leather was attached with dome-headed brass tacks arranged to give a simple decorative border.

The windows, which could be lowered on braided straps, were framed with a smooth finished gilt timber subframe. Around the windows and along the top, just beneath the roof, there was a carved gilt wood trim.

**Figure 23.20** The chair, after treatment.

The leather was chemically stable and reasonably strong. The patent leather finish was in good condition except for a few vertical lines on the sides, apparently caused by a liquid running down. However, it had split in a number of places both on the flat sides and on the concave segments of roof. There was some distortion associated with the splits.

The gilt areas were in good condition except for some fairly light soiling and some worn areas on the highest points.

The textiles on the inside were heavily soiled and one of the straps for lowering the window had broken at the point where it joined the sash frame.

### 23.6.2 Repairs

The leather was repaired using three different techniques. The method chosen for each repair depended on two factors: the availability of access to the back of the leather, and whether or not the leather was flat, as on the vertical surfaces, or curved, as on the roof. In addition, three different repair materials were used, leather for larger repairs to flat areas, Reemay for smaller ones on flat areas, and cast GRP (Glass Reinforced Plastic/fibreglass) for the curved areas on the roof. The only areas where there was any access to the back of the leather without going through the split were in those places where the interior fabric had come away partially. This did not give full access, but it did allow some unwanted materials used during the repair work to be removed when the work was finished.

Where there was access to the back, a card sandwich was used. This was used with both leather and Reemay as a repair material. The sandwich was made up with stiff card on the outsides, two layers of silicone release paper and the repair material in the middle. A tracing of the split was made on Melinex and this was used to pierce pairs of holes along the line of the split. Loops of strong thread were passed through the holes and tied loosely. The back piece of card, silicone release paper and the repair material were inserted through the split and then drawn up into position using the threads. Adhesive was then inserted between the repair material and the back of the leather. This was a mixture of acrylic dispersions, Lascaux 498 HV and 360 HV, in a 3:1 ratio and was chosen as it does not grab quickly and thus allows time for the repair to be manoeuvred into position. Once the position was correct, the outer silicone paper and card were slid into place down the thread which was cut and then retied close to the leather.

280  *Conservation of leather and related materials*

**Figure 23.21** The repair sandwich held together with wedges.

Wedges were then inserted in the loops to pull them tight and hold the sandwich tightly together (*Figure 23.21*). Once dry, the thread was cut and pulled out through the card. The card and silicone release paper were removed from the front and back. In places where the card on the inside would drop into an inaccessible space, it was found helpful to tie a thread to the corner prior to insertion and bring this out through a suitable gap. This allowed the card to be retrieved.

In areas where the card could not be readily removed from the inside, a low pressure suction box was used (*Figures 23.22–23.24*). This was simple in construction. It was a sealed box with a perforated steel face on one side. The box was connected to a variable speed vacuum cleaner. The repair material, usually leather, was slipped behind the split and held in position with a thread brought out through the split. Once the repair was in place, and the Lascaux adhesive inserted, the low pressure box was placed over the area being repaired. This sucked the original leather onto its front face ensuring it was flat, and the air being drawn through the split held the repair onto the back of the leather. It was then left under low suction to dry. There was a small risk that adhesive would be drawn out onto the front face of

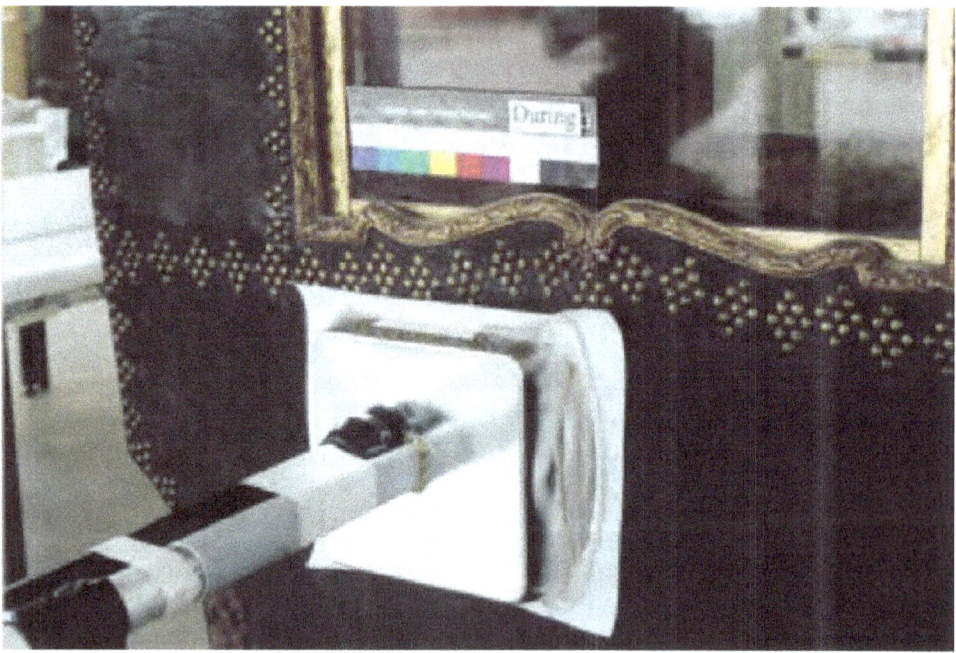

**Figure 23.22** Low pressure suction box.

the suction box. To avoid any risk of the leather sticking to the metal, a sheet of Reemay was taped over the perforated surface. The advantage of this method was that no unneeded materials remained on the back, so completely blind repairs were possible. However, the system with card does ensure that the patch is in good contact with the back. Both hold the leather flat during the repair which helps to ensure a minimum of distortion.

The roof required a radically different approach as the leather was curved and there was no access to the back (*Figure 23.25*). The varying curvature made a low pressure box impracticable, and the card sandwich will not work without a flat surface. A curved repair piece was required which could be inserted. Moulded leather was considered but it was not felt to be stiff enough to hold against the underside. Instead, a glass fibre reinforced polyester (GRP) repair piece was made to fit each split. These were stiff enough to be held in place, but slightly flexible, allowing them to adjust slightly to fit. To obtain the correct shape, plaster of Paris was cast onto an undamaged section of roof with a similar curvature. While this was being done, the surface of the leather was protected with cling film. Fortunately, a second chair with an identical framework and shape was in the workshop. Although this too needed some roof repairs, it provided an extra opportunity to cast plaster of Paris to

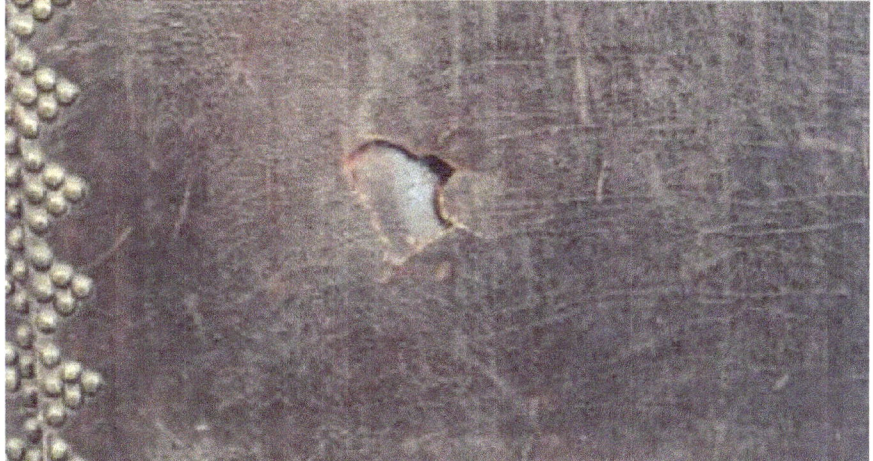

**Figure 23.23** Damage before repair using low pressure suction box.

**Figure 23.24** Damage after repair using low pressure suction box.

**Figure 23.25** The roof, before treatment.

**Figure 23.26** Positioning the GRP repairs with threads.

match damaged areas. The curvature was then adjusted if necessary. The surface of the plaster was waxed and the GRP was cast onto it using 1 oz/sq.ft (300 g/m$^2$) chopped strand fibreglass mat and Trylon SP701PA polyester lay-up resin. The resulting patch was trimmed and degreased with acetone. The patch had loops of strong thread passed through it on the line of the split (*Figure 23.26*). These were used to hold the patch in position by attaching them to elastic bands suspended from the ceiling (*Figure 23.27*). This held the patch securely against the back of the leather while it was attached with the Lascaux acrylic

*Case histories of treatments* 283

mixture. Contact was improved by inserting wedges and pressing the leather down with sandbags. There was some distortion to the leather so it was humidified by laying Aquatex covered with damp paper towel over the distorted edges. This was done prior to the application of adhesive but with the supporting GRP in place.

### 23.6.3 Cleaning

The surface of the leather was cleaned with a cloth and white spirits. No way was found to remove the vertical runs. It was thought this might be blanching from water and that it might be possible to reform the surface with solvent. However, no suitable solvent could be found for this, or to remove the runs.

The carved gilt wood was lightly cleaned with a 50/50 mixture of water and white spirits emulsified with 1% non-ionic detergent. This cleaned the surface safely but care needs to be taken if water-based cleaning is used on gold in case it is bonded to the wood with a water soluble adhesive. As always, cleaning tests are essential.

The fabric on the interior was cleaned with a soft brush and a low powered vacuum cleaner. The broken strap was repaired by adding new fabric which was both sewn and adhered with the Lascaux mixture. It was then reattached with the original nails.

**Figure 23.27.** The GRP patch held in place with elastic bands. Also shows smaller patch prior to insertion.

**Figure 23.28** The finished roof complete with Beva fills.

### 23.6.4 Gap filling and finishing

The two holes were filled with new vegetable-tanned calfskin dyed with Sellaset dyes. The larger gaps in the splits were filled with pigmented solid Beva 371 applied with a heated spatula (*Figure 23.28*). Care was taken that it was melted right down into the split to ensure a good bond. If the heating is too superficial, a bond only forms on the top surface and it can pull away. The fine gaps in the splits were filled with Aquafil 945, an acrylic filler. This was coloured, prior to application, with artists' quality acrylic paints. This filler dries by evaporation and there is some shrinkage. As a result it is not suitable for larger areas of damage.

The areas where the black finish had been lost and the underlying brown was showing through were inpainted with Sellaset dyes. The fills were inpainted with Winsor and Newton watercolours.

The whole of the outside was polished with National Trust furniture polish. No dressing was applied as this could soften and damage the finish.

## 23.7 Jewellery Box

Private collection
LCC Ref: 593–00. Conservator: **Theodore Sturge**
Date: May 2000

### 23.7.1 Description

The outside of this box was covered with a red, vegetable-tanned, sheepskin skiver. The top was padded and slightly domed. Inside there was a lift-out tray. Both this and the inside of the top and bottom of the box are lined with red velvet. The box is 20.4 × 13.8 × 9.5 cm and is probably twentieth century.

The red skiver was very thin, and over the top it had split and pulled away with about 7 cm² missing. In addition, the surface was scuffed and a strip of adhesive tape had been fixed to the surface and then pulled off. Both these had exposed the underlying brown colour of the leather (*Figure 23.29*).

### 23.7.2 Treatment

There were two aspects to the treatment of the leather: repairs to the splits and filling the gaps. The principal problem was that the leather was very thin, and any repair had to be light enough not to show through. In addition, some of the splits ran along the edge where the leather folded over the top and down the side. The leather on the sides was very well attached. To repair these areas, the leather needed to be eased away from the wood so that a repair material could be inserted behind it. Because the leather was very thin and well attached to the wood, there was a danger that the repair material would show through on the outer surface. Leather was felt to be too bulky so a fine Japanese tissue paper was used for most of the work. Before use, this was coloured with a dilute wash of artists' quality acrylic paint. This was matched approximately to the colour of the leather, so that if a sliver showed through a crack it would be unobtrusive. Had white tissue paper been used it would have caught the eye. To keep the thickness down an adhesive with low bulk was needed. After testing to ensure that the red colour of the leather would not be affected by water, wheat starch paste was selected. If the two surfaces to be joined are clean and reasonably smooth, and both are of natural materials, wheat starch paste can give a very strong bond. In addition, most of its bulk is water so the adhesive adds little to the overall thickness, and does not stiffen it significantly either. The disadvantage of wheat starch paste is that its high water content can overwet the leather unless it is used very sparingly.

Because the top was curved by the padding, it was not practical to repair all the splits with a single piece of tissue paper. Instead, the splits were systematically joined with smaller pieces until there was, in effect, a single flap of leather. The gaps were then filled. The leather had an unusual finish which it was impossible to match. Initially, it was thought that suitably coloured Japanese tissue paper would be more appropriate than a non-matching leather which could look obtrusive. However, in practice, a fine, dyed, vegetable-tanned sheepskin skiver proved a more appropriate choice. Patches of this were placed on the back of the leather so that they filled the gaps. These were attached with an acrylic emulsion

**Figure 23.29** Before conservation.

**Figure 23.30** After conservation.

mix, Lascaux 498 HV and 360 HV, in a 3:1 ratio. Finally, the resultant large flap of leather was secured around the edges using leather or tissue paper, as appropriate, to bridge the gap. Where necessary the leather on the sides of the box was eased away, so that the repair material could be inserted behind it.

The areas where the surface of the leather had been lost, exposing the brown underneath, were inpainted with a Sellaset dye mix. This was done with a fine brush so that it was only applied to the damage. If the dye solution spreads onto the sound surface it often does not 'take' and 'bronzing', a visual effect similar to oil on water, forms. This can be very difficult to remove although isopropyl alcohol may be effective. It can be difficult to match the colour exactly. If the dye is too bright it will make the inpainting obtrusive. It is better if it is a little duller in colour than the original so that the eye is drawn to the original rather than the inpainting.

A final finish of leather balm, a wax emulsion, was applied to the outside (*Figure 23.30*).

## 23.8 Dining Chairs

Private collection
LCC Ref: 539–99 and 521–98. Conservators: **Ian Beaumont and Theodore Sturge**
Date: November 1999

### 23.8.1 Description

Three sets of chairs are discussed in this case study. Their needs were very different, as were their treatments. Two of the sets, both Victorian, belonged to a single owner. One was in relatively poor condition and it was to be used on a daily basis. The owner was clear that these should be re-covered to give a usable set of chairs. The second set was to be used less regularly and although worn, conservation was a viable long-term option. The third group came from a larger set of late eighteenth century chairs, the majority of which were in good condition. However, four of the seat covers were worn and split, and the underlying upholstery was in need of attention. These chairs were to be used and a reasonably robust repair was needed that would not affect their overall appearance relative to the untreated chairs in the set.

### 23.8.2 The set of eight chairs for reupholstering

The original vegetable-tanned goatskin appeared to be dark green. However, when the leather was taken off the chairs, hidden unfaded areas were found to be a dark royal blue. The panels on the backs were embossed in gold with the family crest and monogram. The leather was extensively split, there were missing areas, self-adhesive tape had been applied to the surface, and it had been overdressed leaving it feeling damp and sticky (*Figure 23.31*).

**Figure 23.31** Chair prior to reupholstery. Note the extensive use of tape used for previous repairs.

**Figure 23.32** The reupholstered chair.

288  *Conservation of leather and related materials*

*23.8.2.1  Treatment*

The chairs were stripped of their leather and existing upholstery. The underlying upholstery was kept and as much of it as possible was reused during the subsequent reupholstery so that the shape and look of the chairs would be retained.

The leather was replaced with vegetable-tanned hair sheep dyed to match the original royal blue. This was used in preference to goat as larger sizes are more readily available. Its visual appearance and strength are very similar to that of goat, and it is a much more robust material than wool sheep. The crest on the back was reproduced. One of the existing crests was scanned and the artwork re-established by hand. This was used to make an embossing plate. To increase the durability of the gold embossing, it was carried out using looseleaf gold rather than the thinner gold film. This was more time consuming and expensive, but the increased thickness should ensure a longer life. The resultant chairs were suitable for heavy use (*Figure 23.32*).

### 23.8.3  The set of eight chairs repaired without removing the covers

The seats of these Victorian dining chairs were upholstered in vegetable-tanned goatskin which had faded from royal blue to green. The tops were deeply buttoned with leather-covered buttons.

The condition of the chairs varied considerably. Four of them had been treated on a previous occasion by another workshop. There were no details of this treatment available, but it included refinishing with a green-pigmented surface coating. The damage included splits, particularly on the front corners, loss of leather associated with some of the splits, splashes of paint and missing buttons. In addition, there were flaps of loose grain surface and flakes coming away. These were exposing the lighter colour of the underlying corium (*Figure 23.33*).

At some time, the leather on all the chairs had been given a heavy coat of wax polish. The webbing underneath was also starting to break down on most of the chairs.

*23.8.3.1  Treatment*

To ensure that the underlying upholstery was sound, all the chairs had an additional layer of webbing applied to the underside. This was placed over the top

**Figure 23.33**  Chair prior to repair without removal of cover.

of the existing webbing. In this way the seat was made structurally sound but the original was not lost.

The splits were repaired with 0.5 mm vegetable-tanned goatskin. This was dyed with Sellaset dyes, and was used both to repair the splits and to fill any gaps where the leather was missing. This was chosen because it matched the original well, and it was still strong when reduced to this thickness. Some leathers, when reduced in thickness, lose their strength as the longer, stronger, bundles of fibre on the flesh side are removed. If the leather is too thick or springy, it does not blend in so well and the repair becomes cumbersome, may not bond well, and can show through onto the front. The repairs were bonded into place with Hewit Reversible PVA adhesive M218. This was selected for its high tack which assists in the positioning of awkward repairs. The M218 adhesive was also slipped under the flaps and flakes of loose grain surface so that they could be pressed back into place.

**Figure 23.34** Chair after conservation.

**Figure 23.35** Chair prior to removing cover.

The areas where the leather surface had been scuffed or flakes had been lost were light in colour. These were inpainted with Sellaset dyes to reduce the visual impact of the damage. A final finish of SC6000, an acrylic resin and wax mixture, was applied to all the leather. This consolidated the surface, re-established a finish in those areas where it had been lost, and drew the overall appearance together (*Figure 23.34*). SC6000 needs to be used with care. The result can be overshiny or uneven, so careful testing, as with all finishes, is essential. In addition, it contains isopropyl alcohol which may dissolve the existing finish.

### 23.8.4 The four chairs where the covers were removed and conserved

These chairs were part of a large set. The leather on the rest of the chairs, although worn, was still intact and the requirement was for the conserved leather to match the untreated seats. The vegetable-tanned goatskin was heavily worn, and much of the red surface finish was missing. There were also areas, particularly on the corners, where the leather had split and, in some areas, adjacent leather had been lost. The leather was no longer strong enough to stand being restretched onto the new upholstery (*Figures 23.35* and *23.36*).

#### 23.8.4.1 *Treatment*

The leather was carefully removed from the chairs. The dome-headed nails attaching it were difficult to remove without damaging either them or the leather. A decision was made to sacrifice the nails to protect the leather.

The back of the leather was cleaned with Wishab sponges to give a clean surface for repair. Where there were gaps these were filled with new vegetable-tanned goatskin applied to the back (*Figure 23.37*). The new leather was initially dyed with Sellaset dyes. The burgundy colour, and the finish, were broken by lightly sanding the surface with fine glass paper. It was further adjusted by applying a second, darker dye on a cotton wool swab. These were damp, rather than wet, and gave a mottled effect. The splits where there was no loss were repaired

**Figure 23.36** Detail of damage prior to removal of cover.

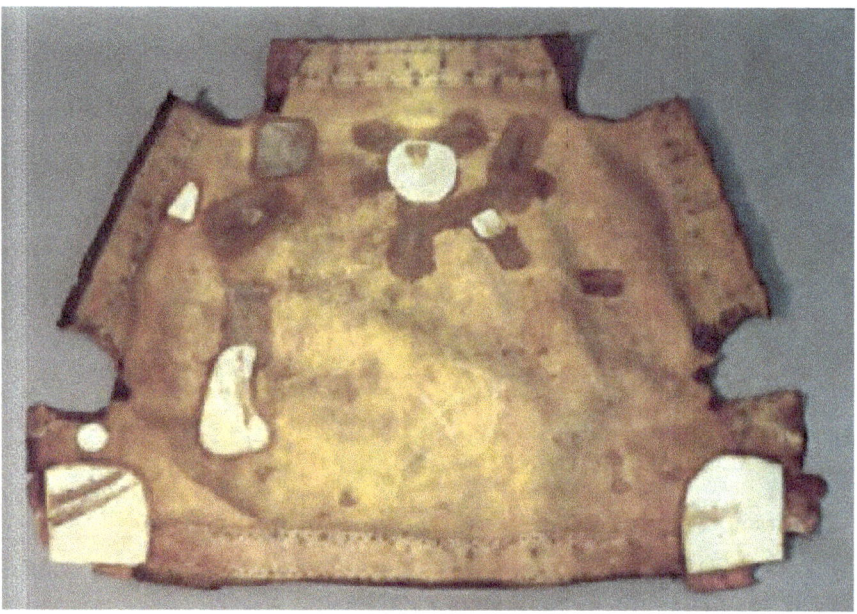

**Figure 23.37** The back during repair showing the leather repairs in the areas where the leather was missing and the Cerex repairs to the splits.

with localized patches of fine Cerex, a non-woven, nylon fabric (*Figure 23.37*). The entire back of the leather was then covered with more Cerex. Cerex is quite soft and flexible and can be worked onto an uneven surface. However, to aid application, it was applied in large strips. These went over the whole surface, including the leather and localized Cerex repairs, and extended beyond the tacking edges (*Figure 23.38*). This surplus was left in place to give the upholsterer an edge to pull on. It was trimmed off after the leather had been replaced on the seat. Cerex was chosen in preference to Reemay or a woven fabric, because of its softness and low bulk. Both the patches and the full lining were applied with a mixture of Lascaux acrylic dispersions, 498 HV and 360 HV, in a 3:1 ratio. This was flexible, wetted out the Cerex well and could be brushed out to a thin layer. The lined leather was still soft and flexible enough to be replaced on the chairs, but had sufficient strength to withstand some use. The overall appearance had not changed relative to the untreated chairs in the set. Because the chairs had to match the rest of the set, the worn surface was not restored in any way. No dye was applied to the areas where colour was lost and no surface finish or dressing was added (*Figure 23.39* and *23.40*).

### 23.8.5 Overview

These three treatments were very different, but were each legitimate in their own context. None of the items belonged to a museum or was to be displayed in an historic house. Instead, they all came from private owners, and their needs were not the same. Indeed, two of the sets came from a single owner and her requirements varied. Unless they are in an environment such as a museum, chairs usually need to be functional, and the treatment should reflect this. Although replacing the leather on chairs as a routine treatment is to be discouraged, there are occasions when it may be necessary, and a purist conservation philosophy is inappropriate. Factors which need to be considered include the rarity and value of the chair, the use to which the chair is to be put, the room setting in which it is to be used, and the viability of preserving the existing leather. The cost of the work may also be relevant. The cost of conservation often exceeds that of re-covering, and may significantly exceed the value of the chair itself. Where a chair has been in a family for a long time, the value of the object relative to the cost of conservation may not be an issue. But if the chairs have been bought recently, and there are no emotional ties, a more pragmatic approach by the owner is likely.

**Figure 23.38** The finished back with the full Cerex lining.

292  *Conservation of leather and related materials*

**Figure 23.39**  The finished chair.

**Figure 23.40**  Detail of repaired seat.

## 23.9 Alum Tawed Gloves, having belonged to Oliver Cromwell

Private collection
LCC Ref: 607.00. Conservator: **Ian Beaumont**
Date: March 2001

### 23.9.1 Description

The gloves date from the mid-to late seventeenth century, and are made from alum-tawed lambskin, the traditional glovemaking leather. The leather has been stained a buff colour on the grain, leaving the suede flesh side white. The gloves have been sewn inside out and then turned right side out. The cuffs of the gloves have been pierced to look like lace work. This may have been done by hand but it is more likely that the leather has been cut with a blade shaped to the lace design. There are traces of an adhesive on the turned face of the cuff. This would suggest that there had been something bonded to it at some time. Under the microscope the fibres of what could have been a pink or blue textile could be seen. The gloves measure 220 mm in length and 95 mm in width.

### 23.9.2 Condition

While the general condition of the gloves was quite good, they had sustained some damage (*Figure 23.41*). Most of this was confined to the cuff area of the gloves. This was associated with the very delicate nature of the lace work.

The left-hand glove had a large area of the underside of the lace work cuff missing. This also included quite a large area of the glove itself, underneath the turned cuff. There was also a small area missing opposite this large missing area. There had been some attempt in the past at repairing the damage with a cotton thread that had been wound around the leather lace work.

The right-hand glove was in better condition in that it was complete. There were though many breaks in the leather lace work.

The gloves were very flat having been stored in the same position for a long time and in both cases the cuffs were very misshapen.

Although the gloves were soiled this was dirt that they had acquired during their useful life and not while in storage.

### 23.9.3 Treatment

#### 23.9.3.1 *Removal of old repairs*

It was necessary to remove previous restoration before treatment could commence. The cotton thread was removed from around the leather lace work and where it had been stitched through the glove itself.

#### 23.9.3.2 *Reshaping*

The leather was humidified in order that the gloves could be reshaped. This was done using a Gore-Tex membrane which allows water vapour to enter the leather without exposing it to water droplets. The lace work was repositioned and allowed to dry in the correct position.

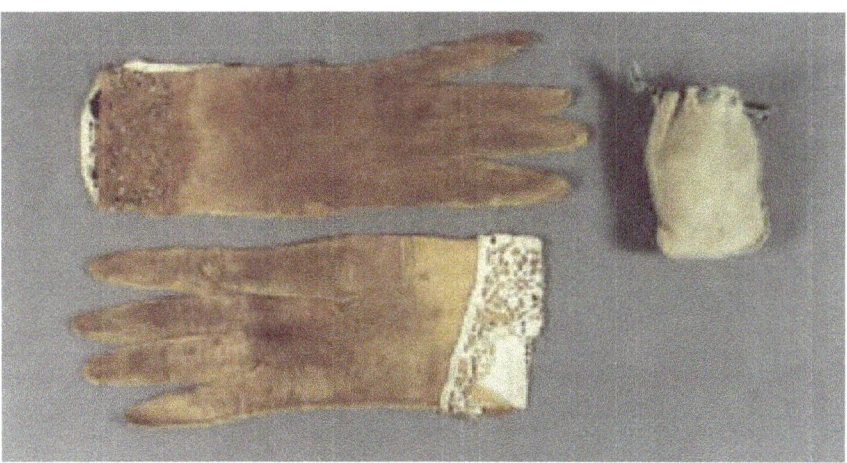

**Figure 23.41**  Gloves before treatment.

294  *Conservation of leather and related materials*

### 23.9.3.3  *Repairs*

The breaks in the lace work were repaired using very small pieces of alum-tawed lambskin leather, which had been skived down to a very fine substance. This was bonded across the breaks with wheat starch paste. After drying, the repairs were in painted with Windsor and Newton watercolours (*Figure 23.42*).

The large missing area was replicated from a similar piece of alum-tawed leather, which had been dyed to match the original with Windsor and Newton watercolours. This was also bonded into place with wheat starch paste.

### 23.9.3.4  *Second reshaping*

The gloves were humidified a second time to allow the cuffs to be turned and to re-establish their original turn-back folds.

### 23.9.3.5  *Mounts*

Perspex mounts were made to support the gloves, the fingers of which were filled with acid-free rolls to keep them in a more natural shape. The mounts are not just for display, they can and should be used for handling and storage purposes (*Figure 23.43*).

## 23.9.4  Future care

On no account should any kind of leather dressing be applied to the gloves. The leather is supple and in good condition. They should not be handled unless absolutely necessary. Should they need dusting this should be carried out with a pony hair brush. They should be kept in a stable environment and not exposed either to high or to low humidities. High humidity is especially dangerous as this can encourage mould growth.

**Figure 23.42**  Replicated leather lace work being fitted into place.

*Case histories of treatments* 295

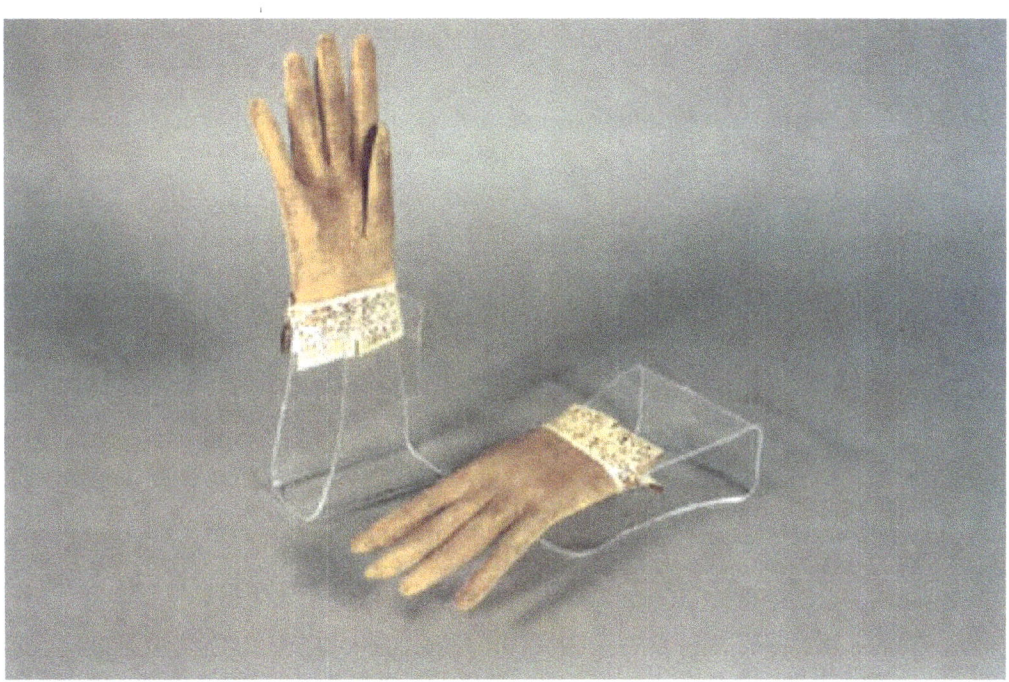

**Figure 23.43** Finished and mounted gloves.

**Figure 23.44** Detail of replicated leather area.

## 23.10 Court Gloves

Private collection
LCC Ref: 435–97. Conservator: **Theodore Sturge**
Date: August 1998

### 23.10.1 Description

There were two pairs of full length, white leather gloves. They were probably mid-twentieth century. One pair was finished with the grain side out and the other was suede finished. There were buttons at the wrist (*Figure 23.45*). On the inside were maker's marks in purple ink.

Both pairs of gloves were soiled. The grain-out pair had a crease down either side and this was particularly dirty. There were some small splits between the fingers adjacent to the stitching, where the leather had given way.

### 23.10.2 Treatment

These gloves may originally have been designed to be washed. However, in gloves of this type the leather has often aged to a point where this is no longer safe. Examples have been seen where washing has left them split, heavily distorted and beyond rescue. However, non-polar solvents are usually safe. If there are, as here, any maker's marks present, they must be tested very carefully to see if these will run. In addition, the solubility of the buttons in the cleaning solvent should be checked.

Most of the heavy soiling along the fold was cleaned off with a soft pencil eraser. However, the amount of cleaning that was possible was limited by the fragility of the grain surface, which tended to lift if too much work was done on it.

Both pairs were cleaned by immersing them in odourless mineral spirits. This was chosen in preference to white spirit because it does not have a lingering smell as it dries. However, it is not as efficient a solvent for some elements of the soiling as white spirits, as it contains less aromatic hydrocarbons and its effect was rather limited, especially on the sueded pair. As a result, on this pair, white spirits was used as well. This still did not clean very well and, as a last resort, a small amount of non-ionic detergent was added to the white spirits. The detergent is a concentrated solution in water, so this added a small amount of water. The polar water, combined with the detergent, worked well and much of the dirt was removed. It must, however, be stressed that the use of water in this way bears a high risk and all concerned should be aware that the object could be in jeopardy.

When white gloves like these are immersed in solvent they become translucent and orange/yellow in colour. Faults in the skin may become visible, and imprints such as the purple maker's mark show through to the outside. In all cases where gloves such as these have been cleaned in this way at the Leather Conservation Centre, the opacity of the leather has returned on drying, but again, a small test is advisable.

After cleaning, the gloves were dried flat. They were laid on absorbent paper towel which drew most of the solvent out of the leather. In the early stages of drying the paper was changed frequently.

The splits were repaired with very fine white silk fabric. This was pre-coated with a mixture of Lascaux acrylic dispersions, 498 HV and 360 HV, in a 3:1 ratio. When this was dry, the patch was cut out with the bias along the line most likely to be stretched. The Lascaux can be used as a heat-sensitive adhesive. In this case it was fixed into position on the inside with an electric-heated spatula. A heat-sensitive adhesive was used because it enabled the spread of the adhesive to be controlled. It also gave a very rapid bond. This made the rather awkward location inside the gloves easier to work on as the repair could be held in place for the short period while it cooled.

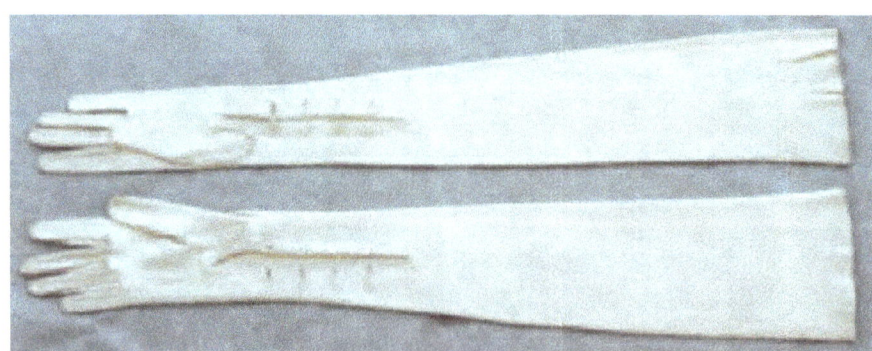

**Figure 23.45** The grain-out gloves after treatment.

## 23.11 Mounting of a Collection of Flying Helmets

Museum collection
LCC Ref: 793–03. Mounts made by **Ian Beaumont**
Date: September 2003

### 23.11.1 Description

A collection of military flying helmets and accessories from the twentieth century collected from British and foreign sources. These were made from a variety of leather types. It was requested that after conservation each be supplied with a storage mount. These were to be inexpensive, interchangeable where possible, chemically and physically stable and had to provide adequate support to each object. Space restrictions also had to be borne in mind.

The simple storage mounts were designed and made from Plastazote foam as shown in Figures 23.46–23.53.

### 23.11.2 Mount instructions

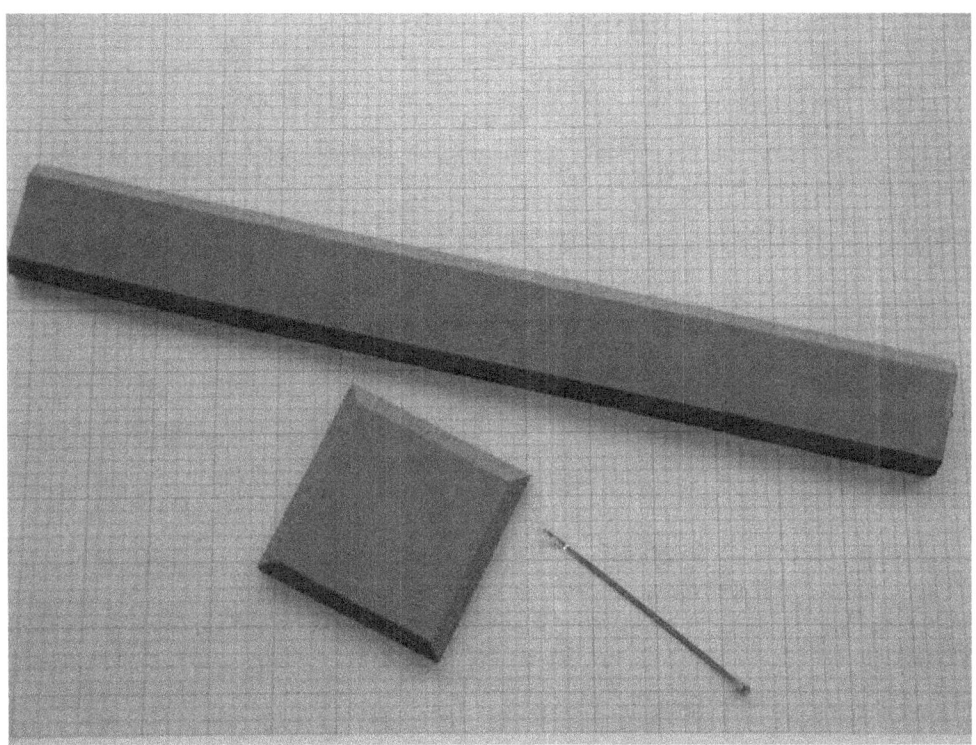

**Figure 23.46** The components for the mount made from Plastazote foam and a cable tie.

**Figure 23.47** The Plastazote is cut to shape and size on a small hobby band saw. This is not essential and can be done with a knife. The chamfered edge on the base is for aesthetic reasons but is not essential. The chamfered edge of the main support softens the edge where it comes into contact with the helmet.

**Figure 23.48** The mount is folded over and placed on the centre of the base so that the profile can be transferred onto the base and the location hole cut out.

*Case histories of treatments* 299

**Figure 23.49** Two holes are punched through the mount near the base.

**Figure 23.50** The cable tie is threaded through the holes tightened and the excess tie cut off.

300  *Conservation of leather and related materials*

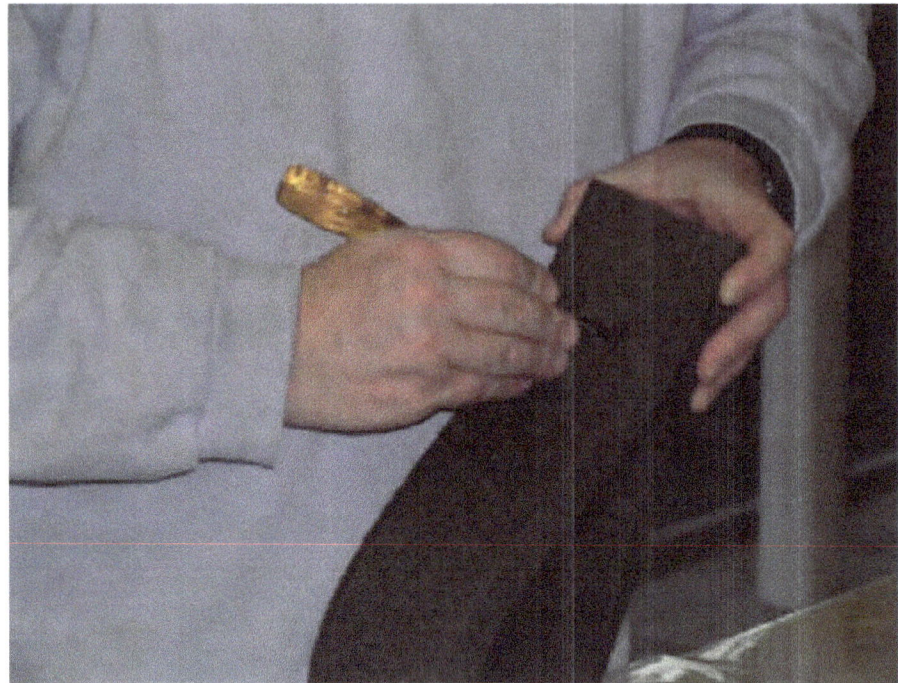

**Figure 23.51** The mechanism is then pushed into the foam to hide it from view. The mount can then be pushed into the base; this can be bonded if necessary.

**Figure 23.52** The flying helmet can the be placed on the mount.

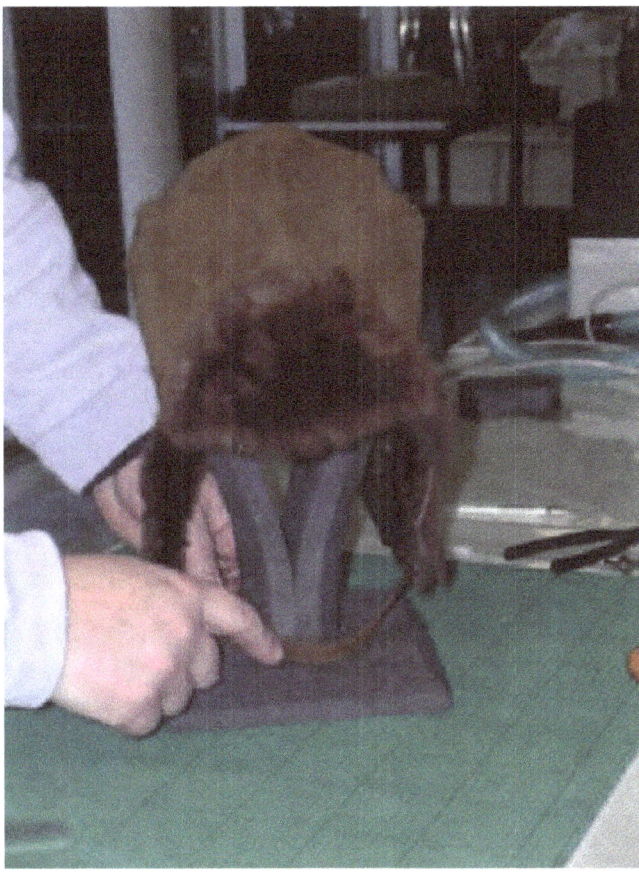

**Figure 23.53** The flying helmet on its completed mount.

302 *Conservation of leather and related materials*

## 23.12 Leather Components from Panhard et Levassor Automobile. 1899

Museum collection
LCC Ref: 737–02. Conservator: **Yvette Fletcher**
Date: May 2003

### 23.12.1 Description

Leather components from a 1899 Panhard automobile, including the hood, the rear seat assembly with loose squab cushion and the fall from the front seat. All the leather is black vegetable-tanned cattle hide in various weights (*Figure 23.54*).

#### 23.12.1.1 Hood

The hood is attached to a painted ferrous metal and wooden frame, has a blue wool textile lining and measures 950 × 1160 × 1350 mm when erected. The front edge is of heavy patent finished carriage leather with brass beading. There is a loose flap on

**Figure 23.54** View of Panhard with hood in place.

**Figure 23.55** Hood before conservation.

**Figure 23.56** Hood after conservation.

the back of the hood, which is buckled into place and can be rolled up out of the way with two leather straps from the inside. There are also four buckles around the edge of the back of the hood, and two more inside each of the sides. There is a length of black painted copper alloy at the junction between the top of the hood and the back on the outside.

#### 23.12.1.2 *Rear seat*

The seat assembly is on a wooden base with an iron frame and a brass rail and trim. The seat back is covered in leather front and back, and measures 695 × 150 mm. There is a trim around the edge of the wooden portion of heavy patent-finished carriage leather, bound in a finer leather, with a brass beading around the middle.

The loose squab cushion, measuring 650 × 365 × 100 mm, is of leather with a textile lining on the base, and has leather piping and buttons in a square 3 × 5 arrangement. It has been reupholstered in the past using foam, and the buttons were possibly replaced at this time.

#### 23.12.1.3 *Seat fall*

The fall from the front seat of the car is of leather with a nailed top edge (associated metal work not available) measuring 800 (min.)–845 (max.) × 350 mm. It was originally lined in black-coated textile which has been removed and replaced with black leather cloth.

#### 23.12.1.4 *Front seat*

The front seat squab cushion is a modern leather reproduction. The original leather upholstered back of the seat is black leather with deep buttoning.

### 23.12.2 Condition

#### 23.12.2.1 *Hood*

The hood leather had been heavily oiled, which may affect the success of some of the treatments. Generally, the hood was sound but dusty and dirty, with an area of paint damage on the interior lower right-hand corner. The leather had worn at the folds with small associated tears, and there had been some loss of surface. There were several large tears, one small tear at the front of the hood and a few small holes. Two buckles were missing from the hood back where the buckle straps had broken, and the two studs on the exterior back top edge (which the straps on the interior hook onto to hold up the back flap) were missing. There was a small sewn leather repair on the back flap of the hood. One of the interior strap ends was also missing. Where the leather was wrapped around the metal frame, there had been some losses and shrinkage. The lower edge of the hood flap was torn and there were small losses where it attaches to the studs on the metal frame. There were residues of metal polish around the brass beading on the front edge. The wool textile lining was in a generally good condition if a little dusty, but was faded on what would be the top fold when the hood is down.

#### 23.12.2.2 *Rear seat*

The seat back was in good condition, with a little cracking to the surface at the outside edges of the upholstered front. The leather edging around the wooden seat was somewhat cracked, as is the nature of this type of coated leather. The front edge was folding inwards a little. There was some wear at the corners and the edges. Generally the edging leather was in good condition, but had been disfigured by

**Figure 23.57** Before conservation – inside of squab cushion with metal sprung unit and padding.

brass polish around the beading and wax polish in the stitching.

The squab cushion was in very poor condition (*Figure 23.57*). The foam upholstery, which was a relatively recent inappropriate addition, had forced the cushion apart and had caused the seams to fail. It had also caused the leather to split alongside those seams which were holding. The corners were in particularly bad condition, with splits, losses and old repairs.

### 23.12.2.3   Seat fall

The seat fall was in a very fragile condition and had also been heavily oiled, which may affect the success of some of the treatments. There was a waxy residue in the stitching. There were two large areas of weakness in the leather with two associated holes, and an old, stuck repair of a leather patch at the top edge. The original stitching had failed in some areas. The leather had torn along a length of the stitching on the bottom right-hand corner. There were tears and small losses along the nailed top edge.

### 23.12.2.4   Front seat

The back of the seat had several small splits and holes and there were areas of surface loss on the leather.

The end (9–10 cm) of one of the straps used to hold the seat in place was missing. One buckle and strap was found under the seat, the second was missing.

### 23.12.3   Treatment

It is understood that the original leather on the automobile will be for museum display only, and that a project is under way to make replica leather components for use out of doors.

After testing to find suitable cleaning systems and solvents, it was felt that Leather Groom (a foaming proprietary leather cleaner and conditioner) was most suitable and all the leather was cleaned to remove dust and dirt, and some of the excess leather dressings.

#### 23.12.3.1   Hood

1. The holes and tears were patched using an archival quality, vegetable-tanned, calf leather which had been dyed to match the original. These patches were adhered using a reversible PVA adhesive. Any small losses and the edges of larger repairs were infilled using pigment-coloured Beva, a thermoplastic wax/resin adhesive which was melted into place using a heated spatula. There was no reason to disturb the sewn leather patch in the back flap, as this is perfectly sound and is part of the history of the object.
2. The broken buckle straps were removed on the back of the hood (for return to the museum) and were replaced using two new buckles. As it appeared that earlier replacement straps had been sewn through the blue lining, the new buckles were also stitched to the hood in this way, thereby removing the necessity of unstitching the lining.
3. The holes in the bottom of the back flap where it attaches to the studs on the frame were repaired using the small patches of leather and the pigment-coloured Beva. However, as it was felt that these holes should not be used to attach the hood to the lower metal bar, small loops of black elastic were sewn to the hood lining to stretch over the metal studs.
4. The brass polish deposits proved difficult to remove from the leather on the front edge and it was felt that any treatment used to effect complete removal would be too invasive. The waxy/oily deposits from the grain of the hood leather were removed as much as possible by mechanical means.
5. The loss of surface was made less obtrusive using leather dyes mixed to the correct colour.
6. The hood was given a light surface coating of SC6000, an acrylic wax blend, to protect the leather and improve its appearance.
7. The two side covers were cleaned and conserved using the same treatments.
8. The lining was lightly vacuum cleaned to remove surface dust.
9. The missing buckles and studs were replaced.

#### 23.12.3.2   Rear seat

##### 23.12.3.2.1   Back

Leather dyes mixed to the correct colour were used to disguise any small cracks. A light surface coating of SC6000 was applied to protect the leather.

**Figure 23.58.** After conservation – squab cushion.

#### 23.12.3.2.2 *Leather edging*

The brass polish deposits proved difficult to remove from the leather edging and it was felt that any treatment used to effect complete removal would be too invasive.

#### 23.12.3.2.3 *Squab cushion (Figure 23.58)*

1. The cushion was disassembled and the inside lightly vacuum cleaned.
2. The sides were lined with archival quality, vegetable-tanned, calf leather which had been dyed to match the original. This lining was first sewn in using pre-existing holes and then adhered using a reversible PVA adhesive.
3. Missing edging was replaced were it was possible to do so without causing further damage to the original.
4. Any areas of loss were then filled using a pigment-coloured Beva and leather dyes were painted on areas of surface loss to improve the appearance.
5. The bottom of the hessian lining surrounding the spring box had been stamped with a large maker's mark. This area had several areas of loss and in order to conserve the maker's mark for the future, it was felt that a new lining was necessary. Therefore the original was unstitched and a new piece of hessian inserted under the original. The areas of loss were couched to the new lining and the opening was also stitched with couch stitching, using a polyester thread. The original lining had been stitched using large stitches which went under the metal bar and which increased the tension of the covering. As the lining is fragile, it was felt that this tensioning would cause further damage and therefore was not replaced.
6. The cushion was reassembled. A felt pad which did not appear to be original and which had caused a great deal of tension on the leather covering was not refitted (but has been returned to the museum). A small amount of horse hair was tucked around the sides. The bottom hessian backing was restitched using the original stitching holes.
7. The whole surface was given a light coating of SC6000 to protect the leather and to improve its appearance.

#### 23.12.3.3 *Seat fall*

1. It was not felt necessary to remove the leather cloth lining, as this is causing no active damage to the skirt. Although it is a later addition, it is part of the history of the object.
2. The two sides were unstitched and the seat fall was relined using lightweight spun-bonded polyester (Reemay). Because Reemay is spun bonded rather than woven, it has no nap or grain and it can move with the leather's own dimensional changes without imposing any of its own on the leather. The Reemay was adhered using a 3:1 mix of Lascaux acrylic adhesives 498 HV and 360 HV. The sides were resewn using the pre-existing stitching holes.
3. The nailed edge was also strip lined with Reemay.
4. Any losses were infilled using pigment-coloured Beva, a thermoplastic wax/resin adhesive which was melted into place using a heated spatula. The tear at the stitching was similarly repaired. The areas of surface loss were painted with colour-matched leather dyes.
5. The seat fall was given a light surface coating of SC6000 to protect the leather and improve its appearance.
6. The seat fall was then reattached to the car using the existing holes.

#### 23.12.3.4 *Front seat*

1. The front seat squab cushion is a modern leather reproduction and is in good condition. However, the bottom hessian cover was torn and therefore this was replaced, being sewn into place using the original stitching holes.
2. A button was missing from the seat, and this was replaced.
3. The back of the seat had some small splits and holes and this was patched using archival quality, vegetable-tanned, calf leather which had been dyed to match the original. These patches were adhered using a reversible PVA adhesive.
4. The areas of surface loss were painted with colour-matched leather dyes.
5. The seat was given a light surface coating of SC6000 to protect the leather and improve its appearance.

6. On the broken strap, a strip of matching leather, shaped and hole punched using the second strap as a template, was fixed to the broken end of the original strap using a reversible PBA adhesive.
7. The found buckle strap was refitted. A second buckle strap was made and fitted.

### 23.12.4 Future care

The leather should be lightly dusted using a lint-free cloth as necessary. No dressings or cleaners should be used.

The original holes in the lower edge of the hood should not be used to fix the hood to the metal bar. Elastic loops have been fitted for this purpose. If possible, the buckles should not be undone.

The hood should be kept in an open position at all times.

The seat apron should be moved with care. It is understood that the apron has to be lifted to gain access when the car is used (approximately 12 times per year) and this has been taken into account when selecting the treatment. However, the apron is fragile and unnecessary usage should be avoided.

The rear squab seat was in a very fragile state and although it has been lined and repaired carefully, it would be preferable if the seat was not used for seating.

## 23.13 Altar Frontal. 1756

Private collection
LCC Ref: 404–96. Conservator: **Ian Beaumont**
Date: May 1998

### 23.13.1 Description

This altar frontal from Little Malvern Court, Worcestershire, is made in traditional gilt leather. The geometric design is picked out with black lines and a translucent red glaze and the background is an off white paint. It is the same design as that in the Marble Dining Room at Ham House, Richmond upon Thames, but the colours are different. The leather at Ham House was probably made by John Hutton in 1756. The Ham House leather is a wall-hanging, and only has the geometric pattern, but this altar frontal has an oval central motif added. The overall size is 153.5 × 83.5 cm. and it was attached with tacks to a timber frame made from 6 × 1.6 cm battens. The paint surface and the silver leaf were in generally good condition except for some surface soiling, scuffing and scratching. The central motif appeared to have been varnished at some time. This was rather uneven. It is possible that this motif had been repainted at some time.

The leather itself, although chemically stable with no signs of red rot, was badly damaged. There were several major splits and substantial sections had torn away along the tacking edge (*Figures 23.59* and *23.60*). There was considerable distortion associated with the tears and some shrinkage. In the centre of the bottom edge the torn leather had been reattached to the timber frame, probably with animal glue.

### 23.13.2 Treatment

The leather was removed from the frame by removing the tacks and releasing the adhesive on the bottom edge. The leather was laid out flat on a table and humidified by laying Aquatex, a semi-permeable membrane, over the leather with a layer of damp blotting paper over the top. This allowed water vapour, but not liquid water, to penetrate the leather. The leather relaxed and the edges of the splits came back together. Gentle pressure was used to hold it flat but care was needed because the paint was quite hard and brittle, and had caused localized distortion. It had pulled the leather out of shape along the lines of the design so that there was a series of slight domes in the surface. These were not removed as it was feared that flattening them could crush the paint.

**Figure 23.59** Before treatment.

**Figure 23.60** Detail before treatment.

Once the splits were lined up they were held in place with localised facings of Japanese tissue paper and wheat starch paste applied to the front (*Figure 23.61*). Wheat starch was chosen because it could be very easily removed at a later stage, and the tissue paper was used because it had little bulk and was unlikely to leave an impression. In addition the torn edges of the paper feathered off without a hard edge.

The splits were repaired from the back with Reemay, a non-woven polyester fabric, and an acrylic dispersion mixture, Lascaux 498 HV and 360 HV, in a 3:1 ratio (*Figure 23.62*). The Reemay was chosen because it has a similar structure to the leather, does not impose a weave, and the repair is not too bulky. It will also take up the shape of the uneven surface. The adhesive was applied to the leather and the Reemay, then they were brought together and the Reemay stippled gently into place. It was then covered with silicone release paper of the type known as parchment paper for cooking, and blotting paper. This was held in place and in shape with sandbags. The blotting paper takes the water away from the adhesive, speeding the drying of the adhesive and reducing the wetting of the leather. In the early stages of drying it needs to be changed frequently.

The original plan was to line the leather completely with Reemay. The intention was to considerably increase the strength of the entire object to protect it against future damage. However, once work started, the potential fragility of the paint layer suggested that this might present too high a risk. Application of adhesive introduces water which leads to expansion then contraction of the leather, and the stippling of the adhesive into the Reemay would also apply stresses. While this risk was acceptable for the areas which had to be repaired, it was deemed unacceptable for those areas where it was not essential. Once the repairs were dry, the facing was removed with a small amount of moisture. At this point it was found that the wheat starch paste had pulled at the paint surface and some of the fine lacquer on the surface was starting to lift. This did not cause significant long-term problems, but it did suggest that wheat starch paste should be used with caution in future. An alternative, which has been used elsewhere, is to apply small patches on the back along the join to hold it in place, and then apply the full repair, incorporating the small repair patches. However, this does not hold it as well as a continuous strip of facing along the front of the split.

The entire frontal was strip lined with new vegetable-tanned calfskin which had been retanned with 1% aluminium to improve its longevity (*Figure 23.63*). This was chosen in preference to Reemay because it gives a more robust edge to reattach to the wooden frame. Reemay combined with leather is very strong, but on its own it can distort and pull out of shape. Because the altar frontal had shrunk slightly the edges were not going to go quite back into place. As a result, the reattachment was,

Case histories of treatments 309

**Figure 23.61** Japanese tissue paper facing along split.

**Figure 23.62** Reemay repairs.

**Figure 23.63** The finished repairs complete with the strip lining. Note the distortion caused by the painted pattern in the left half.

on at least part of the edge, through the strip lining rather than the original leather so the extra strength of a leather strip lining was needed. Before the leather was replaced, the frame was loose lined with a black polyester fabric. This gave the back a tidy appearance, and also gave some support to the leather in case it should receive knocks in the future.

The paint surface was given a light overall clean with a soft brush and a vacuum cleaner. The central motif, which had been revarnished, had this layer of varnish reduced, but not fully removed, with isopropyl alcohol (*Figure 23.64*). This reduced the visual impact of the extra varnish. Isopropyl alcohol is a polar solvent. The yellow varnish over the silver on gilt leather is usually soluble in polar solvents and as a general rule they should be avoided. If they are used extreme care must be taken, along with very careful testing. The solvent can penetrate right through an overlying paint layer and dissolve the varnish underneath, at which point, the paint starts to move about on a bed of liquid varnish.

Gap filling was carried out in three ways. Around the edges there were areas where the leather had shrunk and the strip lining was showing. The strip lining was on the back surface of the original and was thus set back. It also had an inappropriate surface appearance. New vegetable-tanned calfskin with a more suitable grain structure was skived down so that it was very slightly thinner than the original leather, and it was then dyed with Sellaset dyes to an unobtrusive colour. This was then used as an inlay in the missing areas (*Figure 23.65*). The larger gaps, where the leather had split, were filled with solid Beva 371 applied with a heated spatula. The fine cracks along splits were filled with a filler based on Encryl E, an acrylic dispersion. The basic recipe for this is 85% Encryl E, 9% wax FF, 1% thickener A and 5% glass microspheres, all by weight. This is white but pigments can be added. It is somewhat fluid and tends to shrink, and is improved by adding extra microspheres. It needs to be pressed well into the cracks with a spatula. Because it sets by evaporation it tends to shrink so it is only suitable for very fine cracks.

Inpainting on the Beva, acrylic fills and areas of loss on the paint layer was carried out with Winsor and Newton artists' quality watercolours, which were selected from those which are known to be resistant to fading (*Figures 23.66* and *23.67*).

No varnish or other finishing material was applied to the painted surface. As far as could be ascertained, nothing had been applied since the leather had been made. The exception to this was the central motif which had been varnished. The paint surface was fragile but sound, so nothing was needed to consolidate it. Cleaning had not affected the surface and the colour was still saturated, so a varnish was not needed to re-establish the appearance. A varnish would have brightened the finish and provided some protection.

Case histories of treatments 311

**Figure 23.64** Cleaning.

**Figure 23.65** Inlay around the edge.

However, old varnishes applied for these reasons in the past are now discoloured and proving problematic. They are very difficult, or impossible to remove because the solvents required usually dissolve the original yellow varnish which had been applied over the silver. Although it is believed that the varnishes in use now will be readily removable in the future, we cannot be certain, so caution is desirable.

**Figure 23.66** Major split in *Figure 23.60* after treatment.

**Figure 23.67** The finished frontal.

## 23.14 Gilt Leather Screen

Historic house
LCC Ref: 594–00. Conservator: **Ian Beaumont**
Date: May 1998

### 23.14.1 Description

This six-fold chinoiserie screen is probably English and dates from the late seventeenth or early eighteenth century (*Figure 23.68*). It was covered with gilt leather which had been made in the traditional manner. The embossed cover strips were gilt leather too. Unusually, these were found on the back of the joins between the folds, as well as on the front.

There had been some previous restoration work carried out on the screen. This included lining with a fine fabric and the insertion of wooden panels in the back of the framework. This afforded considerable protection to the leather but added significantly to the weight. The painted surface was generally in good condition, but there were 11 substantial splits.

The leather had been reattached to the frames with tacks. Many of these passed through the fabric lining alone and this had torn, allowing the edge to pull away. This may have been caused by the screen being stored at some time in a dry environment. If screens get too dry the leather can become very tight and damage is possible. The cover strips were in very poor condition. They had severe red rot and had lost most of their strength (*Figure 23.69*).

### 23.14.2 Treatment

To allow the damage to be worked on, the individual folds were separated out. When the cover strips were removed they fell apart. The leather on four of the folds was removed completely, while on the other two it was retained in position. The decision to remove only some of the leather was based on the level of repair needed. By taking it right off, it was possible to bring the edges of the major splits back together. It also made carrying out the repairs easier.

**Figure 23.68** The screen after treatment.

**Figure 23.69** Detail. Before treatment.

**Figure 23.70** Detail. After treatment, complete with new cover strips.

The earlier lining was still securely bonded to the leather. Where the leather had torn, the fabric had split along the same line. Rather than remove the lining, the leather and old lining were treated as a single layer. The splits were repaired with Reemay, a non-woven polyester fabric, and an acrylic dispersion adhesive, a mixture of Lascaux 498 HV and 360 HV in a 3:1 ratio. The Reemay and leather were held flat with sandbags as the adhesive dried. The edges were strip lined with Reemay and the Lascaux mix to give a secure edge to reattach to the frames. This was done with stainless steel staples rather than tacks as they are less likely to split the wood and are easier to remove should this be necessary.

The fine cracks left where the splits had been brought back together were filled with Aquafil 945, an acrylic-based filler. This is white, so the correct colour was obtained by adding artists' quality acrylic paints. These were used in preference to dry pigments because they mixed in more readily. The larger gaps were filled with pigmented solid Beva 371 applied with an electric-heated spatula.

The inpainting was carried out with Winsor and Newton artists' quality watercolours chosen from those with good stability when exposed to light.

The cover strips were no longer viable for regular use. They could have been conserved for static display of the screen in a museum, but they were too fragile for the flexing and knocks involved with active use. New cover strips were made to imitate the gilt leather. For this a goatskin with a silver-coloured finish was used as a base material. Initially, the silver layer was wiped over with swabs of toluene. This took some of the surface off and gave an aged appearance. Winsor and Newton acrylic gloss varnish was used as a new surface finish. To give the golden colour Winsor and Newton calligraphy inks were added along with a small amount of black Winsor and Newton artists' quality watercolour. Fumed silica was added to reduce the gloss. The original cover strips were embossed, but replicating this would have required the production of an embossing die. This would have added considerably to the cost and could not be justified (*Figure 23.70*).

## 23.15 Gilt Leather Wall Hangings, Levens Hall

Historic house
LCC Ref: 591–99. Conservators: **Theodore Sturge, Ian Beaumont, Ruth Rodriquez and Elise Blouet**
Date: March 2000

### 23.15.1 Description

This leather is on the main staircase of Levens Hall, Cumbria. There are 60 individual panels of gilt leather mounted on three sides of the staircase. The full-sized pieces are approximately 61 × 73 cm (*Figure 23.71*).

There are two designs on the staircase. The first, running up the bottom flight of stairs and on the mid landing, dates from the first decade of the eighteenth century and was probably manufactured in the southern Netherlands or France. The other design, alongside the upper flight of stairs, was probably produced in the Netherlands in the second quarter of the eighteenth century.

The individual pieces of leather were originally sewn together, and were probably hung up in a similar way to a tapestry. During conservation, stitch holes and the remains of thread were found (*Figure 23.72*). The leather pieces are now mounted separately onto softwood subframes to which they were attached with close-nailed tacks. The individual panels thus formed are then fitted into an oak framework to give a panelled effect.

The leather appears to have been remounted in its present position in 1838. This date was written on the plaster behind the leather (*Figure 23.73*) and an inscription, in pencil, on the softwood subframe for panel 56 reads 'John Bolton aged 17 1838 son of John Bolton of Levens Park Joiner 1838'. This is one of the earliest examples of this type of reuse and remounting.

The leather had been rather crudely revarnished, possibly in 1838 when it was installed in this location. This has now yellowed. The decorative surface was in good condition except for some relatively minor scuffs, cracks and scratches. The leather was in a stable state with no sign of red rot, but there was considerable physical damage with splits and holes and associated distortion. However, very little of the leather was actually missing.

Some parts of the walls had been affected by damp. This had been cured but it had led to rot in some of the wood. The oak framework had been weakened but was still intact and usable with care. A number of the softwood subframes on which the leather was directly mounted were completely rotten and had to be replaced. The leather was unaffected.

**Figure 23.71** The conserved leather on the stairs.

**Figure 23.72** The remains of the original stitching dating back prior to the leather's reuse in 1838.

**Figure 23.73** Inscription on wall behind the leather.

### 23.15.2 Treatment

The method of installing the leather in the oak framework varied. On two of the walls, the leather, on its subframes, had been installed as the oak framework was assembled. As a result it could not be removed from the front. On the third wall, the oak beading holding the leather in place was pinned in place from the front allowing easy removal. However, the damage caused to the plaster by the damp in the past needed further attention, so the oak framework, complete with the leather, was taken down by unscrewing it from the wall. All those sections of leather which needed treatment were then removed from the oak.

There were a large number of old patches on the back of the leather (*Figure 23.74*). Many of these appeared to relate to faults in the leather during its original preparation prior to it being made into gilt leather. Others appeared to have been applied later, probably in 1838, at the time the leather was mounted in the oak framework. A few of these were made with fragments of gilt leather. Most of the old repairs were sound, but a small number were coming away. These were reattached as necessary as the other repairs were carried out.

There were three levels of treatment needed to repair the splits in the leather. The most badly damaged leather, *Figure 23.75*, was removed from the subframes, *Figure 23.76*, and given a full lining. In less serious cases a strip lining was sufficient, while the least damaged had localized repairs applied without removing it from the subframes.

Where necessary, the gilt leather was relaxed with moisture vapour to allow it to be reshaped. Where the leather had been removed from the subframes this was carried out by placing the leather in a chamber with an ultrasonic humidifier. Once humid, it was placed between boards to reshape it. Where localized humidification was needed for leather still on its subframe, Aquatex, a semi-permeable membrane,

**Figure 23.74** Old leather patches on the back and a more recent repair with timber.

was placed over the leather and then damp blotting paper was laid over this. This allowed the water vapour to pass into the leather, but the liquid water was kept away by the membrane.

**Figure 23.75**  A badly split section of leather.

The new repairs were almost all carried out using Reemay, a non-woven polyester fabric available in a variety of thicknesses. The Reemay was chosen for a number of reasons. As a non-woven fabric it does not have a weave and structure of its own which it imposes onto an object. It is also quite thin. This was a distinct advantage over, say, leather or polyester sailcloth, because once the leather was put back onto its subframes it had to fit back into the rebates in the oak framework. It was quite a tight fit, and had the increase in thickness been substantial, this would have led to significant problems. The Reemay also worked well with the adhesive used, an acrylic dispersion mixture of Lascaux 498 HV and 360 HV in a 3:1 ratio. It wetted out well and went into good contact with the leather. Reemay also comes on a roll so its use is easy and, being thin, it can be joined by overlapping the edges without any special treatment of the cut edge to feather it off. Its only significant disadvantage is that on its own it is not very strong. When the repaired leather was replaced on the subframes there was no problem if the staples could be put through the leather and the Reemay, but if there was not enough leather the staples went through the Reemay alone. This led to two problems. If the staples were put in with too much force, they could cut the Reemay, and if the leather was too tight, the Reemay could distort and pull away from the staples. A number of solutions to this were explored. All involved increasing the strength along the tacking edge with a secondary strip lining. The following were tried: a fine polyester sailcloth, leather, an

**Figure 23.76**  The leather in Figure 23.75 removed from its frame.

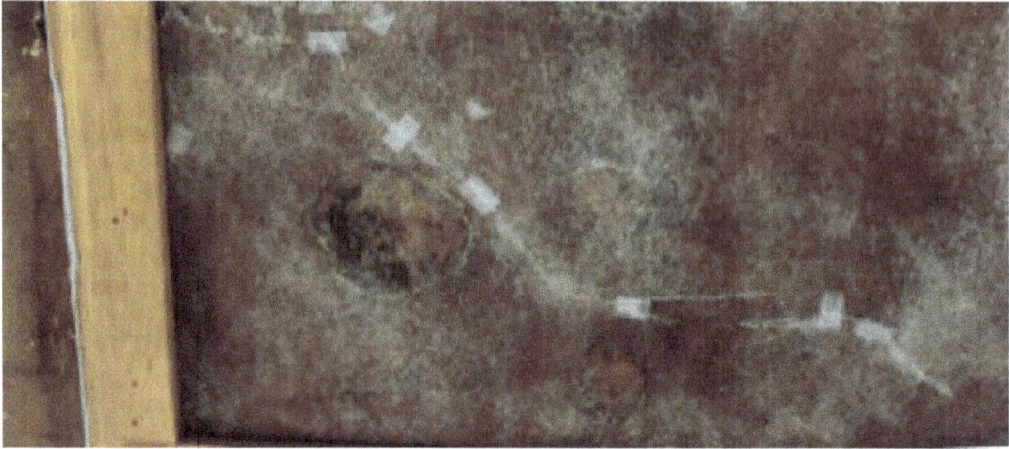

**Figure 23.77** Patches of Japanese tissue paper along the join to hold it in place as the lining is applied. The Reemay has been applied over them.

extra layer of the thin Reemay and a thicker Reemay. Although the increase in strength was greatest with the sailcloth and leather, these were relatively bulky and more likely to cause problems when putting the leather back in the oak frame. The addition of a layer of a heavier Reemay gave a significant increase in strength without too much bulk. It was also easy to apply as it was fully compatible with the thinner Reemay. Provided care was taken not to split it as the staples were inserted, it provided the best all-round alternative.

The most badly damaged leather had distorted, and on these pieces the major splits did not always lie comfortably in their correct alignment. It was necessary to hold them in position during treatment. Two methods were used. First, Japanese tissue paper was applied to the front of the leather with wheat starch paste. This was applied as a strip along the split. There were two potential problems with this. The moisture in the wheat starch paste could be absorbed by the leather, leading to localized expansion, which threw the joint out of line, and it was slow to dry. After the repair was complete, the facing was removed with a small amount of moisture. The alternative method again used Japanese tissue paper, but in this case it was attached with Beva film to the back of the leather. Small patches, to form links between the two sides of the split, were applied with a heated spatula along the line of the join (*Figure 23.77*). These had the advantage that they were quick to apply and did not involve water, so there was no distortion. The disadvantage was that it did not hold the whole join completely in line in the same way as a facing on the front.

**Figure 23.78** Applying adhesive to the leather and the Reemay.

When the lining was applied these patches were retained and incorporated into it.

The Lascaux adhesive was used for all the repairs, including the linings, and reattaching the loose earlier leather repairs. It was applied with a brush to both the back of the leather and to the Reemay (*Figure 23.78*).

**Figure 23.79** Stippling the Reemay into place.

The two were then brought together and the Reemay stippled into place (*Figure 23.79*). The excess adhesive could be removed by laying paper towels over the wet Reemay, and then rolling lightly with a paint roller (*Figure 23.80*). This is not essential but the result is neater and aesthetically more pleasing. In the case of a full lining, the additional strip lining of thicker Reemay was applied while the adhesive from the first layer was still wet. Silicone release paper was then placed over the adhesive and, in case any has come through the splits, under the leather. This was covered with layers of blotting paper to absorb the moisture, and a board was placed on top to keep everything flat. The blotting paper absorbs the water and in the initial stages of drying it needs to be changed frequently along with the silicone release paper. Once the adhesive is dry enough to hold the Reemay on the leather during careful handling, the leather can be turned over and any excess that has come through splits and holes onto the front can be cleaned away.

The use of water-based adhesives needs to be approached with care. There are times when the distortion they produce cannot be controlled and the result is unsatisfactory. In this case the Reemay was able to accommodate the changes in size of the leather during treatment, and by drying under gentle pressure a good, undistorted, result was obtained. Caution is needed when pressure is used, as it can lead to a flattening of the original moulding of the leather.

Where the damage was less serious, localized repairs were carried out, again mainly with Reemay. If necessary, short sections of the tacked edges were freed so that, say, a corner could be repaired. Some of the very small holes were repaired with vegetable-tanned calfskin which had been retanned with 1% aluminium to increase its longevity. The small patches were cut to size and the edges skived away to give a smooth join with the original. These repairs were similar in style to those applied during the production of the leather and, in some cases, replaced these where they had been lost.

Once the leather had been repaired, it was cleaned. Removal of the yellowed overvarnish was not an economic proposition. Instead, a method of surface cleaning was needed that would remove soiling engrained in the surface while, if possible, leaving a good surface. The presence of the original yellow varnish over the silver ruled out polar solvents as this would have been dissolved. Non-polar solvents on their own had little effect, as did moisture. However, in this case,

320  *Conservation of leather and related materials*

**Figure 23.80**  Removing excess adhesive with paper towel and a roller.

**Figure 23.81**  Cleaning with a cloth.

**Figure 23.82** A half cleaned section.

tests showed that a 1:1 mixture of water and white spirits with 1% non-ionic detergent added worked well. This was applied with soft cloths rather than cotton wool swabs (*Figure 23.81 and 23.82*). This was for two reasons. By using a cloth, the problem of the cotton wool snagging on the surface was avoided. This can lead to numerous fragments of cotton wool being caught on the surface and these are time consuming to remove. In addition, the work was in the nature of a general clean of the surface rather than an in-depth, detailed, cleaning as might be used when removing old varnish from an easel painting. There was old wax on the surface and the dirt was inbedded in this. The cleaning removed much of this. After cleaning, the surface was buffed gently with a clean, dry, cloth. This brought a light sheen, probably associated with the remains of the wax on the surface. The end result was a good, but not spectacular, improvement to the finish.

The very fine cracks where the leather had split were filled with Aquafil 945, a flexible, acrylic, resin-based filler to which a small amount of pigment had been added. The larger cracks and gaps were filled with pigmented, solid Beva 371 applied with an electric-heated spatula (*Figure 23.83*). Where appropriate, this was modelled to match the moulding of the leather.

An ideal filler for leather has yet to be found. The solid Beva 371 is a compromise. It is tough and flexible and sticks reasonably well. However, it can pull away from the edges of the filled area and its texture is not ideal. Careful application to ensure that it melts right down into the gap and bonds at the lower levels, rather than just at the surface, helps. A number of methods can be used to improve the surface. If it is cleaned with a solvent such as xylene, this will cut the surface back to leave a less shiny finish. Another possibility is to take a piece of silicone release paper, screw it up and flatten it several times to give it a finish similar to that of the grain surface of leather. If this is laid over the fill and worked over with a heated spatula, this 'craquelure' is transferred to fill and the surface may be substantially improved.

The fills were initially painted with a mica-based, gold, artists' quality acrylic paint. This was then overpainted with Winsor and Newton watercolours. Scratches and other damage to the leather were also inpainted with watercolours. Where necessary, gum arabic was added to increase the gloss. The watercolours were carefully chosen from those known to be light stable (*Figures 23.84–23.86*).

A final surface finish was not needed as the cleaning had left it with a good surface.

When the leather was put back onto the walls, some of the new plaster work that had to be replaced because of the damp penetration was not completely dry. This raised the humidity behind the leather and

322  *Conservation of leather and related materials*

**Figure 23.83**  Gap filling with Beva 371 and a heated spatula.

**Figure 23.84**  The back of the finished section.

*Case histories of treatments* 323

**Figure 23.85** The front of the finished section.

**Figure 23.86** The back of a completed section of panelling showing two full linings and smaller repairs.

some of it became slack. Those panels which were untreated or fully lined responded evenly, but those where the repairs were localized reacted unevenly, and an outline of the repair was, in a few places, visible. Ideally the reinstatement of the leather would have been delayed a few weeks but the hall was about to reopen to the public for the summer. Instead, some of the leather was left out of the panelling to let the air circulate. When the house was revisited some months later, the walls had dried out and the leather had settled down to its correct tension. Although, in this case, there was no long-term problem, it did highlight a potential pitfall. Localized repairs and strip linings may create a piece of leather which will respond unevenly to changes in relative humidity and the result can be distracting. However, the possible problems with this need to be balanced against the more intrusive treatment of a full lining.

## 23.16 Phillip Webb Settle. 1860–65

Private collection
LCC Ref: 262–93. Conservators: **Aline Angus, Theodore Sturge and Aki Arponen**
Date: January 1994

### 23.16.1 Description

This settle was designed by Philip Webb and was probably made around 1860–65. It was owned by William Morris. The settle, made in ebonized wood, had a curved canopy and it was the leather panels on this upper section which were treated. The settle is 204 cm high, 199 cm long and 54 cm deep (*Figure 23.87*).

The lower half of the canopy had three large vertical panels of leather 42 cm high × 59 cm wide. These had been painted with large flowers and the background was stamped with small circles. This leather was in quite good condition with a pH of 3.5, indicating chemical stability. The punching had cut into the surface and there was some associated lifting and loss of the surface.

The upper curved part had six smaller panels of leather 45 cm × 59 cm. These were of gilt leather, and instead of the large pattern on the lower panels, the design consisted of rows of squares with an alternating tile-like pattern. One of the designs had a circular motif while the other had a symmetrical, stylized, leaf design. This leather was in very poor condition. It had a pH of 2.8 and had deteriorated to a point where it had little strength. In the areas where the condition was poorest, the paint surface was very fragile and there was a significant amount of loss. The leather in these areas was brittle and distorted. There had been some previous work carried out which included some repainting of the gold and a small amount of adhesive.

The leather sections were attached to the timber with dome-headed nails around the edge of each piece. It was also attached with adhesive. As a result, the leather had to be treated in situ on the timber.

### 23.16.2 Treatment

The upper, gilt leather panels were too fragile to clean without consolidation. Prior to consolidation and cleaning, the solubility of the decorative surface was tested to ensure that the solvents used would not affect it. The decorative layer, and to an extent the underlying leather, was consolidated with 1% Pliantex, a flexible acrylic resin, in toluene. This was

**Figure 23.87** The settle in Kelmscott Manor.

also applied to the lower panels, but after cleaning. All the leather was cleaned with mineral spirits. This was chosen because the paint, varnish, and Pliantex were all unaffected by it. The only adverse affect of the cleaning was that the overpainting on some of the gilt leather became more visible. This overpaint was not removed as there were no solvents which would remove it without affecting the underlying original paint and varnish.

The loose leather was laid back down onto the underlying timber. It was humidified by laying Aquatex, a semi-permeable membrane, over the surface and then covering this with damp blotting paper and light boards (*Figure 23.88*). The very brittle gilt leather did not lie down very well. It relaxed partially, but did not go completely flat. Resin based adhesives proved unsatisfactory, but wheat starch paste was found to be viable. The paste has a high water content and this was used to further relax the leather during the reattachment. The paste was worked to a very smooth consistency, and sufficient water was added to allow it to pass through a syringe. It was injected behind the damage and any space was filled (*Figure 23.89*). This brought the paste into intimate contact with the back of the leather. Some of the water was absorbed by the leather which became soft. As the leather was pressed carefully into place the excess paste was extruded and removed (*Figure 23.90*). The leather finished up in very good contact with the wood and an excellent bond formed (*Figure 23.91 and 23.92*). Good contact is essential as wheat starch paste is not a gap filling adhesive. When the leather is in a very poor state, as this was, wheat starch paste can lead to a significant darkening. In this case there was some slight darkening, but most of it was hidden by the paint layer and that which was visible was deemed to be an acceptable price to pay for securing the leather. Some alternatives were considered. The most obvious alternative was Beva 371. The gel could have been injected into and behind the areas of damage, allowed to dry and a bond could have been formed with a heated spatula. This would probably have held it together, but it would have left a significant bulk of adhesive in the object. However, it would have been impossible to remove in the future, should the need arise, and would have severely restricted the options for further work. The wheat starch paste too would be impossible to remove from leather as weak as this, but it would be less of a hindrance to further work.

There were some smaller areas on the lower panels which needed to be secured. The wheat starch paste was tried but the leather was unsuitable. When an attempt was made to press it down into place, the punched decorative circles started to lift away,

**Figure 23.88** Humidification with Aquatex.

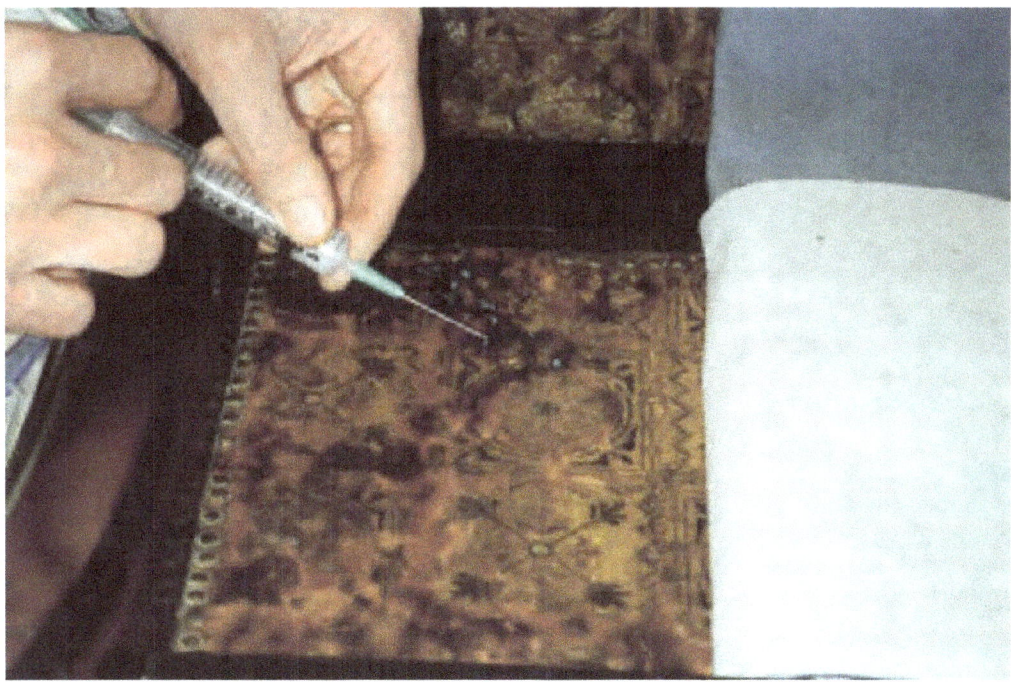

**Figure 23.89** Injecting wheat starch paste.

**Figure 23.90** Excess paste being removed with gentle pressure.

**Figure 23.91.** Detail before treatment.

**Figure 23.92.** Detail after treatment.

exposing the underlying corium. The paste had to be carefully withdrawn and an alternative sought. It proved impossible to find an adhesive which could be introduced in the same way as the paste but without disrupting the surface. Instead, the edges of all the loose areas were secured by inserting a very small amount of adhesive underneath. This ensured that there was nothing lifting which could be caught and damaged during, for example, dusting. A 3:1 mixture of the acrylic dispersions Lascaux 498 HV and 360 HV was used instead of the paste.

On the upper gilt leather panels the golden yellow varnish over the silver (or possibly tin) had, in part, flaked away. This gave a distracting speckled appearance to the surface. This was inpainted with artists' quality Winsor and Newton watercolours. These were carefully selected from those known to be resistant to fading. This gave a consistent appearance. No attempt was made to repaint the design where this had been lost. This was unnecessary for the interpretation of the object, and the areas of loss did not distract from the overall appearance.

When the above work was finished, the decorative surfaces were still quite fragile and vulnerable. A final coat of 2% Pliantex in toluene was applied to consolidate it further and to fix the inpainting.

## 23.17 Gilt Leather Wall Hangings at Groote Schuur, Cape Town

Historic house
LCC Ref: 842–04. Conservators: **Ian Beaumont, Yvette A. Fletcher and Roy Thomson**
Date: March 2004

### 23.17.1 Description

Groote Schuur (Big Barn) (*Figure 23.100*) was purchased by Cecil Rhodes in the late nineteenth century and rebuilt following a disastrous fire in the early twentieth century. For many years it was Cape Town's official Presidential Residence. Both sections of the main reception room are decorated with a frieze of gilt leather. Just over half of the leather panels are of original seventeenth-century manufacture. The remainder are late nineteenth-century replications.

#### 23.17.1.1 Seventeenth century gilt leather wall panels

Eleven wall panels of embossed gilt leather, probably made in the Low Countries in the late seventeenth century. The leather is vegetable-tanned calfskin and the intricate design of fruit and flower swags and garlands with ribbons, birds, apes and other fauna was applied in the traditional manner.

#### 23.17.1.2 Nineteenth century gilt leather panel

One panel of embossed gilt leather made to the same design as the seventeenth century panels.

### 23.17.2 Condition

#### 23.17.2.1 Seventeenth century gilt leather wall panels

The panels had been removed from the north and west wall corner as there had been severe damp problems at this corner (*Figure 23.93*). The backs of the panels had large quantities of paint and plaster attached to them. Underneath this layer were remains of a contact adhesive and animal glue (*Figure 23.94*). Many of the panels had original patches on the reverse, often used to cover holes or thin/weak areas. Some of the panels showed evidence of previous treatments and repairs, such as patching, overpainting and nail holes where they had been fixed to the wall.

A panel from the west wall had a block of wood attached to it. Three panels from the north wall had significant areas of loss of leather.

**Figure 23.93** Wall after replastering ready for leather to be rehung.

Many of the panels had splits and holes. This was particularly evident at the edges where they had shrunk and torn away from the nailing which had been used to fix them to the wall (*Figure 23.95*).

In general the paint surface was in good condition with only minor losses.

#### 23.17.2.2 Nineteenth century gilt leather panel

This panel had been unstitched and removed from the wall for historical analysis. Some of the stitch holes on this and the adjoining panels had perforated leaving small splits along the stitch fold. The bottom edge of the panel had tears where the panel had shrunk and torn away from the nailing fixing it to the wall. On the panel immediately above the panel in question, the stitching joining it to the left side (facing) panel was also broken.

At first sight the panel appeared in very good condition, the back was clean and showed no evidence of having been adhered to the wall, nor of previous treatments/repairs. The painted surface was overall in good condition with only minor losses and flaking of paint. However, small losses

330  *Conservation of leather and related materials*

**Figure 23.94**  Back of panel 4 showing contact adhesive, paint and plaster.

**Figure 23.95**  Front of panel 4 before cleaning.

**Figure 23.96** Cleaning backs of panels, out of doors.

were evident in the deeper areas of embossing and this is consistent with a condition known as red rot which is a result of acidic damage to certain vegetable-tanned leathers, particularly those of nineteenth century origin. This weakening of the leather may also be the reason that the stitching holes are breaking through and the acidity could also be causing the sewing thread to become fragile.

### 23.17.3 Conservation treatment

#### 23.17.3.1 *Seventeenth century gilt leather wall panels*

The plaster and paint were scraped from the back of the panels using a small metal spatula.

Various methods were considered to remove the contact adhesive. A technique was required that would not damage the leather or the painted surface in any way. It was therefore decided to use toluene which was brushed onto the contact adhesive through a buffer of blotting paper, allowed to soak in for only a few seconds (to prevent any potential damage) and then the softened adhesive could be rubbed away using a vinyl glove (*Figure 23.96*). In the absence of a fume cupboard, it was most fortunate that this work could be done in the open air.

The cleaned panels were then humidified using a semi-permeable membrane and dampened blotting paper. This method allows water vapour to transfer into the fibre structure of the leather to soften it without potentially damaging liquid water to come into direct contact with the leather. The leather was then flattened as far as possible under boards and sandbags and allowed to dry.

Where edges joined to another panel, lengths of archival quality vegetable-tanned calfskin were adhered to the sides with an overlap to allow for shrinkage and to facilitate joining the panels together. The top and bottom edges of the panels were also strip lined in this way to allow for fixing to the wall. The lengths of leather were skived along the edges to reduce the bulk and to prevent the edges from showing through in the future. A water reversible PVA adhesive was used to attach the strips.

Any holes and splits were repaired by attaching patches of the archival quality calf leather to the reverse with a water reversible PVA adhesive. Any edges which were damaged were strip lined, again using the archival quality leather (*Figure 23.97*). The repairs were infilled using pigment-coloured Beva, a thermoplastic wax/resin adhesive which was melted into place using a heated spatula.

The panels were joined together in pairs (lower and upper panels) using the strip lining and PVA adhesive and these pairs were then fixed to the wall using staples through the overlapping strips of leather. Each pair of panels was fitted to the adjoining panel in the same way (*Figure 23.98*).

Those areas that had been filled with Beva were inpainted to match the leather and continuity of the design. It was decided in consultation with the client that the strips of leather should be painted to match the background of the original panels but the design need not be inpainted (*Figure 23.99*). Acrylic paints were used for this purpose.

Finally, the panels were given a very light coating of micro-crystalline wax.

#### 23.17.3.2 *Nineteenth century gilt leather panel*

A length of archival quality vegetable-tanned calfskin was adhered to the bottom edge of the panel with an overlap of approximately 3 cm. A water reversible PVA adhesive was used to adhere the strip of leather. The inside edges of the strip were skived to prevent the edges of the leather from showing through the panel in the future.

As there was broken stitching on the panel above and, necessarily, those to the left side (facing) of

332  *Conservation of leather and related materials*

**Figure 23.97** Panels, cleaned and patched.

**Figure 23.98** Some panels back on wall.

**Figure 23.99**  Section of wall after completion.

both panels, a further strip of leather was attached in the same way from top to bottom of these two panels.

The detached panel was stitched to the panels to its right and above using the original stitch holes. A waxed linen thread was used.

These two panels were then adhered to the strip of leather using a water reversible PVA adhesive. The bottom of the panel was fixed to the wall using staples through the overlapping strip of leather.

Any new leather which was showing through was painted to match the background or the existing panels.

### 23.17.4  Future care

The seventeenth century panels are now in a stable condition and should remain so provided care is taken. As with all leather objects the panels are sensitive to changes in humidity. The nineteenth century leather is affected by red rot and its condition should be monitored.

The leather will not require feeding or dressing. Application of oils or fats to the decorated surface will soften the paint and varnish surface causing irreparable damage. An occasional light dusting with a lint-free cloth, or soft brush should be sufficient.

**Figure 23.100**  Groote Schuur (Great Barn), built for Cecil Rhodes and for many years Cape Town's official Presidential residence.

# Index

Acids, role in deterioration, 52
   acid hydrolysis, 38
Acrylics, 127
Adhesives, 108, 124, 126–7
   ethnographic objects, 189
   freezing tests, 165
   fur repairs, 160–1
   leather bookbindings, 228
   parchment glue, 193
   parchment repair, 217–20
   *See also* Animal glue
Aldehyde tanning, 31
Altar frontal case history, 307–12
Alum tawed gloves, case history, 293–5
Aluminium alkoxide, 122
Aluminium tanning, 29
   detection of aluminium, 59
Amino acids, 44–7
   deterioration mechanisms, 45–7
   general characteristics, 44–5
   side chains, 45
Ammonia, 125
Animal glue, 108, 124, 127, 192
   bone glue, 193
   fish glue, 108, 192, 194
   hide glue, 193
   rabbit skin glue, 193
   skin glues, 192
   *See also* Adhesives
Aquatic skins, 173–8
   conservation, 178
   ethnographic objects, 175–8
   fish skin preparation, 174
   structure and identification, 174–5
Aqueous cleaning, 125
Archaeological leather, 244–61
   condition assessments of treated leather, 260
   dry leather, 251–7
      condition, 251–3
      consolidation, 257
      on-site retrieval, 253–4
      present-day treatments, 256–7
      recording procedures, 254–6
      reshaping, 257
   mineralized leather, 257–60
      condition, 257
      on-site retrieval, 259
      recording, 259
      treatment, 259–60
   old collections, 260

   purpose of treatment, 260–1
   retreatments, 260
   storage, 260
   wet leather, 245
      composites and special items, 250–1
      condition, 245–6
      freeze-drying, 248–9, 251
      marine leather, 250
      past treatments, 247–8
      present-day treatments, 248–51
      preservation before treatment, 246–7
      reshaping and reconstruction, 249–50
      solvent dehydration techniques, 247–8
*Australopithecus habilis*, 66
Automobile components, case history, 302–6
Autoxidation, 39–40
Awl, 103

Back stitch, 110
Bating, 77
Bavon leather lubricants, 121, 122, 248, 251
Beating, fur-skins, 153
Bedacryl 122X, 122
Beva 371, 123, 127, 127–8
Bird skins, *See* Feathered skins
Birds, taxidermy techniques, 131–2
   bind up, 131
   problems, 132
   soft stuff, 131
Bleaching, 80
Bloom, 24
Board slotting, *See* Leather bookbindings
Boards, 109
Bone folder, 104
Bone glue, 193
Bookbindings, *See* Leather bookbindings
Boot, 111
Bound edge, 112
Bound seam, 111
Box stitch, 110
Brain tanning, 31
British Long Term Storage Trials (BLTST), 36–7

British Museum Leather Dressing, 121, 129
Brooklyn seam, 111
Brosser (round) seam, 111
Butt stitch, 110
Butted edge, 111

Cable stitching, 110–11
Calfskin, 14
Carbon monoxide, 40
Carriages, 120
Cars, 120
   case history, 302–6
Case histories:
   altar frontal, 307–12
   alum tawed gloves, 293–5
   court gloves, 296
   *cuir bouilli*, 98–101
   dining chairs, 287–92
   dog whip, 268–70
   fire bucket, 271–3
   fireman's helmet, 274–5
   flying helmets, mounting, 297–301
   gilt leather screen, 313–14
   gilt leather wall hangings:
      Groote Schuur, 329–33
      Levens Hall, 315–24
   gold state coach, 265–7
   jewellery box, 285–6
   leather lion, 276–8
   Panhard et Levassor automobile, 302–6
   Philip Webb settle, 325–8
   sedan chair, 279–84
Casting, 128
Cattle skin, 12–13
Chain stitch, 110
Chamois leather, oil tanning, 30–1
Chrome tanning, 27–9, 80–1
   fur-skins, 152–3
   history, 74
Clam, 105
Cleaning, 125
   archaeological leather, 256
   ethnographic objects, 187
   furs, 160
   parchment, 210–11
Clearing, 80
Clicking knife, 104
Closed seam, 111

335

Clothing, 117
Cold storage, furs, 167
Collagen, 4–10
  bonding between molecules, 6–8
    covalent intermolecular bonding, 7–8
    salt links, 6–7
  bonding within molecule, 6
  colloidal nature, 42
  deterioration mechanisms, 43–7
    amino acids, 44–7
    peptides, 43–4
  fibril structure, 8–9, 11
  shrinkage temperature, 9–10
Compass, 104–5
Condensed tannins, 25–6, 75
  detection of, 59
Connolly's Leather Food, 121
Conservation, 113
  aquatic skins, 178
  bird skins, 181–2
  carriages, 120
  cars, 120
  clothing, 117
  *cuir bouilli*, 97–101
    case study, 98–101
    damage caused by old treatments, 98, 99–101
    original treatments, 98
    stability, 97–8
  display versus storage, 114
  exotic skins, 172
  feathered skins, 181–2
  finish, 115
  furs, 159–65
    case histories, 161–5
    cleaning, 160
    repair methods, 160–1
  handling by the public, 114–15
  harness, 118–19
  luggage, 117
  objects in use, 113–14
  preventive conservation, 115–16
    environment, 115–16
    pests, 116
    storage and display, 116
  saddles, 117–18
  screens, 119–20
  sedan chairs, 120
  shoes, 117
  treatment levels, 114
  wall hangings, 119–20
  *See also* Case histories; Conservation treatments
Conservation treatments:
  adhesives, 126–7
  archaeological leather, 260–1
    dry leather, 257–9
    mineralized leather, 259
    purpose of treatment, 260–1
    retreatments, 260
    wet leather, 247–51
  casting, 128
  consolidation techniques, 128
  dressings, 128–9
  dry cleaning, 124
  finishes, 128–9

humidification, 125–6
moulding, 128
parchment, 209–20
past treatments, 121–4
  1982 Jamieson survey, 121–2
  1995 survey, 122–3
  2000 list, 123–4
  2003 Canadian Conservation Institute (CCI) survey, 124
proprietary leather cleaners, 125
repair materials, 126
solvent cleaning, 125
surface infilling materials, 127–8
wet cleaning, 125
*See also* Case histories; *Specific treatments*
Consolidation techniques, 128
  archaeological leather, 256–7
  inks and pigments, 216–17
  leather bookbindings, 230–2, 233
  weak parchment, 215–16
Corner stitch, 110
Cortex, hair, 151
Court gloves, case history, 296
Craft guilds, 82–3, 142
Crease, 104
Crease iron, 106
Crocodile skin, 171
  *See also* Exotic skins
Crosses, of fur, 157–8
Crust leathers, 75–6
*Cuir bouilli*, 94–101
  changes undergone by leather in process, 97
  conservation, 97–101
    case study, 98–101
    damage caused by old treatments, 98, 99–101
    original treatments, 98
    stability, 97–8
  leather moulding techniques, 94
  origins of technique, 94–7
Curriers, 86
Currying, 245
Cuticle, hair, 150–1

Damage, *See* Deterioration
Decay, resistance to, 1
Deerskins, 15
Denaturation, 52
Detergent, 125
Deterioration:
  agents of, 37–43
    acid hydrolysis, 38
    heat, 41
    metals and salts, 40–1
    oxidation, 38–40
    water, 41–3
  archaeological leather, 253
  determination of, 59–60
    fat content, 63–4
    moisture content, 63
    pH, 61–2
    pH difference, 62–3
    shrinkage temperature, 60–1
    sulphate content, 63
  furs, 159

mechanisms, 36–54
  acids, 52
  collagen, 43–7
  fats, oils and waxes, 51
  perspiration, 52
  shrinkage temperature as measure of deterioration, 52–3
  vegetable tannins, 47–50
parchment, 203–9
Dining chairs, case history, 287–92
Dirt, 116
Display, 116
  ethnographic objects, 190
  furs, 166–7
  parchment, 209
  versus storage, 114
Dividers, 104–5
Dog whip case history, 268–70
Double-sided tape, 108
Dressings, 115, 128–9
  archaeological leather, 256
Drills, for leather bookbindings, 229–30
Dropping, fur-skin joining, 155–6
Dry cleaning, 124
Dyeing, 81
  feathers, 181
  fur-skins, 148–9, 153–4
    killing, 154
    mordanting, 154
  leather bookbindings, 225
Dyestuffs, 75, 245

Edge shave, 105–6
Ellagic acid, 48–9
Ellagitannins, 23
Environment, 115–16
Ethnographic objects, 184
  conservation, 186–90
    adhesives, 189
    cleaning, 187–8
    condition, 187
    cosmetic repairs and infills, 190
    display, 190
    mending, 189
    mounts/internal supports, 188–9
    poisons, health and safety issues, 186–7
    pre-treatment examination, 186
    repair supports, 189
    reshaping, 188
    sewing, 189
    storage, 190
  construction techniques, 185
  decoration, 185–6
  ethical aspects, 184
  fish skins, 175–8
  tanning methods, 185
  uses, 184–5
Ethylene glycol, 123
Ethylene vinyl acetate (EVA), 127
Excavation:
  dry leather, 253–4
  mineralized leather, 257–9
Exotic skins, 170–3
  conservation, 172
  history, 170

skin preparation, 171
tanning and dressing, 170–1
 light fastness, 172
 water resistance, 172
uses of, 170–1

Faced edge, 111
Facteka A, 121–2
Fancy stitching, 110
Fatliquoring, 81
Fats:
 deterioration assessment, 63
 role in deterioration, 51
Feathered skins, 178–82
 conservation problems, 181–2
 processing, 178–81
'Feeding' leather, 63
FEIN MultiMaster tool, 241–2
Finish, in conservation process, 115, 128–9
Finishing:
 fur-skins, 154
 tanning process, 81
Fire bucket case history, 271–3
Fireman's helmet case history, 274–5
Fish, taxidermy methods, 135–7
 problems, 136–7
Fish glue, 108, 192, 194
Fish skin, 173–4
 conservation, 178
 ethnographic objects, 175–8
 preparation, 174
 structure and identification, 174–5
Fleshing:
 fur-skins, 152
 parchment manufacture, 198
Flying helmets, case history, 297–301
Foam, 109
Fold endurance test, leather bookbindings, 238
Formaldehyde tanning, 31
Free radicals, 38–9
Freeze-drying, 248–9, 251
Freezing:
 adhesive tests, 165
 furs, 167
French seam, 111
Fuller's earth, 123
Fumigation, parchment, 210
Furs:
 care of, 166–7
  freezing, 167
  pest monitoring, 167
  pest treatment, 167
  storage and display, 166–7
 conservation methods, 159–65
  case histories, 161–5
  cleaning, 160
  repair methods, 160–1
 crosses, 157–8
 damage, 159
 dyeing, 148-9, 153–4
  killing, 154
  mordanting, 154
 finishing, 154

hair/fur fibres, 149
 morphology, 150–1
history of use, 141–6
 fashionable furs and dates, 146–7
 skin processing and dyeing, 148–9
husbandry and harvesting, 145–6
making into garments or accessories, 154–7
 dropping or stranding, 155–6
 sewing, 156–7
 skin on skin method, 155
plates, 157–8
pointing, 154
skin dressing, 151–3
 beating, 153
 chrome, 152–3
 fleshing, 152
 oiling, 153
 pickling, 152
 soaking, 152
 unhairing, 152
species identification, 158–9
terminology, 148

Gallic acid, 48–9
Gallotannins, 23
Gelatine, 124, 128, 193–4
 parchment repair, 218
Gilt leather, 88–90
 production technique, 89
 publications, 90–3
 screen, case history, 313–14
 wall hangings, case histories:
  Groote Schuur, 329–33
  Levens Hall, 315–24
Glass bristle brushes, 124
Glove, 111
Gloves, 117
 case histories, 293–6
Glue, See Adhesives
Glutaraldehyde tanning, 31
Glycerol, 248
Goatskins, 14
Gold state coach case history, 265–7
Goldbeaters' skin, 195
 parchment repair, 219
Groomstick, 124
Guilds, 82–3, 142
Gut membrane, 195

Hair fibres, 149
 morphology, 150–1
  cortex, 151
  cuticle, 150–1
  medulla, 151
Hammer, 105
Handling, by the public, 114–15
Harness, 118–19
Head knife, 103–4
Heat, role in leather deterioration, 41
Hide, 148
 raw hide, 185
Hide glue, 193
Humidification, 125–6
 parchment, 211–15
Hydrolysable tannins, 23–5, 75
 detection of, 59

Hydroxyproline, 5
Hysteresis, 42

Identification:
 aquatic skins, 174–5
 furs, 158–9
Industrial revolution, 86
Infestations:
 furs, 167
  monitoring, 167
  treatment, 167
 parchment, 206, 208
  mould, 210
Infilling, 127–8
 ethnographic objects, 190
 parchment, 218–19
Inks, 206–8
 consolidation, 216–17
Inseam, 111
Invasol S, 123
Isinglass, 194
Isopropanol alcohol, 123

Japanese tissue paper, 126, 127, 234–5
 solvent-set book repair tissue, 232–4
Jean seam, 111
Jewellery box case history, 285–6
Joint tacket system, 229

Keratin, 149–50
Killing, fur-skins, 153–4
Klucel G, 128, 226–7, 230–2
 application, 227
Knives, 103–4

Lapped/overlay, 111
Lascaux acrylics, 127, 128
Latex, 108
Leather:
 definition, 3
 gilt leather, 88–90
  publications, 90–3
 moulding techniques, 94
  See also Cuir bouilli
 properties of, 1–3
  fibre weave and movement, 20–1
  skin fibre structure influence, 19–20
 See also Archaeological leather; Collagen; Mammalian skin structure
Leather bookbindings, 225–42
 binding solutions, 225–30
  adhesives, 228
  board attachment, 228
  board treatment, 228
  drills, 229–30
  facing degraded leather, 227–8
  helical oversewing, 228–9
  joint tacket, 229
  Klucel G, 226–7
  offsetting, 228
 board slotting, 236–42
  board slotting machine, 238
  board treatment, 236
  dyeing, 241
  reattachment, 237–8
  scientific analyses, 238–39

Leather bookbindings (*contd*)
  text block treatment, 237
  variation on the board slotting machine, 241
  leather consolidants, 230–2
    STEP Leather Project, 230
  solvent-set book repair tissue, 232–4
  split joints on leather bindings, 234–5
    adding a cloth inner hinge, 235
  split-hinge board reattachment, 235–7
Leather Conservation Centre, Northampton, UK, 37
Leather Groom, 125
Leather lion case history, 276–8
Leatherworkers:
  social status of, 82–7
    social separation, 84–7
  trade guilds, 82–3
Leatherworking, 246
  tools, 103–8, 256
Light:
  role in deterioration, 39, 115–16
  taxidermy specimen care, 137
Liming, 77
  parchment manufacture, 198, 201–2
Lipids, autoxidation of, 39–40
Lock stitch, 110
Luggage, 117

Mammalian skin structure, 12
  calfskin, 14
  deerskins, 15
  directional run of fibres, 19
  fibre structure influence on leather properties, 19–20
  goatskins, 14
  grain surface patterns, 17
  mature cattle skin, 12–13
  pigskins, 15–17
  sheepskins, 14–15
  suede surfaces, 19
  variation with location in skin, 19
  *See also* Collagen
Mammals, taxidermy techniques, 132–5
  problems, 134–5
Marine leather, 251
Medieval period:
  fashionable furs, 146
  tanning, 68–73
Medulla, hair, 151
Metals, role in leather deterioration, 40–1
Microcrystalline wax, 125, 129
Mineral spirit, 125
Mineralized leather, 257–60
  condition, 257–9
  on-site retrieval, 259
  recording, 259
  treatment, 259–60
Moccasin, 111

Mock stitching, 110
Moisture content, 64
Mordanting, 154
Mould, parchment, 210
Moulding, 128
Mounts:
  ethnographic objects, 188–9
  flying helmets, case history, 297–301

N-methoxymethyl nylon, 124
National Trust Furniture Polish, 129
Neanderthal Man, 66
Needles, 107–8
Neoprene, 108
Neutralfat SSS, 122
Nineteenth century:
  fashionable furs, 146
  tanning, 73–7
Nitrogen dioxide, 40

Oak bark, 86
Oil tanning, 30–1
Oils:
  deterioration assessment, 63
  fur-skin dressing, 153
  role in deterioration, 51
Open (flat) seam, 111–12
Opodeldoc, 123
Overlay, 111
Oxazolidine tanning, 31
Oxidation, 38–40
  atmospheric pollutants, 40
  autoxidation of lipids, 39–40
  free radicals, 38–9
  light, 39
Ozone, 40

Panhard et Levassor automobile components, case history, 302–6
Papers, 109
Paraloid B67, 128
Paraloid B72, 127
Parchment, 2, 200
  characteristics, 203–9
  conservation treatments, 209–20
    cleaning methods, 210–11
    consolidation of inks and pigments, 216–17
    consolidation of weak parchment, 215–16
    humidification and softening, 211–15
    mould and fumigation, 210
    repairs and supports, 217–20
  deterioration, 203–9
  display, 209
  format, 208–9
  manufacture of, 198–9, 200–3
    drying, 198–9
    liming, 198, 201–2
    soaking, 198
    temporary preservation, 198
    unhairing and fleshing, 198
  storage, 209
  use of, 200–3

Parchment glue, 193
Paring knives, 104
Pelt, 148
Peptides, 43
  formation in deteriorated leather, 43–4
Perspiration, 52
Pests, 116
  furs, 167
    monitoring, 167
    treatment, 167
  parchment, 208
pH, 61
  deterioration indication, 61–2
  pH difference, 62–3
Philip Webb settle, case history, 325–8
Photolysis, 39
Photoxidation, 39
Pickling, 77–80
  fur-skins, 152
  in taxidermy, 131
Pigments, 206–7, 245
  consolidation, 216–17
  *See also* Dyeing
Pigskins, 15–17
Piped seam, 112
Pique, 112
*Pithecanthropoids*, 66
Plastination, 251
Plates, of fur, 157–8
Plexisol, 122
Pliancreme, 122
Pliantex, 122, 123, 128
Pliantine, 123, 129
Plumage, *See* Feathered skins
Pointing, fur-skins, 154
Poisons, ethnographic objects, 186–7
Polar organic solvents, 125
Polyester sailcloth, 126
Polyethylene glycol (PEG), 248–9
  PEG 400, 123, 249
Polyvinyl acetate (PVA), 108, 127
Post-medieval period, 68–73
Post-tanning, 81
Prehistoric times, 66–8
Preparation, 109
Preservatives, taxidermy, 140
Pretanning, 77–80
Pricking iron, 107
Prix-seam, 112
Pseudo leathers, 1–2

Quilting, 110

Rabbit skin glue, 193
Race, 105
Raw hide, 185
Raw/cut edge, 112
Reinforcements, 108–9
Relative humidity, 42–3, 115
  taxidermy specimen care, 137
Renaissance wax, 123
Repairs:
  archaeological leather, 257
  ethnographic objects, 189

furs, 160–1
parchment, 217–20
repair materials, 126
sedan chair case history, 279–84
solvent-set book repair tissue, 233
Reptile skin, See Exotic skins
Reshaping:
    archaeological leather, 249–50, 257
    ethnographic objects, 188
Revolving hole punch, 105
Rounding, 110
Rubber cement, 124
Rubber solution, 108

Saddle soap, 123, 128
Saddle stitch, 109–10
Saddles, 117–18
Sausage casings, 197
SC6000, 128, 129
Screens, 119–20
    case history, 313–14
Seal skin, 144
Seams, 111–12
Sedan chairs, 120
    case history, 279–84
Settle, case history, 325–8
Sewing, 109–10
    ethnographic objects, 189
    furs, 156–7
    leather bookbindings, 228–9
    See also Stitching
Sheepskins, 14–16
Shoes, 117
    archaeological specimens, 250–2
Shrinkage temperature, 2, 9–10, 22–3
    as measure of deterioration, 52–3, 60–1
    determination of, 60–1
Silked seam, 112
Silking, 219
Skin glues, 192–3
Skin on skin method, 155
Skin, See Aquatic skins; Feathered skins; Mammalian skin structure
Skiving, 109
Skiving knives, 104
Smoke sponges, 124
Smoking, 185
Social status of leatherworkers, 82–7
Sodium carboxy methyl cellulose (CMC), 127
Soluble nylon, 124
Solvent cleaning, 125
    ethnographic objects, 188
    furs, 160
    limitations on solvent transport, 254
Solvent dehydration techniques, 247–8
Solvent-set book repair tissue, 232–4
Spanish (gilt) leather, 88–90
Species identification:
    aquatic skins, 173–8
    furs, 158–9

Spiral bladed cylinder knives, 74
Splitting machine, 74
Spot tests, 59
Sprung seam, 112
Spun-bonded polyester, 126
Squirrel skins, 143
Starch pastes, 108
State coach case history, 265–7
Steel rule, 104
STEP project, 37, 44
Steramould, 128
Stitch marker, 106–7
Stitching, 109–10
    decorative machine stitching, 110–11
    decorative stitching, 110
    machine stitching, 110
    parchment repair, 217
    See also Sewing
Stoddard solvent, 123, 125
Storage, 116
    archaeological leather, 246, 254, 260
    ethnographic objects, 190
    furs, 166–7
        cold storage, 167
    parchment, 209
    taxidermy specimens, 137–40
    versus display, 114
Stranding, fur-skin joining, 155–6
Stripping, 80
Strop, 104
Suede surfaces, 17
Sulphur compounds:
    deterioration assessment, 63
    role in deterioration, 51–2
    sulphur dioxide, 40
Supports:
    ethnographic objects, 188, 189
    parchment, 217–20
Syntans, 32–4, 80
    auxiliary syntans, 32–3
    combination or retanning syntan, 33
    replacement syntans, 33–4

Tanneries:
    location of, 84–5
    See also Leatherworkers
Tanning, 22–34, 77–81, 245–6
    aldehyde tanning, 31
    determination of tanning process, 58–60
        aluminium, 59
        archaeological leather, 254–6
        condensed tannins, 59
        hydrolysable tannins, 59
        vegetable tannins, 59
    ethnographic skins, 185
    exotic skins, 171–2
    history of, 66–77
        medieval and post-medieval periods, 68–73
        nineteenth century, 73–7
        prehistoric and classical times, 66–8

importance of tanning industry, 82
mineral tanning, 27–31
    aluminium(III) salts, 29
    chromium(III) salts, 27–9, 74, 80–1
    titanium(IV) salts, 29–30
    zirconium(IV) salts, 30
oil tanning, 30–1
post-tanning, 81
pretanning, 77–80
stages, 77–81
syntans, 32–4, 80
    auxiliary syntans, 32–3
    combination or retanning syntan, 33
    replacement syntans, 33–4
vegetable tanning, 23–6, 80
    deterioration mechanisms, 47–50
    history, 68–74
Tanning drum, 74
Taxidermy, 130–40
    birds, 131–2
        bind up, 131
        problems, 132
        soft stuff, 131
    care, 137–40
        light, 137
        relative humidity, 137
        storage, 137–40
        temperature, 137
    fish, 135–7
        problems, 136–7
    mammals, 132–5
        problems, 134–5
    preservatives, 140
    terms, 131
Temperature, 115
    taxidermy specimen care, 137
    See also Shrinkage temperature
Tensile strength test, leather bookbindings, 239
Thonging, 110
Thread, 108
Titanium tanning, 29–30
Toluene, 125
Tools, leathermaking, 103–8
Top stitching, 110
Trade guilds, 82–3, 142
Turned over edge, 112
Turned seam, 112

Unhairing:
    fur-skins, 152
    parchment manufacture, 198

Varnishes, 115
Vegetable tanning, 23–6, 80
    antioxidant ability of tannins, 47
    detection of vegetable tannins, 59
        archaeological leather, 254–5
    deterioration mechanisms, 47–50
    history, 68–75
Vellum, 2
Vulpex, 122, 123

Wadding, 109
Wall hangings, 119–20
  case histories:
    Groote Schuur, 329–33
    Levens Hall, 315–24
Water:
  types of in leather, 41–2
  water activity, 42

Waterlogged leather, *See* Archaeological leather
Waxes, role in deterioration, 51
Webb, Philip, settle case history, 325–8
Wet cleaning, 125
Wheat starch paste, 127
White spirit, 123, 125

Worshipful Company of Skinners of London, 142
Woven fabric materials, 109

Xeroradiographic imaging, 99

Zirconium tanning, 30

For Product Safety Concerns and Information please contact our EU
representative  GPSR@taylorandfrancis.com
Taylor & Francis Verlag GmbH, Kaufingerstraße 24, 80331 München, Germany

www.ingramcontent.com/pod-product-compliance
Lightning Source LLC
Chambersburg PA
CBHW081757300426
44116CB00014B/2155